CALCIUM-BINDING PROTEINS AND CALCIUM FUNCTION

We wish to express our grateful appreciation to the following for generous support:

Abbott Laboratories, Ross Laboratories Division
Armour Pharmaceutical Company
Cornell University:
    Division of Biological Sciences
    Division of Nutritional Sciences
    New York State College of Veterinary Medicine
Hoffmann-LaRoche, Inc.
Merck, Sharp & Dohme Co., Inc.
National Institute of Arthritis, Metabolic and Digestive Diseases
National Institute of Dental Research
Norwich Pharmacal Company
  (Div. of Morton-Norwich Products, Inc.)
Procter and Gamble Company

Calcium-Binding Proteins and Calcium Function

Proceedings of the International Symposium on Calcium-Binding Proteins and Calcium Function in Health and Disease, June 5-9, 1977.
Sponsored by the Divisions of Biological and Nutritional Sciences, and the New York State College of Veterinary Medicine, Cornell University

Editors:

R.H. Wasserman, R.A. Corradino, E. Carafoli,

R.H. Kretsinger, D.H. MacLennan, and F.L. Siegel

NORTH-HOLLAND·NEW YORK
NEW YORK · AMSTERDAM · OXFORD

ELSEVIER NORTH-HOLLAND, INC.
52 Vanderbilt Avenue, New York, New York 10017

ELSEVIER/NORTH-HOLLAND BIOMEDICAL PRESS
Jan Van Galenstraat 335, P.O. Box 1527
Amsterdam, The Netherlands

The logo appearing on the title page and
incorporated into the cover design was designed
by Dr. Klaus Müller, Laboratorium für Biochemie,
Eidgenössische Technische Hochschule Zürich, Politecnico
Federale Svizzero Zurigo, Zürich, Switzerland.

Library of Congress Cataloging in Publication Data

International Symposium on Calcium-Binding Proteins
     and Calcium Function in Health and Disease, 2d,
     Cornell University, 1977.
     Calcium-binding proteins and calcium function.

     Proceedings of the 1st meeting are entered under
title: Calcium binding proteins, 1974.
     Bibliography: p.
     Includes indexes.
     1.   Calcium in the body—Congresses.
2.   Protein binding—Congresses.   I.   Wasserman,
Robert Harold.   II. Cornell University. Division
of Biological Sciences.   III.   Cornell University.
Division of Nutritional Sciences.   IV.   The
New York State College of Veterinary Medicine.
V.   Title   [DNLM: 1.   Protein Binding
—Congresses.   2.   Calcium—Metabolism—Congresses.
3.   Calcium—Physiology—Congresses. QU55 I641c
1977]
QP535.C2156   1977   599′.01′9245   77-23899
ISBN 0-444-00245-6

Manufactured in the United States of America

CONTENTS

LIST OF PARTICIPANTS

Abramson, J.   University of Rochester, Rochester, New York
Acuto, O.   Swiss Fed.Inst.Tech., Zurich, Switzerland
Adelstein, R.S.   NIH-Nat. Heart & Lung Inst., Bethesda, Md.
Allen, L.H.   University of Connecticut, Storrs, Connecticut
Armbrecht, H.J.   Cornell University, Ithaca, New York
Barker, W.C.   Georgetown Univ. Med. Cntr., Washington, D.C.
Bennick, A.   University of Toronto, Toronto, Canada
Bidlack, J.M.   University of Rochester, Rochester, New York
Blinks, J.R.   Mayo Medical School, Rochester, Minnesota
Bocckino, S.   Rutgers Med. School, Piscataway, New Jersey
Bredderman, P.J.   University of Tennessee, Knoxville, Tennessee
Bronner, F.   University of Connecticut Health Cntr, Farmington
Brooks, R. R.   Norwich Pharmacal Co., Norwich, New York
Bruns, D.E.   Wash. Univ. Sch. of Med., St. Louis, Missouri
Bruns, E.   Wash. Univ. Sch. of Med., St. Louis, Missouri
Bugg, C.   University of Alabama, Birmingham, Alabama
Campbell, K.P.   University of Rochester, Rochester, New York
Carafoli, E.   Swiss Fed.Inst.Tech., Zurich, Switzerland
Cheung, H.C.   University of Alabama, Birmingham, Alabama
Chiesi, M.   Swiss Fed.Inst.Tech., Zurich, Switzerland
Choih, Sun-Jin   9906 Kell Ave.S., Bloomington, Minnesota
Clagett, C.O.   Penn State, University Park, Pennsylvania
Coffee, C.J.   University of Pittsburgh, Pittsburgh, Pennsylvania
Cormier, M.J.   University of Georgia, Athens, Georgia
Corradino, R.A.   Cornell University, Ithaca, New York
Cox, J.A.   University of Geneva, Geneva, Switzerland
Daniel, D.S.   3051 St. Paul Blvd., Rochester, New York
de Bernard, B.   Istituto di Chimica Biologica, Trieste, Italy
Drabikowski, W.   Nencki Inst., Warwaw, Poland
Einspahr, H.M.   University of Alabama, Birmingham, Alabama
Eldefrawi, A.T.   University of Maryland Sch.Med., Baltimore, Maryland
Eldefrawi, M.E.   University of Maryland Sch.Med., Baltimore, Maryland
Ellis, K.O.   Norwich Pharmacal Co., Norwich, New York
Ettienne, E.M.   University of Mass. Med.Sch., Worcester, Mass.
Fallon, R.J.   New York University Med. Cntr., New York, N.Y.
Feher, J.J.   Cornell University, Ithaca, New York
Fitts, R.H.   Marquette University, Milwaukee, Wisconsin
Foster, C.J.   Rutgers Med. School, Piscataway, New Jersey
Franciskovich, P.P.   S.I.U. School Med., Springfield, Illinois
Frazier, P.D.   NIH-Dental Res., Bethesda, Maryland
Freund, T.S.   Fairleigh Dickinson University, Hackensack, N.J.
Fullmer, C.S.   Cornell University, Ithaca, New York
Furie, B.   Tufts Univ. Sch.Med., Boston, Mass.
Gallop, P.M.   Children's Hosp. Med. Cntr., Boston, Mass.
Gillis, J.M.   University of Louvain, Brussels, Belgium
Golub, E.E.   University of Connecticut Health Cntr., Farmington
Green, N.M.   Nat. Inst. for Med. Res., London, England
Gunter, K.K.   University of Rochester, Rochester, New York
Gunter, T.E.   University of Rochester, Rochester, New York
Halloran, B.P.   University of Chicago, Chicago, Illinois
Hauschka, P.V.   Children's Hosp. Med. Cntr., Boston, Mass.
Hay, D.I.   Forsyth Dental Clinic, Boston, Mass.
Haynes, D.H.   University of Miami, Miami, Florida
Head, J.F.   Boston University Med. Cntr., Boston, Mass.

Hofmann, T.	University of Toronto, Toronto, Canada
Jeng, A.Y.	University of Rochester, Rochester, New York
Katz, A.	University of Connecticut Health Cntr., Farmington
Kawakami, M.	University of Toronto, Toronto, Canada
Kolbeck, R.C.	Medical College of Georgia, Augusta, Georgia
Koo, S.I.	Cornell University, Ithaca, New York
Krall, A.R.	Med. Univ. of S. Carolina, Charleston, S. Carolina
Kretsinger, R.H.	University of Virginia, Charlottesville, Virginia
Lamar, E.	University of Rochester, Rochester, New York
Leach, R.M., Jr.	Penn State University, University Park, Pennsylvania
Leavis, P.	Boston Biomed. Res. Inst., Boston, Mass.
Levine, B.A.	University of Oxford, Oxford, England
Lian, J.B.	Children's Hosp. Med. Cntr., Boston, Mass.
Lindsay, R.	1237 West Rd., Hilton, New York
Lippert, J.	Rochester Inst. of Tech., Rochester, New York
Long, M.M.	University of Alabama, Birmingham, Alabama
MacLennan, D.H.	University of Toronto, Toronto, Canada
Mahendran, C.	Mt. Sinai Sch. Med., New York, N.Y.
Makinose, M.	Max-Planck Inst. Med., West Germany
Marche, P.	Univ.de Res. Sur Le Metab.-INSERM, France
McGowan, E.B.	State University of New York, Brooklyn, New York
Merle, Y.	State University of N.Y. at Buffalo, Buffalo, N.Y.
Messer, H.H.	University of Minnesota, Minneapolis, Minnesota
Mircheff, A.K.	University of California Med. Sch., Los Angeles, CA
Moriuchi, S.	University of Tokyo, Tokyo, Japan
Mullins, L.J.	University of Maryland, Baltimore, Maryland
Murphy, T.	University of Rochester, Rochester, New York
Myers, M.	400 Kendrick Rd., Rochester, New York
Navickis, R.	University of Illinois, Urbana, Illinois
Nelsestuen, G.	University of Minnesota, St. Paul, Minnesota
Nelson, D.J.	Clark University, Worcester, Mass.
Ohnishi, S.T.	Hahnemann Medical College, Philadelphia, Pennsylvania
Oku, T.	Cornell University, Ithaca, New York
Patchornik, A.	Weiszman Institute, Rehovoth, Israel
Pechere, J.-F.	CNRS, Montpellier, France
Pemrick, S.M.	Mt. Sinai Med. Sch., New York, N.Y.
Peterlik, M.	University of Vienna, Vienna, Austria
Potter, J.	University of Cincinnati Coll.Med., Cincinnati, Ohio
Powers, L.	Bell Laboratories, Murray Hill, New Jersey
Prendergast, F.G.	Mayo Foundation, Rochester, Minnesota
Price, P.A.	University of California, San Diego, California
Racker, E.	Cornell University, Ithaca, New York
Reuben, J.	University of Penn. Sch. Med., Philadelphia, PA
Riddle, C.V.	Temple University Med. Sch., Philadelphia, PA
Roche, R.S.	University of Calgary, Alberta, Canada
Routledge, L.M.	University of Cambridge, Cambridge, England
Russell, R.G.G.	University of Sheffield, Sheffield, England
Ryan, T.E.	20 Ellicott Ave., Batavia, New York
Scott, T.L.	University of Rochester Med. Cntr.,Rochester, N.Y.
Shamoo, A.E.	University of Rochester, Rochester, New York
Sheppard, H.	Hoffmann-LaRoche, Inc., Nutley, New Jersey
Siegel, F.L.	University of Wisconsin, Madison, Wisconsin
Simon, W.	Swiss Fed. Inst. Tech., Zurich, Switzerland
Smith, Q.T.	University of Minnesota, Minneapolis, Minnesota
Sobieszek, A.	NIH-Heart, Lung & Blood Inst., Bethesda, Maryland
Suttie, J.	University of Wisconsin, Madison, Wisconsin
Tada, M.	Osaka University Med. Sch., Osaka, Japan
Taylor, A.N.	Baylor Coll. Med. Sch., Dallas, Texas
Thomasset, M.	INSERM, U.120, LeVisinet, France

Tivol, W.H.   University of Rochester, Rochester, New York
Traverso, H.   Children's Hospital Med. Cntr., Boston, Mass.
Tripp, M.J.   Oregon State University, Corvallis, Oregon
van Eerd, J.-P.   State University of Groningen, The Netherlands
Vanaman, T.C.   Duke University Med. Cntr., Durham, N. Carolina
Veis, T.   Northwestern University Med. Sch., Chicago, Illinois
Vincenzi, F.F.   University of Washington, Seattle, Washington
Wasserman, R.H.   Cornell University, Ithaca, New York
Watterson, D.M.   Duke University Med. Cntr., Durham, N. Carolina
Weeds, A.   MRC Lab. Mol. Biol., London, England
Whitaker, M.   Kings College, Strand, London, England
Wolff, D.J.   Rutgers Med. Sch., Piscataway, New Jersey
Wortsman, J.   SIU School of Medicine, Springfield, Illinois
Yamamoto, T.   Osaka University, Osaka, Japan
Zimmer, J.F.   Cornell University, Ithaca, New York

     Participants from Cornell University include the following: C. Akey,
S. Akiyama, D. Axelrod, B. Baird, L. Ballas, J. Belt, C. Blincoe, M.J. Boros,
W.D. Bowen, M. Brindak, M.J. Bunk, M. Caffrey, M.W. Chapman, D.E. Cohn,
M.J. Cooper, J.K. Crane, F.C. Davis, W.L. Dills, Jr., W. Dukinsky, J. Eisemann,
N. Epstein, G.W. Feigenson, M. Finley, G. Fisher, E. Frelier, M. Fuchs, H. Furr,
E.L. Gasteiger, V. Gildow, M. Glade, L. Heppel, R. Hinnen, F.M. Hoffmann,
D.A. Holowka, F.A. Kallfelz, S.Y. Kang, H. Kasprzak, J.P. Leonard, D. Levy,
E. London, M.A. Longino, R. Martinelli, B. MacKinnon, C.C. McOsker, E.C. Melby,
Jr., R.R. Minor, J. Nagy, R. Noble, E. Nothnagel, K. Obendorf, R.D. O'Brien,
A.H. Parsons, S. Pilch, N. Rackovsky, M. Salpeter, R. Schwartz, D. Simmons,
R. Sleigh, J.B. Smith, P. Slagsvold, D. Sogin, J. Spence, P.C. Sternweis,
A.C. Storer, G. Struve, J.A. Swartzman, H. Tripathi, R. Tripathi, G. Villegas,
R. Villegas, L. Weissberger, M.M. White, D.V. Wilson, R. Wissler, N.M. Wolfman,
J.F. Wooton, A. Zimniak, P. Zimniak.

PREFACE

The present volume represents the proceedings of the International Symposium on Calcium-Binding Proteins and Calcium Function in Health and Disease held at Cornell University from June 5–June 9, 1977. This is the second conference devoted exclusively to this specialized topic, the first held in Jablonna, Poland, in 1973. In the four years since the Jablonna conference, a substantial increase in new and exciting information has accumulated, and the view of several investigators in the field was that another international meeting on this topic would prove useful. Since the exact function of many of the calcium-binding proteins is not known at present, it was proposed that the cross-fertilization and stimulation from the information provided at the Symposium and contained in this volume will aid such an objective.

Calcium-binding proteins, as a class, defy a simple all-inclusive definition, particularly since it is recognized that most proteins will bind calcium under the appropriate conditions. A calcium-binding protein, as such, might be so designated if it has one or more of the following properties:

1. High, or in some cases, moderate to low, affinity site(s) for calcium.
2. Low affinity site(s) but high binding capacity for calcium.
3. The biochemical action of the protein (or enzyme) is stimulated or inhibited by calcium.
4. The stability or lability of the protein is affected by the presence or absence of calcium.
5. The known or hypothesized function of the protein is dependent upon its capacity to bind calcium.
6. The binding of calcium by the protein is required for interaction with other biological entities (e.g., biomembranes).

The aforementioned properties are not mutually exclusive since items (1) and (2) would undoubtedly be a necessary condition for their function, the latter emphasized in the other characteristics. Although these calcium-binding proteins are diverse in origin and function, their commonality is an ability to interact with calcium and related cations and, as a consequence, their biological behavior is modified and/or their biological function becomes manifest.

The Symposium, for the most part, was organized around the organs and tissues of vertebrates from which the various calcium-binding proteins were derived. Included also were proteins from diverse extracellular and intracellular sources, and from non-vertebrate organisms. The opening session was devoted to physical chemical concepts and techniques that bear on defining the characteristics of these important macromolecules. The order of presentation of material in this volume differs minimally from that of the Symposium.

With gratitude, the fine cooperation of the other members of the Organizing Committee, comprised of Dr. E. Carafoli, Dr. R. H. Kretsinger, Dr. D. H. MacLennan and Dr. F. L. Siegel, is acknowledged. Sincere appreciation is extended to Dr. R. A. Corradino, Dr. C. S. Fullmer, Dr. K. Moffat and Dr. H. J. Armbrecht of the Local

Committee who were instrumental in assuring the smooth running of the Symposium. The chairmen of the various sessions, who we sincerely thank, were, as follows: Drs. W. Drabikowski and R. H. Kretsinger (Underlying Physical Chemical Concepts); Drs. F. L. Siegel and M. E. Eldefrawi (Calcium-Binding Proteins in Nervous Tissue); Drs. D. H. MacLennan and A. Katz (Sarcoplasmic Reticulum); Drs. D. J. Coffee and A. Sobieszek (Calcium-Binding Proteins in Muscle); Drs. F. Bronner and P. Gallop (Vitamin D- and Vitamin K-Dependent Calcium-Binding Proteins); Drs. R. S. Roche and B. de Bernard (Extracellular Calcium-Binding Proteins); and Drs. E. Carafoli and M. J. Cormier (Intracellular Calcium-Binding Proteins).

Our thanks are extended to others in my laboratory who unselfishly devoted time and effort to the operation of the Symposium. However, special gratitude and thanks are extended to Mrs. Norma Jayne, who served as secretary to the conference and without whose efforts the Symposium would not have been possible.

R. H. WASSERMAN

Ithaca, New York
June 12, 1977

UNDERLYING PHYSICAL CHEMICAL CONCEPTS

Calcium Chemistry and Its Relation to Protein Binding

R.J.P. Williams
Inorganic Chemistry Laboratory
South Parks Road, Oxford, OX1 3QR
England.

The relevance of calcium chemistry to the understanding of calcium interaction with proteins lies in the clarification of four properties through the study of model complexes (i) the strength of calcium binding to the protein backbone and side-chains; (ii) the selectivity of this binding relative to that of other cations, mainly of magnesium and to a lower degree of sodium and potassium, of the transition metal ions including zinc and cadmium, and of lead; (iii) the conformational constraints which this binding generates in a ligand (iv) the kinetic constraints, that is the on/off rates of the ion reactions and the internal mobility of the complex. We can then write a general equation for the steps we wish to understand

$$\text{Ca} + \text{L} \underset{k_{-1}}{\overset{k_1}{\rightleftharpoons}} \text{CaL'} \underset{k_{-2}}{\overset{k_2}{\rightleftharpoons}} \text{CaL''} \quad \text{etc} \ldots \ldots \quad (1)$$

where k_1 is the rate constant for the first binding step, k_2 is the rate constant of the first conformational adjustment after binding, k_3 the rate constant of the second conformational adjustment and so on. The rate constants for the reversals of these steps are k_{-1}, k_{-2}, k_{-3} etc. The first binding constant is $K_1 = k_1/k_{-1}$ and the free energy change of each sequential conformational adjustment is $-\Delta G = RT \ln K_n = RT \ln k_n/k_{-n}$ where n = 2, 3 etc. Inspection of a molecule at any instant in time will give the conformation of one of the forms ML', ML'' etc but a time averaged picture even over a short period of time may give a blurred image depending upon the rates k_1 to k_n and k_{-1} to k_{-n}. Thus we are concerned with fluctuations of conformation as well as the conformations themselves. Action may demand any one of the possible states.

All the evidence available at present suggests that the binding of calcium to proteins is via oxygen-atom coordination.[1] The oxygen-atoms involved fall into two classes - anionic sites due to carboxylate or phosphate groups and neutral oxygen due to carbonyl, ether, or alcohol groups some of which may be from the main peptide chain but most of which will be due to protein side-chains. For the most part the complex ion chemistry of small calcium ligands has concentrated on units made from carboxylate residues linked by neutral nitrogen, not oxygen, centres as in substituted glycines which are based on $N(CH_2CO_2H)_{1-3}$. Despite the fact that nitrogen ligands on the whole are poor ligands for calcium, for example ethylenediamine is a very weak ligand, the nitrogen in chelates of the type of nitriloacetic acid, $N(CH_2CO_2H)_3$, do assist binding to a degree not very different from carbonyl, ether or alcohol oxygen. Thus the ligands based on acetate substituted amines can be used as models for the behaviour of protein ligands.

Overall Binding Strength

While there is a succession of steps in the formation of many complexes and the rates and relative stabilities of each step are important there is also the overall thermodynamic stability of the system of complexes which is represented by $RTlnK_o$. This is the measured stability constant of the complex of a given composition, here M:L is one to one.

$$K_o = \left\{ \frac{[ML']}{[M][L]} \quad \cdots\cdots\cdots\cdots (2) \right.$$

where ML' is any complex from ML', ML", etc. When we are concerned with competition between cations this constant is of the greatest influence. It is the stability constant reported in text-books on complex ions or indeed by protein chemists. For the proteins of interest in biology log K_o must lie in the region from 2 to 8 approximately. Smaller values than K_o=100 will be generated by sites of little consequence since the free calcium ion concentration does not exceed 3×10^{-3} in biology. Thus we need only examine model chelate ligands which bind in this range of stability constant. Before doing so however we are required to examine the solvent in which the ligand and the calcium are found. The simplest equilibria are for reactions in water (aq)

$$Ca^{2+}(aq) \quad + \quad L(aq) \rightleftharpoons CaL(aq) \quad \cdots\cdots (3)$$

Biological ligands are not confined to aqueous media though in large part calcium ions must be so confined. There is then the possibility that we should be examining too the equilibrium

$$Ca^{2+}(aq) \quad + \quad L(non-aq) \rightleftharpoons CaL(non-aq) \quad \cdots (4)$$

when we could be dealing with a membrane bound protein. In a slightly superficial way we can redefine the problem by taking the reaction in two steps

$$Ca^{2+}(aq) \quad + \quad L(aq) \rightleftharpoons CaL(aq) \rightleftharpoons CaL(non-aq) \quad \cdots (5)$$

when all non-aqueous chemistry will be covered by the consideration of the second step of (5), that is by partition coefficients. We must be very careful in the examination of the partition coefficient however as it could involve a very large conformational rearrangement of the ligand.

We shall inspect first simple equilibria in water. The general importance of aqueous solvation itself will not be summarised since for such material reference can be made to early reviews.[1] Table I provides a set of examples of binding constants in model complexes. From the Table we draw the deductions (1) for a rigid frame cryptate ligand it is possible to bind calcium in the required above range of binding strengths even with a neutral ligand.[2]

Table I Binding Constants of Some Metal Ions with Carboxylate Ligands

| Ligand | Binding Constant $\log K_0$ (ML) | | | | | Note |
	Ca^{2+}	Mg^{2+}	Cd^{2+}	Zn^{2+}	Pb^{2+}	
Imidodiacetate	2.7	2.9		7.1		(a)
Dipicolinate	4.4	2.7	5.7	7.0	8.7	(a)
Nitrilotriacetate	6.4	5.4	9.8	10.7	11.5	
EDTA	10.7	8.9	16.5	16.5	17.5	(b)
EGTA	10.7	5.4	16.7	14.5	14.6	(b)
Acetate	0.7	0.8	1.5	1.4	2.4	(d)
Malonate	2.5	2.8	3.0	2.8	3.1	(c)
Succinate	2.0	1.2	2.1	1.6	2.8	(c)
Citrate	4.8	3.2	4.0	4.8	6.5	(d)
Cryptate^2	8.0	6.5			11.5	(e)

Notes

(a) The constant here shows how a ligand with a fixed chelate bite, dipicolinate, can bind large cations more strongly but small cations more weakly than a non-rigid ligand.

(b) The calcium binding is little affected by the size change of the ligand but magnesium binding is adversely affected.

(c) The further apart the carboxylates on a flexible chain the poorer the binding.

(d) Lead has a very high affinity for oxygen anion donor groups which is not understood. There is also a general increase in binding from extra donor centres.

(e) See Table 7 of reference (2).

Table II Ca^{2+} protein-binding

Protein	Coordination no.	No. of $-CO_2-$	Log Stability Constant	Reference
Parvalbumin	(1) $6?+H_2O$	4	7	(6)
	(2) 8	4	7	
Thermolysin	6-8	2-3	4?	(7)
Nuclease	$6?+H_2O$	3	3	(8)
Concanavalin A	6?	2	3	(9)
Lysozyme	(?)	1-2	1.5	(10)

(2) for more open ligands which have flexible arms, e.g. $N(CH_2CO_2H)_3$, binding in the required range requires a minimum of two carboxylate groups. One carboxylate plus neutral donor atoms is not known to give $\log K_o > 2$ in water. Given present day knowledge of calcium binding to proteins in water we can be confident that indeed at least two negative groups, for example carboxylates, are required for binding of calcium with $\log K_o > 2$, see Table II. Thus the cryptate situation[2] has not been found in aqueous protein calcium complexes as yet. The reason would appear to be as follows.

Proteins are built from substituted α-amino-acids. The required centres of co-ordination for calcium are then the oxygens of the carbonyl groups of the peptide bond or the oxygens of the side-chains. Most of the peptide carbonyl groups are linked to -NH groups by hydrogen-bonds to provide stability to the protein-fold and it is then very difficult to create a cryptate like cavity. (Obviously this is not the case for small cyclic peptides to which we return below). It is presumably for this reason that there is rather little strong binding of sodium or potassium to aqueous proteins. The side-chains of proteins are based on the units $\alpha\text{-}CH\text{-}\beta\text{-}CH_2\text{-}(CH_2)_n\text{-}X$ where X is the ether, alcohol or carboxylate group and $\underline{n}$ can be 0-3. The $\beta\text{-}\overline{CH}_2$ group provides great flexibility as in $N(CH_2.CO_2H)$, and this very flexibility requires that the group X as a monodentate centre must have a reasonably high affinity for calcium if the centre is to bind calcium with the required strength and despite the fact that the complex formed will be a chelate. Ether, carbonyl or alcohol oxygen in simple ethers, amides and alcohols, X = neutral oxygen, bind inadequately as monodentate centres. On the other hand when X = carboxylate or phosphate binding is sufficiently strong to give the possibility of strong complexes in chelates such as $N(CH_2CO_2H)_3$. Thus selection by calcium of liganding atoms of the protein is based upon binding strengths of individual groups X which have to be effective against a background of the hydration energy of the aquated calcium ion and of the loss of configurational entropy of the unbound chelating protein. Protein side-chains such as carboxylates of glutamic and aspartic acids usually have a high degree of configurational entropy for they often lie on the surface of the protein in a non-constrained fashion. Thus several centres are required to give $\log K_o > 3$.

We shall take it that in general adequate binding to give $\log K_o$ from 3 to 8 can be provided by a chelating agent even of considerable configurational entropy provided that some two or four carboxylate groups are available and which may be assisted to a limited degree by additional binding to uncharged oxygen (nitrogen) atoms. Phosphate of serine phosphate can substitute for carboxylate. Before leaving $\log K_o$ we need to make two further observations.

The first concerns competition between cations for binding sites.

Competition between Cations

Here we summarise our previous papers.[1] Competition for calcium sites is largely from magnesium ions. Discrimination in favour of calcium is usually based on (1) the absence of nitrogen-atom donors for such donors favour magnesium binding relatively. (2) the small size of the magnesium ion which militates against multidentate binding of bulky ligands and gives water a relatively favourable binding energy for this cation (3) the demand of magnesium for a well-defined geometry of its ligands, octahedral symmetry. It would appear that proteins do not readily provide the necessary well organised (stereochemically) binding site which would give high stability binding sites for magnesium. The best suitable selective configuration for it which is known to occur would appear to be two carboxylates plus one imidazole nitrogen in a geometry where magnesium retains three water molecules. e.g. in phosphomutases. It is notable that calcium binds relatively poorly to such a site but given a larger number, 5 or 6, donor centres its binding is of sufficient strength to some protein sites even to overcome the bias of concentration of the free ions $[Mg^{2+}]/[Ca^{2+}] > 1,000$. The sites always have at least three carboxylate residues, compare Table I.

Of course magnesium can be taken up in a non-protein cryptate of fixed geometry and cavity size very effectively e.g. in chlorophyll. This brings out a very striking contrast with calcium biochemistry - we do not know of a similar biological cryptate which has been designed for the calcium ion and which is retained by proteins. On the other hand there are ionophores for calcium which are clearly closely related to such cryptates. These ionophores often differ from the sodium/potassium ionophores in that they are usually open chain anions of limited conformational mobility but like them have a large number of closely related (in space) oxygen atoms. They usually bind calcium with but a poor binding strength, $\log K_0 = 3$, but as they are lipid soluble it is hard to give a true constant separate from the partition coefficient. Again the discrimination against magnesium is not very strong. Synthetic cryptates of much greater selectivity are known.[2] In many ways the manner in which proteins bind calcium ions can be said to be intermediate between cryptate ligand binding and the binding by very flexible multidentate model ligands such as ethylenediaminetetracetate as we shall show by reference to the control exerted upon binding by the stability of the protein fold, see below.

The competition between calcium ions and cations such as those of the transition metals, zinc, cadmium and lead is dominated by three factors. (a) the competing ions are in much lower concentration. (b) the competing ions are largely removed by their relatively very high affinity for nitrogen and sulphur donor sites. Calcium does not have this affinity. (c) Overall in biological systems the number of chelating sites exceeds the number of trace metal ions available. Thus only under abnormal conditions does a metal such as lead, cadmium or copper represent a competitive hazard in calcium biochemistry.

Protein Fold Energies

A protein is a very complicated ligand and it can exert additional constraints upon binding which are not seen to the same degree in any model complexes. The protein through its fold energy can force groups together which repel one another but this is not a rigid constraint as in a cryptate. It is therefore hard to predict from its sequence the ability of an apo-protein to provide a fixed site for a given cation. We do not know and I would say that we can not know from sequence studies the degree to which calcium sites are preformed in the apo-proteins and we can not therefore assess the configurational entropy changes on chelation. A second point of difference between small ligands and proteins is that a protein is not able to bring liganding atoms into any desired interaction configuration using the binding energy of the cation for this change. It can be that the required bond distances have to be made against the total fold energy of the protein. Herein lies the possibility for a protein to generate an extremely specific interaction with a given cation for it may fold appositely in the presence of one cation and not another. Although an apo-protein may be far from rigid internally it can be sufficiently restricted to have a relatively small configurational entropy and yet only a limited number of configurations may be achievable which can act as a cation binding site. Thus ligand atoms may only come to bonding distances with cations of a given size. The selectivity, related to the entatic state hypotheses, implies that the whole protein may feel the binding of a cation to a local part of it. This cooperativity of the whole protein is of the very essence of the control function of proteins. It is in this area that NMR spectroscopy is proving so valuable for although it does not have the structural precision of X-ray crystal studies it can look at minute changes on binding of ions to proteins revealing the running changes of conformation, (see Levine and Williams, this volume). If the sense of this paragraph is correct then the binding affinity of a protein site for calcium can only be understood when the fold energy of the protein is understood in both the apo-protein and the cation-bound protein states.

The Acidic Nature of Calcium-Binding Proteins

There is now a very general point about the calcium-binding proteins. They have far too many acidic groups for the number of calcium ions they are said to bind. A glance at the major binding proteins whether they contain glutamic, aspartic or γ-carboxyglutamic acid residues shows that the proteins remain extremely negatively charged even after binding some calcium ions. Such proteins would be expected to be either random coil or very loosely held together in the absence of compensating positive charge, i.e. especially as apo-proteins. We shall show that this is the case later in the symposium. A simple conclusion is that on binding calcium (or any other positive charge) the proteins will in-

evitably undergo very considerable conformational rearrangement and that this rearrangement will be extensively distributed throughout the proteins. Part of the
binding energy of any cation is not from the ligands but is a general electrostatic binding.

It is particularly instructive to consider some of the calcium binding proteins as a hydrophobic control core, which is somewhat structured even in the absence of calcium, and a large negative charge network on the surface which is
roughly spherical. Binding of calcium then causes a contraction of the whole
sphere and superimposed on the general contraction are a large number of specific
local conformational changes.

The Special Calcium Site: γ-carboxy-glutamic acid

A separate description of the unit $-CH(CO_2H)_2$ is necessary for it appears to
provide a special biological calcium site. The unit has two ionised carboxylate
groups at pH = 7 and these two units are held close together by covalent bonds so
avoiding the problem of the compensatory fold energy of the protein normally
needed for such an enforced juxtaposition of negative charge. Secondly the two
groups are fixed very closely together and there is then little configurational
entropy of the group. Compare methyl malonic acid $CH_3.CH(CO_2H)_2$. Two such carboxylate groups bind considerably more strongly than those in a unit such as
$CH_2(CH_2CO_2H)_2$. However the γ-Glu unit is not good enough to bind calcium unless
the free calcium approaches 10^{-3}M. Moreover the unit by itself does not discriminate between magnesium and calcium. As there is hardly any part of a
biological system in which $\left[Ca^{2+}\right] > \left[Mg^{2+}\right]$ the specially created binding agent is
not functional directly in a calcium-specific manner. Only where two such
groups are brought together in a peptide chain such as γ-Glu.x.x. γ-Glu is calcium selectivity assured. See above for the failure of magnesium to bind such a
site. However it should be noted that it is now very hard to visualise that the
site could be preformed before calcium binding for there must be intense repulsion between the four carboxylate groups. γ-Glu binding of calcium would
apparently involve a compulsory conformational change much as in other proteins.

Successive Conformational Changes

The very nature of the calcium sites in proteins means that there will be
conformational changes on calcium binding for the volume of the protein is reduced on binding. We have observed this effect even on binding of calcium to
lysozyme,[3] not a true calcium binding protein. The question as to how many
steps there are before the final binding condition is obtained, see equation (1),
will be a property of the individual proteins. There can be no useful parallel
with model ligands but one or two small points are worth noting. Rearrangement
about the $-CH_2$ group of a carboxylate anion is unlikely to provide a set of con-
formers. In the binding of calcium to model ligands it is only those ligands
which have considerable constraints on their own mobility, e.g. cryptates, which

appear to show a succession of binding steps. Eigen and Hammes[4] in particular
drew attention to the slow step by step stripping of water from an ion such as
calcium which could be required if the cation is to enter into a hole formed by
such a ligand. As we have indicated above the proteins fall between cryptate
ligands and very mobile open chain ligands and we can expect well-defined steps
in the formation of some protein calcium complexes. It is quite impossible to
predict the relative stabilities of the different complexes ML'. ML", etc. and
therefore their contributions to K_o, equation (2).

Rate Constants

The discussion of the rates of the steps in the formation of ML', ML", etc.
follows directly from the above paragraph. If the on-rate for the step ML' is
diffusion controlled then the off-rate depends on the binding constant and the
on diffusion rate constant. The first step of protein binding is likely to be
fast. The subsequent steps can not be estimated as protein energy changes are
involved.

Fluctional Molecules

While investigating the complexes of dipicolinic acid with the lanthanide ions
we have observed that the dipicolinate anion can rotate about the metal-N axis.[5]
The rotation is more restricted in the case of the bis dipicolinate complexes and
we would suppose that it does not occur in the nine-coordinate tris dipicolinate
complexes. Although we have no evidence of such easy motion in other complexes
we suspect that it must occur generally. There is then the question as to the
state of the calcium complexes - do they have a fixed conformation? While there
is no experimental evidence the experience with the lanthanide cations suggests
that in the absence of high coordination the ligands may be mobile on the surface of the cation. In calcium protein complexes the calcium ion appears to
have a low coordination number. Relative motion of the ligands on the surface
of the calcium sphere is to be expected.

More Complex Equilibria

The article has looked at one to one complexes. There are many calcium proteins which are of the composition $Ca_x L$ and calcium can cross-link proteins CaL_y.
The major new factors which are introduced are the cooperativities involved in
the steps. Given that the protein conformation changes on binding each calcium,
every step must be cooperative or anti-cooperative and each protein will behave
differently. The ability of calcium to cross-link proteins is very difficult to
discuss as yet for there is too little experimental evidence.

Bone Formation

The ultimate form of cooperativity is seen in ionic crystals. We should
therefore enquire closely into the role of protein in the initiation of bone
formation since given cooperativity it is the nucleation of growth which is be-
set by the greatest energy barrier. At the moment we can but speculate that

there are proteins for this initiation. These proteins should be of high
negative charge density at the surface and the charges should be exposed in a
rather fixed array. In this way there will be formed a patterned grid of calcium ions which can then be bridged by phosphate or carbonate anions. The low
configurational entropy of the γ-Glu group makes it ideal for this function.
The crystals which form are not those of conventional salts.

Some Biological Consequences

I wish to summarise some views about calcium in biology which I have formed
from the above study of models and of proteins.

(1) Although any particular short sequence of a protein e.g. the so-called E-F
hand fold, could give a potential calcium-binding site this is not certain since
the energy of complex formation depends on the total sequence and the fold
energy as well as on a local sequence. This local/global energy interaction in
proteins gives bond energy and rate specificity to calcium sites of an unusual
kind and is determined by genetics.

(2) From (1) binding constants for single sites are modulated by proteins in
the range log K = 2 to 8. The extreme values are largely controlled by the
number of carboxylate groups at the binding site, but precise values are regulated by the overall fold energies.

(3) There are many calcium binding proteins with different functions and the
binding constant is tailored to the function e.g. triggering, transport, buffering.

(4) Calcium ion binding must produce conformational changes and often these
are global rather than local.

(5) Global conformation changes readily permit connection to triggering and
energised transport for the protein becomes an allosteric protein, whether this
be through-bond or a through-space (electrostatic) effect.

(6) By the use of slightly different proteins, parvalbumins and troponin C for
example, in suitable concentrations in cells calcium ion concentration changes
can be used to work triggers, but prevented from excessive alterations. This is
calcium ion buffering, which needs to be elaborated in terms of the binding constants and stoicheiometries.

Acknowledgement

R.J.P.W. is a member of The Oxford Enzyme Group.

References

1. Williams, R.J.P. (1976) in Calcium in Biological Systems Symposium XXX Soc.Expt.Biology, Cambridge Univ.Press p.1-18.

2. Lehn, J.-M, (1973) Structure and Bonding, 16, p.1-69.

3. Dobson, C.M. and Williams, R.J.P. (1977) in Metal-Ligand Interactions in Organic Chemistry and Biochemistry, eds. B.Pullman and N.Goldblum, Reidel Pub.Co. Dordrecht Holland, Part I p.255-282.

4. Eigen, M. and Hammes, G.G. (1963) Adv.Enzymology 25, p.1-43.

5. Alsaadi, B., Rossotti, F.J.C. and Williams, R.J.P. (1977) Chem.Comm. Chem.Soc. London, submitted.

6. Moews, P.C. and Kretsinger, R.H. (1975). J.molec.Biol., 91, p.201-228.

7. Edelman, G.M., Cunningham, B.A., Reeke, G.N., Becker, J.W., Waxdal, M.J. and Wang, J.L. (1972). Proc.natn.Acad.Sci. USA, 69, p.2580-2584.

8. Cotton, F.A., Brier, C.J., Day, V.W., Hazen, E.E. and Larson, S. (1971). Cold Spring Harb.Symp.quant.Biol., 36, p.243-255.

9. Matthews, B.W., Weaver, L.H. and Kester, W.R. (1974). J.biol.Chem., 249, p.8030-8044.

10. Campbell, I.D., Dobson, C.M. and Williams, R.J.P. (1975). Proc.R.Soc. Lond. A, 345, p.41-59.

11. Inoue, S. and Okazati, K. (1977) Scientific American, 236, 83-92.

Crystal Structures of Calcium Complexes
of Amino Acids, Peptides and Related Model Systems

Howard Einspahr and Charles E. Bugg
University of Alabama in Birmingham
Institute of Dental Research
University Station
Birmingham, Alabama 35294

INTRODUCTION

   Despite the many fundamental roles that calcium ions play in biological systems, relatively little is known about the structural chemistry of calcium. Since calcium does not seem to form stable covalent compounds, there has been a tendency to assume that calcium complexes must be stabilized by interactions that lack stereospecificity. However, it is clear that many biological processes display specificities for calcium that must be attributable to highly selective calcium-binding sites. The available structural evidence indicates that such specificity is largely due to the overall spatial arrangements of suitable ligands at calcium-binding sites, but little is known about the exact geometrical factors that are involved.

   Crystallographic, spectroscopic and chemical data all indicate that the most common ligands at the calcium-binding sites on proteins are the peptide-carbonyl groups and the carboxyl groups from glutamic- and aspartic-acid residues. We have reviewed the Ca--carboxyl and Ca--carbonyl interactions in 60 crystal structures (Table I) in an effort to identify those factors that might influence the geometry of calcium interactions with these functional groups. To the best of our knowledge, this set includes all published crystal structures of calcium salts and complexes of amino acids and peptides, plus several such crystal structures that have not yet been published. In addition, we have included a number of other crystal structures in which calcium ions are complexed with carboxyl and carbonyl groups.

   In this paper we briefly review the geometries of Ca--carboxyl and Ca--carbonyl interactions as found in these 60 crystal structures. At this stage we have not yet fully analyzed the detailed geometrical constraints that seem to accompany these interactions. Consequently, a more thorough analysis of the calcium interactions in these crystal structures will be published

13

Table I. The crystal structures that are included in this survey.

Ca(Tartrate)-4H$_2$O
Ambady, G.K. (1968) Acta Cryst. **B24**, 1548-1557.

CaCl$_2$(Sarcosine)$_3$
Ashida, T., Bándo, S. & Kakudo, M. (1972) Acta Cryst. **B28**, 1560-1565.

Ca(5-Keto-D-Gluconate)-2H$_2$O
Balchin, A. A. & Carlisle, C. H. (1965) Acta Cryst. **19**, 103-111.

Ca$_2$(Ethylenediaminetetraacetate)-7H$_2$O
Barnett, B. L. & Uchtman, V. A. (1977) Personal Communication.

CaNa(Nitrilotriacetate)
Barnett, B. L. & Uchtman, V. A. (1977) Personal Communication.

Ca(Barbital)-3H$_2$O
Berking, B. (1972) Acta Cryst. **B28**, 98-113.

Ca(Acetate)(Thioacetate)-3H$_2$O
Borel, M. M. & Ledesert, M. (1975) J. Inorg. Nucl. Chem. **37**, 2334-2335.

Ca(Hydrazinecarboxylate)$_2$-H$_2$O
Braibanti, A., Manotti Lanfredi, A. M., Pellinghelli, M. A. & Tiripicchio, A. (1971) Acta Cryst. **B27**, 2261-2268.

Ca(Hydrazinecarboxylate)$_2$
Braibanti, A., Manotti Lanfredi, A. M., Pellinghelli, M. A. & Tiripicchio, A. (1971) Acta Cryst. **B27**, 2448-2452.

Ca(Malate)-2H$_2$O
Bränden, C.-I. & Söderberg, B.-O. (1966) Acta Chem. Scand. **20**, 730-738.

CaK(Co(Dithiooxalate)$_3$)-4H$_2$O
Butler, K. R. & Snow, M. R. (1975) Acta Cryst. **B31**, 354-358.

Ca(Oxalate)-H$_2$O
Cocco, G. & Sabelli, C. (1962) Atti Soc. Tosc. Sc. Nat., Serie **A**, 3-12.

Ca(Trans-1,2-Diaminocyclohexane-N,N'-Tetraacetatoaquoferrate(III))$_2$-8H$_2$O
Cohen, G. H. & Hoard, J. L. (1966) J. Amer. Chem. Soc. **88**, 3228-3234.

CaBr(Lactobionate)-4H$_2$O
Cook, W. J. & Bugg, C. E. (1973) Acta Cryst. **B29**, 215-222.

Ca(Benzo-15-Crown-5)$_2$(3,5-Dinitrobenzoate)$_2$-3H$_2$O
Cradwick, P. D. & Poonia, N. S. (1977) Acta Cryst. **B33**, 197-199.

CaBr(Glucuronate)-3H$_2$O
DeLucas, L., Bugg, C. E., Terzis, A. & Rivest, R. (1975) Carbohyd. Res. **41**, 19-29.

Ca(Urea)$_4$(Sulphate)
DeVilliers, J. P. R. & Boeyens, J. C. A. (1975) J. Cryst. Mol. Struct. **5**, 215-226.

Ca(Glutamate)-3H$_2$O
Einspahr, H. & Bugg, C. E. (1974) Acta Cryst. **B30**, 1037-1043.

Ca(Glutamate)2-4H$_2$O
Einspahr, H. & Bugg, C. E. (1977) Unpublished data.

Ca(Aspartate)(Ascorbate)-H$_2$O
Einspahr, H. & Bugg, C. E. (1977) In preparation.

CaCl(Glutamate)-H$_2$O
Einspahr, H., Gartland, G. L. & Bugg, C. E. (1977) Acta Cryst. In press.

Ca(Hydroxyprolinate)$_2$-5H$_2$O
Eriks, K. & Kadlec, R. J. (1977) Private communication.

Ca(Cycloheptanecarboxylate)-5H$_2$O
Flapper, W. M. J., Verschoor, G. C., Rutten, E. W. M. & Romers, C. (1977) Acta Cryst. **B33**, 5-10.

Ca(Arabonate)$_2$-5H$_2$O
Furberg, S. & Helland, S. (1962) Acta Chem. Scand. **16**, 2373-2383.

Ca(Glucarate)-4H$_2$O
Glusker, J. P., Minkin, J. A. & Casciato, C. A. (1971) Acta Cryst. **B27**, 1284-1293.

Ca(Hibiscus-ate)-4H$_2$O
Glusker, J. P., Minkin, J. A. & Soule, F. B. (1972) Acta Cryst. **B28**, 2499-2505.

CaNa(Galacturonate)$_3$-6H$_2$O
Gould, S. E. B., Gould, R. O., Rees, D. A. & Scott, W. E. (1975) J. Chem. Soc. Perkin **II**, 237-242.

Ca(Ethylacetate)$_2$(PO$_2$F$_2$)$_2$
Grunze, H., Just, K.-H. & Wolf, G.-U. (1969) Z. Anorg. Allg. Chem. **365**, 294-300.

CaH$_2$(Homophthalate)-5H$_2$O
Gupta, M. P. & Prasad, N. (1976) Acta Cryst. **B32**, 3257-3261.

Ca(Fumarate)-3H$_2$O
Gupta, M. P., Prasad, S. M., Sahu, R. G. & Sahu, B. N. (1972) Acta Cryst. **B28**, 135-139.

Ca(Ascorbate)$_2$-2H$_2$O
Hearn, R. A. & Bugg, C. E. (1974) Acta Cryst. **B30**, 2705-2711.

Ca(Diphenyl propionate)$_2$-1/2(Ethanol)
Hollander, F. J., Templeton, D. H. & Zalkin, A. (1973) Acta Cryst. **B29**, 1295-1303.

Ca2(Mellitate)-9H$_2$O
Jandacek, R. J. & Uchtman, V. A. (1977) Personal communication.

Ca(Urea)(Nitrate)$_2$-3H$_2$O
Lebioda, L. (1972) Roczniki Chem. **46**, 373-385.

Ca(Urea)$_4$(Nitrate)$_2$
Lebioda, L. (1977) Private communication.

CaBr$_2$(Urea)$_6$
Lebioda, L. & Stadnicka, K. (1977) Private communication.

Ca$_2$Sr(Propionate)$_6$,Paraelectric Phase
Maruyama, H., Tomiie, Y., Mizutani, I., Yamazaki, Y., Uesu, Y., Yamada, N. & Kobayashi, J. (1967) J. Phys. Soc. Japan **23**, 889.

Ca(Terephthalate)-3H$_2$O
Matsuzaki, T. & Iitaka, Y. (1972) Acta Cryst. **B28**, 1977-1981.

Ca(2-Keto-D-Gluconate)$_2$-3H$_2$O
Mazid, M. A., Palmer, R. A. & Balchin, A. A. (1976) Acta Cryst. **B32**, 885-890.

Ca(Glycerate)$_2$-2H$_2$O
Meehan, E. J., Einspahr, H. & Bugg, C. E. (1977) Unpublished data.

CaCl$_2$(Glycylglycine)
Meulemans, R., Piret, P. & Van Meerssche, M. (1971) Bull. Soc. Chim. Belges **80**, 73-81.

CaCl$_2$(Glycine)$_2$-4H$_2$O
Natarajan, S. & Mohana Rao, J. K. (1976) Curr. Sci. **45**, 490-491.

CaBr$_2$(Glycine)$_3$
Natarajan, S. & Mohana Rao, J. K. (1976) Curr. Sci. **45**, 793-794.

Ca(Formate)$_2$
Nitta, I. & Osaki, K. (1948) X-Sen **5**, 37-42.

Ca(D-Glucoisosaccharate)$_2$
Norrestam, R., Werner, P.-E. & Von Glehn, M. (1968) Acta Chem. Scand. **22**, 1395-1403.

Ca(Co(Aspartate)$_2$)$_2$-10H$_2$O
Oonishi, I., Shibata, M., Marumo, F. & Saito, Y. (1973) Acta Cryst. **B29**, 2448-2455.

Ca$_2$(Co(Aspartate)$_2$)$_4$-15H$_2$O
Oonishi, I., Sato, S. & Saito, Y. (1975) Acta Cryst. **B31**, 1318-1324.

CaBr$_2$(Diacetamide)$_4$
Roux, J. P. & Boeyens, J. C. A. (1970) Acta Cryst. **B26**, 526-531.

Ca(Diacetamide)$_5$(Perchlorate)$_2$
Roux, J. P. & Kruger, G. J. (1976) Acta Cryst. **B32**, 1171-1175.

CaH(Citrate-3H$_2$O
Sheldrick, B. (1974) Acta Cryst. **B30**, 2056-2057.

Ca(Serinate)2-2H$_2$O
Sicignano, A., Gandhi, S. & Eriks, K. (1977) Private communication.

Ca(Oxalate)-2H$_2$O
Sterling, C. (1965) Acta Cryst. **18**, 917-921.

CaBr(Pantothenate)
Sternglanz, H., DeLucas, L., Einspahr, H. & Bugg, C. E. (1977) Unpublished data.

Ca(Dipicolinate)-3H$_2$O
Strahs, G. & Dickerson, R. E. (1968) Acta Cryst. **B24**, 571-578.

Ca(D-Glucarate)-4H$_2$O
Taga, T. & Osaki, K. (1976) Bull. Chem. Soc. Japan **49**, 1517-1520.

Ca(Oxydiacetate)-6H$_2$O
Uchtman, V. A. & Oertel, R. P. (1973) J. Amer. Chem. Soc. **95**, 1802-1811.

CaCl$_2$(Glycylglycylglycine)-3H$_2$O
Van Der Helm, D. & Willoughby, T. V. (1969) Acta Cryst. **B25**, 2317-2326.

Ca(Antipyrine)$_6$(Perchlorate)$_2$
Vijayan, M. & Viswamitra, M. A. (1968) Acta Cryst. **B24**, 1067-1076.

CaH(Nitrilotriacetate)-2H$_2$O
Whitlow, S. H. (1972) Acta Cryst. **B28**, 1914-1919.

Ca(2,4,6,8-Cyclooctatetraene-1,2-Dicarboxylate)-2H$_2$O
Wright, D. A., Seff, K. & Shoemaker, D. P. (1972) J. Cryst. Mol. Struct. **2**, 41-51.

later. It is our hope that the rather superficial analysis described here will serve to underscore the fact that Ca--ligand interactions can involve considerable geometrical specificity, and that this analysis may help to stimulate the further experimental and theoretical studies that are needed to properly interpret these interactions.

CALCIUM---CARBOXYL INTERACTIONS

Forty-eight of the crystal structures that we examined display calcium--carboxyl interactions. These structures exhibit 190 crystallographically independent Ca--carboxyl interactions. For the purpose of our analysis, we divide the interactions into the four general classes depicted in Figure 1. Those calcium interactions with only one oxygen atom of a carboxyl group are classified as unidentate; there are 95 such examples. Those interactions involving chelation by the pair of oxygen atoms from the carboxyl group are classified as bidentate, and there are 23 examples of this type. Another class includes those carboxylic acids that chelate calcium ions through a carboxyl-oxygen atom acting in concert with a suitable ligand (oxygen or nitrogen) that is bonded to the α-carbon atom. This mode of chelation, labelled alpha, is responsible for 36 of the calcium--carboxyl interactions included in our survey. The alpha-chelation pattern is found in almost all cases where a suitable ligand is in the alpha position. The fourth class of calcium--carboxyl interactions that we recognize is labelled complex, since it consists of chelation patterns that are more complicated than those of the three other classes. The complex-chelation patterns involve one or more

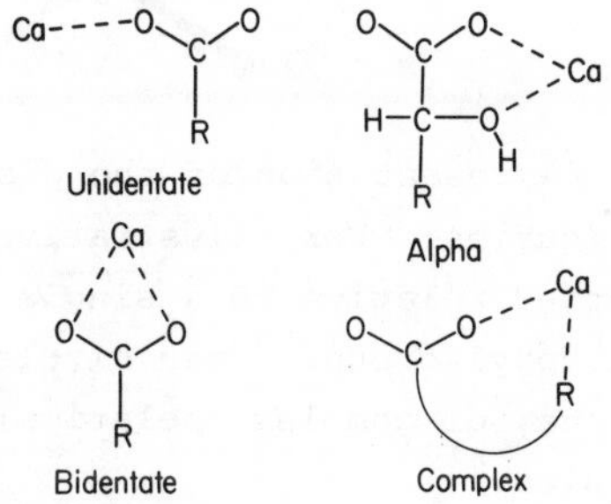

Figure 1. The major types of calcium--carboxyl interactions.

15

ligands acting in concert with a carboxyl group to chelate the calcium ion. We find 13 examples of such complex patterns. Most of the complex examples include a ligand in the alpha position and follow the general trend that is observed for the simple alpha chelates.

The overall geometrical features of the calcium--carboxyl interactions can be seen from the stereoscopic drawing in Fig. 2, which shows a composite of calcium positions, as viewed perpendicular to a carboxyl group. In this drawing, we have plotted all calcium positions relative to a single (the upper) oxygen atom of the carboxyl group. The unidentate-coordinated calcium ions are depicted as solid black circles. All other calcium ions (alpha, complex, and bidentate examples) are represented by open circles. The alpha and complex examples lie to the left of the vertical C-O bond. All of the alpha-chelated calcium ions fall within the tight clump that is situated between the oxygen atom and the R-group. The few open circles that lie to the right of this clump, but to the left of the vertical C-O bond, correspond to calcium ions in complex-chelation patterns that do not involve an α-substituent. The bidentate-chelated calcium ions lie in the region between the two oxygen atoms of the carboxyl group.

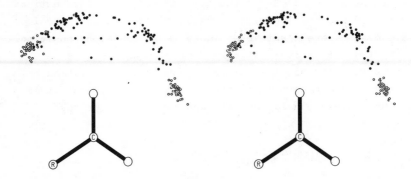

Figure 2. A stereoscopic representation of the distribution of observed Ca--carboxyl geometries. For illustrative purposes, calcium positions are plotted relative to a single oxygen atom (the upper one) of the carboxyl group. Open circles represent examples of bidentate, alpha, or complex chelation; filled circles represent unidentate examples.

Examination of Fig. 2 reveals a number of interesting features about the geometry of the Ca--carboxyl interactions in these crystal structures:

(1) There is little tendency for Ca ions to lie colinear with the C-O bond.

(2) Most of the Ca ions lie near the plane of the carboxyl group.

(3) The unidentate examples cluster on both sides of the C-O bond, with most Ca ions lying at C-O--Ca angles of about 140-160° on the side toward the R group, and 120-150° on the side toward the second oxygen atom of the carboxyl group.

(4) The bidentate-chelated calcium ions are confined to a restricted range by interactions with both oxygen atoms of the carboxyl groups; these calcium ions occur at C-O--Ca angles of about 90°.

(5) For the most part, the alpha- and complex-chelation patterns confine the Ca ions to a tight region at C-O--Ca angles of about 110-130°.

(6) Interactions of calcium ions with a single oxygen atom of the carboxyl group are much more common than bidentate interactions with the pair of oxygen atoms.

Close examination of the Ca--O distances for these interactions with carboxyl groups reveals several pertinent features. Except for a few examples, the Ca--O distances lie in the range 2.3-2.6 Å. There is a clear relationship between binding mode and Ca--O distance: the distances for unidentate-bound Ca ions tend to be shorter than those for the alpha and complex examples, which, in turn, are somewhat shorter than those for bidentate-bound Ca ions. The unidentate examples display an average Ca--O distance of 2.37 Å; the average for the alpha and complex examples is 2.41 Å; and the bidentate examples have an average Ca--O distance of 2.54 Å. As might be expected, there is a correlation between Ca--O distance and coordination number (the total number of ligand atoms that are in the coordination polyhedron around the calcium ion). The general trend is for Ca--O distances to increase with increasing coordination number. The average distances for Ca contacts with carboxyl-oxygen atoms are about 2.35 Å for 6-fold coordination, 2.38 Å for 7-fold coordination, and 2.45 Å for the 8-fold and 9-fold coordination patterns.

The 60 crystal structures contain a total of 70 crystallographically independent Ca ions. Of these 70 examples, 14 display

6-fold coordination polyhedra, 21 have coordination numbers of 7, 29 show 8-fold coordination patterns, and 6 Ca ions have 9-fold coordination polyhedra. The distribution of carboxyl-oxygen atoms among these various coordination schemes is roughly similar to this overall distribution. The majority of the carboxyl groups fall into 7-fold and 8-fold patterns; the 8-fold patterns seem to be slightly preferred. One interesting finding is that bidentate, alpha, and complex examples were not present in 6-fold coordination polyhedra.

The Ca--carboxyl interactions display a rather strong correlation between Ca--O distance and C-O--Ca angle. The general trend is for Ca--O distance to decrease as the C-O--Ca angle increases. The longest Ca--O distances correspond to the bidentate examples, all of which display C-O--Ca angles in the range 80-100°. As noted earlier, the alpha and complex examples tend to cluster between C-O--Ca angles of about 110° and 130° and to display Ca--O contact distances that are somewhat shorter than those for the bidentate examples. The unidentate examples, which cover a relatively wide range of C-O--Ca angles, are also consistent with the trend. The net result is an overall decrease in Ca--O distance in going from C-O--Ca angles of 90° to ones that approach 180°.

CALCIUM INTERACTIONS WITH CARBONYL GROUPS

Of the 60 crystal structures that we examined, 17 involve calcium interactions with carbonyl groups. A total of 56 crystallographically independent Ca--carbonyl interactions are observed. These interactions have been subdivided into the two categories depicted in Fig. 3. Those calcium interactions with only the carbonyl-oxygen atom are classified as unidentate; we find 26 interactions of this type. The second category of interactions includes all those examples where the Ca ion is

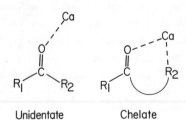

Unidentate   Chelate

Figure 3. The two classes of Ca--carbonyl interactions.

chelated through one or more additional ligands acting in concert
with the carbonyl-oxygen atom.   There are 30 examples of
Ca--carbonyl interactions of the chelate type.   Most of the chelate
examples are compounds in which the second ligand lies
close to the plane of the carbonyl group, a feature that may
tend to bias our results.

The overall geometrical features of calcium--carbonyl
interactions can be seen in the stereoscopic drawing shown in
Figure 4.   The drawing shows a plot of Ca ion positions relative
to the oxygen atom of the carbonyl group.   In this drawing, R_1 is
the substituent that displays the shorter bond to the carbon atom
of the carbonyl group; and, because most of the chelating groups'
included in this survey (for example, diacetamide) have the shorter
C-R bond lying between the ligand atoms, the chelated Ca ions lie
to the right of the vertical C-O bond.   These chelate examples are
clustered at C-O--Ca angles of 110°-140°.   The unidentate-bound
calcium ions, on the other hand, form a relatively diffuse,
umbrella-shaped cap around the carbonyl-oxygen atom.   Unlike the
carboxyl examples, there is no particular orientation of unidentate-
bound Ca ions that seems to be preferred, although there is a general
tendency for them to lie within a torus-shaped region defined by

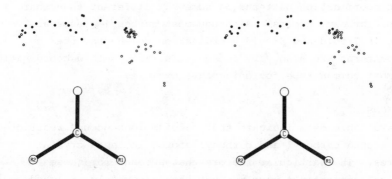

Fig. 4.   A stereographic representation of the distribution of
observed Ca--carbonyl geometries.   For illustrative purposes, the
calcium ions are plotted relative to the oxygen atom of a single
carbonyl group.   Open circles represent chelate examples; filled
circles represent unidentate examples.

C-O--Ca angles of about 120-160°.

The majority of both unidentate and chelate examples display C-O--Ca angles between 130° and 150°, with most clustered near 135°. As in the case of carboxyl groups, there is no obvious tendency for the calcium ions to be colinear with the C-O bond, and few examples occur at C-O--Ca angles greater than 150°. All of the chelated calcium ions display angles below 140°.

From Fig. 4, it can be seen that many of the Ca ions are considerably displaced from the plane of the carbonyl group. While there may be a slight tendency for the calcium ions to lie near the carbonyl plane, an appreciable percentage show displacements greater than 1 Å. However, the majority of Ca ions lie within 0.6 Å of the carbonyl plane.

The Ca--carbonyl distances follow the same general trends as those found for Ca--carboxyl interactions. The overall average of Ca--O distances in the carbonyl case is about 2.39 Å. The average for unidentate interactions is 2.34 Å, and the average for chelate interactions is 2.44 Å. As with the carboxyl groups, the Ca--O distances increase as the coordination number increases. Likewise, there is a clear correlation between C-O--Ca angle and Ca--O distance, with distance decreasing as angle increases.

The distribution of carbonyl groups among the various possible coordination patterns is somewhat different than that found for carboxyl groups. Although most carbonyl groups are involved in 7-fold and 8-fold coordination polyhedra, the 6-fold patterns are also very common. In fact, 6-fold coordination is the most common type for unidentate examples.

CONCLUSIONS

The available data indicate that calcium ions occupy restricted regions around carboxyl and carbonyl groups in these crystal structures. The particular factors that account for these geometrical constraints have not yet been adequately analyzed, but the observed arrangements suggest that calcium--carboxyl and calcium--carbonyl interactions may involve more stereo-specificity than is generally assumed.

Supported by N.I.H. grants DE-02670, CA-12159, and CA-13148. We thank Drs. Barnett, Eriks, Jandacek, Lebioda, and Uchtman for unpublished data.

Lanthanides as Spectroscopic and Magnetic Resonance Probes

Jacques Reuben*

Departments of Biochemistry & Biophysics and Biology,
University of Pennsylvania
Philadelphia, Pennsylvania 19104, U.S.A.

INTRODUCTION

Calcium ions play a key role in a great variety of biological processes and there is a growing interest in elucidating the mechanism of their action and the mode of their interaction with proteins and other constituents of living systems. However, as is often the case, the importance of studying these aspects of calcium action has been found to be inversely proportional to the ease of doing so, mainly because Ca^{2+} is an ion with a closed electronic shell, devoid of spectroscopic properties that could be used to monitor the interaction with its environment. Therefore, the suggestion[1] that lanthanides could be used as probes for calcium binding sites was met with great interest. The possibility of using spectroscopic and magnetic resonance methods in the study of lanthanide complexes with proteins was explored and the first encouraging reports[2-7] appeared in 1970 and 1971. Particularly important was the finding that Nd^{3+} would accelerate the rate of the trypsin catalyzed conversion of trypsinogen more efficiently than Ca^{2+}.[8]

The literature entries on the general subject of the use of lanthanides as probes and on their interaction with systems of biological interest is now well into the hundreds. A number of review articles have already appeared. Glasel has summarized the chemical shift effects of lanthanides on the nuclear magnetic resonance (NMR) spectra of molecules of biological interest.[9] Nieboer

---

* On leave from the Isotope Department, the Weizmann Institute of Science Rehovot, Israel.

has treated the subject _vis_ _a_ _vis_ the coordination chemistry of lanthanides in model systems. [10] Mikkelsen has reviewed the use of lanthanides as calcium probes in biomembranes. [11] This author has written a short summary[12] as well as a chapter extensively covering the literature and critically reviewing the field. [13] This paper is intended to provide merely a key to the literature. Given in the next section is a brief outline of the different physical techniques. It is followed by a summary of the calcium binding proteins, the interaction of which with lanthanides has been studied.

The lanthanides are the series of elements with atomic numbers between 57 (La) and 71 (Lu) in which the 4f electronic shell is progressively filled. Since virtually all the states from $4f^0$ to $4f^{14}$ are available (promethium has no stable isotopes and $4f^4$ is practically missing) the lanthanides offer a remarkable multitude of spectroscopic and magnetic properties that may be useful in probing their interaction with the environment. Their stable oxidation state in aqueous solution is +3. The tripositive ions have similar chemical properties with only gradual variations along the series resulting from the lanthanide ionic radius contraction. The ionic radius of Ca^{2+} (0.99A) is well within the range of ionic radii of La^{3+} (1.061A) and Lu^{3+} (0.848A).

PHYSICAL METHODS

Optical Absorption: The spectral changes in the hypersensitive transitions of Nd^{3+} in the 500-600 nm region that are observable upon complex formation have been used by Birnbaum et al. to study the interaction with proteins. [3,14] The changes are small and the recording of a difference spectrum is required. Similar difference spectra are obtained upon binding to proteins and to simple amino acids. It is, therefore, difficult to identify the binding site from such measurements. However, the method is well suited as an analytical technique.

The ultraviolet absorption spectra of proteins are perturbed when lanthanides bind in the vicinity of aromatic chromophores. The magnitude of the perturbation can be used for analytical purposes.[6]

Fluorescence: Both Tb^{3+} and Eu^{3+} are useful as fluorescence probes. Upon binding to proteins the fluorescence of Tb^{3+} is often enhanced due to energy transfer.[6] The enhancement can be used to monitor the binding of Tb^{3+} and the competition of other cations as well.[15] Concomitant with the enhancement, quenching of the protein fluorescence is observed, which along with the wave length of maximum enhancement can be used to identify the protein residue involved in energy transfer.[6,15] Quenching of Tb^{3+} fluorecence may occur as a result of dipole-dipole interaction with another nearby paramagnetic ion. In this way, the distance between two metal-ion binding sites on the same protein molecule can be determined.[6,16,17]

The circular polarization of luminescence reflects the asymmetry of the ligand field in the excited electronic state of the Tb^{3+} ion and can be used qualitatively to observe similarities or differences between the metal-ion binding sites of different proteins[18] and to detect heterogenous binding[19] (cf. also ref. 20).

Proton Relaxation Rates: The longitudinal relaxation rate of water protons in solutions of Gd^{3+} is much faster than that in pure water owing to the magnetic dipolar interaction between the unpaired electronic spin of the ion and the proton magnetic moments of the water molecules in the first hydration sphere. The magnitude of the effect is directly proportional to the number of water molecules attached to the cation and inversely proportional to the sixth power of the distance between the two magnetic moments. It also depends upon the correlation time modulating the interaction (a combination of the rotational correlation time, the electron spin relaxation time, and the

23

mean residence time of a proton in the hydration sphere) and upon the resonance frequency. When Gd^{3+} binds to macromolecules a net enhancement in the water proton relaxation rate is observed, which can be used to monitor the binding.[4,5] The enhancement results from a change of the dominant correlation time, an effect that more than compensates for the loss of water molecules accompanying the binding. From a detailed analysis of the frequency and temperature dependence of the relaxation rates an estimate for the number of waters of hydration of the bound Gd^{3+} ion can be obtained, provided that the ion-proton distance is known.[5,19] Conversely, nuclear relaxation rates can be used to estimate distances of substrate or inhibitor nuclei from a Gd^{3+} binding site on an enzyme.[21] However, this approach requires an almost absolute homogeneity of the system with respect to the binding of both Gd^{3+} and the inhibitor (or substrate). If this requirement is not satisfied, erroneous conclusions will result.[22]

Relaxation of Lanthanum-139: Lanthanum-139 is a nucleus with a spin quantum number $I = 7/2$ possessing an electric quadrupole moment. The interaction of the latter with intramolecular electric field gradients modulated by molecular reorientation is responsible for the relaxation of such nuclei. Recent studies have shown that the relaxation of ^{139}La can be used to determine molecular tumbling times of macromolecules.[23]

NMR Chemical Shifts: Paramagnetic lanthanides (except for Gd^{3+}) induce large chemical shifts in the NMR absorptions of ligands in their vicinity. One of the main origins of the shifts is the dipolar interaction between the nuclear spin and the anisotropic magnetic moment of the lanthanide. The shifts depend both on the ion-nucleus distance, r, and on the angles between the vector r and the principal magnetic axes and can be used in structure determinations.[2,9,24]

<u>EPR of Gadolinium (III)</u>: The electron paramagnetic resonance (EPR) spectrum of Gd^{3+} in aqueous solution consists of a broad single line. The line-width changes upon binding to proteins.[7] Recent studies have shown that these changes reflect on the ligand-field symmetry at the binding site.[25]

<u>Mossbauer Spectroscopy</u>: Most of the lanthanides have isotopes suitable for Mossbauer spectroscopy. However, the technique has not been explored in detail with regard to systems of biological interest. In the only published report the spectrum of the Eu^{3+}-transferrin complex has been examined.[26] It contains little information since at the temperature of the experiment (4.2K) the ground singlet level of Eu^{3+} is the only one populated and it is unaffected by the environment.

<u>X-ray Crystallography</u>: All of the lanthanides give rise to anomalous scattering components of x-ray radiation that are useful for phase determinations in crystallographic studies of macromolecules. The isomorphous replacement of calcium by lanthanides in thermolysin has been studied in detail by x-ray crystallography.[27]

INTERACTION OF LANTHANIDES WITH CALCIUM BINDING PROTEINS

Complex formation between a lanthanide ion and a protein may result in spectral perturbations of the following general types: modification of the lanthanide spectrum, alteration of the protein spectrum, and changes in the proton relaxation times of the water of hydration. The magnitudes of perturbations of this kind as well as some common biochemical techniques have been used to quantitate the binding of lanthanides to proteins. The effect of lanthanide substitution on the activity of a number of calcium binding proteins has also been studied. A listing is given in Table 1.

The ionic radius dependence of the stability constants of the lanthanide-trypsin complexes exhibit a striking similarity to those of the nitrilotriacetate (NTA) chelates, indicating a major contribution from the free entropy of complexation. The temperature dependence of the dissociation constants

TABLE 1

CALCIUM BINDING PROTEINS STUDIED BY THE

LANTHANIDE PROBE METHOD

Protein	Activity*	Spectr. Study	Refs.
Aequorin	+		28
α-Amylase	±	+	20, 29-31
Concanavalin A	+	+	20, 32, 33
Factor X, Bovine	-		34
Parvalbumin		+	20, 35
Phospholipase A_2	+	+	20, 36
Staph. Nuclease	-	+	37
Thermolysin	+	+	16, 17, 20, 27
Troponin		+	20
Trypsin	±	+	8, 15, 19, 21, 22, 38

*   Activity or inhibition are marked by the + and - signs respectively

of the trypsin and bovine serum albumin complexes of Gd^{3+} also reveal large free entropies. With the assumption that this is the result of dehydration, their values can be used to estimate the number of water molecules remaining on the complexed Gd^{3+} ion. These estimates are in good agreement with the results of a detailed analysis of the frequency dependence of the water proton relaxation rates.[13,19]

With regard to the effect of lanthanide substitution on enzymatic activity, no generalizations can be offered at this time. It is advantageous to use the lanthanides as probes with enzymatically active systems. If the lanthanides are inhibitory, the mechanism of inhibition has to be elucidated before mean-ingful conclusions can be drawn from spectroscopic and magnetic resonance studies. An example of a successful application of the lanthanide probe approach is the positive identification of the calcium binding site on the mo-lecular model of trypsin.[19]

REFERENCES

1. Williams, R. J. P. (1970) Quart. Rev. 24, 331.

2. Morallee, K. G., Nieboer, E., Rossotti, F. J. C., Williams, R. J. P.,
   Xavier, A. V., and Dwek, R. A. (1970) Chem. Commun. 1132.

3. Birnbaum, E. R., Gomez, J. E., and Darnall, D. W. (1970) J. Am. Chem.
   Soc. 92, 5287.

4. Dwek, R. A., Richards, R. E., Morallee, K. G., Nieboer, E., Williams,
   R. J. P., and Xavier, A. V. (1971) Eur. J. Biochem. 21, 204.

5. Reuben, J. (1971) Biochemistry 10, 2834.

6. Luk, C. K. (1971) Biochemistry 10, 2838.

7. Reuben, J. (1971) J. Phys. Chem. 75, 3164.

8. Darnall, D. W., and Birnbaum, E. R. (1970) J. Biol. Chem. 245, 6484.

9. Glasel, J. A. (1973) Current Research Topics in Bioinorganic Chemistry,
   Lippard, S. J., Ed., John Wiley and Sons, New York, pp. 383-413.

10. Nieboer, E. (1975) Structure and Bonding 22, 1.

11. Mikkelsen, R. B. (1976) Biological Membranes, Vol. 3, Chapman, D.,
    and Wallach, D. F. H., Eds., Academic Press, New York, pp. 153-190.

12. Reuben, J. (1975) Naturwissenschaften 62, 172.

13. Reuben, J. (1977) Handbook on the Physics and Chemistry of Rare Earths,
    Gschneider, K. A., Jr., and Eysing, L., Eds., North-Holland, Amster-
    dam, in press.

14. Birnbaum, E. R., and Darnall, D. W. (1973) Bioinorg. Chem. 3, 15.

15. Epstein, M., Levitski, A., and Reuben, J. (1974) Biochemistry 13, 1977.

16. Berner, V. G., Darnall, D. W., and Birnbaum, E. R. (1975) Biochem.
    Biophys. Res. Commun. 66, 763.

17. Horrocks, W. DeW., Jr., Holmquist, B., and Vallee, B. L. (1975) Proc.
    Nat. Acad. Sci. USA 72, 4764.

18. Gafni, A., and Steinberg, I. Z. (1974) Biochemistry 13, 800.

19. Epstein, M., Reuben, J., and Levitzki, A. (1977) Biochemistry, in press.

20. Brittain, H.G., Richardson, F.S., and Martin, R.B. (1976) J. Am. Chem. Soc. 98, 8255.

21. Abbott, F., Gomez, J.E., Birnbaum, E.R., and Darnall, D.W. (1975) Biochemistry 14, 4935.

22. Epstein, M., and Reuben, J. (1977) Biochem. Biophys. Acta 481, 164.

23. Reuben, J., and Luz, Z. (1976) J. Phys. Chem. 80, 1357.

24. Marinetti, T.D., Snyder, G.H., and Sykes, B.D. (1977) Biochemistry 16, 647.

25. Reed, G.H., private communication.

26. Spartalian, K., and Oosterhuis, W.T. (1973) J. Chem. Phys. 59, 617.

27. Matthews, B.W., and Weaver, L.H. (1974) Biochemistry 13, 1719.

28. Izutsu, K.T., Felton, S.P., Siegel, I.A., Yoda, W.T., and Chen, A.C.N. (1972) Biochem. Biophys. Res. Commun. 49, 1034.

29. Levitzki, A., and Reuben, J. (1973) Biochemistry 12, 41.

30. Darnall, D.W., and Birnbaum, E.R. (1973) Biochemistry 12, 3489.

31. Steer, M.L., and Levitzki, A. (1973) FEBS Lett. 31, 89.

32. Sherry, A.D., Newman, A.D., and Gutz, C.G. (1975) Biochemistry 14, 2191.

33. Barber, B.H., Fuhr, B., and Carver, J.P. (1975) Biochemistry 14, 4075.

34. Furie, B.C., and Furie, B. (1975) J. Biol. Chem. 250, 601.

35. Donato, H., Jr., and Martin, R.B. (1974) Biochemistry 13, 4575.

36. Hershberg, R.D., Reed, G.H., Slotboom, A.J., and deHaas, G.H. (1976) Biochemistry 15, 2268.

37. Furie, B., Eastlake, A., Schecter, A.N., and Anfinsen, C.F. (1973) J. Biol. Chem. 248, 5821.

38. Gomez, J.E., Birnbaum, E.R., and Darnall, D.W. (1974) Biochemistry 13, 3745.

NMR Studies of Various Calcium-Binding Proteins

B.A. Levine*, R.J.P. Williams*, C.S. Fullmer** and R.H. Wasserman**
*Inorganic Chemistry Laboratory
Oxford University, England

** Dept. of Physical Biology
Cornell University
Ithaca, N.Y. 14850

NMR studies of various calcium binding proteins

Calcium mediated processes which link external stimuli to intracellular events involve Ca(II) binding proteins whose function is not only closely related to the triggering event but also to the control through transport and buffering of the changes in calcium ion concentration. Elucidation of the function of these proteins requires not only a knowledge of their structures in the calcium bound form but also of the extent of dynamic fluctuations of groups in the various conformational states adopted by each protein in calcium free and bound states. Whereas X-ray diffraction studies can yield a static structural model of a globular protein they can not define more mobile structures. Structural studies in solution are required in order to obtain both static and dynamic parameters. NMR spectroscopy undoubtedly provides the most detailed information about protein conformational states in solution since signals are detected from most atoms in the molecule. Structural determination by NMR is obtained by assignment of the resonances of specific amino acid side chains. These resonances can then be used as probes to investigate the running changes in the protein conformation which occur in response to a variation in the solution conditions.[1,2,3]

Assignment of spectra

Each resonance in the NMR spectrum is defined by several spectral parameters; (i) its intensity, which is a direct measure of the number of contributing nuclei; (ii) its chemical shift (resonance energy), which is determined by the chemical environment of the nuclei and therefore on the structure and conformation of the molecule; (iii) the multiplicity (fine structure) of the signal, which yields the nuclear spin/spin coupling to the immediately neighbouring nuclei and (iv) the relaxation properties seen in part in the resonance lineshape. The last parameters contain information relating to the dynamics of a group since the processes which contribute to the observed lineshape include effects of exchange between magnetically inequivalent sites and relaxation (motional) effects which influence the lifetime of the nuclei in their excited states.

Classification of any resonance to a certain type of chemical group, $-CH_3$ or
$-NH$, is often based on major chemical shift differences and intensity data
(e.g. aromatic sidechain protons appear in the range 6-8.5ppM, methyl groups
ca -1 to 2ppM, solvent exchangeable peptide $-NH$ resonances 7-9.5ppM) (Figure 1)
Assignment to a type of amino acid (e.g. leucine or phenylalanine) may be

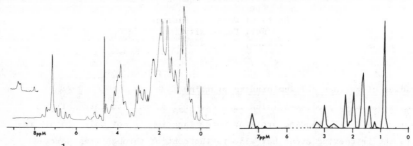

Figure 1. ^{1}H NMR spectrum of bovine Vit-D dependent CaBP. Native Ca(II)-
band protein (left), showing $-NH$ resonances on an expended scale;
simulated random coil (right).

achieved by perturbation of spin/spin coupling (double resonance effects) since
the multiplicity of a resonance, its intensity and the resonance energy separation between coupled resonances characterize signals deriving from groups
on particular amino acids.

The NMR spectrum of a globular protein with a preferred conformation in
solution typically shows many nuclei which experience superimposed relatively
minor changes in resonance energy (secondary shifts) due to variations in local
environment defined by the native protein conformation rather than a solvent environment as in the case of a random coil protein (Figure 1). The measurement
of the secondary shifts is therefore linked to the definition of the protein
structure.

The major effect we shall discuss here is the secondary shift of a resonance
from a nucleus, A, due to the proximity of a ring current, B, of an aromatic
group. The observed secondary shift varies as $(3\cos^2\theta-1)/r^3$ where r is the distance between A and B and θ is the vector direction A→B relative to a perpendicular to the plane of the benzene ring. Monitoring these shifts therefore
permits us to follow changes in disposition of A relative to B. We shall also
use both shift and relaxation probes.[4,5] Such an NMR study is greatly helped if
the structure of one state of the system has been defined by X-ray structure
determination for then changes of state are easily followed by NMR.

We have studied the binding of calcium to proteins from four different
classes (1) those related to activation (Troponin-C), (2) those thought to be
involved in calcium ion buffering (parvalbumin), (3) proteins related to calcium
transport (Vitamin-D dependent CaBP from bovine intestine) and (4) extracellular
enzymes (lysozyme and phospholipase A2).

CONFORMATIONAL STUDIES IN SOLUTION

I. Lysozyme

Hen egg white lysozyme (molec.wt. ca. 14,000) is an enzyme consisting of a single polypeptide chain. This protein has several very weak single carboxylate metal binding sites, e.g. Asp-101, Asp-87 and one stronger site in the active site cleft between two carboxylate groups Glu-35 and Asp-52. (ca. 0.8-1nm apart). Many resonance assignments to specific residues have been achieved through chemical modification coupled with the effects of binding of paramagnetic probes (lanthanide cations and spin labels), inhibitors and protons.[4] Though not required for activity binding of calcium and the lanthanides occurs predominantly at the active site carboxylates.[5] Conformational studies using the lanthanides as probes (in place of calcium) have enabled the whole of the active site cleft to be mapped out relative to the bound paramagnetic probe. There is a close similarity between this solution structure and the X-ray crystal structure.[6] There are three tryptophan residues in the active site and the NMR data shows that this region of the protein possesses a certain degree of mobility, largely oscillation of tryptophan-62, and rotation of groups such as valine and leucine.

In the protonated state, Glu-35 (pK_a>6) lies close to the aromatic group of Trp-108. NMR data show that ionization of this carboxylate in a pH titration or on metal binding results in a local conformational change in the active site region, the carboxylate moving into free solution away from the tryptophan ring. The displacement of a proton from this carboxylate residue plays a key role in the proposed catalytic mechanism of lysozyme.[6] Addition of metal ions causes a movement together of the two active site carboxylates as Glu-35 leaves the vicinity of Trp-108. The lack of exchange broadening shows that this movement is fast (>10^3sec^{-1}). This conformational change has been suggested as a model for an ion gating mechanism[7] whereby the entrance to a calcium channel consists of two carboxylates held in a particular constraint (as in the Trp-108/Glu-35 complex) which is altered upon interaction of the calcium with this site.

The interior of the protein is mobile as shown by the high flipping rates of tyrosine and probably phenylalanines. The slow decrease of the ring current secondary shifts with temperature shows that the protein expands slowly until at about 75°C when it denatures to a random coil with no secondary shifts. Metal binding causes a small tightening of all the protein.

It is very likely that other globular proteins which are calcium requiring enzymes, e.g. phospholipase A and bacterial nucleases, are affected by calcium binding in a similar, relatively small, way.

II. Parvalbumins

Parvalbumins form a class of closely homologous globular proteins of low molecular weight (ca. 11,5000) which contain two calcium binding sites and have a high phenylalanine content (8-10 residues). Most of the phenylalanine

31

residues are internal and are conserved in the sequence of parvalbumins from different species. Although chemically well characterized, the biological role of the parvalbumins has not yet been established, though they may act as calcium ion buffers in muscle cells (Pechere, this volume).

Most of the aromatic proton resonances in the spectra of several different parvalbumins have been resolved and coupling patterns have been found using multiple resonance techniques [8] It has been established that most, if not all, of the aromatic residues are undergoing rapid flips about the C_β-C_γ bonds, thereby demonstrating extensive mobility of the internal structure of these proteins. A feature of the NMR spectrum is the detection of resonances from para and meta protons of three different phenylalanine residues which suffer large upfield shifts from their primary positions. This shows that many aromatic residues are close together and that the respective orientation between residues is such that the para and/or meta protons of one side chain are pointing towards the π electron distribution of another residue. These conclusions indicate that the solution conformation of the calcium bound form is the same as that given by the X-ray structure[9], the internal core of the protein consisting of a cluster of eight phenylalanine residues with no π-π stacking.

Proton NMR studies have provided evidence for expansion of the structure on increase of temperature and also dimerization by formation of a disulphide bridge. Interestingly the X-ray structure shows that the single cysteine residue of each monomer is not exposed on the surface. The dimerization process can only occur by a conformational fluctuation which exposes to solvent the sulfhydryl groups of two molecules. ^{13}C NMR studies[10] have also been used to investigate the intramolecular motions of the hydrophobic core of these proteins and the conformational change resulting from the removal of one of the calcium ions (the E-F hand). The findings confirm the observations discussed earlier that the whole protein may be cooperative in that it may feel the binding of a cation to a local region of its sequence. In fact there is evidence that the calcium free protein approaches a random coil and successive binding of calcium ions causes major conformational changes. It is not until two calcium ions are bound that the protein has a relatively tight globular structure as seen in crystals.

III. Troponin-C (TN-C)

Similar data to that given for parvalbumin have been obtained by NMR studies on this protein which is involved in the contractile response to calcium. Four calcium ions bind to the protein which resembles a random coil in the metal free state. Binding takes place in two stages, the first two calcium ions promoting the fold of the protein as seen by induced ring current shifts and changes in the backbone α-CH_α resonances. The second two calcium ions tighten this fold. Troponin-C and parvalbumin both show that calcium binding constants are related differently to protein fold energies and not just to ligand binding strengths at

the preformed site. Details of studies on TN-C have been published[11],[12] and
point to the following additional observations:-

(a) There is internal rotation of aromatic residues but this is restricted
in some cases at room temperature in the presence of calcium.

(b) The protein has a smaller coefficient of expansion than parvalbumin.

(c) Proton binding is cooperative in the metal free state.

(d) The conformational rearrangement resulting from the binding of the first
two calciums may be characterized by a rate constant for the calcium on/off reaction which is of the order of 10 sec^{-1} or less and is therefore unlikely to be
involved in regulation. The on/off rate constants are relatively faster for the
second two calcium ions (10^1-10^2 sec^{-1}). This binding reaction is modified in
the presence of the other two component proteins of troponin.

We concentrate now on the interaction with troponin-I and troponin-T. For
many resonances in the spectrum of TN-C the linewidth is governed largely by the
overall rotation of the protein molecule. Correlation between linewidth and
molecular tumbling is not however expected to hold in case of resonances of surface groups for there is a high probability of considerable local mobility for
these residues, e.g. lysine, arginine and histidine. The mobility of exposed
groups, independent of the motion of the protein as a whole, has also been indicated by relaxation time studies on lysozyme.[13] The mobility at the surface
of a protein is of importance in the understanding of protein-protein recognition, e.g. in the troponin trimer complex. Binding of the last two calciums
to TN-C influences the surface state of the protein and these conformational
effects are observed to be transmitted to the other adjacent components. The
structure of TN-C is altered to different extents upon complexation with TN-I
and TN-T, the reduction in mobility of lysine and arginine resonances of TN-I
and TN-T demonstrate the interaction of basic regions of these proteins with
acidic sites on TN-C. A general point to notice is that the conformation of an
acidic protein can be altered by binding to a basic one and we shall be interested to see if the resulting conformational change alters calcium-binding to
the acidic component.

IV Vitamin-D dependent Ca(II) binding protein

The work to be described has been carried out in collaboration with Professor
R.H. Wasserman. The bovine intestinal calcium binding protein has a molecular
weight of 9800, two calcium binding sites of affinity >ca 10^6 and is involved in
the Vitamin D-dependent calcium translocation process (see papers by Wasserman
and Fullmer, this volume). No crystallographic data are as yet available.

The NMR spectrum is shown in Figure 2 for the protein in the absence of calcium, one calcium and two calcium bound states. The aromatic resonances derive
from one tyrosine and five phenylalanines, the sequence of the protein from
mammalian species lacking in histidine and tryptophan residues.[14] The resonances

of the tyrosine residue are shown in the figure. The spread in chemical shift

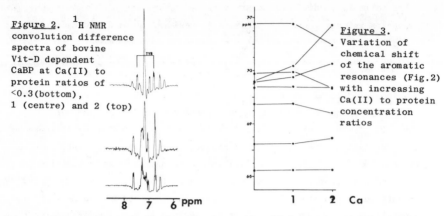

Figure 2. ^{1}H NMR convolution difference spectra of bovine Vit-D dependent CaBP at Ca(II) to protein ratios of <0.3(bottom), 1 (centre) and 2 (top)

Figure 3. Variation of chemical shift of the aromatic resonances (Fig.2) with increasing Ca(II) to protein concentration ratios

values of the resonances indicates the degree to which these aromatic groups interact with one another upon binding calcium. In the calcium free state per-turbation of the resonances of the tyrosine and one phenylalanine residue is ob-served. Binding of one calcium leads to a perturbation of a ring shifted iso-leucine methyl resonance and an increased interaction of the aromatic residues whose resonances are all markedly perturbed in the two calcium state, Figure 3. The consecutive course of folding of the polypeptide chain may be summarized as:

Ca(II) free	⇌	one Ca(II)	⇌	two Ca(II)
little overall fold;		Tyr/Phe fold;		well defined protein
local tyrosine/		4 Phe residues		fold involving both
phenylalanine fold;		fold influences		Tyr/Phe and 4 Phe
4 Phes weakly folded				strongly folded

The first calcium bound is held tenaciously and the on/off rate constant must be extremely slow (ca. 10 sec^{-1}). The spectral changes resulting from the bind-ing of this calcium suggest that the site, partly preformed, is stabilized in the presence of calcium. Evidence for this comes from the very high pK_a(>12) obser-ved for the hydroxyl group of the tyrosine residue in the absence of calcium and the increased thermal stability of the protein in the one calcium state. Calcium binding at the first site may also stabilize and/or in part preform the second calcium site as indicated by the increased interaction of four of the phenyl-alanine residues. The second calcium is held less firmly and the on/off rate constants are >10^2 sec^{-1}. Binding at the second site completes the protein fold initiated at the first binding step and results in a well defined, thermally stable conformation. There is further evidence for cooperativity between the two binding sites, the conformational change induced by binding of the second calcium influencing the local environment of the tyrosine and the methyl groups of an isoleucine residue. Further addition of calcium did not result in any de-

34

tectable spectral changes. Thus the reported binding of calcium to sites with affinity ca 10^2 [14] must occur to exposed carboxylates and does not modify the overall protein conformation.

The stability of the protein fold is also indicated by a study of the exchange of peptide -NH resonances. Four -NH protons resonances are observed in the NMR spectrum of the two calcium bound protein (Figure 1) which are found to be resistent to exchange with the solvent, D_2O, under varying pH conditions even after prolonged heating above $80^{\circ}C$. These resonances are exposed to solvent upon removal of the second calcium by back titration with EGTA. The protein conformation does not however place severe packing constraints on the aromatic groups which are found to possess rotational mobility in the two calcium state.

The overall conformation of the protein is not yet known but using the NMR data we have at present we can obtain a working outline model. This is based on the presence of two calcium binding regions in the protein sequence. Each site is taken to consist of a helix-loop-helix configuration, about 30 residues in length. The trial structure (Figure 4) shows tyrosine to be in a hydrophobic pocket containing an isoleucine and phenylalanine residue and is consistent with the observed stabilization by calcium of tge interaction between the remaining aromatic groups in the sequence. This model is consistent with sequence data available (Fullmer & Wasserman, this volume). Further details of the structure can be obtained by a thorough assignment of the resonances in the NMR spectrum and the use of paramagnetic probes to map out the environment of the calcium binding sites.[15]

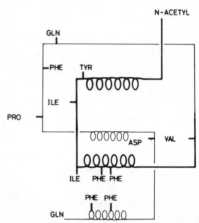

Figure 4. Proposed outline structure of bovine Vit-D dependent CaBP

CONCLUSIONS FROM CONFORMATIONAL STUDIES

The observations in all the calcium proteins we have studies are summarized in Table I. We see that for the enzymes which have relatively weak calcium binding the sites and overall fold are virtually preformed so that the only notable conformational changes promoted by calcium are in the local active site region. The on/off reactions are fast. The calcium buffering, triggering and transport proteins are very different. Calcium binding occurs in steps. The first step (one or two calcium ions) results in a major conformational change. The on/off reactions are slow. The second step gives a general and often considerable tightening of the whole protein. The on/off reactions are now generally faster.

Thus we may suppose that it is the second step which is the active (triggering) mode in most circumstances. Note that the calcium ion binding in both steps generally affects the whole of the protein. The conformational change may arise both via through bond effects or electrostatic energy changes. In all the proteins in Table I the binding constants reflect conformational energies and are therefore modulated by the energy of the protein fold.

TABLE I Summary of the conformational effects induced by the binding of calcium to various proteins studied

	Ca(II)-free	First calcium ion(s) bound	Second calcium ion(s) bound
Lysozyme	Highly structured	1 Ca(II) Structure tightens; Small effect; Exchange rate $>10^4s^{-1}$	2 Ca(II) No effect; very weak binding
Phospholipase A2	Highly structured	1 Ca(II); local conformational change	
Phosphatase	Highly structured	1 Ca(II) small effect	
Parvalbumin	Lowly structured	1 Ca(II); large conformational change; Exchange rate $>10s^{-1}$	1 Ca(II) Some tightening of conformation; Local structural changes; Exchange rate $10^1-10^2s^{-1}$
Troponin-C	Lowly structured	2 Ca(II) Large conformational change; Exchange rate $>10s^{-1}$	2 Ca(II) Some tightening of conformation; Local structural changes; Exchange rate $10^1-10^2s^{-1}$
Vit. D-	Lowly structured	1 Ca(II) Large conformational change; Exchange rate $ca10s^{-1}$	1 Ca(II) Tightening of the protein fold; Exchange rate $>10^2s^{-1}$

The binding of ions other than Calcium

Binding of magnesium to troponin-C (overall K $\sim 10^3$) is less specific than for calcium. Magnesium ions are observed to bind to the first two Ca(II) sites only but do not induce a conformational change identical to calcium. Two separate magnesium binding sites have also been detected. Cadmium binding occurs to all four Ca(II) sites and results in structural changes similar to calcium. The binding of protons to TN-D also mimics that of calcium in a qualitative sense.

Acknowledgements

This work is supported by the Medical Research Council. The studies on troponin were carried out in close collaboration with Dr. D. Mercola and are supported by a grant from the Muscular Dystrophy Group of Great Britain. R.J.P.W. is a member of the Oxford Enzyme Group.

REFERENCES

1. Campbell, I.D., Dobson, C.M. and Williams, R.J.P. (1975) Proc. Roy. Soc. Lond. A345, 23-40

2. Campbell, I.D., Dobson, C.M., Williams, R.J.P. and Xavier, A.V. (1973) J. Mag. Res. 11, 172-181

3. Campbell, I.D., Dobson, C.M., Williams, R.J.P. and Wright, P.E. (1975) FEBS Letts., 57, 96-99

4. Cassels, R., Dobson, C.M., Poulsen, F.M. and Williams, R.J.P. (1977) Europ. J. Biochem. (submitted)

5. Dobson, C.M. and Williams, R.J.P. (1977) in "Metal-Ligand Interactions in Organic Chemistry and Biochemistry" B. Pullman and N. Goldblum (eds), Reidel Publishing Company.

6. Imoto, T., Johnson, L.N., North, A.C.T., Phillips, D.C. and Rupley, J.A. (1972), in "The Enzymes" Vol.III 3rd edition, Boyer, P.D. (ed), Academic Press, N.Y.

7. Williams, R.J.P. (1976) in "Calcium in Biological Systems", 30th Symposium of the Society for Experimental Biology, C.U.P.

8. Cave, A., Dobson, C.M., Parello, J. and Williams, R.J.P. (1976) FEBS Letts., 57, 96-99

9. Moews, P.C. and Kretsinger, R.H. (1975) J. Mol. Biol., 91, 201-228

10. Nelson, D.J., Opella, S.J. and Jardetzky, O. (1976) Biochemistry, 15 5552-5560

11. Levine, B.A., Mercola, D. and Thornton, J.M. (1976) FEBS Letts., 61, 218-222

12. Levine, B.A., Mercola, D., Thornton, J.M. and Goffman, D. (1977) J. Mol. Biol. (in press)

13. Campbell, I.D., Dobson, C.M., Ratcliffe, R.G. and Williams, R.J.P. to be published.

14. Wasserman, R.H., Corradino, R.A., Fullmer, C.S. and Taylor, A.N. (1974) Vitamins and Hormones, 32, 299-324

15. Since this paper was typed, experiments using Mn(II) as a relaxation probe have provided results consistent with the model.

37

The Role of Calcium as a Conformational Lock in the Structure of Thermostable Extracellular Proteases

R. S. Roche and G. Voordouw*

Biopolymer Research Group
Department of Chemistry
University of Calgary
Calgary, Alberta, Canada T2N 1N4

INTRODUCTION

As far as can be discerned from the literature, one of the first reports to suggest that calcium may play a role in the structure and function of extracellular proteases is that published by Delezenne[1] in 1905 concerning the accelerating effect of calcium salts on the activation of pancreatic juices. The rationalisation of the early observations of Delezenne and others in terms of the autoactivation and autolysis of trypsinogen and the stabilisation of the resultant trypsin by calcium has been reviewed by J. H. Northrop and his coworkers[2]. It has been established that trypsinogen binds two calcium ions one of which ($K_d \sim 10^{-1.8}$ M)[3] is essential for the activation process and is lost in the formation of trypsin; the other higher affinity calcium ($K_d \sim 10^{-3.2}$ M)[4] binds at a site which is common to both proteins and which protects trypsin against autolytic[5], thermal and chemical degradation[6]. The elegant structural studies of trypsinogen[7], trypsin[8] and in particular the high affinity calcium binding site[9,10,11] now provide a basis for a discussion of the earlier observations in considerable molecular detail. A similar stabilising role seems to be played by bound calcium ions in the structure of a number of other proteases[12].

In this paper we review attempts to elucidate the role of bound calcium in stabilising the structure of *five* calcium-binding extracellular proteases from non-mammalian sources. Three of the proteins discussed are serine proteases: Thermomycolase (or Thermomycolin) (TMN) from the thermophilic fungus *Malbranchea pulchella* var. *sulfurea*; Subtilisin Novo (or BPN') (SN) and Subtilisin Carlsberg (SC) both from *B. Subtilis*. The remaining two proteases are of the neutral zinc metalloendopeptidase class: Thermolysin from the thermophile *B. thermoproteolyticus* and Neutral protease A from *B. Subtilis*.

On the basis of criteria to be discussed below we have classified Subtilisin Carlsberg, Thermomycolase and Thermolysin as thermostable[13]. Although, in the case of the latter two proteins, this classification is fully consistent with the

---

* Department of Biochemistry, The Agricultural University, Wageningen, The Netherlands.

known thermophily of the organisms from which they are excreted it must be stressed that the dividing line between "thermostable" and "non-thermostable" chosen by us is purely arbitrary and has only operational validity. Indeed, all of the above proteases are stabilised against autolysis and thermal denaturation to some extent by bound calcium[13,14]. However, even for sequence homologous proteins such as the two subtilisins and the zinc metalloendopeptidases Thermolysin and Neutral protease A the differences in stability are quite marked. How can these differences be rationalised? It could quite legitimately be argued that, because of the subtle origin of these differences, the latter question is too naive since a proper answer requires a theory of protein structure much more highly developed than current theories. That rather subtle factors are operative in determining the stability of certain structures is made very clear in the case of some point mutants of bacteriophage T4 lysozyme, as discussed by Matthews[15]. In the case of proteins stabilised by bound calcium the situation is somewhat clearer and some progress has been made in developing strategies to answer the above question.

One generally applicable approach is to elucidate in as much detail as possible the thermodynamics of calcium binding to a given protein and the associated changes in the equilibrium and kinetic properties of the molecule. Although this approach is essentially phenomenological it can be formulated in terms of a model which draws heavily on the current *microscopic* view of globular proteins as fluctuating or "stochastic" systems rather than the "platonic" time-averaged structures revealed by X-ray crystallography[16]. When the latter structures are available, however, a molecular mechanism can, in principle, be derived for the stabilising role of calcium. In the case of thermostable calcium-binding proteins Thermolysin is the only one for which detailed 3-D structural information on the protein and its calcium binding sites is available[17,18]. It therefore provides a unique opportunity for a discussion in molecular detail of the role of calcium in the structure of thermostable proteins. Recently, we have reviewed in detail the more general question of the structural and functional role of a variety of metal ions bound at the active and calcium binding sites of thermolysin[19] As we shall attempt to demonstrate here, the availability of structural data alone, while a necessary prerequisite for a discussion of possible molecular mechanisms, is not sufficient without the insights provided by the phenomenological approach. Because of its thermodynamic foundation the latter enjoys complete generality and is, in principle, applicable to the rationalisation of the behavior of any calcium binding protein. Hopefully, the present discussion will stimulate the application of our calcium-binding model to a broader range of calcium-binding proteins.

A MODEL FOR THE MODULATION OF PROTEIN STRUCTURE BY BOUND CALCIUM

Although globular proteins have structures which essentially exclude water from their interior the evidence now available suggests that their core is far from impenetrable and that as a result of random fluctuations on the nanosecond to microsecond time scale small molecules such as O_2 may penetrate deeply into the molecule. The latter view of the protein molecule as a stochastic system has been reviewed by Weber[20], Careri[21,22] and Ikegami[23]. Formally, the protein may be thought of in terms of a multidimensional hyperspace the coordinates of which determine its equilibrium and dynamic properties[24]. Fluctuations between microscopic states, which may correspond to local minima in free energy, give rise to conformational changes in the molecule. The latter can be either subtle changes associated with ligand (substrate) binding or gross changes associated with folding-unfolding (denaturation) processes.

Although a phenomenological description of protein folding-unfolding processes in terms of a "two-state" model is adequate for many purposes[25] it is now generally accepted that the folding of a protein must involve a cooperative sequence of many discrete steps;conversely for unfolding (denaturation)[24]. The latter processes can formally be regarded as arising from highly correlated fluctuations in the hyperspace of the protein. If the individual kinetic processes involved can be regarded as microscopically reversible then the kinetics and equilibria between fluctuational (conformational) states can be described using the formalism developed by Tanford[26], Weber[27] and Laidler and Bunting[28]. Thus we can better describe, for example,the "two state" equilibrium between the native (N) and denatured (D) states of a protein as involving fluctuations of the multistate manifold connecting the two via intermediates (X) on the unfolding pathway. It should be stressed, however, that the formalism is a general one and can be applied to processes connecting any two states in the protein hyperspace. The following is a general description of a calcium-modulated protein structure:

$$
\begin{array}{ccccc}
N_i(i=1,1) & \underset{}{\overset{k^o_{ij}}{\rightleftharpoons}} & X_j(j=1,m) & \underset{}{\overset{k^o_{jg}}{\rightleftharpoons}} & D_g(g=1,n) \\
K_N \Updownarrow & & K_X \Updownarrow & & K_D \Updownarrow \\
N_iCa_p(i=1,1') & \underset{}{\overset{k_{ij}}{\rightleftharpoons}} & X_jCa_q(j=1,m') & \underset{}{\overset{k_{jg}}{\rightleftharpoons}} & D_gCa_r(g=1,n')
\end{array}
$$

where $(1 \leq 1 \leq m \leq n)$ and $(1' \leq m' \leq n')$ represent the number of fluctuational (conformational) states in the free and calcium occupied manifolds of N_i (native), X_j ("activated" unfolding intermediate) and D_g (denatured) states respectively; $(1 \leq p \leq q \leq r)$ or $(1 \leq p \geq q \geq r)$ represent the calcium binding stoichiometry for each state respectively; $(K_N \geq K_X \geq K_D)$ represent the experimentally determinable calcium binding constants for each state respectively and $(k^{o'}_{ij}, k^{o''}_{ij},$

etc.) are the equilibrium constants for the fluctuational isomerisations between microstates. Fluctuations within the sub-manifolds $(N_i; X_j; D_g)$ and $(N_iCa_p; X_jCa_q; D_gCa_r)$ can be represented as follows:

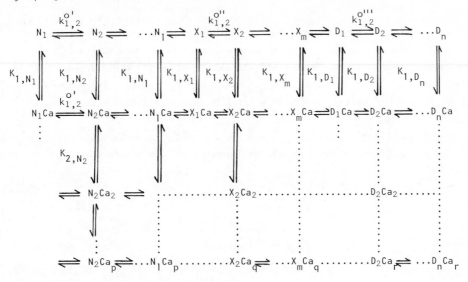

Figure 1

Multistate Fluctuation Manifold

In terms of the above formalism, it can be seen that the binding of calcium ions will stabilise the native structure if $(K_N > K_X > K_D)$ and $(p \geq q \geq r)$. Experimental evidence suggests that the first condition is always satisfied; the high affinity sites usually being found in the native structure. However,it is possible that because in general there are many low affinity sites in the manifold D_g (i.e. $r > p,q$) that even when $(K_N > K_X > K_D)$ calcium may act as a stabiliser at low concentrations and a denaturant at high concentrations[29]. It is also clear that if calcium binding to N_i involves a significant conformational change in the protein the number of fluctuational states in N_iCa_p may be reduced $(1 < 1')$ and as a result the properties of the protein (enzyme activity; protein-protein interactions; hydrogen isotope exchange rates, etc.) may be modulated by changes in the concentration of free calcium ion in the medium. In this sense calcium acts as a conformational lock on the native structure (N_i).

As we have shown elsewhere[19,30] the parameters of the model described in Figure 1 can be related to observable quantities in a straightforward way. The measured calcium-binding constant K_N and the observed rate constant, k_{obsd}, for a kinetic process the rate of which is dependent on the conformation of the molecule and hence on the statistical weight of a particular microstate,which may in turn be modulated by the binding of calcium, is given by

$$k_{obs} = k_I \frac{\Sigma_{N_{i+z}}}{\Sigma_{N_i}} \quad ; \quad z = 1,(1-1); \quad i = 1,1 \tag{1}$$

where Σ_{N_i} and $\Sigma_{N_{i+z}}$ are respectively the binding polynomials*[19,31] for the states N_i and N_{i+z}; and k_I is the number average intrinsic rate constant[30] for the kinetic process in the absence of calcium. Equation (1) would account for calcium modulated enzyme activity in the case where N_i and N_{i+z} are both members of the manifold of native structures but differ in enzyme activity and calcium binding affinity.

In the case of autoproteolysis, where the unfolded protein serves as substrate for its native conformation, the calcium dependent autolysis rate constant will be given by

$$k_{autolysis} = k_I(autolysis) \frac{\Sigma_{substrate}}{\Sigma_{Native\ fold}} \quad . \tag{2}$$

Where $\Sigma_{substrate}$ and $\Sigma_{Native\ fold}$ are respectively the calcium binding polynomials for the unfolded substrate molecule and native enzymatically active structure and $k_I(autolysis)$ is the intrinsic rate for autolysis in the absence of calcium. When the substrate molecules are highly unfolded members of the manifold D_g (g =1,n) and presumably, therefore, virtually void of bound calcium $(K_D \ll K_N)$ the binding polynomial $\Sigma_{substrate} = \Sigma_D = 1$. In the latter circumstance, as predicted by eq. (2), the native fold is maximally protected against autolysis by bound calcium. In the case of calcium activated zymogens one can envisage the intermediate situation where the autocatalytic cleavage of the activation peptide is modulated by calcium $(\Sigma_{zymogen} \approx \Sigma_{enzyme})$ and further autolysis is prevented because $\Sigma_{substrate} = 1 \ll \Sigma_{enzyme}$ in accord with equation (2).

In the case of calcium modulated thermal denaturation where the chain unfolding pathway passes through a critically activated intermediate which is a member of the manifold X_j (j = 1, m) the observed rate constant is given by

$$k_{thermal\ denaturation} = k_{I(thermal\ denaturation)} \frac{\Sigma_{X_j}}{\Sigma_{N_i}} \quad . \tag{3}$$

If a reversible folding-unfolding equilibrium can be established between states $N_i (i = 1, 1)$ and $D_g (g = 1, n)$ then the apparent equilibrium constant in the N $\leftrightarrow$ D conversion, K, is given by

---

$$* \quad \Sigma_{N_i} = \prod_{\alpha=1}^{p} (1 + K_\alpha N_i [Ca^{2+}])$$

$$K = K_I \frac{\Sigma_D}{\Sigma_N} \qquad\qquad (4)$$

where K_I is the intrinsic equilibrium constant in the absence of calcium ion and Σ_D and Σ_N are the calcium binding polynomials for the denatured and native states respectively.

In order to apply equations (1) - (4) to the rationalisation of calcium modulated protein behavior, either thermodynamic or kinetic, it is essential to obtain the binding polynomials of the various states. If k_{obsd} as a function of pCa^{2+} and Σ_{Native} have been measured then a self-consistent description of the system is usually obtainable[19,30].

We have shown previously[13] that is is possible to separate intrinsic and calcium ion contributions to the total kinetic thermal stability of calcium stabilised proteins. For the latter purposes we have defined a thermostable protein as one which has total activation free energy for thermal denaturation, $\Delta F_T^{\ddagger} \geq 25$ kcal mol^{-1} at $70^{\circ}C$ under conditions (pH, pCa^{2+}, ionic strength, etc.) which maximises its total kinetic thermal stability. The calcium contribution to $\Delta F_T^{\ddagger}$ we have shown resides in $\Delta(\Delta F_{Ca^{2+}})$ the difference between the calcium binding free energy to N and X. Calcium stabilises the native fold when $\Delta(\Delta F_{Ca^{2+}})$ is *negative*.

MATERIALS AND METHODS

Thermolysin (lot no. 54C-0211), S. Carlsberg (lots no. 33C-3270 and 94C-0245) and S. Novo (BPN') (lot no. 13C-2360) were from Sigma Chemical Co. Neutral protease A from *B. Subtilis* purified by affinity chromatography was obtained through the courtesy of Dr. K. A. Walsh, University of Washington, Seattle. Thermomycolase was purified as described previously. Methods for enzyme assay: measurement of calcium binding isotherms; thermal and autolytic degradation; calcium dependent spectroscopic and other equilibrium properties have all been described previously[13,30,36].

RESULTS AND DISCUSSION

Calcium Binding Stoichiometry and Polynomials

The calcium binding stoichiometry and binding polynomial for each of the five proteases we have studied are summarised in Table 1.

TABLE 1

Protein	p	Σ_N	K_1	$K_{1,2}$	K_3	K_4
TMN-DIP	1	(e)	5×10^5 M^{-1} [a]	-	-	-
SN-DIP	1	(e)	$>10^{11}$ M^{-1} [a]	-	-	-
SC-DIP	1	(e)	1.1×10^6 M^{-1} [a]	-	-	-
TLN [d]	4	(f)	-	2.8×10^9 M^{-2}	3.2×10^6 M^{-1} [b]	($K_3=K_4$)
				1.6×10^8 M^{-2}	5.0×10^4 M^{-1} [c]	($K_3=K_4$)
NPA [d]	3	(g)	-	8.3×10^8 M^{-2}	5.0×10^4 M^{-1} [c]	-

Abbreviations: TMN: Thermomycolin (Thermomycolase); SN: Subtilisin Novo (BPN);
SC: Subtilisin Carlsberg; TLN: Thermolysin; NPA: Neutral Protease A; DIP:
di-isopropylfluorophosphoryl.
(a) 25°C, in 0.01 M Tris, pH 7.50 + NaCl to bring Ionic Strength to 0.1 M.
(b) 25°C, in 0.01 M TEP, 0.1 M NaCl, pH 9.00. See reference 36.
(c) 6°C, in 0.01 M TEP, 0.1 M NaCl, pH 9.00. See reference 19.
(d) Under the conditions described in (c) TLN and NPA are present as the zinc
free apo-enzyme. All binding isotherms were determined by the gel-
filtration technique described previously.
(e) $\Sigma_N = (1+KC)$
(f) $\Sigma_N = (1+K_{1,2}C^2)(1+K_3C)(1+K_4C)$
(g) $\Sigma_N = (1+K_{1,2}C^2)(1+K_3C)$ where $C = [Ca^{2+}]$

Thermomycolase (Thermomycolin) has been characterised as a globular protein
and to be stabilised by the single bound calcium. This protein was isolated
from the thermophilic fungus *Malbranchea pulchella* var. *sulfurea* by our
colleague G. M. Gaucher and his student P. Ong. As a serine protease of fungal
origin it is of considerable theoretical interest from the point of view of the
evolution of this class of proteases particularly in the light of the
convergence of function seen in Chymotrypsin and Subtilisin Novo (BPN')[32].
Preliminary sequence work suggests that Thermomycolase may be closer to the
bacterial Subtilisins[33]. Although the latter have been discussed extensively
in the literature no previous report that they bind calcium can be found. The
value for $K_{1,N}(>10^{11}$ $M^{-1})$ seems rather high for a calcium binding site (see
dicussion of Williams[34]) and may be an artefact of the gel-filtration technique
which requires equilibration within 45 minutes. It is thus conceivable that the
above value is the result of the very slow dissociation of a site of much lower
affinity.

The binding polynomial for Thermolysin has been discussed extensively elsewhere[19,36]. The most notable feature of the binding of calcium to this protein
is the complete cooperativity in the binding of two of its four calcium ions.
We have suggested[19,30] that the most probable location for the cooperatively
binding calcium ions is at site S(1)-S(2) the calcium-binding double-site in

which the ions are only 3.7 Å apart. We have also found that two calcium ions
bind to Neutral Protease A with complete positive cooperativity (Table 1). The
latter observation strongly suggests that a calcium binding double-site similar
to the one found in Thermolysin may be present in this molecule. This suggestion
is made highly plausible by the rather striking sequence homology between the
two molecules in the region associated with the S(1)-S(2) site in Thermolysin[35]:

```
       186              190
TLN  -Trp-Glu-Ile-Gly-Glu-Asp-Val-Tyr-

NPA  -Trp-Asp-Ile-Gly-Glu-Asp-Ile-Thr- .
```

Calcium Dependent Equilibrium Properties

The extent to which the binding of calcium modifies the normal conformational
fluctuations will be reflected in the perturbation of a number of equilibrium
properties such as sedimentation coefficient, intrinsic viscosity, extinction
coefficient and molar ellipticity. If $Y_i(1)$ is the value of a given property
for state $N_i Ca$ and $Y_i(0)$ is the value for N_i, the the average $\overline{Y}$ for the native
state, N, is given by:

$$\overline{Y} = \frac{\sum\limits_{i=1}^{1'} Y_i(1)[N_i Ca] + \sum\limits_{i=1}^{1} Y_i(0)[N_i]}{\sum\limits_{i=1}^{1'} [N_i Ca] + \sum\limits_{i=1}^{1} [N_i]} \tag{5}$$

The $[N_i]$ values can be expressed in terms of $[N_1]$ and the isomerisation equilibrium constants $k^{o'}_{i,i+1}$ (Figure 1) as follows:

$$[N_\alpha] = \prod\limits_{i=1}^{\alpha-1} k^{o'}_{i,i+1}[N_1] \tag{6}$$

Similarly, $[N_1]$ can be expressed in terms of $k^{o'}_{i,i+1}$, K_N, $[Ca^{2+}]$ and C_p, where

$$K_N = \frac{\sum\limits_{i=1}^{1'} [N_i Ca]}{[Ca^{2+}] \sum\limits_{i=1}^{1} [N_i]}$$

and C_p = total protein concentration = $\sum\limits_{i=1}^{1'} [N_i Ca] + \sum\limits_{i=1}^{1} [N_i]$, as follows:

$$[N_1] = \frac{C_p}{(1+K_N[Ca^{2+}])} \left\{ \sum\limits_{i=1}^{1} \prod\limits_{i=1}^{\alpha-1} k^{o'}_{i,i+1} \right\}^{-1} \tag{7}$$

by combining (6) and (7) we can express $[N_\alpha]$ as a function of C_p, K_N, $[Ca^{2+}]$ and

45

the $k_{i,i+1}^{o'}$:

$$[N_\alpha] = \frac{C_p}{(1+K_N[Ca^{2+}])} \left\{ \frac{\prod\limits_{i=1}^{\alpha-1} k_{i,i+1}^{o'}}{\sum\limits_{i=1}^{1} \prod\limits_{i=1}^{\alpha-1} k_{i,i+1}^{o'}} \right\} \qquad (8)$$

Similarly we can derive for $[N_\alpha Ca]$:

$$[N_\alpha Ca] = \frac{C_p K_N[Ca^{2+}]}{(1+K_N[Ca^{2+}])} \left\{ \frac{\prod\limits_{i=1}^{\alpha-1} k_{i,i+1}^{o''}}{\sum\limits_{i=1}^{1} \prod\limits_{i=1}^{\alpha-1} k_{i,i+1}^{o''}} \right\} \qquad (9)$$

from (5), (8) and (9) we obtain for $\overline{Y}$

$$\overline{Y} = \frac{K_N[Ca^{2+}]}{(1+K_N[Ca^{2+}])} \overline{Y}(1) + \frac{1}{(1+K_N[Ca^{2+}])} \overline{Y}(0) \qquad (10)$$

Where $\overline{Y}(1)$ and $\overline{Y}(0)$, averaged over $N_i Ca$ and N_i, have been identified as:

$$\overline{Y}(1) = \sum_{i=1}^{1'} \left\{ \left(\frac{\prod\limits_{i=1}^{\alpha-1} k_{i,i+1}^{o''}}{\sum\limits_{i=1}^{1'} \prod\limits_{i=1}^{\alpha-1} k_{i,i+1}^{o''}} \right) \right\} Y_\alpha(1)$$

$$\overline{Y}(0) = \sum_{i=1}^{1} \left\{ \left(\frac{\prod\limits_{i=1}^{\alpha-1} k_{i,i+1}^{o'}}{\sum\limits_{i=1}^{1} \prod\limits_{i=1}^{\alpha-1} k_{i,i+1}^{o'}} \right) \right\} Y_\alpha(0)$$

It is useful to express the difference $\Delta\overline{Y} = \overline{Y} - \overline{Y}(1)$ in terms of the maximum difference $\Delta\overline{Y}_{max} = \overline{Y}(0) - \overline{Y}(1)$ in the observed equilibrium parameter, $\overline{Y}$, as follows:

$$\Delta\overline{Y} = \frac{\Delta\overline{Y}_{max}}{(1+K_N[Ca^{2+}])} = \frac{\Delta\overline{Y}_{max}}{\Sigma_N} \qquad (11)$$

It is evident from (11) that the calcium dependence of any equilibrium property is expressible in terms of the binding polynomial determined by a direct method. It is also clear that, because the observation of $\overline{Y}$ necessarily involves an averaging of $\overline{Y}$ over the manifolds N_i and $N_i Ca$, the calcium coupled effect may appear to involve only "two states". Clearly, such a view is oversimplified. Rather, one should think in terms of the modulation of fluctuations of the

46

molecule by bound calcium ion (Figure 1).

It is, of course, not possible to predict the magnitude of $\Delta\bar{Y}_{max}$ for a given equilibrium property. In the case of the muscle proteins these changes have been shown to be quite large, being detectable with relatively insensitive parameters such as the sedimentation coefficient. In contrast, the five proteins examined by us show minor calcium-dependent changes in their equilibrium properties detectable only by spectroscopic means.[19,30]

Calcium Dependent Kinetic Effects

It is clear from the discussion of the previous section that large energetically unfavorable fluctuations, because of their low statistical weight, will not be revealed by measurement of equilibrium properties. However, if highly reactive conformational states are populated by random fluctuations in the manifolds N_i, X_j or D_g then they will be detected in rate processes. If the fluctuations leading to highly reactive species are strongly modulated by bound calcium then the rate constant for the reaction will be dependent on pCa^{2+} in accord with equations (1) - (3).

In the case of TMN, SN and SC which each bind only *one* calcium ion equations (1) - (3) reduce to the simple form

$$k = k_{intrinsic} \frac{(1+K_R[Ca^{2+}])}{(1+K_N[Ca^{2+}])} \tag{12}$$

where K_R is the equilibrium calcium binding constant for the reacting conformation of the protein. Data for the autolytic loss of enzyme activity of TMN, SN and SC were fitted to (12)[14] and the parameters so obtained are summarised in TABLE 2.

TABLE 2

Protein	k_I	K_N	K_R
TMN	$6.2 M^{-1}sec^{-1}$	$2.2\times10^5 M^{-1}$	0
SN	$(0.43\pm0.04) M^{-1}sec^{-1}$ [a]	K_N =	K_R
SC	$0.230 M^{-1}sec^{-1}$	$1.91\times10^6 M^{-1}$	$7.64\times10^5 M^{-1}$

(a) The average of k_{obs} over the range $4.0<pCa^{2+}<8.0$. The rate of autolysis is independent of pCa^{2+} in the latter range.

The results summarised in Table 2 clearly reveal that in the case of TMN the most readily autolysed species has lost its calcium affinity ($K_R=0$) which suggests either that only highly unfolded species are autolysed at an appreciable rate or that the calcium binding site is associated with a part of the molecule essential for enzyme activity. SN, in contrast, is not stabilised by its bound

47

calcium ion against autolysis (cf. its thermal denaturation). This implies that catalytically essential parts of the molecule are lost before the calcium binding site is destroyed. SC is only partly stabilised by its single bound calcium ion. It will be interesting to see if the contrasting behaviour of SN and SC can be accounted for in molecular terms when their calcium sites have been characterised structurally.

Because the calcium binding sites of Thermolysin have been structurally characterised it is possible to offer a more definitive interpretation of its autolysis behaviour. We have found[30] that the autolysis rate constant is given by

$$k_{autolysis} \ (TLN) \ = \ \frac{k_I}{\Sigma_N} \ = \ \frac{k_I}{(1+K_{1,2}C^2)(1+K_3C)(1+K_4C)} \tag{13}$$

since $K_3C \gg 1$ and $K_4C \gg 1$, equation (13) reduces to

$$k_{autolysis} \ (TLN) \ = \ \frac{k_I(K_3K_4)^{-1}}{C^2(1+K_{1,2}C^2)}$$

The data are best fitted by $\log\{k_I(K_3K_4)^{-1}\} = -8.64$ and $K_{1,2} = 3.0 \times 10^8 M^{-2}$.[30] It is clear that all four calcium ions stabilise Thermolysin against autolysis. Since $K_R = 0$ in equation (13) we can also infer that autolysis requires complete unfolding of the molecule to conformations in the manifold D_g with negligible calcium-binding affinity.

In contrast to the latter results the calcium-dependent thermal denaturation kinetics for Thermolysin at 70°C are best rationalised by the following equation

$$k_{thermal} = k_I \frac{(1+K_{1,x}C)}{(1+K_{1,N}C)} \tag{14}$$

with $\log(k_I/k_N) = -6.51$ and $K_X = 600 M^{-1}$.[30] Equation (14) implies that only <u>one</u> of its four calcium ions stabilises Thermolysin against purely thermal degradation at 70°C. We have previously reviewed the evidence which suggests that this critical ion is either Ca(3) or Ca(4).[19]

We have shown in an earlier communication that is is possible to separate the calcium contribution to the total kinetic thermal stability of a given protein from the intrinsic contribution of the polypeptide fold itself.[13] The results for the five enzymes which we have studied are discussed in detail elsewhere and are summarised in TABLE 3.[13]

TABLE 3

Protein	$\Delta F,^{\ddagger}$ (kcal/mol)	$\Delta F_T^{\ddagger}$	$-\Delta(\Delta F_{Ca^{2+}})$	pH
S.Novo (BPN')	22.4±0.3	24.6±0.2	2.2	6.60
S. Carlsberg	22.5±5.5	25.5±1.0	3.0	7.20
TMN	24.1±0.5	26.7±0.1	2.7	6.80
TLN	20.5±0.2	28.6±0.4	8.1	6.00
NPA	22.0±0.4	22.9±3.0	0.8	6.50

The above results suggest that calcium ion binding at specific sites in the structures of globular proteins can contribute significantly to their stabilisation. Perhaps, the most striking example and the one for which a molecular mechanism can be suggested is Thermolysin. Although it is the least intrinsically stable, the large contribution, 8.1 kcal/mol at 70°C, to the total kinetic thermal stability from only *one* of its four calcium ions, as revealed by the analysis based on our model, makes it the most thermostable of the given proteases studied.

REFERENCES

1. Delezenne, M. C. (1905) C. R. Soc. Biol., 59, 476.
2. Northrop, J. H., Kunitz, P., and Herriot, R. M. (1948) "Crystalline Enzymes", Columbia U.P., 2nd Ed.
3. McDonald, M. R. and Kunitz, M. (1941) J. Gen. Physiol., 25, 53.
4. Abita, J. P., Delaage, M., Lazdunski, M. and Savrda, J. (1969) Eur. J. Biochem., 8, 314.
5. Gabel, D. and Kasche, V. (1973) Acta Chem. Scand., 27, 1971.
6. Delaage, M. and Lazdunski, M. (1967) Biochem. Biophys. Res. Commun., 28, 390.
7. Kossiakoff, A. A., Chambers, J. L., Kay, L. M. and Stroud, R. M. (1977), Biochemistry, 16, 654.
8. Stroud, R. M., Kay, L. M. and Dickerson, R. E. (1974) J. Mol. Biol., 83, 185.
9. Chambers, J. L. and Stroud, R. M. (1977), Acta Crystallog., (in Press).
10. Bode. W. and Schwager, P. (1975) J. Mol. Biol., 98, 693.
11. Epstein, M. (1976) "The Calcium Binding Site of Bovine and Porcine Trypsin as Probed by Lanthanides", Ph.D. Thesis, Weizmann Inst. of Science, Rehovot, Israel.
12. Matthews, B. W., Weaver, L. H. and Kester, W. R. (1974) J. Biol. Chem., 249, 8030 and references cited therein.
13. Voordouw, G., Milo, C. and Roche, R. S. (1976) Biochemistry, 15, 3716.
14. Voordouw, G. (1975) "The Role of Bound Calcium Ions in the Structure of Proteolytic Enzymes", Ph.D. Thesis, University of Calgary.
15. Matthews, B. W. (1976) in "Enzymes and Proteins from Thermophilic Microorganisms", Zuber, H. Ed. Proc. Int. Symp., Zurich, July 28-Aug.1, 1975, Experentia Suppl. 26, Birkhauser Verlag, Basel.
16. Weber, G. (1975) Adv. Protein Chem., 29, 1.
17. See reference 12.
18. Matthews, B. W. and Weaver, L. H. (1974), Biochemistry, 13, 1719.
19. Voordouw, G. and Roche, R. S.,C.R.C. Crit. Rev. Biochem., (in Press).
20. See reference 16.
21. Careri, G. (1974) in "Quantum Statistical Mechanics in the Natural Sciences", Eds., S. L. Mintz and S. M. Widmayer, Plenum, New York.
22. Careri, G., Fasella, P. and Gratton, E. (1975) C.R.C. Crit. Rev. Biochem., 3, 141.
23. Ikegami, A. (1977) Biophys. Chem., 6, 117.
24. Anfinsen, C. B. and Scheraga, H. A. (1975) Adv. Prot. Chem., 29, 205.
25. Baldwin, R. L. (1975) Ann. Rev. Biochem., 44, 453.
26. Tanford, C. (1970) Adv. Prot. Chem., 24, 1.
27. Weber, G. (1972) Biochemistry, 11, 864.
28. Laidler, K. J. and Bunting, P. S. (1973) "Kinetics of Enzyme Action", O.U.P.
29. von Hippel, P. H. and Wong, K. Y. (1965) J. Biol. Chem., 240, 3909.
30. Voordouw, G. and Roche, R. S. (1975) Biochemistry, 14, 4659, 4667.
31. Schellman, J. A. (1975) Biopolymers, 14, 999.
32. Robertus, J. D., Alden, R. A.,Birktoft, J. J., Powers, J. C. and Wilcox, P. E. (1972), Biochemistry, 11, 2439.
33. Gaucher, G. M. and Stevenson, K. J. (1976), Methods in Enzymol., 45, Chap. 34.
34. Williams, R. (1976) in "Calcium in Biological Systems", Symp.30, Soc.Exp.Biol.
35. Pangburn, M. K., Levy, P. L., Walsh, K. A. and Neurath, H. in reference 15.
36. Voordouw, G. and Roche, R. S. (1974), Biochemistry, 13,5017.

Calcium Ionophores

Wilhelm Simon, Werner E. Morf, and Daniel Ammann
Department of Organic Chemistry
Swiss Federal Institute of Technology
Universitätstr. 16, CH-8092 Zuerich
Switzerland

INTRODUCTION

   Ionophores or ion carriers are molecules of rather small relative molar mass which have the capability of complexing ions and of transporting these across lipophilic membranes by carrier translocation (see Fig. 1). In order for a ligand to behave as an ionophore, several aspects have to be considered[1-4]:

a) A carrier molecule should be composed of polar and nonpolar groups (see section 2).

b) The carrier should be able to assume a stable conformation (see section 4) that provides a cavity, surrounded by the polar groups, suitable for the uptake of a cation, while the nonpolar groups form a lipophilic shell around the coordination sphere. These groups must ensure sufficiently large lipid solubility for ligand and complex. This is one reason why classical electrically charged complexing agents such as EDTA will not behave as carriers in membrane systems.

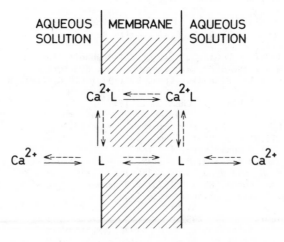

Fig. 1. Schematic representation of the mechanism of a carrier-mediated Ca^{2+}-transport through a membrane

c) Among the polar groups of the ligand sphere, there should be pre-
   ferably 5-8, but not more than 12 coordinating sites such as oxygen atoms (see section 3).

d) High selectivities are achieved by locking the coordinating sites
   into a rigid arrangement around the cavity. Such rigidity can be
   enhanced by the presence of bridged structures or hydrogen bonds.
   Within one group of the periodic system, the cation that best
   fits into the offered cavity is preferred (see sections 3 and 4).
   Ideally, all cations should be forced into accepting the same
   given number of coordinating groups (see section 3).

e) Notwithstanding requirement d), the ligand should be flexible
   enough to allow a sufficiently fast ion-exchange. This is possible only with a stepwise substitution of the solvent molecules by
   the ligand groups. Thus, a compromise between stability (d) and
   exchange rate (e) has to be found.

f) To guarantee an adequate mobility, the overall dimensions of a
   carrier should be rather small but still compatible with high lipid solubility.

SIMPLE BINDING SITES FOR Ca^{2+}

Attractive binding sites are ligand atoms which are capable of
competing with water molecules in the complexation of Ca^{2+}. An estimate of the effectiveness of different sites may be obtained by
simple model calculations. The estimated energies of interaction of
different group IA and IIA ions with one water molecule are given in
Table 1, column 2. They were calculated by considering the ion-di-
pole interactions, ion-induced dipole interactions, as well as contributions due to repulsion (see Refs. 5, 6) and are in excellent
agreement with data obtained by much more sophisticated computational
techniques[7] (column 3) as well as with experimental values[8] (column
4). In columns 5 to 7 the increments of the interaction energies are
given (relative to column 2) which result from changes in the dipole
moment (column 5), the polarizability (column 6), and the radius
(column 7) of the ligand site. The data indicate that an increase in
the dipole moment or the polarizability as well as a decrease in the
radius of the ligand atom increase the stability of the hypothetical
complexes. The effect is especially large for small and multiple
charged cations. Therefore, small and polar binding sites generally
tend to prefer Ca^{2+} over Na^+ and Ba^{2+}. An extreme situation is found

Table 1

INTERACTIONS BETWEEN BINDING SITE MODELS AND CATIONS

Ion (Radius, Å)	Interaction energy [kcal/mol] for the complex Ion-H_2O			Change in interaction energy [kcal/mol] obtained for			
	Calculated values using the data of Table 2[*]	Values from Ref. 7	Values from Ref. 8	Increase in dipole moment by 0.5 D	Increase in polarizability by 1Å^3	Increase in ligand radius by 0.1Å	Substitution of dipole moment by charge $-e_O$
1	2	3	4	5	6	7	8
Li^+ (0.68)	-34.1	-34.1	-34.0	-6.8	-6.1	+3.8	-123
Na^+ (0.98)	-24.4	-23.2	-24.0	-5.2	-3.6	+2.3	-110
K^+ (1.33)	-17.5	-16.2	-17.9	-3.9	-2.0	+1.4	-98
Rb^+ (1.49)	-15.3		-15.9	-3.5	-1.6	+1.2	-93
Cs^+ (1.65)	-13.5		-13.7	-3.1	-1.3	+1.0	-89
Mg^{2+} (0.78)	-75.4			-12.4	-20.3	+8.9	-236
Ca^{2+} (1.06)	-54.1			-9.7	-12.5	+5.5	-214
Sr^{2+} (1.27)	-43.5			-8.2	-9.0	+4.0	-199
Ba^{2+} (1.43)	-37.4			-7.3	-7.1	+3.2	-190

* The basic formula[5] was

$$E = - \frac{x-2}{x} \frac{ze_O p}{r^2} - \frac{x-4}{x} \frac{\alpha (ze_O)^2}{2r^4} \, ,$$

with ionic charge ze_O, dipole moment p, polarizability α, ion-ligand distance r, and repulsion coefficient x = 12.

for anionic sites (last column in Table 1) where these effects are amplified. The data presented in Table 1 refer to interactions of one ion with one ligand molecule in the absence of other solvating species (gas phase) and are intended to give semiquantitative information. Some molecular parameters for a series of possible binding sites are compiled in Table 2.

EFFECT OF COORDINATION NUMBER

The often observed variation in the coordination number of group IA and IIA cations is a consequence of the radius-ratio-effect, which was introduced into discussions of ion-packing in crystals[10,11] and of ionic hydration[5]. If we consider for example coordination

Table 2

MOLECULAR PARAMETERS FOR SOME BINDING SITES[9]

Molecule	Permanent dipole moment [D]	Polarizability* [Å^3]	Van der Waals radius of ligand atom [Å]
H_2O	1.85**	1.46**	1.38**
NH_3	1.47	2.26	1.50
H_2S	0.92	3.67	1.85
PH_3	0.55	4.28	1.90
CH_3–O–CH_3	1.30	0.65***	
CH_3–NH–CH_3	1.03	0.94***	
CH_3–S–CH_3	1.50	3.06***	
CH_3–C(=O)–CH_3	2.88	0.84***	
CH_3–S(=O)(=O)–CH_3	4.49		
CH_3–C(=O)–O–CH_2CH_3	1.78		
CH_3–C(=O)–N(CH_3)CH_3	3.81		

* Calculated from molar refraction data.

** Values used in Table 1 (column 2) and in Ref. 5.

*** Polarizability increment of ligand atom.

53

spheres of 8 (cubic), 6 (octahedral), and 4 (tetrahedral) oxygen atoms, forming cavities for the uptake of a cation, we find minimal cavity radii of 1.0, 0.6, and 0.3 Å, respectively[1,2,5]. This is consistent with the usually observed coordination number n of about 8 for Ca^{2+} (1.06 Å) and only 6 for Mg^{2+} (0.78 Å)[11]. Such a situation is represented in Fig. 2 for the hydration of A cations. The calculated values ΔG_H^o in Fig. 2 indicate that the most stable aquo-complexes of alkali metal ions and Mg^{2+} have a coordination number of 6 whereas the larger alkaline earth cations prefer n = 8. If n could be fixed to 8 throughout, Mg^{2+} would heavily be discriminated relative to the situation with a free choice of n. For alkali ions a coordination with n = 8 would discriminate Li^+ heavily and would also give some destabilization of the other ions (see Fig. 2). These results clearly demonstrate that a coordination with oxygen and n = 8 is especially attractive for complexing Ca^{2+}.

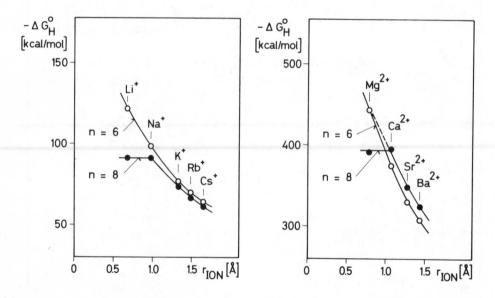

Fig. 2. Calculated free energies of hydration ΔG_H^o for alkali ions (left) and alkaline earth ions (right). Values for cubic and octahedral coordination geometry are given as a function of the ionic radius[5].

POLYDENTATE LIGANDS

It is a well known fact that complexes of polydentate ligands have enhanced stability over their unidentate counterparts (chelate

54

effect) and complexes of macrocyclic ligands are, as a rule, more stable than those of noncyclic polydentate ligands (macrocyclic effect)[3,12-17]. According to Adamson[14], the chelate effect is largely a consequence of the asymmetry of the standard reference state. This has the mathematical consequence (1):

$$\log K \text{ (n-dentate)} = \log \beta_n \text{ (unidentate)} + \text{(n-1)} \log 55.5 \tag{1}$$

where K is the stability constant of the 1:1-cation:polydentate ligand complex in water, and β_n is the cumulative stability constant of the 1:n-complex of the unidentate analogues (see also refs. 14, 17).

Stabilization effects beyond what has been discussed may be due to a reduction in translational and/or rotational entropy of the free polydentate ligand.

In macrocyclic and especially in macropolycyclic ligands, repulsions e. g. between binding sites are already built in[3] and, in addition, optimal solvation of the free ligand sites may be prevented so that the formation of the complexes is further favoured.

Due to favoured conformations of a multidentate ligand of a given constitution, the coordination shell (i. e. coordination number and cavity radius) may be predetermined and therefore selectivity may be induced even by non-macrocyclic molecules (see section 3). Indeed, the noncyclic neutral carriers 4, 5 and 6 (Fig. 4) are capable of inducing extremely high selectivity for Ca^{2+} in certain membranes.

SELECTIVITY OF NEUTRAL CARRIERS IN MEMBRANES

The selectivity of complex formation of a given neutral carrier molecule in water or water-like solvents need not to be identical to the ion selectivity of the corresponding carrier membrane, e. g. in transport experiments. The problems involved have been discussed in detail elsewhere[2,18-21]. Accordingly, the free energy of transfer of an ion from the aqueous solution into the membrane phase includes the term (2)[1-3]:

$$\Delta G_B = - \frac{(ze_o)^2}{2(r_{Ion} + s)} \left(1 - \frac{1}{\varepsilon}\right) \tag{2}$$

which describes the interaction of the cation:carrier complex of radius r_{Ion} + s and charge ze_o with the membrane phase of dielectric constant ε. The parameter s signifies the average thickness of the ligand shell around the cation and is important for the discrimination between cations of the same radius but different charge. Fig. 3 indicates that small values of s and a high ε lead to a preferential uptake of Ca^{2+} relative to Na^+ into the membrane (other parameters kept constant).

The description of systems with variable cation:carrier complex stoichiometry becomes rather involved[19,20] and yields no direct correlation between complex stability constants and the extraction or transport behavior of the corresponding membranes.

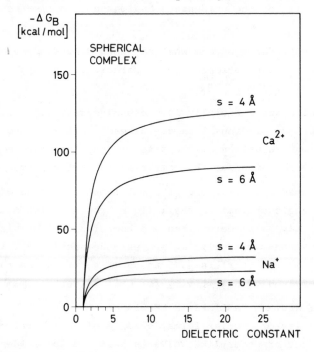

Fig. 3. Free energy of the electrostatic interactions between cationic complex and membrane-solvent. The values ΔG_B were estimated for two metal ions of nearly the same size but of different charge, for two values of s, and for a varying dielectric constant of the membrane medium (see Eq. (2)).

EXAMPLES OF Ca^{2+} IONOPHORES

Some naturally occuring molecules (1 to 3) which were reported to be Ca^{2+} ionophores[11,22,23], as well as synthetic carriers[24,25] for Ca^{2+} (4 to 6) are presented in Fig. 4. The antibiotics X-537A (1) and A23187 (2) are monocarboxylic acids, give rise to anions and therefore are expected to interact strongly with divalent cations (Table 1).

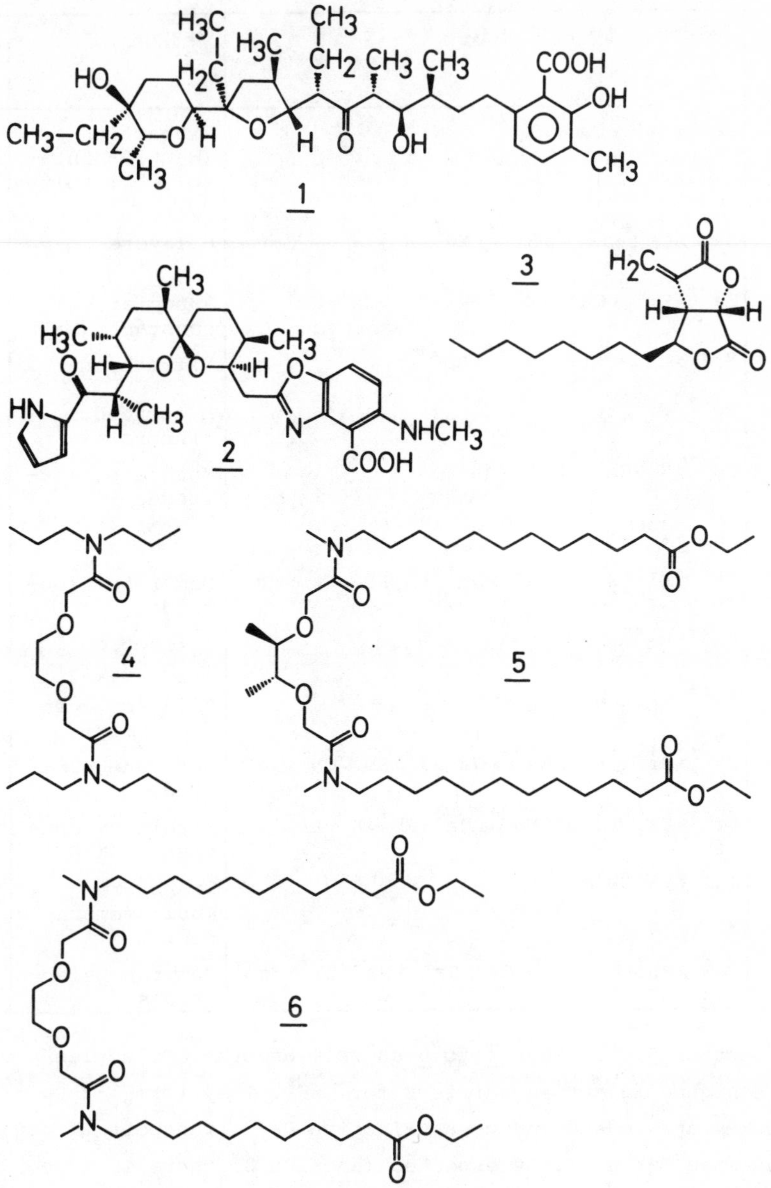

Fig. 4. Structures of some Ca^{2+} ionophores

Avenaciolide (3) contains oxygen functional groups of rather high dipole moment (Table 2) leading to a preferential interaction with small divalent cations. As suggested earlier[23], several such ligand molecules will be necessary to form the required coordination shell

Table 3

Li-gand (Fig.4)	Selectivity Sequence	Method	Lite-rature
1	Ba^{2+}, Sr^{2+}, Ca^{2+}, Mg^{2+}	Extraction (H_2O/30% buta-nol, 70% tolue-ne)	29
	Ba^{2+}>Ca^{2+}>Mn^{2+}>Sr^{2+}≫Mg^{2+}	Bilayers	30
	Ba^{2+}>Sr^{2+}>Ca^{2+},Mg^{2+}>K^+~Rb^+~Cs^+>Na^+	Association constants (MeOH)	31
2	Mn^{2+}>Ca^{2+}~Mg^{2+}>Sr^{2+}>Ba^{2+}	Extraction	32
	Mg^{2+}, Ca^{2+} ?	Mitochondria transport	33
	Ba^{2+}>K^+>Ca^{2+}>Na^+>Mg^{2+}	Membrane elec-trodes	34
3	?		
4	Ca^{2+}>Li^+>Sr^{2+}~Ba^{2+}>Na^+~K^+>Rb^+~Cs^+>Mg^{2+}	Membrane elec-trodes	24
5	Ca^{2+}>Ba^{2+}~Sr^{2+}≫Mg^{2+}	Extraction (H_2O/CH_2Cl_2)	19
	Ca^{2+}>Ba^{2+}~Na^+~Mg^{2+}>K^+	Bilayers (con-ductivity)	35
	Ca^{2+}>Sr^{2+}>Li^+>Ba^{2+}>Na^+>K^+>Rb^+~Cs^+>Mg^{2+}	Membrane elec-trodes	36
	Sr^{2+}>Ba^{2+}>Ca^{2+}>Mg^{2+}>Na^+>Rb^+~K^+~Li^+	Stability con-stants (EtOH)	19
6	Ca^{2+}≫Mg^{2+}; Ca^{2+}>Na^+	Transport (bulk membra-nes)	37,38
	Ca^{2+}>Rb^+>Ba^{2+}>K^+~Cs^+>Sr^{2+}>Na^+>Li^+>Mg^{2+}	Membrane elec-trodes	39

(see section 3). Ligands 4 to 6 as well as other molecules[24] have been designed to act as carriers for Ca^{2+}. They form 1:2-Ca^{2+}:carrier complexes and indeed induce a selective Ca^{2+} transport through membranes (see Table 3). As expected (section 3) there is a coordination of Ca^{2+} by 8 oxygen atoms (see Fig. 5). In contrast to complexes of 1 and 2, which are electrically neutral, those of 3 to 6 are charged (see Fig. 5); therefore effects due to the thickness, s, of the ligand shell around the cation (Eq.(2)) become important. For this reason s was kept small to obtain the Ca^{2+}-selective carriers 4, 5 and 6. Details of the X-ray analysis of the 1:2-Ca^{2+}:carrier complex of 4 are given in Fig. 6[28].

58

$(X-537 A)_2 Ba \cdot H_2O$ $(A \ 23187)_2 Ca \cdot 2(C_2H_5OH)$ $L_2 Ca \cdot Cl_2$

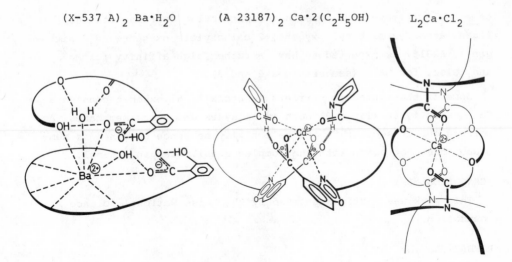

Fig. 5. Schematic representation of the complexes of ligands 1, 2, and 4 (Fig. 4) with selected divalent cations[26-28].

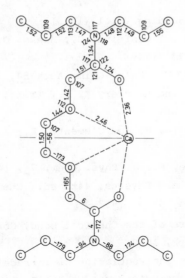

Fig. 6. Molecular geometry for the Ca^{2+} complex of ligand 4 (crystallographic 222 symmetry; only one ligand L is shown). On the upper half bond lengths and angles are indicated; on the lower half torsion angles are given[28].

Selectivity sequences obtained in different systems are given in Table 3. It appears that the so-called Ca^{2+}-ionophore 1 is actually a Ba^{2+}-ionophore. The selectivity sequences found for 2 are, somewhat, controversial. Due to the particular coordination shell (6 ligand atoms: 2 carbonyl oxygens, 2 carboxylate oxygens and 2 nitrogens), A23187 is expected to have a rather high affinity for Mg^{2+} and, also, for Ca^{2+} (see sections 2 and 3).

Some of the synthetic carriers (especially 5) exhibit considerable Ca^{2+} selectivity (Table 3) both in bilayers and thick membranes. As pointed out in section 5, the selectivity sequence obtained in membranes does not agree with the complex stabilities in ethanol.

ACKNOWLEDGEMENT

This work was partly supported by the Swiss National Science Foundation.

REFERENCES

1. Morf, W. E. and Simon, W., Helv. Chim. Acta 54, 2683 (1971).
2. Simon, W., Morf, W. E. and Meier, P. Ch., Struct. Bonding 16, 113 (1973).
3. Lehn, J.-M., Struct. Bonding 16, 1 (1973).
4. Diebler, H., Eigen, M., Ilgenfritz, G., Maass, G., and Winkler, R., Pure Appl. Chem. 20, 93 (1969).
5. Morf, W. E. and Simon, W., Helv. Chim. Acta 54, 794 (1971).
6. Muirhead-Gould, J. S. and Laidler, K. J., Trans. Farad.Soc. 63, 944 (1967).
7. Kistenmacher, H., Popkie, H. and Clementi, E., J. Chem. Phys. 59, 5842 (1973).
8. Džidić, I. and Kebarle, P., J. Phys. Chem. 74, 1466 (1970).
9. Handbook of Chemistry and Physics, 44th ed., Chemical Rubber Publishing Co., Cleveland, Ohio, 1963.
10. Pauling, L., The Nature of the Chemical Bond, Cornell University Press, Ithaca, New York, 1960.
11. Williams, R. J. P., in The Regulation of Intracellular Calcium (Duncan, C. J., ed.), Cambridge University Press, England, 1976, p. 1f.
12. Martell, A. E. and Calvin M., The Chemistry of the Metal Chelate Compounds, Prentice Hall, New York, 1952, p. 149f.
13. Schwarzenbach, G., Helv. Chim. Acta 35, 2344 (1952).
14. Adamson, A. W., J. Amer. Chem. Soc. 76, 1578 (1954).
15. Cabbiness, D. K. and Margerum, D. W., J. Amer. Chem. Soc. 91, 6540 (1969).

16. Callear, A. B., Fleming, I., Ottewill, R. H., Twigg, M. V.,
    Warren, S. G. and Prince, R. H., Chemistry and Industry, 80
    (1977).

17. Hancock, R. D. and Marsicano, F., J. Chem. Soc. Dalton, 1096
    (1976).

18. Morf, W. E., Wuhrmann, P. and Simon, W., Anal. Chem. $\underline{48}$, 1031
    (1976).

19. Kirsch, N. N. L. and Simon, W., Helv. Chim. Acta $\underline{59}$, 357 (1976);
    Kirsch, N. N. L., Diss. Nr. 5842 ETH, Zurich, 1976.

20. Simon, W., XVIth Solvay Conference on Chemistry, Brussels,
    22-26 November 1976.

21. Eisenman, G., Membranes, Vol. 2, Chaps. 2 and 3, Marcel Dekker,
    New York, 1973.

22. Pressman, B. C., Ann. R. Biochem. $\underline{45}$, 501 (1976).

23. Truter, M. R., in The Regulation of Intracellular Calcium
    (Duncan, C. J., ed.), Cambridge University Press, England,
    1976, p. 19f.

24. Ammann, D., Bissig, R., Güggi, M., Pretsch, E., Simon, W.,
    Borowitz, I. J. and Weiss, L., Helv. Chim. Acta $\underline{58}$, 1535 (1975).

25. Ammann, D., Bissig, R., Cimerman, Z., Fiedler, U., Güggi, M.,
    Morf, W. E., Oehme, M., Osswald, H., Pretsch, E. and Simon, W.,
    Proceedings of the International Workshop on Ion Selective
    Electrodes and on Enzyme Electrodes in Biology and Medicine
    (Kessler, M., Clark, L. C., Jr., Lübbers, D. W., Silver, I. A.,
    Simon, W., Eds.), Munich, Berlin, Vienna, Urban & Schwarzen-
    berg, 1976.

26. Johnson, S. M., Herrin, J., Liu, S. J. and Paul, I. C., Chem.
    Comm., 72 (1970).

27. Chaney, M. O., Jones, N. D. and Debono, M., J. of Antibiotics
    $\underline{29}$, 424 (1976).

28. Neupert-Laves, K. and Dobler, M., Helv. Chim. Acta, 1977, in
    press.

29. Pressman, B. C., Fed. Proc. $\underline{32}$, 1698 (1973).

30. Célis, H., Estrada-O., S. and Montal, M., J. Membr. Biol. $\underline{18}$,
    187 (1974).

31. Degani, H., Friedman, H. L., Navon, G. and Kosower, E. M.,
    Chem. Comm., 431 (1973).

32. Pfeiffer, D. R., Reed, P. W. and Lardy, H. A., Biochemistry
    $\underline{19}$, 4007 (1974).

33. Wong, D. T., Wilkinson, J. R., Hamill, R. L. and Hong, J.-S.,
    Archs. Biochem. Biophys. 156, 578 (1973).
34. Covington, A. K. and Kumar, N., Anal. Chim. Acta 85, 175 (1976).
35. Amblard, G. and Gavach, C., Biochim. Biophys. Acta 448, 284
    (1976).
36. Ammann, D., Güggi, M., Pretsch, E. and Simon, W., Anal. Letters
    8(10), 709 (1975).
37. Wuhrmann, P., Thoma, A. P. and Simon, W., Chimia 27, 637 (1973).
38. Simon, W., Morf, W. E., Pretsch, E. and Wuhrmann, P., Proceedings
    of the International Symposium on Calcium Transport in Contrac-
    tion and Secretion held at Bressanone, Italy (Carafoli, E.,
    Clementi, F., Drabikowski, W. and Margreth, A., Eds.), Amster-
    dam - Oxford, North-Holland Publishing Company, New York, Else-
    vier Publishing Company, Inc., 1975.
39. Ammann, D., Pretsch, E. and Simon, W., Anal. Letters 5(11), 843
    (1972).

Evolution of the Informational Role
of Calcium in Eukaryotes

Robert H. Kretsinger
Department of Biology
University of Virginia
Charlottesville, Virginia, 22901

INTRODUCTION

The literature on calcium in biological systems is vast and cannot be reviewed here. Most of the material related to this article is documented in my reviews.[1,2] I suggest that the five postulates listed below provide a valuable framework for summarizing past results and designing future experiments. I intentionally state the postulates in a general form. Subsequently I discuss each in turn, provide documentation, and explore variations on, or exceptions to, the general theme stated in the postulate.

1) All resting eukaryotic cells maintain the concentration of free Ca^{2+} within the cytosol between 10^{-7} and 10^{-8} M.

2) The sole function of Ca^{2+} within the cytosol is to transmit information.

3) The target of Ca^{2+}, functioning as a second messenger, is a protein in the cytosol.

4) Calcium modulated proteins contain EF-hands.

5) Cells initially extruded calcium so they could use phosphate as their basic energy currency; $Ca_3(PO_4)_2$ is insoluble.

DISCUSSION

1) <u>All resting eukaryotic cells maintain the concentration of free Ca^{2+} in the cytosol between 10^{-7} and 10^{-8} M</u>. Although this generalization is widely accepted, it is an extrapolation from a limited number of examples. The most direct measurements come from observations of cells that have been injected with the Ca^{2+} sensitive luminescent protein aequorin. Ca^{2+} sensitive dyes, such as murexide and arsenazo III, have also been used <u>in vivo</u>, but since the negative logs of their Ca^{2+} dissociation constants, $pK_d(Ca)$, are in the range 5 to 6, they can give only an upper limit to resting levels of free Ca^{2+} in the cytosol. Many eukaryotic cells and nearly all prokaryotes are difficult or impossible to inject.

Less direct measurements include: i) Treating cells that have a defined Ca^{2+} dependent response, such as contraction or exocytosis, with a divalent specific ionophore like A23187. If the cells are bathed in higher Ca^{2+} concentrations the characteristic response occurs; if they are bathed in a Ca^{2+} free medium, there is ideally no response. A frequent artifact arises from the ionophore's having entered the cell and released Ca^{2+} from internal stores. ii) Injecting known quantities of Ca^{2+} directly into the cell. This is technically difficult, and the Ca^{2+} buffering capacity of the cell cannot be measured. iii) Testing sequestering capacity. Internal vesicles or membrane systems, such as the sarcoplasmic reticulum (S.R.) of skeletal muscle, can pump down the free Ca^{2+} of the bathing solution to 10^{-7} M. Mitochondria have a similar Ca^{2+} accumulating ability; however, they probably function as intermediate, as opposed to final steady state, Ca^{2+} storage sites. Bacteria, which lack mitochondria and S.R., normally extrude Ca^{2+} through their cell membranes as shown by Ca^{2+} accumulation by inverted membrane vesicles.[3]

The measurement of total cell calcium is not a good indicator of free Ca^{2+} in the cytosol since cells have so many calcium storage sites. Table 1 summarizes a few of the more reliable determinations of free Ca^{2+} in the cytosol:

TABLE 1

EXAMPLES OF DETERMINATIONS OF FREE Ca^{2+} IN THE CYTOSOL

Cell Organism ref.	Technique	pCa
Erythrocytes human[15]	ATPase(Ca)	>7
Muscle (various)	aequorin	∿7
	ATPase(Ca) S.R.	>7
Axon squid[16]	aequorin	7.7
	arsenazo III	7.3
(17)	inject Ca	>7
Neuron Helix[18]	aequorin	7.3
Photoreceptor Limulus[19]	aequorin	∿7
Salivary gland Chironomus[6]	aequorin	>6.3
Egg Medaka[20]	aequorin	∿7
	A-23187	∿7
Physarum[7]	aequorin	∿7

2) The sole function of Ca^{2+} within the cytosol is to transmit information.
The implication of these first two postulates is that for any eukaryotic cell at
"rest" the cytosol concentration is pCa 7 to 8. Following an appropriate stimulus, Ca^{2+} enters from either the extracellular medium or from an internal store.
This causes a transitory increase in Ca^{2+} (pCa 4 to 5), followed by a pumping
down to pCa 7. Hence the target, the molecule with which Ca^{2+} interacts, has
a $pK_d(Ca)$ between 7 and 5. If it bound more tightly than $pK_d \simeq 7$, it could
never release Ca^{2+}, or relax. If it bound Ca^{2+} more weakly than $pK_d \simeq 5$, it
would never bind Ca^{2+}, or activate.

Not only does the target of second messenger calcium have a $pK_d(Ca)$ between
5 and 7 but also any molecule in the cytosol whose $pK_d(Ca)$ lies between 5 and 7
is by definition modulated by Ca^{2+}.

If Ca^{2+} had any function in the cytosol other than serving as a second messenger, then the entity that would bind calcium would have to have a $pK_d(Ca)$ greater
than 7. I know of no evidence that any cytosol molecule can win in a tug of war
for Ca^{2+} with the calcium pumps.

The justification for this postulate, as well as postulates 3 and 4, is that
there are no known exceptions.

3) The target of Ca^{2+}, functioning as a second messenger, is a protein in the
cytosol. The validity of this postulate depends on postulate 1), that the resting cytosol concentration is low, pCa $\simeq$ 7 to 8, and that the Ca^{2+} concentration
rises to pCa 5 following a stimulus. Neither the phospholipid, nucleic acid nor
carbohydrate of the cytosol have both the Ca^{2+} affinity and the selectivity to
bind Ca^{2+} at 10^{-5} M in the presence of cytosol levels of free Mg^{2+}, about 10^{-3} M.
Proteins have evolved this affinity and selectivity simply because they can
place six to eight oxygen atoms, each carrying partial or full negative charge,
in exact position to form a cavity to accept a cation of crystal radius 0.99 A
as opposed to the 0.65 A of Mg^{2+}. Of course Ca^{2+}, as a second messenger, may
modify a membrane via a membrane bound protein. As noted in the previous section the $pK_d(Ca)$ for these calcium modulated proteins must lie between 7 and 5,
that is between resting and excited levels of free pCa.

I suggest that even though there may be a charge transfer carried by the trans-ransmembrane flux of Ca^{2+}, the target of Ca^{2+} is not some ill defined charge detector but a protein that binds Ca^{2+} specifically and strongly. It is difficult to estimate the actual flux of Ca^{2+} following a cell stimulus. As a rough estimate, a cell of diameter 7 μ has a volume: $(3.5\cdot10^{-4}cm)^3\cdot4\cdot\pi/3 = 1.8\cdot10^{-10}cm^3$; at rest it contains $1.8\cdot10^{-10}cm^3\cdot10^{-7}M\cdot6\cdot10^{23}$ ions/$10^3cm^3\cdot M = 1.1\cdot10^4$ free Ca^{2+} ions. At 10^{-5} M it would contain $1.1\cdot10^6$ ions, that is an increase of $1.09\cdot10^6$ ions. Even with the uniquely high Ca^{2+} buffering capacity provided by the parvalbumins of fish muscle,[4] upper limit 1.0 mM and 2 Ca^{2+} bound/protein, only $1.1\cdot10^8 + 1.1\cdot10^6 = 1.1\cdot10^8$ Ca^{2+}ions would be required to saturate the buffer and change the free pCa from 7 to 5. The surface area of this hypothetical 7 μ cell would be $4\cdot\pi\cdot(3.5\cdot10^{-4}cm)^2 = 1.5\cdot10^{-6}cm^2$. The estimate of $1.1 \times 10^8 Ca^{2+}/1.5\cdot10^{-6}$ cm^2 ($1.2\cdot10^{-10}$ moles Ca^{2+}/cm^2) would in fact exceed the inward Na^+ current during a nerve spike.[5] However this estimate of Ca^{2+} is an upper limit and is surely several decades too large. Few cells have such high Ca^{2+} buffering capacity. Further Rose and Lowenstein[6] have conclusively shown in _Chironomus_ salivary gland that a Ca^{2+} pulse is localized to a volume about 1/100 of the cell. A localized calcium response such as contraction in _Physarum_[7] or cleavage furrow formation[8] can occur without bringing the entire cytosol to pCa ≃ 5.

TABLE 2

POSSIBLE CALCIUM MODULATED PROTEINS

Protein	Function	$pK_d(Ca)$	n	$M_r\cdot10^{-3}$
Phosphorylase b kinase (2.7.1.38)[21,22]	Phosphorylase b → a	6	1	55
Tyrosine hydroxylase (1.14.16.2)[23]	Dopamine biosynthesis	5	?	40
Aequorin[24]	Bioluminescence ??	7	3	28
Luciferin Binding Protein[24]	Bioluminescence ??	7	?	20
ATPase(Ca)	Calcium pump (SR)	7	1	102
Spasmin[25]	_Vorticellid_ contraction	7	2	20
Tubulin[26] (or associated proteins)[27]	Spindles, cilia, etc.	4?	?	110/αβ
S-100	??	5	2	$21/\alpha_2\beta$
L-1[28]	??	5	2	14

'4) <u>Calcium modulated proteins contain EF-hands</u>. The EF-hand conformation, first observed in the crystal structure of carp muscle calcium binding parvalbumin, consists of two turns of α helix, a twelve residue loop containing six calcium coordinating ligands, and two more turns of helix. The helix, loop, and helix are related to one another like the forefinger, middle finger and thumb of a hand (figure 1).

When I first presented this EF-hand homology argument,[9] its appeal was primarily intuitive. It seemed to make evolutionary sense that once Nature had found a good target for Ca^{2+}, she would usually prefer to vary her basic theme than to start anew. That is, divergent rather than convergent evolution seemed more appealing; however such an argument is not readily subjected to critical analysis. The following considerations and corollaries are associated with this homology postulate:

i] Nature has devised various ways of coiling a protein about a calcium ion. The various extracellular proteins - concanavalin A, thermolysin, <u>Staphylococcus</u> nuclease, trypsin, phospholipase A_2 and the γ-carboxyglutamic acid containing proteins -- all bind calcium, and are neither homologous to one another nor to the EF-hand proteins. Their calcium affinities tend to be lower, $pK_d(Ca)$ 3 to 5, and they function in environments having higher Ca^{2+} concentrations, pCa 3 to 2. However their basic architectures could certainly be used to achieve $pK_d(Ca)$ values of 7. The EF-hand conformation is certainly not required for high Ca^{2+} affinity or selectivity.

ii] In addition to affinity and selectivity, one might anticipate that Ca^{2+} should express its presence by inducing a significant conformational change in its target. This has been observed for parvalbumin, TN-C and modulator protein.[10] Solano and Coffee[11] cleaved parvalbumin at Arg-75 thereby generating two fragments -- 1 to 75 and 76 to 108 -- each of which contains one EF-hand. Although the intact molecule binds two Ca^{2+} with $pK_d(Ca) \simeq 7$, the isolated EF-hands do not bind calcium. Chemical modification of the invariant Arg-75 greatly reduces Ca^{2+} affinity. Not only does the binding of Ca^{2+} alter the conformation of the protein, but also changes in the protein conformation alter the affinity of the protein for Ca^{2+}.

iii] The EF-hand modulated protein is most easily described as being itself an enzyme or being an activator associated with an enzyme, as troponin and light chains are associated with myosin ATPase. This concept must be broadened to

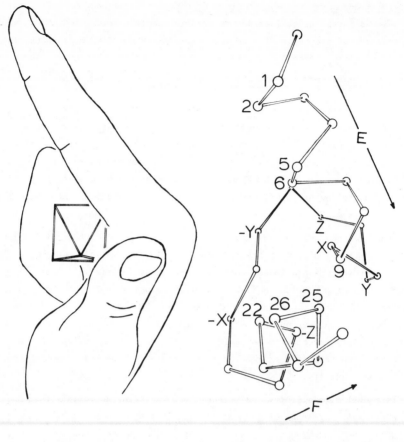

The following data table accompanies the figure (conserved positions with their residue frequency of occurrence):

	1	2	5	6	9	10	12	14	15	17	18	19	20	21	25	26	29
label	n		n	n	n	X	Y	Z		n	−Y	−X	−Z		n	n	n
	Glu 12	Phe 4	Ala 5	Phe 7	Phe 5	Asp 12	Asp 8	Asp 7	Gly 14	Ilu 12	Asp 5	Glu 5	Glu 11	Phe 7	Met 5	Met 7	Leu 4
	Asp 3	Leu 4	Met 3	Leu 3	Gly 3		Asn 4	Asn 3		Val 3	Asn 2	Asp 3		Leu 7	Val 5	Leu 3	Ilu 2
	Ilu 3		Leu 2	Ilu 3	Val 2		Glu 2	Ser 2			Glu 2	Glu 2			Leu 2	Val 2	
	Val 2		Val 2	Met 2	Leu 2						Ser 2	Ser 2					

Fig. 1 The α-carbon drawing is from the crystal structure of carp muscle calcium binding parvalbumin. The indicated positions are highly conserved in the two EF-hands of parvalbumin, the four EF-hands each of TN-C, myosin light chains, and modulator protein and the one EF-hand of VitDCaBP. The frequency of occurrence, based on these 15 types of EF-hand, are listed at each conserved position. The inner (designated "n") residues of both E and F α-helices tend to be hydrophobic. The Ca^{2+} coordinating residues (X, Y, Z, −X, and −Z) have oxygen-containing side chains. At position 15 Gly usually occurs at a sharp turn. The Ilu, almost invariant at position 17, appears to hold the calcium binding loop to the hydrophobic core of the molecule.

include calcium buffer and/or transport proteins. The functions of parvalbumin and of the vitamin D induced calcium binding protein (VitDCaBP) are not yet established; however, they do not seem to be directly associated with enzyme activation.

iv] In the course of evolution an EF-hand may have lost its ability to bind Ca^{2+}, yet retained its conformation. The various myosin light chains each consist of four EF-hand regions; however, they bind only one or no Ca^{2+} ions. Whereas skeletal TN-C consists of four EF-hands and binds four Ca^{2+}, cardiac TN-C has lost two ligands in the first of its four EF-hands and binds three Ca^{2+}.[12] Relative to the "standard" EF-hand of figure 1 the first third of parvalbumin has two residues deleted, while the second EF-hand, in sequence, of the alkali extractable light chain has two residues inserted. The loop that connects the first pair of EF-hands to the second pair is three residues shorter in modulator protein than in TN-C. The amino acid sequence of the calcium binding protein from intestinal mucosa[13] indicates one EF-hand. The other half of the molecule may have evolved from a different gene.

Several other proteins (Table 2) of the cytosol, including membrane associated proteins exposed to the cytosol surface, bind Ca^{2+} tightly $pK_d(Ca) > 5$ and appear to be associated with a process involving Ca^{2+} as a second messenger.

In an evolutionary sense, cells first had to lower their intracellular free Ca^{2+} before they could use the pCa_{out}/pCa_{in} gradient to transmit information. Certainly the most exciting prediction is that the calcium pump itself contains an EF-hand.

5) Cells initially extruded calcium so they could use phosphate as their basic energy currency; $Ca_3(PO_4)_2$ is insoluble. Near pH 7, if $[Ca^{2+}]$ were at ocean (or plasma) levels of 10^{-2} M, then $[HPO_4^{2-}] + [H_2PO_4^{-}]$ could not exceed $10^{-7.7}$ M. At physiological levels of phosphate ~10^{-2} M, $[Ca^{2+}]$ could not exceed $10^{-5.2}$ (calculated from Sillén[14]). Magnesium phosphate is soluble. Magnesium is the counterion associated with ATP and other biological phosphorylated compounds. Further, most enzymes dealing with phosphorylated substrates require magnesium and conversely most magnesium-requiring enzymes have phosphorylated substrates. An interesting corollary to this $Ca_3(PO_4)_2$ postulate is that since prokaryotes also have phosphate based energy systems, they too should actively extrude Ca^{2+}. Whether this involves an ATPase(Ca) pump or an ATPase(Na) pump with

subsequent Ca^{2+} for Na^+ or for H^+ exchange is not known.[3] Nor is it known whether any prokaryotes have been clever enough to use their pCa_{out}/pCa_{in} gradients for information transfer.

SUMMARY

The preceding five postulates -- 1) cytosol $pCa \simeq 7.5$. 2) only information. 3) protein target. 4) EF-hands. 5) $Ca_3(PO_4)_2$ -- are intentionally stated in very comprehensive, hopefully comprehensible, terms. They do not directly concern extracellular events, such as calcification, cell adhesion or enzyme stabilization. Inevitably exceptions will be found; details must be altered. After all, penguins don't fly; Ø-X DNA is single stranded. Nonetheless I suggest that these postulates provide a conceptual framework for interpreting an immense body of phenomenological data. Further, they should focus our design and interpretation of experiments so that within a few years a better general theory can replace this one.

REFERENCES

1. Kretsinger, R.H. (1976) "Calcium-Binding Proteins." Ann. Rev. Biochem. 45, 239-266.

2. Kretsinger, R.H. and Nelson, D.J. (1976) "Calcium in Biological Systems." Coord. Chem. Rev. 18, 29-124.

3. Tsuchiya, T. and Rosen, B.P. (1976) "Calcium Transport Driven by a Proton Gradient in Inverted Membrane Vesicles of Escherichia coli." J. Biol. Chem. 251, 962-967.

4. Baron, G., Demaille, J. and Dutruge, E. "The Distribution of Parvalbumins in Muscle and in Other Tissues." FEBS Lett. 56, 156-160.

5. Meves, H. and Vogel, W. (1973) "Calcium Inward Current in Internally Perfused Giant Axons." J. Physiol. 235, 225-265.

6. Rose, B. and Loewenstein, W.R. (1976) "Permeability of a Cell Junction and the Local Cytoplasmic Free Ionized Calcium Concentration: A Study with Aequorin." J. Memb. Biol. 28, 87-119.

7. Ridgway, E.B. and Durham, A.C.H. (1976) "Oscillation of Calcium Ion Concentrations in Physarum polycephalum." J. Cell Biol. 69, 223-226.

8. Hollinger, T.G. and A.W. Schuetz (1976) "'Cleavage and Cortical Granule Breakdown in Rana Pipiens Oocytes Induced by Direct Microinjection of Calcium." J. Cell Biol. 71, 395-401.

9. Kretsinger, R.H. (1975) "Hypothesis: Calcium Modulated Proteins Contain EF-hands" in Calcium Transport in Contraction and Secretion E. Carafoli (ed.) North-Holland Publishing Co. pp. 469-478.

10. Liu, Y.P. and Cheung, W.Y. (1976) "Cyclic 3':5'-Nucleotide Phosphodiesterase: Ca^{2+} Confers More Helical Conformation to the Protein Activator." J. Biol. Chem. 251, 4193-4198.

11. Solano, C. and Coffee, C.J. (1976) "Preparation and Properties of Carp Muscle Parvalbumin Fragments A (Residues 1→75) and B (Residues 76→108) Biochim. Biophys. Acta. 453, 67-80.

12. van Eerd, J.-P and Takahashi, K. (1976) "Determination of the Complete Amino Acid Sequence of Bovine Cardiac Troponin C." Biochemistry 15, 1171-1180.

13. Hofmann, T., Kawakami, M., Morris, H., Hitchman, A.J.W., Harrison, J.E. and Dorrington, K.J. "The Amino Acid Sequence of a Calcium Binding Protein from Pig Intestinal Mucosa." in this volume.

14. Sillén, L.G. (1961) "The Physical Chemistry of Sea Water." in Oceanography (ed. M. Sears) pp. 549-581. American Association for the Advancement of Science (Publication no. 67) Wash. D.C.

15. Porzig, H. (1973) "Calcium-Calcium and Calcium-Strontium Exchange across the Membrane of Human Red Cell Ghosts." J. Membrane Biol. 11, 21-46. (Bern)

16. Dipolo, R., Requena, J., Brinley, F.J., Mullins, L.J., Scarpa, A. and Tiffert, T. (1976) "Ionized Calcium Concentrations in Squid Axons." J. Gen. Physiol. 67, 433-467.

17. Kusano, K., Miledi, R. and Stinnakre (1975) "Postsynaptic Entry of Calcium Induced by Transmitter Action." Proc. R. Soc. Lond. B 189, 49-56.

18. Meech, R.W and N.B. Standen (1975) "Potassium Activation in Helix Aspersa Neurones under Voltage Clamp: A Component Mediated by Ca Influx." J. Physiol. 249, 211-239.

19. Brown, J.E. and Blinks, J.R. (1974) "Changes in Intracellular Free Calcium Illumination of Invertebrate Photoreceptors." J. Gen. Physiol. 64, 643-665.

20. Ridgway, E.B., Gilkey, J.C. and Jaffe, L.F. (1977) "Free Calcium Increases Explosively in Activating Medaka Eggs." Proc. Nat. Acad. Sci. 74, 623-627.

21. Khoo, J.C. (1976) "Ca^{2+}-Dependent Activation of Phosphorylase by Phosphorylase Kinase in Adipose Tissue." Biochim. Biophys. Acta. 422, 87-97.

22. Keppens, S., Vandenheede, J.R. and DeWulf, H. (1977) "On the Role of Calcium as Second Messenger in Liver for the Hormonally Induced Activation of Glycogen Phosphorylase." Biochim. Biophys. Acta. 496, 448-457.

23. Murrin, L.C. and Roth, R.H. (1976) "Dopaminergic Neurons: Effects of Electrical Stimulation on Dopamine Biosynthesis." Mol. Pharmacol. 12, 463-475.

24. Cormier, M.J., Hori, K. and Anderson, J.M. (1974) "Bioluminescence in Coelenterates." Biochim. Biophys. Acta. 346, 137-164.

25. Routledge, L.M., Amos, W.B., Gupta, B.L., Hall, T.A. and Weis-fogh, T. (1975) "Microprobe Measurements of Calcium Binding in the Contractile Spasmoneme of a Vorticellid." J. Cell Sci. 19, 195-201.

26. Solomon, F. (1977) "Binding Sites for Calcium on Tubulin." Biochemistry 16, 358-363.

27. Schliwa, M. (1976) "The Role of Divalent Cations in the Regulation of Microtubule Assembly. In Vivo Studies on Microtubules of the Heliozoan Axopodium Using the Ionophore A23187." J. Cell Biol. 70, 527-540.

28. Alema, S., Calissano, P. and Giuditta, A. (1974) "Studies on a Calcium Binding, Brain Specific Protein from the Nervous System of Cephalopods." in Calcium Binding Proteins, eds. W. Drabikowski, H. Strzelecka-Golaszewska and E. Carafoli (Elsevier) pp. 739-749.

Evolutionary Relationships
among Calcium-Binding Proteins

W. C. Barker, L.K. Ketcham, and M.O. Dayhoff
National Biomedical Research Foundation
Georgetown University Medical Center
3900 Reservoir Road, N.W.
Washington, D.C. 20007

INTRODUCTION

We have constructed an evolutionary tree of the troponin C superfamily from the available sequences of myosin light chains, cardiac and skeletal muscle troponins C, parvalbumins, calcium-dependent modulator protein, and vitamin D-induced calcium-binding protein. Myosin (the major component of thick filaments in the contractile system) is a complex molecule consisting of heavy chains and at least three types of light chains: LC-1 and LC-3 may help to maintain the structural integrity of the myosin head,[1] LC-2 may be involved in calcium regulation of muscle contraction. Troponin C (TnC) confers calcium sensitivity on the contractile process and parvalbumin (PV) may play a role in the transport of calcium ions from myofibrils to sarcoplasmic reticulum in vertebrate fast skeletal muscle, providing an essential relaxing action in rapidly contracting muscles.[2] The modulator protein (MP), also called phosphodiesterase activator protein, confers calcium control on hydrolysis of cyclic AMP and GMP in many tissues. The vitamin D-induced calcium-binding protein (CBP) is somehow involved in storage or transport of calcium in the intestinal mucosa, and may be found in other tissues also.

METHOD

Matrices of estimated point mutations between sequences are derived from alignments of sequences. For each possible topology, branch lengths that fit the matrix most closely are calculated. The topology selected is the one having the minimum total number of mutations. Our computer methods for deriving trees and for detecting distant relationships are described elsewhere.[3]

RESULTS AND CONCLUSIONS

The amino acid sequences of these proteins indicate a history of internal gene duplications. PV has two homologous regions, each binding a calcium ion, and a third region that does not bind calcium and that is not detectably homologous to the other two. CBP also has two homologous regions but binds only one calcium ion. The TnC, MP, and myosin light chains have four homologous regions. Myosin LC-1 chain has 41 additional amino-terminal residues. LC-1 and LC-3 are not known to bind calcium; LC-2 binds one calcium ion, whereas MP and skeletal muscle TnC bind

four calcium ions and cardiac muscle TnC binds three. These proteins have
descended from a calcium-binding ancestor about 39 residues long (see Fig. 1).
Early in evolution two internal duplications produced a chain four times the
length of the common ancestral chain. By aligning half chains of the longer
sequences with the CBP and PV sequences and constructing topologies that separately reflect the evolution of both halves of the longer chains, we determined
where to place the trunk of the tree and where to place the PV and CBP branches.
These diverge together from the second half of TnC, and then diverge from each
other. Therefore, although these proteins are similar in length to the doubled
ancestor, they seem to have arisen by loss of the amino-terminal portion of the
redoubled ancestor.

Because the bovine cardiac TnC diverged before chicken and rabbit skeletal
TnCs, a gene duplication is represented rather than a species divergence. The
divergences of ancestral myosin light chain, TnC, and PV are more ancient than
the divergence of cardiac from skeletal muscle TnC. The rate of change of TnC
is estimated to be 1.6 PAMs/100 my from the divergence of chicken and rabbit
skeletal muscle TnC. If TnC has been changing at this unusually slow rate since

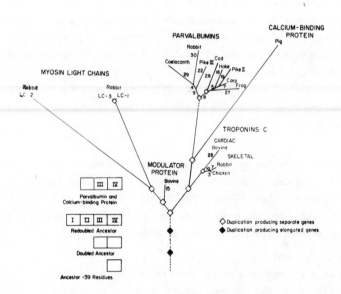

Fig. 1. Evolutionary tree of troponin C superfamily. The tree is a composite of
the best topologies for the parvalbumins, for the sequences with four homology
regions, and for half-chains of these compared with parvalbumin and calcium-
binding protein. The sequences used are published except for intestinal calcium-
binding protein (T. Hofmann, these Proceedings), and modulator protein (T.C.
Vanaman, personal communication).

the divergence of the cardiac and skeletal forms, the gene duplications that allowed the specialization of cardiac and skeletal muscle may have occurred a billion years ago. At least two duplications of the PV gene must be postulated to explain the observed topology. The estimated rate of change of PV is 5.6-6.9 PAMs/100 my, considerably faster than that of TnC.

On the other major branch of the tree, we see that the first duplication produced the genes for MP and for the ancestral myosin light chain. That MP is the most slowly changing protein of this family is indicated both by the very short branch length on the tree and by the fact that the halves of the MP sequence are more similar to each other than are the halves of any of the other sequences. Presumably its function also corresponds most closely to the function of the common ancestor of these proteins. Next occurred a duplication that gave rise to the two main types of myosin light chain. These, the parvalbumins, and the calcium-binding protein are changing faster than TnC or MP. Therefore the proteins with the more conserved primary structure retain more of the functional ability to bind calcium than the faster-evolving proteins of this superfamily.

We also searched the calcium-binding regions of TnC and other proteins against the Protein Sequence Data Bank of the *Atlas of Protein Sequence and Structure*, using a scoring matrix based on mutation data. The results conformed to postulated evolutionary relationships rather than to proposed similarities in calcium affinity. Testing of the following sequenced calcium-binding proteins against each other for evolutionary relationships using the program RELATE[3] revealed no unsuspected relationships: myosin LC-2 chain, parvalbumin, skeletal and cardiac troponins C, intestinal calcium-binding protein; gamma-carboxy-glutamic acid-containing protein; prothrombin (residues 1-323), prothrombinase (residues 1-140); trypsin, proteases A and B; thermolysin; staphylococcal nuclease; deoxyribonuclease; prophospholipase A2; colipase II; alpha casein; and concanavalin A.

REFERENCES

1. Kendrick-Jones, J. and Jakes, R., Trends Biochem. Sci. 1, 281-284, 1976.
2. Pechere, J.-F., Derancourt, J., and Haiech, J., FEBS Lett., 75, 111-114, 1977.
3. Dayhoff, M.O., in *Atlas of Protein Sequence and Structure*, ed. Dayhoff, M.O., Vol. 5, Suppl. 2, pp. 1-8, National Biomedical Research Foundation, Washington D.C., 1976.

(Supported by NIH grants HD 09547 and RR 05681.)

Geometry of Calcium: Water Interactions in Crystalline Hydrates

Howard Einspahr, Charles E. Bugg, and William J. Cook
University of Alabama in Birmingham
Institute of Dental Research
University Station
Birmingham, Alabama 35294

As part of a study of the geometrical factors involved in
calcium--ligand interactions, we have surveyed the calcium--water
arrangements in crystal structures of hydrated calcium salts. Data
were obtained from published crystal structure analyses that were
sufficiently accurate to reveal experimental positions for the
water-hydrogen atoms, yielding some 100 examples of calcium--water
interactions. The primary geometrical parameter examined was the
acute angle between the calcium--water-oxygen (OW) vector, and the
vector joining the OW position with the mean position of the two
water-hydrogen (HW) atoms. This angle (θ) was used to judge the
degree to which a Ca ion is colinear with the water dipole. We
also monitored the Ca--O distance (R) and the dihedral angle (ϕ)
between the water plane and the plane defined by the Ca position
and the water dipole.

A scatter plot of θ <u>versus</u> R is shown in Figure 1. Although
the distribution of sample points is somewhat diffuse, there is
a clear correlation between θ and R. For those examples displaying
the shorter Ca--OW distances, the Ca ion tends to be more nearly
colinear with the water dipole; for examples with longer Ca--OW

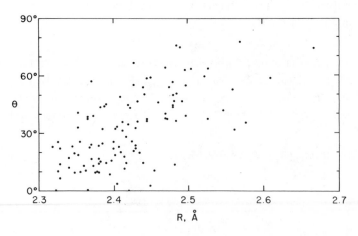

Fig. 1. A scatter plot, θ <u>versus</u> R, of the Ca--water examples.

distances, the tendency toward linearity is relaxed. Scatter
plots (not shown) of ϕ versus R and θ versus ϕ revealed an
additional relationship: for examples with longer Ca--O distances
(and larger θ), the Ca ion tends to be equidistant from the water
hydrogens; for examples with shorter Ca--O distances (and smaller
θ), the restriction disappears.

Fig. 2(a) is a histogram showing numbers of examples versus
deviations from linearity (θ). This overall distribution shows
a primary maximum at about 15° and a secondary one at about 40°.
This secondary maximum seems to be due to water examples that
form a total of 4 or more hydrogen bonds and bonds to cations;
as shown in Fig. 2(a), the secondary maximum disappears if these
water molecules are omitted from the sample. When the histogram
is properly modified to show density of examples versus deviations
from linearity (Fig. 2(b)), the distribution is approximately
Gaussian with a maximum at $\theta=0°$.

It is concluded that the equilibrium configuration of the
calcium--water system is one in which the Ca ion is colinear with
the dipole moment vector of the water molecule. This configuration
may be rationalized in electrostatic terms, with both Ca--OW
attraction and Ca--HW repulsions playing key roles.

Supported by N.I.H. grants DE-02670, CA-12159, and CA-13148.

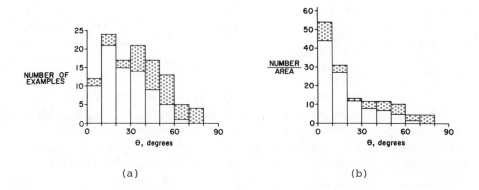

(a)   (b)

Fig. 2. The distribution of θ values. The unpatterned portion
of each histogram represents a subset of examples with water
coordination numbers of 3 or less. (a) The raw histogram,
population versus θ. (b) The corrected histogram, population
density versus θ.

The Calcium-Binding Site of Trypsin

Michael Epstein,[‡] Jacques Reuben,[§] and Alexander Levitzki[¶]
From the Departments of Biophysics and Isotope Research
The Weizmann Institute of Science, Rehovot, Israel

The recently growing recognition of the fundamental involvement of calcium ions as bioregulators and activators in a wide variety of biological processes enhances the importance of probing calcium binding to proteins using the lanthanides as probes. The establishment of methods for the application of the lanthanides is thus necessary. In the present study a systematic approach to the use of lanthanides as probes for Ca^{2+} was developed. The approach was guided by the experimental system chosen, which was the specific Ca^{2+} binding sites in bovine and porcine trypsin. The three dimensional structure of bovine trypsin has been known since 1971 (1), but information concerning its Ca^{2+} binding site was not available. The methodologies developed in this work enabled us to identify the Ca^{2+} binding sites in both proteins as well as to characterize them. Furthermore, these approaches seem to be of general significance in probing Ca^{2+} binding sites using lanthanides.

In this study it was found that the trivalent lanthanide ions replace Ca^{2+} stoichiometrically and mimic its function. The fluorescence of Tb^{3+} is tremendously enhanced when bound to bovine and porcine trypsin, depending on the excitation wavelength (up to 7.10^4 fold). The approximately fourfold difference between the Tb^{3+} fluorescence enhancement factor for the two enzymes reflects a difference between the corresponding metal binding sites. The difference in the relative intensities of the three main Tb^{3+} emission lines between the Tb^{3+}-bovine and Tb^{3+}-porcine complexes also reflect a difference between these two binding sites.

The Tb^{3+}-fluorescence was used to measure the binding constants of a series of metal ions to porcine and bovine trypsin. These binding constants imply that a metal ion, for which octahedral coordination is preferred, has a higher affinity for the enzyme. Within the lanthanide series, the ionic radius significantly effects the affinity of the ions to the enzyme. The binding constants of lanthanide ions vary over an order of magnitude from 180 M^{-1} to 2200 M^{-1}. It is also

---

[‡]Present address: Section of Biochemistry, Molecular and Cell Biology, Cornell University, Ithaca, New York 14853.

[§]Present address: Department of Biochemistry and Biophysics, University of Pennsylvania School of Medicine, Philadelphia, Pennsylvania 19174.

[¶]Present address: Department of Biological Chemistry, Hebrew University, Jerusalem, Israel.

concluded that the coordination number changes from 9 for the light lanthanides up to Nd^{3+} to 8 for the heavier ones. The fact that the ionic radius is important in determining the metal ion affinity for the enzyme is also indicated by the lack of binding of Sr^{2+} and Mg^{2+}, which have coordination properties similar to those of Ca^{2+}.

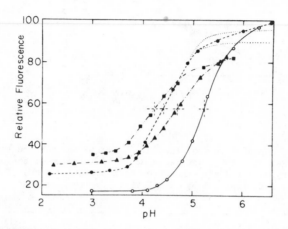

Fig. 1. The stability of the Tb^{3+}-trypsin complexes as a function of pH. Unbuffered solutions of 2.5×10^{-5} M trypsin containing $TbCl_3$ were brought to the desired pH using dilute HCl, and the Tb^{3+} fluorescence was measured. o – o porcine trypsin with 10^{-2} M $TbCl_3$. $\blacktriangle - \blacktriangle$ bovine trypsin with 3.6×10^{-2} M $TbCl_3$. $\bullet - \bullet$ bovine trypsin with 10^{-2} M $TbCl_3$. $\blacksquare - \blacksquare$ bovine trypsin with 4.10^{-3} M $TbCl_3$. The shift of the titration midpoint with the Tb^{3+} concentration indicates a H^+-Tb^{3+} competition.

The pH dependence of the Tb^{3+} fluorescence (Fig. 1) leads to the assignment of carboxylate side chains as primary ligands to the metal ion in both trypsins. Two modified treatments of the pH dependence data (a double logarithmic presentation and the slope of the titration curve at its midpoint) indicate participation of at least two carboxylates in the metal chelation. This is corroborated by proton release measurements upon metal binding.

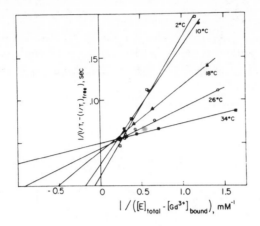

Fig. 2. A double reciprocal plot of the Gd^{3+} titration by porcine trypsin, monitoring the water PRR. The τ_{null} was measured of 100 µl samples containing 0.5 mM Gd^{3+} and porcine trypsin at a varying concentration up to 4.5 mM. The aqueous solutions were buffered at pH 6.3 by .2 M PIPES. For the details of the calculation see text.

The magnetic relaxation enhancement measurements of water protons, PRR, caused by Gd^{3+} binding to porcine trypsin (Fig. 2), indicate that 2 water molecules are left in the coordination sphere of the bound Gd^{3+}.* This finding means that 6 ligands of the Gd^{3+} coordination sphere are donated by the protein. Four of them are probably the oxygen atoms of two carboxylates and two are neutral ligands, such as carbonyl oxygens. Lanthanide chemistry and crystallography support the involvement of such carbonyls in metal ligation.

---

*Brief description of the calculation: From Fig. 2 one obtains the $1/T_1$ of bound Gd^{3+} (refer to equations given by Koenig and Epstein 1975, J. Chem. Phys. 63, 2279). From the steeply positive frequency dependence of the PRR of trypsin-bound Gd^{3+} (Epstein and Koenig, unpublished) as apposed to the intense frequency dependence of the PRR of free Gd^{3+} (Koenig and Epstein, the above reference) one learns that the correlation time τ_c is largely dominated by τ_e, the electron relaxation time. τ_e has been calculated independently from esr spectra of Gd^{3+}-trypsin, and turned out to be faster by one order of magnitude than the other main contribution to τ_c, namely the rotation time of trypsin as determined by fluorescence polarization (Yguerabide, et al. 1970, J. Mol. Biol. 51, 573). The value which has to be most accurately determined is the distance Gd^{3+}-proton (of a bound water molecule). This was taken from crystallographic data of the lanthanide-oxygen distance (Hoard et al. 1965, J. Am. Chem. Soc. 87, 1612), and both the H-O-H angle and O-H distance of water (Chidambaram 1962, J. Chem. Phys. 36, 2361). Thus the relaxation time of water in the coordination sphere of Gd^{3+} $T_1 = 6.6 \times 10^{-7}$ sec. This, together with the measurement and the known Gd^{3+} concentration yield q = 2.2 for the number of water molecules left on the bound Gd^{3+} ion.

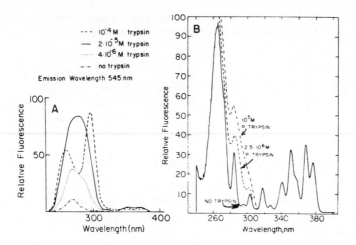

Fig. 3. Enhancement of Tb^{3+} fluorescence by porcine trypsin. Corrected excitation spectra of 0.1 M TbCl$_3$ (A) and 0.01 M TbCl$_3$ (B) in presence and absence of porcine trypsin at the concentrations shown. In B, the sensitivity is enhanced compared to A, and at wavelengths higher than 310 nm the spectra overlap. t = 25°C, pH = 6.3 in 0.2 M PIPES. The maximal enhancement is at 295 nm.

The fluorescence excitation spectra of trypsin-bound and free Tb^{3+}, Fig. 3, and the long (millisecond range) lifetime of the Tb^{3+} fluorescence imply that energy transfer occurs from a protein tryptophan to the Tb^{3+} ion. (Whenever tryptophan dominates the protein absorption spectrum of the Tb^{3+} binding protein, the excitation spectrum of the enhanced Tb^{3+} fluorescence peaks at 295 (2)).

Using the above mentioned information the metal binding site was identified on the three dimensional model of bovine trypsin (3). The binding site is composed of the carboxylates of Glu 70 and Glu 80 and the peptide carboxyl oxygen of Asn 72 and Val 75. Facing this site at a distance of about 7 Å is the indole ring of Trp 141. Recent x-ray diffraction maps of the bovine enzyme and its Ca^{2+} complex have identified this same site as the Ca^{2+} binding site (4).

A different conclusion concerning the identification of the metal binding site was reached on the basis of the effect of Gd^{3+} on the magnetic relaxation of active-site bound inhibitor protons (5). In our study we could show that these enhancement effects were in fact due to Gd^{3+} bound to low affinity lanthanide binding sites, different from the primary specific Ca^{2+} binding site (6).

REFERENCES
1. Stroud, R.M., Kay, L.M. and Dickerson, R.E. (1971) Cold Spring Harbor Sympt. Quant. Biol. 36, 125.
2. Brittain, H.G., Richardson, F.S., and Martin, R.B. (1976) J. Am. Chem. Soc. 98, 8255.
3. Huber, R., Kulka, D., Bode, W., Schwager, P., Bartels, K., Deisenhofer, J. and Steigemann, W. (1974) J. Mol. Biol. 89, 73.
4. Bode, W. and Schwager, P. (1975a) FEBS Letters, 56, 139.
5. Abbott, F., Gomez, J.E., Birnbaum, E.R. and Darnall, D.W. (1975b), Biochemistry, 14, 4935.
6. Epstein, M. and Reuben, J. (1977) Biochim. Biophys. Acta 481, 164.

Natural Abundance Carbon-13 Magnetic Resonance Study of Phospholipase A$_2$ (α) from *Crotalus adamanteus*†

Donald J. Nelson[†], William C. Hutton[*] and Michael A. Wells[**]

Dept. of Chemistry[†], Clark Univ., Worcester, MA., USA 01610

Dept. of Chemistry[*], Univ. of Virginia, Charlottesville, VA., USA 22901

Dept. of Biochemistry[**], Univ. of Arizona, Tucson, AZ., USA 85724

SUMMARY

Natural abundance carbon-13 nuclear magnetic resonance spectroscopy is used to investigate the monomer-dimer transition in phospholipase A$_2$(α) from the venom of Crotalus adamanteus, and to study the effect of calcium addition to the dimeric protein. While relatively minor spectral perturbations accompany the monomer-dimer transtion, major conformational alterations, including some involving arginine and tyrosine side chains, occur upon addition of calcium to dimeric phospholipase A$_2$(α).

INTRODUCTION

Carbon-13 NMR provides one of the most productive approaches to the study of the static configuration and dynamic properties of a protein[1]. In the current preliminary study carbon-13 NMR is applied to the calcium-activated venom phospholipase A$_2$(α) from Crotalus adamanteus. Phospholipase A$_2$(α) is a heat-stable, esterolytic enzyme having activity on 1,2-diacyl-sn-phosphoglycerides, liberating specifically the fatty acid esterified at the glycerol C-2 position. Phospholipase A$_2$(α) is composed of two identical 15,000 molecular weight subunits, each containing a binding site for either the activator ion, calcium, or any of the competitive inhibitors: strontium, barium, zinc or cadmium[2]. Tyrosine and tryptophan residues have been associated with specific cation-induced ultraviolet and fluorescence spectral perturbations and arginine modification results in complete loss of enzyme activity.

MATERIALS AND METHODS

Phospholipase A$_2$(α) from Crotalus adamanteus was isolated and assayed according to the procedures of Wells[3]. Carbon-13 magnetic resonance spectra were taken on a JEOL PS-100 P/EC 100 spectrometer operating at 23.5 kG. Routinely 8K data points were used over a frequency range of 5 kHz. The recycle time was 1 second, employing at 45° flip angle. Chemical shifts are reported in parts per million from dioxane as an external reference. Typically, phospholipase A$_2$(α) concentrations of about 10 mM (in monomer units) were employed.

[†]Acknowledgment is made to the Donors of the Petroleum Research Fund, administered by the American Chemical Society, for the support of this research.

RESULTS AND DISCUSSION

The carbon-13 spectrum of monomeric phospholipase $A_2(\alpha)$ is shown in Figure 1. The large number of overlapping resonances at various locations precludes a complete interpretation of the entire spectrum; however, useful information can be obtained from resonance bands derived from a limited number of carbon atoms. Consider, for example, the region from -90 to -80 ppm, which contains resonances derived solely from five arginine ζ-carbons (guanido) and seven tyrosine ζ-carbons (phenolic). Since the guanido carbon resonates at slighly lower field and is generally much less sensitive to environmental perturbations than the tyrosine ζ-carbon, the sharp peak at about -88 ppm should be dominated by the arginine resonances. Figure 2a, b and c illustrate the arginine ζ/tyrosine ζ-carbon region under a variety of conditions. Dimerization induces some spectral changes in the arginine/tyrosine region; however, it appears that major conformational alterations accompany calcium addition to the dimeric protein, most probably involving the rearrangement of a number of tyrosine side chains. The carbon-13 NMR difference spectrum (Figure 3) further illustrates that relatively minor structural rearrangements accompany dimerization; however, the arginine/tyrosine difference peak (labeled "b") is prominent downfield and the aspartic β-carbon and glutamic γ-carbon peaks, as expected, are prominent upfield.

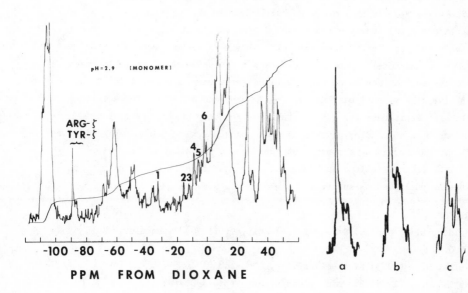

Fig. 1. Natural abundance carbon-13 NMR spectrum of monomeric phospholipase $A_2(\alpha)$. (80,000 transients).

Fig. 2. Arg-ζ/Tyr-ζ region of phospholipase $A_2(\alpha)$.
a. pH=2.9 (monomeric).
b. pH=8.1 (dimeric, -calcium).
c. pH=8.0 (dimeric, +calcium).

83

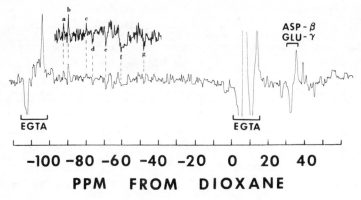

Fig. 3. Carbon-13 NMR difference spectrum. The spectrum
of dimeric phospholipase $A_2(\alpha)$ at pH=8.1 (-calcium) was
subtracted from that of the monomeric protein at pH=2.9.

An interesting and unexpected result of the phospholipase $A_2(\alpha)$ spectroscopy is
the appearance of the peaks labeled 1 through 6 in Figure 1. These six resonance
bands, which are still present in the spectrum following exhaustive dialysis of
the protein, do not derive from any amino acid residue; however, they are compatible with the presence of tightly bound carbohydrate.

Preliminary carbon-13 nuclear relaxation measurements (i.e., spin-lattice relaxation time, T_1, and nuclear Overhauser enhancement, NOE) on monomeric
phospholipase $A_2(\alpha)$ indicate the presence of significant degrees of motional
freedom in the protein. For example, the alanine β-carbon resonance band has
$T_1 \simeq 170$ msec and NOE$\simeq$1.6, the lysine ε-carbon resonance band has $T_1 \simeq 185$ msec and
and NOE$\simeq$1.6 and the resonance band containing the phenylalanine δ and ε carbons
and the tyrosine γ and δ carbons has $T_1 \simeq 170$ msec and NOE$\simeq$1.3. These initial
results are significant for future studies since they suggest that dynamic measurements on tight-binding, carbon-13 enriched substrate analogue molecules are
feasible, thus providing a convenient approach to the study of the nature of the
intermolecular interactions and internal carbon motions in the ternary complex,
calcium-phospholipase A_2-substrate.

REFERENCES

1. Nelson, D.J., Opella, S.J. and Jardetzky, O. (1976) Biochemistry, 15, 5552-
    5560.
2. Wells, M.A. (1973) Biochemistry, 12, 1080-1085.
3. Wells, M.A. (1975) Biochim. Biophys. Acta, 380, 501-505.

CALCIUM-BINDING PROTEINS IN NERVOUS TISSUE

Calcium Binding and Regulation in Nerve Fibers

L. J. Mullins and F. J. Brinley, Jr.
Department of Biophysics and Department of Physiology
University of Maryland School of Medicine
Baltimore, Maryland 21201 U.S.A.

INTRODUCTION

The squid giant axon is a sufficiently large structure to allow the measurement of ionized [Ca] inside it using either the photoprotein aequorin or a metachromic indicator, arsenazo III. Such measurements, coupled with analytical determinations of total Ca content, allow the investigator to follow changes in $[Ca]_i$ as more and more Ca is added to the axon by various experimental techniques. The relationship $\Delta[Ca]_i / \Delta$Total Ca defines buffering of Ca by the axon; this relationship is affected by metabolic inhibitors known to prevent mitochondrial accumulation of Ca. The results to be presented describe how it has been possible to ascertain that most Ca inside the fiber is complexed with non-mitochondrial buffers.

RESULTS

Analytical Ca Content of Axoplasm. Squid axons have an advantage for making analytical determinations of Ca that are truly intracellular, in that samples of axoplasm uncontaminated by external structures such as connective tissue can be readily obtained. Analysis of whole tissue is difficult since it necessarily includes extracellular material. It is also possible to measure the Ca content of mitochondria in vivo and to study the way by which these structures control ionized Ca.

The amount of axoplasm usually available for analysis is of the order of a few milligrams so that analytical methods must be developed to cope with a few tens of picomoles of Ca. We have found it satisfactory to use atomic absorption analysis combined with a graphite furnace to analyze for these small amounts of Ca. A program of furnace heating such that the sample is dried, charred, and finally atomized into the optical light path provides adequate control over sample introduction, and a deuterium lamp provides the necessary correction for broadband, non-specific absorption.

We have found that axons freshly isolated from living squid have a Ca content of 50 μmole Ca/kg axoplasm, and that this value is unchanged by storage if the axon is kept in 3 mM Ca, 450 mM Na seawater[1]. Axons do gain Ca if the usual seawater [Ca] of 10 mM is applied and this gain approximates a net flux of 70×10^{-15} mole/cm^{2}sec (fmole). The application to axons of 450 mM choline seawater containing 3-10 mM Ca leads to analytical Ca gains that represent 1000 fmole/cm^{2}sec or more, suggesting that $[Na]_o$ is an important ingredient in the maintenance of a normal Ca content of tissues. This finding supports notions that an inwardly directed electrochemical gradient for Na$^+$ is important in Ca regulation.

If axons are allowed to increase their analytical [Ca] from 50 μmole to 500 μmole/kg axoplasm, corresponding to a 1 hr immersion in 3 mM Ca choline seawater, and then transferred to 3 mM Ca sodium seawater, the Ca content of the axon declines to control levels as the data in Fig 1 show. One infers, therefore, that the axon possesses regulatory mechanisms that allow it to recover from applied Ca loads and that, in the course of 30-40 min, these mechanisms can return total internal Ca to control levels. Note that this conclusion is independent of any assumptions about how intracellular Ca may be distributed between internal compartments. A further point in these "unloading" experiments is that this loss of Ca occurs whether the axon is a normal one or one treated with mitochondrial uncouplers such as FCCP, or a combination of IAA and FCCP. An inference from these findings is that the recovery of normal Ca content by the axon is not dependent on mitochondrial function or on ATP generated by oxidative phosphorylation or glycolysis.

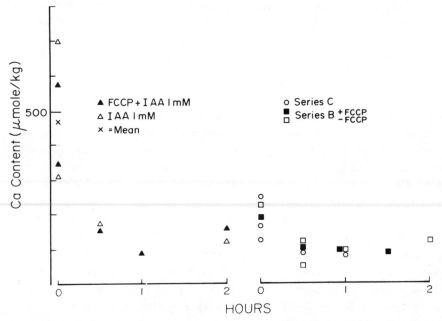

Fig. 1. (left) This shows analytical Ca of axoplasm for axons in 3 mM Ca sodium seawater for 1 hr with either FCCP + IAA or with IAA, followed by 1 hr in 3 mM Ca choline seawater and yields the zero time values. Other axons had the same treatment, followed by 0.5, 1 or 2 hr in 3 mM Ca sodium seawater plus inhibitors (recovery). (right) Axons were kept 0.5 hr in 1 mM Ca choline (series C without inhibitor, series D ± FCCP) and loading is defined by the zero time value. Recovery by other axons kept 0.5, 1, 1.5 or 2 hr in 3 mM Ca sodium seawater is shown on this plot (Requena, Brinley and Mullins, unpublished).

Ionized [Ca]. It has long been recognized that most of the analytical Ca in cells is not ionized but either stored in intracellular compartments or chemically combined with intracellular compounds. Techniques that have been successful in the measurement of ionized Ca are: a) the use of the photoprotein aequorin, which emits light in the presence of

Ca, and b) multiwavelength spectrophotometry using Ca arsenazo III, a dye that can be microinjected into cells, and its association with Ca measured. In principle, aequorin can be injected into cells and if analytical Ca is in a steady state, it will measure the mean [Ca] of the cell. Baker, Hodgkin and Ridgway[2] carried out such experiments and found an ionized [Ca] of 350 nM when total [Ca] was about 400 μmole/kg axoplasm. Aequorin can be confined to a dialysis tube introduced on the axis of an axon and these experiments showed an ionized [Ca] of 20 nM when conditions were such that total [Ca] was 50 μmole/kg axoplasm[3]. These results are not in disagreement; it will be shown later that ionized and total Ca bear a direct relationship to each other. With Ca arsenazo III, the results of DiPolo et al.[3] showed an ionized Ca of 50 nM; we have adopted a mean value for ionized [Ca] of 30 nM.

With ionized [Ca] at 30 nM and analytical [Ca] at 50 μmole/kg, it is clear that most Ca in axoplasm is complexed in some way and that roughly less than 1/1000th of the Ca is in an ionized form. Since mitochondria have been implicated in Ca sequestration, a reasonable guess would be that most of the internal Ca might be contained within these organelles. The experiments of DiPolo et al.[3] showed that there is at most a doubling of light emission when CN is applied to axons from freshly killed squid. On the other hand, the experiments of Baker, Hodgkin and Ridgway[2] showed that if aequorin-injected axons are treated with CN, then after a delay, light emission from aequorin rises to high levels and is reversed by the removal of CN. A resolution of this apparent discrepancy of experimental results is possible by noting that the analytical Ca of the two sorts of axons studied differed by almost a factor of 10. Fresh axons with low analytical Ca do not contain large quantities of Ca releasable by CN, while axons previously loaded with Ca do contain Ca in their mitochondria.

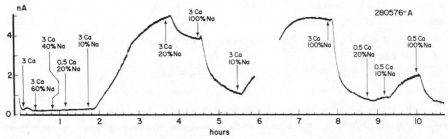

Fig. 2. This is a recorder tracing of light output by aequorin (as photomultiplier current in nanoamps) vs. time. The aequorin was confined to a dialysis tube at the center of an axon. Note that 3 mM Ca 40 percent Na seawater gives a $[Ca]_i$ identical with control (3 mM Ca 100 percent Na) seawater while 20 percent Na or 10 percent Na seawater – choline mixtures give a large (ca. 25-fold) increase in $[Ca]_i$ (from Ref. 1).

Another sort of study has related ionized Ca to external [Ca]; the results of Requena et al.[1] have shown that ionized Ca remains constant with time if $[Ca]_o$ in seawater is 3 mM, just as analytical studies of the Ca content of axoplasm have shown this to remain constant if Ca_o is 3 mM. Reducing the [Na] of seawater causes a large rise in ionized [Ca] and one that is reversed by applying Na seawater to the fiber as the experiment in Fig 2 shows.

89

These findings are in agreement with the idea that it is the energy present in the Na electrochemical gradient that energizes Ca extrusion.

Finally, if an axon is injected with apyrase so that intracellular ATP is reduced to low levels, not only is there no change in ionized [Ca], but if the axon is loaded with Ca, it recovers to its initial value for ionized [Ca] as the results of Fig 3 show. There is, therefore, a strong parallel between studies with aequorin of ionized Ca changes, and analytical changes in total Ca content Such findings make it difficult to postulate a role for ATP in the process of maintaining a low $[Ca]_i$.

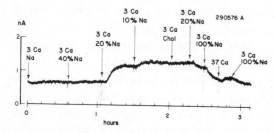

Fig. 3. This is the aequorin light output from an apyrase-injected axon. Compared with Fig. 2, the responses to 20 or 10 percent Na are much smaller (from Ref. 1).

Ca Fluxes. The measurement of unidirectional Ca fluxes with ^{45}Ca has provided some valuable insights into the processes regulating intracellular Ca content. The experiments have been of two sorts: influx measurements in which an axon is immersed in radioactive seawater for an appropriate time, then removed, rinsed and the axoplasm extruded and counted; and efflux measurements in which an axon is injected with ^{45}Ca and the loss of radioactivity followed with time. Influx measurements[4] show that about 100 $fmole/cm^2 sec$ enter a fiber from 10 mM Ca seawater and that this flux is linear with $[Ca]_o$. It has also been shown that Ca influx is raised by increases in $[Na]_i$, and by the substitution of Li for Na in seawater[5]. Parallel measurements of Na efflux that depends on Ca_o show that the factors favoring Ca entry also promote this Na efflux and that about 4 Na leave the fiber for each Ca that enters.

As a technique, Ca efflux from fibers injected with ^{45}Ca suffers from some disadvantages. First, it is not possible to get enough counts into a fiber without the addition of about 100 μmole Ca/kg axoplasm. Since this quantity of Ca is twice that found in fresh axons, a substantial loading of the fiber with Ca necessarily takes place. Second, flux measurements are only meaningful if the concentration of Ca remains constant while the flux analysis is taking place. This seems unlikely to be true for injected fibers, especially when Ca fluxes are not in balance, since the natural buffering of the axon is insufficient to prevent large excursions in [Ca]. This conclusion is inferred from the results of treating a fiber with CN where the ionized Ca and the Ca efflux rise some 40 fold[6]. There are also likely to be large specific activity changes in such axons if influx substantially exceeds efflux.

Much of the criticism of the injected fiber method can be dealt with by using CaEGTA/EGTA buffers to control ionized Ca. While this can be done for one Ca

90

concentration in injected fibers, the technique of internal dialysis[7] allows the continuous control of all dialyzable solutes in the fiber using a dialysis tube inserted into the fiber on axis. The results with internal dialysis show that Ca efflux is about 30 nM[8], that Ca efflux is about half-maximal when $[Ca]_i$ is about $8\,\mu M$[9], that $[Na]_i$ is an inhibitor of Ca efflux with 30 mM Na reducing Ca efflux about half the value when $Na_i = 0$[8,9], and that ATP increases Ca efflux about 10 fold when $[Ca]_i$ is in the physiological range and $[Na]_i$ is 80 mM[10] but has no effect when $[Na]_i$ is zero[11].

Changes in Na_o and Ca_o also affect Ca efflux; the effect of removing both of these cations from seawater is a dramatic fall in Ca efflux if Ca_i is high[12], a 30-50 percent decrease in Ca efflux in injected axons[6,13] or little or no effect when Ca_i is at physiological levels[8,10].

Changes in membrane potential also affect both Ca efflux and Ca influx, the effect of hyperpolarization being to increase Ca efflux and decrease Ca influx[4, 14]; thus Ca fluxes are dependent on six experimental variables, viz. $[Ca]_o$, $[Ca]_i$, $[Na]_o$, $[Na]_i$, $[ATP]_i$ and E_m.

The Energetics of Ca Transport. If the energy of the Na gradient across the cell membrane is to be used to maintain a steady-state Ca_i, it can be shown that the ratio $[Ca]_o/[Ca]_i$ is dependent on the coupling of Na to Ca as follows:

$$\frac{[Ca]_o}{[Ca]_i} = \frac{[Na]_o^2}{[Na]_i^2}, \quad \text{for } n = 2 \qquad\qquad \frac{[Ca]_o}{[Ca]_i} = \frac{[Na]_o^3}{[Na]_i^3} \exp -V_m F/RT, \quad \text{for } n = 3$$

$$\frac{[Ca]_o}{[Ca]_i} = \frac{[Na]_o^4}{[Na]_i^4} \exp -2V_m F/RT, \quad \text{for } n = 4$$

Since in a squid axon the ratio of [Na] across the cell membrane is of the order of 10, a coupling of 2 Na per Ca is equivalent to a Ca ratio of 100, a coupling of 3 is equal to about 10^4, while the actual ratio as found by DiPolo et al.[3] is about 10^5 and hence requires a value of n = 4. An alternate method of stating these relationships is to note that for a steady $[Ca]_o/[Ca]_i$, the flux ratio of Ca is unity and equal to $\exp -(2E_{Na} - E_{Ca} - V_m) 2F/RT$. The experimental data available on Ca fluxes are in reasonable agreement with the idea that both Ca influx and efflux depend on E_{Na}, E_{Ca}, and V_m if allowance is made for the saturation to be expected[15] when $[Na]_o$ is high.

Compartments. The foregoing data, especially that for increases in analytical Ca, makes it possible to relate the entry of a known quantity of Ca to the change in ionized Ca that this entry produces. When this is done in the nM or physiological concentration range, one finds that the entry of sufficient Ca to increase the analytical Ca 1 μM in axoplasm, leads to the increase in ionized [Ca] by 1 nM (a 1:1000 buffering). If large Ca loadings are effected so that ionized [Ca] rises to the μM level, then ionized [Ca] remains stable at 3-5 μM and virtually independent of load[16].

91

When Ca enters a nerve fiber either as a result of stimulation or because Na-free solutions have been applied outside, it can a) be taken up by intracellular organelles such as mitochondria, or b) complex with Ca-binding compounds existing inside the cell, or c) lead to an increase in ionized [Ca] inside the cell. Two sorts of experimental situations should be recognized in attempting to analyze the way that Ca is handled inside the cell. In the first, one considers physiological Ca concentrations which are of the order of 30 nM. No technique exists capable of analyzing the manner by which Ca is handled immediately as it enters the cell. The spectrophotometric method of measurement of Ca arsenazo is poorly sensitive at low $[Ca]_i$ and cannot readily distinguish between a small amount of $[Ca]_i$ at the cell periphery and this same amount of [Ca] uniformly distributed in the axoplasm.

Table 1

ESTIMATES OF CALCIUM BINDING TO AXOPLASMIC CONSTITUENTS

Ligand	Concentration in Axoplasm	Effective Dissociation Constant	Ca Ligand	Reference for Dissociation Constant
	mM	mM	nM	
Aspartate	80	500	5	Tiffert & Brinley Unpub
Glutamate	20	500	1	Assumed equal to aspartate
ATP	4	0.7	171	Assume equal to K_{eff} for MgATP
ADP	0.1	3.3	1	
AMP	1	50	1	De Weer Unpub
Phosphate	3	35	3	17
Ca binding Protein	0.18	0.025	216	18
Ca binding Protein	0.03	0.0005	1800	19
Sum [Ca Ligands] above			2.2 μM	
Mitochondrial Ca Content			2.2 μM	
Axoplasm Ca Content			50 μM	
Non-Mitochondrial Ca			45 μM	

Aequorin is much more sensitive to small amounts of [Ca], but since it is confined to a dialysis tube in the center of the axon, it has a time constant of the order of 1.5 hr for its response to a steady Ca load. The bulk of the evidence with respect to mitochondria in vitro would suggest that these organelles have little ability to accumulate Ca when the ambient [Ca] is less than 1 μM so that at a concentration 30-fold less than this (30 nM), one might expect mitochondria to have little effect in Ca buffering. Such a conclusion fails to recognize that immediately inside the membrane [Ca] may be in the μM range, while in the axon interior it may be in the nM range. This is inferred by the following calculation. The Ca entry in a single action potential with 112 mM Ca_o is 0.14 pmole/cm^2. Since the mean square diffusion distance $\Delta x^2 = D \Delta t$, and the action potential lasts ca. 10^{-3} sec, with D =

10^{-5} cm^{2}sec, $\Delta x^2 = 10^{-8}$ or $\Delta x = 10^{-4}$ cm. We have thus 0.14 pmole/10^{-4} cm^3 = 1.4 μM, a concentration where mitochondria can be expected to buffer strongly. At present, however, buffering in the periphery is not experimentally accessible to measurement.

Experimentally, one finds with aequorin or arsenazo that the rate of increase in [Ca] with an applied Ca load is about 1/1000th of the load when mitochondrial function is intact, while it is 1/20th of the applied load when mitochondria are poisoned with FCCP or with CN. The difference between these values defines a buffer system that is not sensitive to metabolic inhibition and is one that can buffer Ca up to several mmole/kg axoplasm. Table 1 lists a number of compounds that have been shown to complex with Ca and from this tabulation it is clear that even the sum of all the Ca ligands does not complex enough Ca at physiological levels of [Ca] to account for the Ca buffering that is observed in the absence of mitochondrial function.

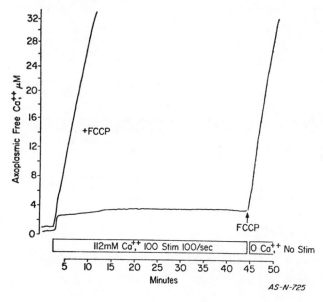

Fig. 4. Ordinate is ionized Ca in axoplasm as measured by arsenazo III; abscissa is Ca net flux into axon produced by nerve stimulation at 100/sec, it is also time of stimulation. Ca load is 50 umole Ca per kg axoplasm per min of stimulation. Record on left is axon treated with FCCP prior to start of stimulation. Record on right shows that axoplasmic [Ca] is constant at about 2.5 uM throughout loading period. At 45 min, the seawater was changed to Ca-free and stimulation stopped. [Ca]$_i$ rose as a result of loss of Ca from mitochondria (from Ref. 16).

With Ca arsenazo, it is possible to explore the range of Ca buffering in axoplasm not accessible to aequorin, since this compound would be destroyed by the high levels of [Ca]$_i$. When [Ca]$_i$ reaches levels of from 3-10 μM, the change in ionized Ca with Ca load is virtually non-existent if mitochondrial function is intact, as Fig 4 shows. In the axons shown, the application of FCCP to inhibit mitochondrial function produces a marked change in Ca$_i$ vs. load so that the regulatory function for [Ca]$_i$ is clearly residing in the mitochondria.

Another sort of experiment that can be performed using either the aequorin technique or that of Ca arsenazo is to poison axons with FCCP and note the change in ionized [Ca]. Typically, this is a 3-fold increase in ionized Ca and when allowance is made for the volume fraction of mitochondria in the axon, it works out that 5 percent of the analytical Ca of an axon is releasable by mitochondrial poisoning. By implication, 95 percent of the analytical Ca of the axon is either complexed with Ca binding proteins in axoplasm, or is in a precipitated and therefore inert form. There is no direct experimental method for distinguishing between these two hypotheses, but the fact that, in the absence of functioning mitochondria, Ca buffering appears to be virtually constant for even large Ca loads argues somewhat in favor of a Ca binding protein that complexes most of the analytical Ca at physiological levels of [Ca]. For large loads, it is clear that mitochondria complex most of the Ca as Fig 4 indicates. It has proven possible to specify an ionized [Ca] for axoplasm solely from total analytical Ca content measurements simply by noting that Ca content and ionized Ca differ by a factor of 1000, as Fig 5 shows[16].

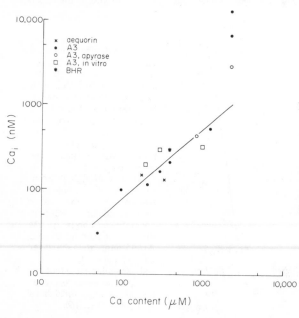

Fig. 5. Ordinate is ionized [Ca] in axoplasm and abscissa is analytical Ca content of axoplasm. Note that the line has a slope close to unity and the ordinate and abscissa differ by 1000. Points have been obtained using aequorin, arsenazo III (A3), and a point from Baker, Hodgkin and Ridgway, 1971 (BHR).

DISCUSSION

Steady-State Ca. The evidence reviewed is persuasive to the view that both the analytical Ca of an axon and the ionized Ca inside the fiber are in a steady state when Ca_o is 3 mM and Ca_i is 30 nM; changes in Ca brought about by a variety of techniques can be reversed by allowing the axon to equilibrate in 3 mM Ca sodium seawater. It is possible to maintain a steady state with respect to either ionized or total Ca in the presence of inhibitors that reduce ATP and/or mitochondrial buffering to low levels. Mitochondrial function is maintained in the absence of ATP (apyrase-injected axons) but is lost by the presence of CN or FCCP.

Transient Ca Fluxes during Metabolic Inhibition. A large decrease in buffering is occasioned by metabolic inhibition; in turn, one must then expect that a given Ca flux generated by transient bioelectric activity will produce a large excursion in $[Ca]_i$ just inside the membrane. In normal axons, however, $[Ca]_i$ is so low that mitochondrial buffering of a transient Ca entry may be expected to be rather slow. This is inferred from the small amounts of Ca that can be released fom mitochondria in a normal, fresh nerve fiber.

REFERENCES

1. Requena, J., R. DiPolo, F. J. Brinley, Jr. and L. J. Mullins (1977) The control of ionized calcium in squid axons. J. Gen. Physiol., in press.
2. Baker, P. F., A. L. Hodgkin and E. B. Ridgway (1971) Depolarization and calcium entry in squid axons. J. Physiol. (Lond.). 218:709-755.
3. DiPolo, R., J. Requena, F. J. Brinley, Jr., L. J. Mullins, A. Scarpa and T. Tiffert (1976) Ionized calcium concentrations in squid axons. J. Gen. Physiol. 67:433-467.
4. Hodgkin, A. L., and R. D. Keynes (1957) Movements of labelled calcium in squid giant axons. J. Physiol. (Lond.). 138:253-281.
5. Baker, P. F., M. P. Blaustein, A. L. Hodgkin and R. A. Steinhardt (1969) The influence of calcium on sodium efflux in squid axons. J. Physiol. (Lond.). 200:431-458.
6. Blaustein, M. P., and A. L. Hodgkin (1969) The effect of cyanide on the efflux of calcium from squid axons. J. Physiol. (Lond.). 200:497-527.
7. Brinley, F. J., Jr., and L. J. Mullins (1967) Sodium extrusion by internally dialyzed squid axons. J. Gen. Physiol. 50:2303-2331.
8. Brinley, F. J., Jr., S. G. Spangler, and L. J. Mullins (1975) Calcium and EDTA fluxes in dialyzed squid axons. J. Gen. Physiol. 66:223-250.
9. Blaustein, M. P., and J. M. Russell (1975) Sodium-calcium exchange and calcium-calcium exchange in internally dialyzed squid giant axons. J. Membr. Biol. 22:285-312.
10. DiPolo, R. (1977) Characterization of the ATP-dependent calcium efflux in dialyzed squid axons. J. Gen. Physiol., in press.
11. DiPolo, R. (1976) The influence of nucleotides on calcium fluxes. Fed. Proc. 35:2579-2582.
12. Blaustein, M. P., J. M. Russell, and P. De Weer (1974) Calcium efflux from internally dialyzed squid axons: The influence of external and internal cations. J. Supramolec. Struct. 2:558-581.
13. Baker, P. F., and P. A. McNaughton (1976) Kinetics and energetics of calcium efflux from intact squid giant axons. J. Physiol. (Lond.). 259:103-144.
14. Mullins, L. J., and F. J. Brinley, Jr. (1975) The sensitivity of calcium efflux from squid axons to changes in membrane potential. J. Gen. Physiol. 65:135-152.
15. Mullins, L. J. (1976) Steady-state calcium fluxes: Membrane versus mitochondrial control of ionized calcium in axoplasm. Fed. Proc. 35:2583-2588.
16. Brinley, F. J., Jr., T. Tiffert, A. Scarpa and L. J. Mullins (1977) Intracellular calcium buffering capacity in isolated squid axons. J. Gen. Physiol., in press.
17. Ringbom, A. (1963) Complexation in Analytical Chemistry. New York: Interscience.

18. Alema, S., P. Calissano, G. Rusca, and A. Giuditta. (1973) Identification of a calcium-binding, brain specific protein in the axoplasm of squid giant axons. J. Neurochem. 20:681-689.
19. Baker, P. F., and W. Schlaepfer (1975) Calcium uptake by axoplasm extruded from giant axons of _Loligo_. J. Physiol. (Lond.). 249:37P-39P.

Divalent Cation Binding Sites of CDR and Their Role in the Regulation of Brain Cyclic Nucleotide Metabolism

Donald J. Wolff, Margaret A. Brostrom and Charles O. Brostrom
CMDNJ-Rutgers Medical School
Department of Pharmacology
Piscataway, N. J.   08854 U.S.A.

INTRODUCTION

In 1972, Wolff and Siegel[1] reported the purification and characterization of a Ca^{2+}-binding protein from brain. This protein was demonstrated subsequently to act as a Ca^{2+}-dependent regulator (CDR) of two functionally related brain enzymes: a cyclic nucleotide phosphodiesterase[2] and an adenylate cyclase[3,4]. Other workers independently purified to homogeneity the protein activator of cyclic nucleotide phosphodiesterase originally reported by Cheung[5] and demonstrated that this protein binds Ca^{2+} [6,7]. More recently, this protein has been shown to possess primary structural features similar to troponin C[8] and to substitute for troponin C in conferring Ca^{2+} dependence on a reconstituted actomyosin system[9].

Despite the differences in nomenclature used by different investigators to describe this protein, it is quite clear that a single specific protein, as judged by a variety of criteria confers Ca^{2+} sensitivity on both the cyclic nucleotide phosphodiesterase and the adenylate cyclase. Discrepancies and incomplete characterization exist in the reported literature, however, with respect to the divalent cation binding properties of this protein. Consequently, our laboratory has undertaken a comprehensive examination of the divalent cation binding properties of CDR by equilibrium dialysis and circular dichroic spectrophotometry to define their characteristics. Putative roles of the divalent cation binding sites of CDR in the regulation of brain cyclic nucleotide metabolism have been inferred from these data.

METHODS

Preparation of the Ca^{2+}-dependent regulator (CDR) protein: CDR was prepared from bovine brain by a procedure described by Wolff et al.[10] resulting in the production of 56 mg/kg wet weight for whole brain.

Circular dichroic spectrophotometry: Circular dichroic spectrophotometric measurements were made at 25° on a Cary model 61 circular dichrometer in a 1 cm quartz cell. Mean residue ellipticities (θ_m) were calculated from the relationship $\theta_m = \theta_{obs}^{\circ} M/100\ 1\ c$; where M is the mean residue weight calculated from the amino acid analysis[1]. CDR samples and buffer (10 mM Tris-Cl, pH 7.4) were freed of divalent cations by passage over columns of Chelex 100 and monitored for cation content by atomic absorption spectrophotometry.

Determination of divalent cation binding by equilibrium dialysis: Dialyses were performed at 25° for 18 hours on a reciprocating shaker in 100 ml polypropylene bottles washed five times with deionized water. CDR and buffer (10 mM Tris-Cl, pH 7.4) were freed of divalent cations by passage over columns of Chelex 100. Following dialysis samples taken from inside and outside the dialysis bag were analyzed by atomic absorption spectrophotometry.

CDR-dependent phosphodiesterase: CDR-dependent phosphodiesterase was prepared from porcine brain as described previously[11] and assayed for activity with standardized concentrations of Ca^{2+} without added chelators as described by Brostrom and Wolff[12].

CDR-dependent adenylate cyclase: CDR-dependent adenylate cyclase was prepared from a washed, particulate fraction from rat brain and assayed as detailed by Brostrom et al.[13].

RESULTS AND DISCUSSION

Characterization of the Ca^{2+}-dependent regulator from bovine brain: CDR was homogeneous by the criteria of gel electrophoresis in 15% acrylamide in the non-denaturing system of Davis[14] or in gels containing 0.1% SDS by the procedure of Laemmli[15]. Calibration of the molecular mass of CDR with standard protein markers provided a value of 18,700 ± 500 which was used in the calculation of the molar quantities of CDR. Determinations of the phosphate content of wet ashed CDR samples by the procedure of Chen et al.[16] established that the preparation contained less than 0.1 moles phosphate/mole CDR.

Circular dichroic spectral studies of CDR: The circular dichroic spectrum of CDR was examined without added divalent cations and in solutions containing Ca^{2+} (25 μM), Mn^{2+} (100 μM) or Mg^{2+} (1 mM) (Figure 1). The far ultraviolet circular dichroic spectrum possessed negative maxima at 222 and 207 nm. In the absence of divalent cations the mean residue ellipticity at 208 nm was -12,000 °cm^2/decimole. The apparent α helical content, estimated by the method of Greenfield and Fasman[17] was 28%. Upon adjustment to 25 μM Ca^{2+}, the mean residue ellipticity decreased to -16,200 °cm^2/decimole, indicating an increase of helical content to 42%. Additional Ca^{2+} produced no further changes. When the divalent cation-free CDR was adjusted to 100 μM with Mn^{2+}, the mean residue ellipticity decreased to -15,600 °cm^2/decimole corresponding to a helical content of 40%. Alternately, the addition of Mg^{2+} (1 mM final) provided a mean residue ellipticity of -15,200 °cm^2/decimole with a helical content of 39%. These concentrations of Mn^{2+} and Mg^{2+} were found to produce maximal changes and were thus saturating. Although a profound change is detectable when CDR interacts with either Ca^{2+}, Mn^{2+}, or Mg^{2+}, the extent of that change is not identical for each of the cations. At saturating concentrations, the conformational change was greater for Ca^{2+} than for either Mn^{2+} or Mg^{2+}.

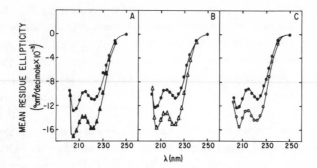

Fig. 1. The far ultraviolet circular dichroic spectrum of CDR. Samples of CDR (24 μg/ml) in 10 mM Tris-HCl, pH 7.4, were prepared free of divalent cations as described in Methods and the spectra examined (A) in the absence (●) and presence (▲) of 25 μM Ca^{2+}; (B) in the absence (●) and presence (△) of 100 μM Mn^{2+}; and (C) in the absence (●) and presence (○) of 1 mM Mg^{2+}.

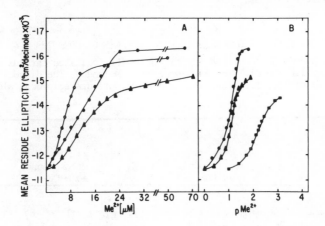

Fig. 2. The divalent cation concentration dependence of the mean residue ellipticity of CDR measured at 222 nm. Divalent cations were removed from CDR by treatment with Chelex-100 as described in Methods. Panel A: The effect of total added divalent cation concentration measured at 3.1 μM CDR plus added Ca^{2+} (○) and at 6.2 μM CDR plus added Ca^{2+} (●) or Mn^{2+} (▲). Panel B: The effect of total added Ca^{2+} (●), Mn^{2+} (▲) or Mg^{2+} (□) on the mean residue ellipticity of CDR (6.2 μM) expressed on a log scale.

The divalent cation specificity of the observed conformational change of CDR was explored by examining the cation concentration dependence of the change of ellipticity measured at 222 nm (Figure 2). A solution containing 3.1 μM CDR required 6.2 μM total added Ca^{2+} to provide a half maximal conformational transition, whereas a solution containing 6.2 μM CDR required 12.4 μM total added Ca^{2+}. At 6.2 μM CDR the half maximal conformational transition was noted at 13.7 μM Mn^{2+} and 140 μM Mg^{2+} respectively. The Mg^{2+} concentration dependence of the conformational change was not markedly influenced by CDR concentration (data not shown).

The observation that the apparent Ca^{2+} concentration dependence of the conformational change was directly proportional to the amount of CDR used in the study supported the interpretation that virtually all of the Ca^{2+} in the system was bound, whereas in the study with Mg^{2+} almost all of the cation was free. Consequently, a comparison of the free concentrations of Ca^{2+} and Mg^{2+} required for interaction with CDR could not readily be obtained from data derived from circular dichroic measurements. The concentration dependence and the specificity of cation binding to CDR were therefore determined by equilibrium dialysis in conjunction with atomic absorption spectrophotometry, utilizing reagents freed of contamination with divalent cations.

TABLE 1

DIVALENT CATION DISSOCIATION CONSTANTS OF CDR

The concentration dependence of binding to CDR of the indicated cations was determined by equilibrium dialysis as described in <u>Methods.</u> The dissociation constants were determined from the abscissal intercepts of reciprocal plots of the binding data.

		Class A		Class B	
Cation	Competing ion	moles bound/ mole CDR	K_d μM	moles bound/ mole CDR	K_d μM
Ca^{2+}	none	3	0.2	1	1
Mn^{2+}	none	3	1.3	1	4
Ca^{2+}	10 μM Mn^{2+}	3	1.5	1	10
Ca^{2+}	1 mM Mg^{2+}	3	3	*	-
Mn^{2+}	1 mM Mg^{2+}	3	20	*	-
Mg^{2+}	none	3	140	1	20

* undetectable

<u>Determination</u> <u>of</u> <u>divalent</u> <u>cation</u> <u>binding</u> <u>to</u> <u>CDR</u> <u>by</u> <u>equilibrium</u> <u>dialysis:</u> The Ca^{2+} concentration dependence of binding to CDR was examined in the absence of other divalent cations (Table 1). Two classes of Ca^{2+} binding sites were observed for CDR at this condition - sites class A binding three moles of Ca^{2+}/mole CDR with a dissociation constant of 2×10^{-7} M, and site class B binding one mole with a dissociation constant of 10^{-6} M.

The concentration dependence of Mn^{2+} binding to CDR in the absence of other divalent cations revealed two classes of binding sites - three moles Mn^{2+} bound with a dissociation constant of 1.3×10^{-6} M, and one mole bound with a dissociation constant of 4×10^{-6} M.

The effect of varying the concentration of free Ca^{2+} at a fixed concentration of Mn^{2+} (10 μM) and CDR was investigated in order to establish the suspected identity of the Ca^{2+} and Mn^{2+} binding sites of CDR. As the free concentrations of Ca^{2+} in the system were elevated, increasing degrees of binding of Ca^{2+} to CDR were detected, accompanied by comparable degrees of dissociation of bound Mn^{2+} from CDR. An exchange of Ca^{2+} for bound Mn^{2+} was therefore indicated. When the Ca^{2+} binding data were plotted in a double reciprocal format (not shown), it was clear that Ca^{2+} binding to CDR in the presence of 10 μM Mn^{2+} occurred at two classes of sites. Three moles bound at sites class A with an apparent dissociation constant of 1.5×10^{-6} M and one mole bound with an apparent dissociation constant of 10^{-5} M at site class B. A plot of the reciprocal of Mn^{2+} binding to CDR as a function of the free inhibitory concentration of Ca^{2+} (not shown) revealed that the exchange of Ca^{2+} for bound Mn^{2+} could be attributed to interactions at two classes of sites: three moles bound to one class with an apparent K_i of 1.5 μM and one mole bound to a second class with an apparent K_i of 16 μM. These data are consistent with a competitive exchange occurring between Ca^{2+} and Mn^{2+} at both site classes A and B.

The intracellular free concentrations of Mg^{2+} in brain are believed to be relatively constant and have been estimated by Veloso et al.[17] to be approximately 1 mM. Since the far ultraviolet circular dichroic spectrophotometric data of Figures 1 and 2 had indicated that Mg^{2+} binds to CDR at this concentration, the effect of 1 mM Mg^{2+} on the concentration dependence of Ca^{2+} or Mn^{2+} binding to CDR was examined. At 1 mM Mg^{2+}, binding of Ca^{2+} to CDR occurred at a single class of sites to the extent of 3 moles/mole with an apparent dissociation constant of 3×10^{-6} M. The binding of Mn^{2+} to CDR at 1 mM Mg^{2+} also occurred at a single class of sites to the extent of 3 moles/mole CDR with an apparent dissociation constant of 2×10^{-5} M.

The binding of Ca^{2+} or Mn^{2+} to site class B of CDR was prevented by 1 mM Mg^{2+} at the concentration ranges studied. This concentration of Mg^{2+} also increased the concentrations of Ca^{2+} or Mn^{2+} required to bind to sites class A approximately fifteen-fold for each ion. These data, however, did not distinguish whether these changes in affinity arose as a consequence of non-competitive inhibition produced at sites class A by the occupancy of site class B by Mg^{2+}, by direct competition of Mg^{2+} with Ca^{2+} or Mn^{2+} at sites class A, or by contributions from both types of interactions. Therefore the effect of variation of Mg^{2+} concentration on Ca^{2+} binding to CDR was examined at 0.2, 0.4, 0.6, and 1.0 μM Ca^{2+}. Under these conditions virtually all of the Ca^{2+} bound to CDR is associated at sites class A, and concentrations of Mg^{2+} required for the dissociation of Ca^{2+} could be examined with minimal interferences from effects produced by cation binding to site class B. These data (not shown), when plotted in a Dixon format as the reciprocal of bound Ca^{2+} versus Mg^{2+} concentration (as inhibitor), provided a series of lines which intersected above the abscissa. The interaction of Ca^{2+} and Mg^{2+} at sites class A was thus competitive and exhibited a K_i for Mg^{2+} of 140 μM.

The Mg^{2+} concentration dependence of binding to CDR in the absence of other divalent cations revealed binding to two classes of sites, three moles bound at sites class A with a dissociation constant of 1.4×10^{-4} M and one mole bound at site class B with a dissociation constant of 2×10^{-5} M.

Studies of Mg^{2+} binding to CDR were conducted in the presence of 2.5, 10, and 25 μM Ca^{2+} (data not shown) at Mg^{2+} concentrations ranging from 10 to 200 μM. Under these conditions one mole of Mg^{2+} bound per mole CDR, and the apparent affinity for Mg^{2+} binding decreased with increasing Ca^{2+} concentrations.

These binding studies establish that two classes of divalent cation binding sites exist on CDR to which either Ca^{2+}, Mn^{2+} or Mg^{2+} may bind. Sites class A bind three moles of cation with a dissociation constant of 0.2 μM for Ca^{2+}, 1.3 μM for Mn^{2+} and 140 μM for Mg^{2+}. Site class B binds one mole of cation with dissociation constants of 1 μM for Ca^{2+}, 4 μM for Mn^{2+}, and 20 μM for Mg^{2+}. Each cation competitively inhibits the binding of the other cations, thus raising their apparent dissociation constants.

Regulatory roles of class A and class B cation binding sites of CDR: The CDR concentration dependence of the cyclic nucleotide phosphodiesterase was examined at 1 mM Mg^{2+} without or with either 8 μM Ca^{2+} or 50 μM Mn^{2+} (Figure 3) to evaluate the relative potencies of the $Mg_{3A}Mg_{1B}CDR$, $Ca_{3A}Mg_{1B}CDR$ and $Mn_{3A}Mg_{1B}CDR$ complexes as activating species. The free Ca^{2+} and Mn^{2+} concentrations in the incubations were present at 2.5 times their respective apparent dissociation constants for binding to CDR as determined in the presence of 1 mM Mg^{2+}. Equal degrees of exchange of Ca^{2+} and Mn^{2+} for Mg^{2+} at sites class A were thereby generated. In the absence of Ca^{2+} or Mn^{2+}, no activation by CDR occurred

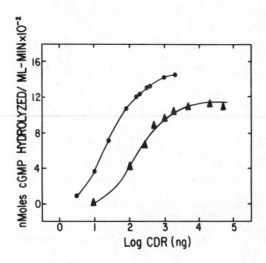

Fig. 3. The CDR concentration dependence of cyclic nucleotide phosphodiesterase activity measured at 1 mM Mg^{2+} and either 8 μM Ca^{2+} (●) or 50 μM Mn^{2+} (▲). CDR-dependent phosphodiesterase was prepared from porcine brain as described previously[11] and assayed in the presence of spectral grade cations, the concentrations of which were determined by atomic absorption spectrophotometry. No chelators were present in the assays which were performed as described previously[12].

demonstrating that the tetramagnesium CDR was inactive. Activation of the phosphodiesterase occurred at ten-fold lower concentrations of CDR with 8 μM Ca^{2+} as compared to 50 μM Mn^{2+}. It is clear, therefore, that the phosphodiesterase discriminated among these species of CDR in favor of the $Ca_{3A}Mg_{1B}CDR$. These data preclude a regulatory role for Mn^{2+} in the control of the cyclic nucleotide phosphodiesterase for the following reasons. First, CDR requires seven-fold higher concentrations of Mn^{2+} than of Ca^{2+} to form equal concentrations of complexed cation. Second, ten-fold higher concentrations of the $Mn_{3A}Mg_{1B}CDR$ complex, as opposed to the $Ca_{3A}Mg_{1B}CDR$ complex, are required to produce comparable degrees of phosphodiesterase activation. Third, the total intracellular content of Ca^{2+} is at least 100-fold higher than that of Mn^{2+} [18]. These effects are potentiating and allow several orders of magnitude of discrimination by the system in favor of Ca^{2+}.

The Ca^{2+} concentration dependence of the activation of cyclic nucleotide phosphodiesterase (Figure 4A) was determined in incubations containing 1 mM Mg^{2+} and a series of fixed ratios of CDR to the catalytic component of the enzyme (CDR-dependent phosphodiesterase). The Ca^{2+} concentrations required for half maximal activation were 6, 3 and 1 μM, respectively, for incubations containing

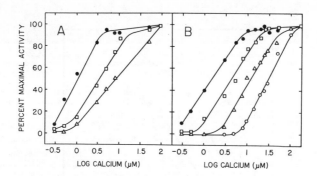

Fig. 4. Panel A: The effect of CDR on the Ca^{2+} sensitivity of porcine brain cyclic nucleotide phosphodiesterase. The Ca^{2+} concentration dependence of the phosphodiesterase activity was determined as described previously[12] in incubations containing fixed concentrations of CDR-dependent phosphodiesterase, 25 μM cGMP, 1 mM $MgSO_4$, either 10 (△), 100 (□) or 1000 (●) ng of CDR. Maximal activities were respectively 395, 1060 and 1051 nanomoles cGMP hydrolyzed/min-ml of enzyme. Panel B: The effect of Mg^{2+} on the Ca^{2+} sensitivity of CDR-dependent phosphodiesterase. The Ca^{2+} concentration dependence of the CDR-dependent phosphodiesterase was determined in incubations containing 33 ng of CDR and fixed concentrations of CDR-dependent phosphodiesterase at 0.3 (●), 1 (□), 3 (△) and 10 (○) mM $MgSO_4$. Maximal activities were respectively 685, 840, 1075 and 998 nanomoles cGMP hydrolyzed/min-ml of enzyme. Cations were of spectrograde quality. Reagents were freed of divalent cations by Chelex-100 pretreatment.

10, 100 and 1000 ng of CDR. These data confirm a previously proposed[12] ordered mechanism for CDR activation of the phosphodiesterase.

$$Ca^{2+} + CDR \rightleftharpoons Ca^{2+} \cdot CDR$$

$$Ca^{2+} \cdot CDR + PDE_{inactive} \rightleftharpoons Ca^{2+} \cdot CDR \cdot PDE_{active}$$

In this model CDR must first undergo an obligatory interaction with Ca^{2+} to form a complex which interacts subsequently with the inactive catalytic component to form an active ternary complex. These complexes are reversibly dissociable. According to this model, as the CDR concentrations in the incubations are progressively increased in excess of the available catalytic component, the Ca^{2+} concentrations required for enzyme activation would shift to lower values, as is observed in Figure 4A. Only when catalytic component is in excess of CDR would the concentration dependence of Ca^{2+} binding to CDR be equal to the Ca^{2+} dependence for enzyme activation.

The Ca^{2+} concentration dependence of binding to CDR has been shown to be competitively inhibited by Mg^{2+} at both the A and B classes of sites. Since phosphodiesterase requires Mg^{2+} for activity, its presence in the assay would be anticipated to influence the Ca^{2+} concentration dependence for enzyme activation.

Measurements of the Ca^{2+} dependence of phosphodiesterase activity conducted at limiting CDR and 0.3, 1, 3 and 10 mM Mg^{2+} required respectively 1.5, 4.5, 10 and 30 µM Ca^{2+} (Figure 4B) for half maximal enzyme activity. At 1 mM Mg^{2+} and limiting CDR the Ca^{2+} concentration producing half maximal activation was 4.5 µM. Binding of Ca^{2+} to CDR at this concentration occurs at sites class A with an apparent dissociation constant of 3 µM, while Mg^{2+} is bound at the B class site. Consequently, these data establish $Ca_{3A}Mg_{1B}CDR$ as the species of CDR which evokes phosphodiesterase activation.

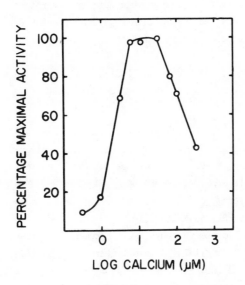

Fig. 5. The Ca^{2+} concentration dependence of CDR-dependent adenylate cyclase of rat brain. Particulate CDR-dependent adenylate cyclase was selectively stabilized to heat inactivation as described previously[13]. The enzyme was freed of Ca^{2+} and CDR by several washes with Ca^{2+}-free. 10 mM imidazole-Cl, pH 7.5, and assayed as described previously[13] with 1 µg CDR, 3 mM Mg^{2+} and varying $CaCl_2$. Ca^{2+} is expressed as the total content of the reactants as determined by atomic absorption spectrophotometry.

The Ca^{2+} dependence of a form of adenylate cyclase from rat brain which is stimulated by CDR has been examined less exhaustively than the Ca^{2+} requirements of the phosphodiesterase. In incubations containing fixed concentrations of CDR (1 µg) and Mg^{2+} (3 mM) the response to increasing concentrations of Ca^{2+} was biphasic with apparent K_m values of 2.5 and 200 µM for activation and inhibition respectively (Figure 5). Similar biphasic responses to Ca^{2+} have been observed for a norepinephrine stimulated adenylate cyclase of C-6 glioma cells[4]. In the C-6 system the Ca^{2+} concentrations producing both activation and inhibition of the adenylate cyclase activity were shifted to progressively lower values as the CDR content of the incubations was increased, suggesting that CDR mediated both the stimulatory and inhibitory effects of Ca^{2+} by formation of reversible complexes of CDR with adenylate cyclase as has been proposed for the phosphodiesterase. The apparent K_d values for dissociation of Ca^{2+} from CDR at 3 mM Mg^{2+} can be calculated from the relationship $Kapp_d = K_d(1 + i/K_i)$ and provide values of 4 and 200 μM respectively for dissociation of Ca^{2+} from the A and B classes of sites. These values closely approximate the apparent K_m values for Ca^{2+}-dependent activation and inhibition, respectively of the adenylate cyclase activity. While it is attractive to speculate that $Ca_{3A}Mg_{1B}CDR$ and $Ca_{3A}Ca_{1B}CDR$ form reversible complexes with the enzyme that are responsible for its activation and deactivation, considerably more detailed information regarding the CDR-dependent adenylate cyclase will be necessary to validate such proposals.

ACKNOWLEDGEMENT

This work was supported in part by U. S. Public Health Service Grants NS 11252 and NS 11340.

REFERENCES

1. Wolff, D. J. and Siegel, F. L. (1972) J. Biol. Chem. 248, 4180-4185.
2. Wolff, D. J. and Brostrom, C. O. (1974) Arch. Biochem. Biophys. 163, 349-358.
3. Brostrom, C. O., Huang, Y.-C., Breckenridge, B. McL. and Wolff, D. J. (1975) Proc. Natl. Acad. Sci. U. S. A. 72, 64-68.
4. Brostrom, M. A., Brostrom, C. O., Breckenridge, B. McL. and Wolff, D. J. (1976) J. Biol. Chem. 251, 4744-4750.
5. Cheung, W. Y. (1967) Biochem. Biophys. Res. Commun. 29, 478-482.
6. Teo, T. S. and Wang, J. H. (1973) J. Biol. Chem. 248, 585-595.
7. Lin, Y., Liu, Y. P. and Cheung, W. Y. (1974) J. Biol. Chem. 249, 4943-4954.
8. Watterson, D. M., Harrelson, W. G., Keller, P. M., Sharief, P. M. and Vanaman, T. C. (1976) J. Biol. Chem. 251, 4501-4513.
9. Amphlett, G. W., Vanaman, T. C. and Perry, S. V. (1976) FEBS Letters 72, 163-168.
10. Wolff, D. J., Poirier, P. G., Brostrom, C. O. and Brostrom, M. A. (1977) J. Biol. Chem. in press
11. Brostrom, C. O. and Wolff, D. J. (1976) Arch. Biochem. Biophys. 172, 301-311.
12. Brostrom, C. O. and Wolff, D. J. (1974) Arch. Biochem. Biophys. 165, 715-727.
13. Brostrom, C. O., Brostrom, M. A. and Wolff, D. J. (1977) J. Biol. Chem. in press.
14. Davis, B. J. (1964) Ann. N. Y. Acad. Sci. 121, 404-427.
15. Laemmli, U. K. (1970) Nature 227, 680-685.
16. Chen, P. S., Toribara, T. Y. and Warner, H. (1956) Anal. Chem. 28, 1756-1758.
17. Veloso, D., Guynn, R. W., Oskarrson, M. and Veech, R. L. (1973) J. Biol. Chem. 248, 4811-4819.
18. Cotzias, G. C. (1962) in Mineral Metabolism (Comar, C. L. and Bronner, F., eds.) Vol. IIB, Academic Press, New York, pp. 414-416.

Structural Homology
between Brain Modulator Protein and Muscle TnCs

T. C. Vanaman, F. Sharief, and D. M. Watterson
Duke University Medical Center, Box 3020
Durham, North Carolina 27710 U.S.A.

INTRODUCTION

An increasing amount of evidence suggests that actomyosin contractile systems, in addition to their documented role in muscle contraction, are the mechanochemical mediators for such vital processes as endo- and exocytosis, inter- and intracellular movement and proliferation in animal systems (for review, see 1, 2). Although it is clear that Ca^{2+} and the cyclic nucleotides are involved in the regulation of these events (for review, see 3-5), their exact roles and the mechanisms through which they act are unclear at present. Studies in this and other laboratories have provided considerable evidence that a ubiquitous, troponin(Tn) C-like protein may serve as the central regulatory element coupling Ca^{2+} to the regulation of cyclic nucleotide metabolism and contractile activity in numerous vertebrate tissues (for review, see 6). This protein has multiple, calcium dependent, regulatory activities including: 1) activation of a specific 3':5' cyclic nucleotide phosphodiesterase (7, 8), 2) activation of adenylate cyclase (9, 10), 3) TnC-like activity with a reconstituted rabbit skeletal actomyosin system (11), 4) activation of erythrocyte membrane ATPase activity (12, 13), and 5) stimulation of erythrocyte membrane Ca^{2+} transport (14). This multiple regulatory potential has led us (15) to term this protein "modulator protein," a term that will be used subsequently in this paper. The amino acid sequence of bovine brain modulator protein, presented in the following sections, is clearly homologous to those reported previously for fast muscle TnCs (16-20). These results indicate that modulator protein and the fast muscle TnCs form a family of structurally and functionally related proteins distinct from the more distantly related myosin light chains and parvalbumins. In addition, the structure, functions, and biological properties of modulator protein suggest that it may serve as a central regulatory element for stimulus-response coupling in smooth muscle and in non-muscle tissues.

METHODS

Modulator protein was prepared according to the procedure of Watterson et al. (15). Performic acid oxidation (21), cyanogen bromide cleavage (22), modification with citraconic anhydride (23), digestion with trypsin (15), digestion with chymotrypsin (24), and digestion with thermolysin (22) were done by previously described procedures. Resultant peptides were isolated using the methods described by Herman and Vanaman (25). The sequence of amino acids was determined by automated and manual Edman degradation of these peptides as previously described (15). Amides were assigned by direct determination of the phenylthiohydantoin derivatives of glutamine and of asparagine using gas liquid chromatography (15) and by amino acid analysis of aminopeptidase digests of peptides (22).

RESULTS AND DISCUSSION

a. Sequence determination

The amino acid sequence of bovine brain modulator protein is presented in Figure 1 aligned with the sequences previously reported for bovine cardiac muscle TnC (16) and rabbit skeletal muscle TnC (17). The sequence shown for the bovine brain protein has been deduced from analyses of peptides derived from the intact protein by a number of different procedures. Tryptic peptides encompassing the entire sequence of the protein were obtained following digestion of the performic acid oxidized protein or of the unmodified protein in the presence of EGTA as previously described (15). The amino acid sequences of these peptides were determined by either automated or manual Edman degradation of the intact tryptic peptides and, where necessary, of thermolysin, chymotrypsin, or cyanogen bromide sub-fragments of individual peptides. The amino to carboxyl terminal order of individual tryptic peptides was established unequivocally by studies of cyanogen bromide peptides derived from the intact protein and of peptides isolated following trypsin cleavage of the citraconylated protein. Sequence assign-

```
                                    10                        20
A.  Ac met ASP asp ile thr lys ala ALA val glu GLN LEU THR GLU GLU GLN lys asn GLU PHE LYS ALA ALA PHE ASP ILE PHE val
B.                          Ac(ala,asx)GLN LEU THR GLU GLU GLN ILE ALA GLU PHE LYS glu ALA PHE SER LEU PHE ASP
C.  Ac ASP thr gln gln ala glu ala arg ser tyr LEU SER GLU GLU met ILE ALA GLU PHE LYS ALA ALA PHE ASP MET PHE ASP

        30                        40                        50
leu gly ALA GLU asp gly cys ILE SER THR LYS GLU GLY LEU GLY lys VAL MET ARG MET LEU GLY GLN ASN PRO THR pro GLU GLU LEU GLN
lys ASP GLY ASX GLY THR ILE THR THR LYS GLU LEU GLY THR VAL MET ARG ser LEU GLY GLN ASN PRO THR glu ala GLU LEU GLX
ala ASP GLY gly GLY ASP ILE SER val LYS GLU LEU GLY THR VAL MET ARG MET LEU GLY GLN THR PRO THR lys GLU GLU LEU ASP

        60                        70                        80
GLU MET ILE ASP GLU VAL ASP GLY SER GLY THR VAL ASP PHE ASP GLU PHE LEU VAL MET MET VAL ARG cys MET LYS ASP ASP
ASX MET ILE ASN GLU VAL ASP ala ASP GLY ASX GLY THR ILE ASP PHE pro GLU PHE LEU thr MET MET ALA ARG lys MET LYS ASP thr
ala ILE ILE GLU GLU VAL ASP GLY SER GLY SER GLY THR ILE ASP PHE GLU GLU PHE LEU VAL MET MET VAL ARG gln MET LYS GLU ASP

        90                        100                       110
SER LYS SER GLU GLU LEU SER GLU ASP leu PHE ARG MET PHE ASP LYS ASN ALA ASP GLY TYR ILE ASP leu GLU GLU LEU LYS
asp         SER GLU GLU GLU ILE arg GLU ala PHE ARG VAL PHE ASP LYS ASP GLY ASN GLY TYR ILE SER ALA ala GLU LEU ARG
ALA LYS LYS SER GLU GLU GLU LEU ALA GLU cys PHE ARG ILE PHE ASP ARG ASN ALA ASP GLY TYR ILE ASP ALA GLU GLU LEU ala

        120                       130                       140
ile MET LEU gln ALA THR GLY GLU thr ILE THR GLU ASP ASP ILE GLU GLU LEU MET LYS ASP GLY ASP LYS ASN ASN ASP GLY ARG ILE
his VAL MET thr asx leu GLY GLU tml LEU THR ASP GLU GLU VAL ASP GLU MET ILE ARG GLU ALA ASN ile ASP gly ASP GLY glx VAL
glu ILE PHE arg ALA SER GLY his VAL THR ASP GLU ILE GLU ser LEU MET LYS ASP GLY ASP LYS ASN ASP GLY ARG ILE

        150
ASP TYR ASP GLU GLU PHE LEU GLU PHE MET lys GLY VAL GLU COOH
ASX TYR GLX GLX PHE VAL GLN MET MET thr ALA lys COOH
ASP PHE ASP GLU PHE LEU lys MET MET glu GLY VAL GLN COOH
```

Fig. 1.   Amino acid sequences of bovine brain modulator protein and muscle TnCs. A. Bovine cardiac muscle TnC taken from van Eerd and Takahaski (16). B. Bovine brain modulator protein; this report. C. Rabbit skeletal TnC taken from Collins (17). Residues shown in capital letters are identical or functionally conservative replacements; those in lower case are assumed to be nonconserved. The numbers refer to positions in the bovine cardiac TnC. Trimethllysine is abbreviated as tml.

ments and overlaps were further confirmed by studies of chymotryptic peptides prepared from the performic acid oxidized protein.

Only residues 1, 2 (alignment position 9, 10; ala, asx), residues 137, 139, and 140 (alignment positions 149, 151, 152; asx, glx, glx) remain to be unequivocally confirmed. However, existing data support the order as shown. Amide assignments have been made unequivocally except for those positions indicated as Asx or Glx. It should be noted that the Asp-Lys-Asp sequences at residues 19-21 (alignment positions 28-30) and 93-95 (alignment positions 105-107) are relatively resistant to trypsin cleavage. Trypsin cleavage also did not occur at the single trimethyllysine which occurs at residue 115 (alignment position 127) in bovine brain modulator protein. In addition, trypsin cleavage is obtained between the methionyl residues at alignment positions 80-81 and following the methionyl-methionine sequence at positions 156 and 157 (between positions 157 and 158). Similar unusual trypsin cleavages have also been observed with muscle TnCs (16).

As can be seen, modulator protein contains 148 residues as compared to 159 and 161 residues for rabbit skeletal and bovine cardiac muscle TnCs respectively. The molecular mass of modulator protein calculated from this sequence is 16,723 daltons, substantially lower than the 17,800 gms/mole previously deduced from physical studies (15). The amino acid composition of bovine brain modulator protein determined from this sequence agrees with that previously reported (15) for the intact protein when those values are normalized to a molecular weight of 16,723 gms/mole as shown in Table I.

b. Homology to fast muscle TnCs.

Alignment of the amino acid sequences of bovine cardiac and rabbit skeletal muscle TnCs with that of bovine brain modulator protein as in Figure 1 demonstrates that the brain protein is closely related in structure to the muscle TnCs. Residues that are identical or functionally conserved among the three proteins are distinguished from non-conservative replacements as noted in the figure legend. Only those residues defined by Dayhoff (26) as functionally conservative and which arise by single nucleotide substitution are considered

TABLE I. AMINO ACID COMPOSITIONS

| | Modulator Protein | | | | Modulator Protein | |
	By[a] Analysis	By Sequence			By[a] Analysis	By Sequence
Lys	7.2	7	Ala		10.7	11
Tml	1.0	1	Half-Cys		0.0	0
His	1.3	1	Val		7.2	7
Arg	6.2	6	Met		9.0	9
Asp	22.2	23	Ile		7.6	8
Thr	11.0	12	Leu		9.4	9
Ser	4.7	4	Tyr		2.0	2
Glu	26.7	27	Phe		7.8	8
Pro	2.0	2	Trp		0.0	0
Gly	10.8	11				148

a. Calculated using a molecular mass of 16,723 M_r

conservative. By aligning the sequence of bovine brain modulator protein (B, middle line) so that residue 1 corresponds to residue 9 in the cardiac TnC (A, top line) and residue 8 in the rabbit skeletal muscle TnC (C, bottom line), maximum homology to both TnCs is maintained throughout the entire linear sequence by introducing only two small gaps in the modulator protein sequence. Therefore, the lower molecular weight of modulator protein noted in the previous section is largely the result of having 7-8 fewer residues at the amino terminus. The fast muscle TnCs also show great variability in amino terminal sequence (17).

The first gap noted above, corresponding to the cysteinyl residue at position 35 in bovine cardiac muscle TnC, is also required for proper alignment of the rabbit skeletal muscle and bovine cardiac muscle TnCs. It appears likely that cysteine 35 in the bovine cardiac muscle TnC has arisen by insertion.

The second gap introduced in the modulator protein sequence corresponds to residues 90-92 (Lys, Gly, Lys) in the bovine cardiac protein; this gap is required for homology with skeletal or cardiac muscle TnC. Due to the large number of overlapping peptides isolated from this region of bovine brain modulator protein, it appears likely that this amino acid sequence is correct. The significance of this gap is further discussed in a following section.

The majority of the modulator protein sequence is either identical to or functionally very similar to those of the muscle TnCs. As tabulated in Table II, the total number of identical plus functionally conserved residues shared by

111

TABLE II. SEQUENCE COMPARISON OF MP AND TnC
(MP Residues No. 1-148)

	Identities	Functionally Conservative Replacements	Sum	Percent
MP vs. Cardiac TnC	76	38	114	77
MP vs. Skeletal TnC	79	35	114	77
Skeletal TnC vs. Cardiac TnC	98	24	122	82
Among all three proteins	65	43	108	73

bovine brain modulator protein and either TnC is 114 out of 148 positions compared, only slightly lower than the total number shared between the two TnCs. A larger number of residues are identical between the two TnCs. However, modulator protein shows more functionally conserved residues when compared to either of the two TnCs. These data show that modulator protein is much more closely related to the fast muscle TnCs than to the alkali light chains of myosin or to the parvalbumins, other members of this calcium binding protein super family (26).

c. Internally homologous (calcium binding) domains in modulator protein.

It has previously been noted (27) that the amino acid sequence of muscle TnCs can be divided into four homologous domains, each of which contains a potential calcium binding site. The amino acid sequence of bovine brain modulator protein also possesses internal homology as shown in Figure 2. Although all four domains are related in sequence, the level of homology is greatest when the first domain (residues 8-40) is aligned with the third domain (residues 81-113) and the second domain (residues 44-76) is aligned with the fourth (residues 117-148). Of the 33 residues compared in each pair of domains, 18 residues are identical and 6 are conservative replacements between domains 1 and 3. Domains 2 and 4 have 14 identical residues and 11 conservative replacements. This level of internal homology appears to be greater than that observed within the TnCs (17).

Based on the crystal structure of parvalbumin and its homology to TnCs, Kretsinger and Barry (28) have predicted a model for the three-dimensional structure of rabbit skeletal muscle TnC in which they identify putative calcium binding residues in each domain. The asterisks in Figure 2 denote the corres-

112

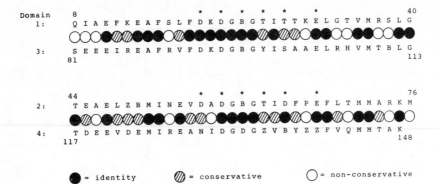

Domain
1:

8
Q I A E F K E A F S L F D K D G B G T I T T K E L G T V M R S L G
* * * * * *
40

3:

S E E E I R E A F R V F D K D G B G Y I S A A E L R H V M T B L G
81
113

2:

44
T E A E L Z B M I N E V D A D G B G T I D F P E F L T M M A R K M
* * * * * *
76

4:

T D E E V D E M I R E A N I D G D G Z V B Y Z Z F V Q M M T A K
117
148

● = identity   ▨ = conservative   ○ = non-conservative

Fig. 2.   Internal homology in modulator protein. Amino acid residues are
abbreviated using the single letter code nomenclature described by
Dayhoff (26). The symbols indicate relatedness between residues
at a given position as described on the figure. The numbers refer
to exact residue position in the sequence of bovine brain modulator
protein.

ponding positions in modulator protein. As is evident, all four domains possess

a potential calcium liganding (side chain oxygen containing) residue at each of

these positions. The amino acid sequence of these putative calcium binding

loops is obviously highly conserved. In addition, the regions of sequence

immediately adjacent to these putative loop structures contain the appropriate

hydrophobic residues for the helices predicted by Kretsinger and Barry (28) as

essential constituents in forming functional calcium binding loops. These data

support the previous observation (15) that bovine brain modulator protein binds

four atoms of calcium per molecule.

It is interesting to note that the three residue gap introduced in the

sequence of modulator protein for proper alignment with the TnC sequences

(Figure 1) occurs immediately before the amino terminus of domain 3 of the

modulator protein (residue 81) shown in Figure 2. This suggests that modulator

protein arose by duplication of a slightly smaller two domain precursor than the

fast muscle TnCs.

SUMMARY AND CONCLUSIONS

Figure 3 presents a model which summarizes the structural features of bovine

brain modulator protein as they relate to the fast muscle TnCs. In addition, it

113

shows the now widely accepted model (for review, see 17, 26, 29) for the evolution

of this class of proteins from a small ancestral precursor containing a single

calcium binding domain. The boxes in this model represent helices directly adja-

cent to the calcium binding regions indicated by loop structures. The fact that

the amino terminal 8-10 residues of muscle TnCs appear to be quite variable as

noted in the figure indicates that their absence in modulator protein is likely

to have little effect on its functional properties. Similarly, the three-residue

difference in the length of the region between domains 2 and 3 in these two pro-

teins should not cause a significant difference in their structures. Therefore,

it is not surprising that modulator protein possesses TnC-like activities as

previously reported (11).

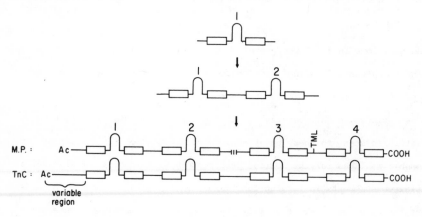

Fig. 3. Schematic representation of modulator protein (MP) and fast muscle TnC
structural features and hypothetical ancestry.

It has recently been reported (30) that rabbit skeletal muscle TnC will fully

activate cyclic nucleotide phosphodiesterase under the appropriate conditions.

However, its affinity for the phosphodiesterase is 600-fold lower than that of

modulator protein. This finding strongly suggests that activation of this phos-

phodiesterase by modulator protein is a specific function of physiological

significance. The presence of a trimethylated lysyl residue in modulator pro-

tein and not in the TnCs appears to be the most obvious significant difference

between these proteins which could account for modulator protein specific func-

tions.

114

Although it has yet to be proven that modulator protein functions in vivo as a TnC for non-fast muscle actomyosins, substantial evidence supporting this possibility is presented in this paper. Its additional potential for providing Ca^{2+} dependent regulation of cyclic nucleotide levels and of calcium sequestration in both smooth muscle and non-muscle tissues suggests that modulator protein may be a central regulator for the complete cycle of stimulus, response, and relaxation in animal cells as shown in Figure 4.

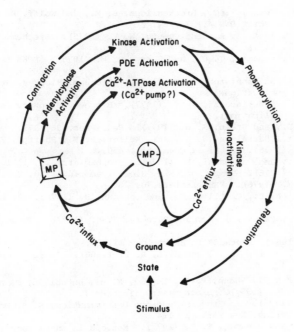

Fig. 4 Hypothetical model for stimulus – response coupling.

ACKNOWLEDGEMENTS

These studies were supported by NIH grant No. 10123. D.M.W. is a recipient of NIH postdoctoral fellowship NS 05132.

REFERENCES

1. Pollard, T. D., and Weihing, R. E. (1974). CRC Critical Reviews in Bio-chemistry 2, 1.
2. Perry, S. V., A. Margreth, and R. S. Adelstein, Ed. (1976). Contractile Systems in Non-muscle Tissues. North Holland Publishing Co., Amsterdam.
3. Rubin, R. P. (1974). Calcium and the Secretory Process, Plenum Press, N. Y.
4. Douglas, W. W. (1974). Biochem. Soc. Symp. 39, 1.
5. Berridge, M. J. (1975). Adv. Cyc. Nuc. Res. 6, 1.
6. Vanaman, T. C., Sharief, F., Awramik, J. L., Mendel, P. A., and Watterson, D. M. (1976). In Contractile Systems in Non-muscle Tissues, S. V. Perry, A. Margreth, and R. S. Adelstein, Eds. Elsevier/North Holland Biomedical Press. p. 165.
7. Cheung, W. Y. (1970). Biochem. Biophys. Res. Commun. 38, 533.
8. Kakiuchi, S., Yamazaki, R., and Nakajima, H. (1970). Proc. Japan. Acad. 46, 589.
9. Brostrom, C. O., Hwang, Y. C., Breckenridge, B. M., and Wolff, D. J. (1975). Proc. Natl. Acad. Sci. (U.S.A.) 72, 64.
10. Lynch, T. J., Tallant, E. A., and Cheung, W. Y. (1975). Biochem. Biophys. Res. Commun. 63, 967.
11 Amphlett, G. W., Vanaman, T. C., and Perry, S. V. (1976). FEBS Letters, 72, 163.
12. Jarrett, H. W., and Penniston, I. T. (1977). Biochem. Biophys. Res. Commun. (Submitted for publication).
13. Gopinath, R. M., and Vincenzi, F. F. (1977). Biochem. Biophys. Res. Commun. (Submitted for publication).
14. MacIntyre, J. D., and Green, J. W. (1977). Fed. Proc. 36, 271.
15. Watterson, D. M., Harrelson, W. G., Jr., Keller, P. M., Sharief, F., and Vanaman, T. C. (1976). J. Biol. Chem. 251, 4501.
16. Van Eerd, J.-P., and Takahashi, K. (1976). Biochemistry 15, 1171.
17. Collins, J. H. (1976). Symp. Soc. Exp. Biol. (Cambridge) 30, 303.
18. Wilkinson, J. M. (1976). FEBS Letters 70, 254.
19. Romero-Herrera, A. E., Castillo, O., and Lehmann, H. (1976). J. Mol. Evol. 8, 251.
20. Van Eerd, J.-P., Capony, J.-P., and Pechère, J.-F. (1977). These Proceed-ings.
21. Hirs, C.H.W. (1967). Methods in Enzymology 11, 95.
22. Vanaman, T. C., Wakil, S. J., and Hill, R. L. (1968). J. Biol. Chem. 243, 6409.
23. Freymeyer, D. K., II, Shank, P. R., Edgell, M. H., Hutchison, C. A., III, and Vanaman, T. C. (1977). Biochemistry (in press).
24. Kasper, C. B. (1975). In Protein Sequence Determination, S. B. Needleman, ed., Springer-Verlag, New York, p. 114.
25. Herman, A. C., and Vanaman, T. C. (1977). Methods in Enzymology (in press).
26. Dayhoff, M. O. (1976). ATLAS OF PROTEIN SEQUENCE AND STRUCTURE, Vol. 5 and Supplement 2. The National Biomedical Research Foundation. Silver Spring, Md.
27. Collins, J. H., Potter, J. D. Horn, M. J., Wilshire, G., and Jackman, N. (1974). In Calcium Binding Proteins, W. Drabikowski, H. Stezelecka-Golaszewska, and E. Carafoli, eds., Amsterdam, p. 51.
28. Kretsinger, R. H., and Barry, C. D. (1975). Biochim. Biophys. Acta 405, 40.
29. Kretsinger, R. H. (1975). In Calcium Transport in Contraction and Secretion, E. Carafoli, F. Clementi, W. Drabikowski, and A. Margreth, eds. Elsevier/North Holland, Amsterdam, p. 469.
30. Dedman, J. R., Potter, J. D., and Means, A. R. (1977). J. Biol. Chem. 252, 2437.

Interactions of Calcium with the Acetylcholine Receptor

Amira T. Eldefrawi and Mohyee E. Eldefrawi
Department of Pharmacology and Experimental Therapeutics,
University of Maryland School of Medicine, Baltimore, Md. 21201, U.S.A.

and

Helga Rübsamen[*] and George P. Hess
Section of Biochemistry, Molecular and Cell Biology
Cornell University, Ithaca, N. Y. 14853, U.S.A.

It has long been known that Ca^{2+} affects acetylcholine (ACh)-induced conductance changes at the postjunctional membrane[1,2]. However, because of the many roles played by Ca^{2+} in the maintenance of membrane structure and function, including the supply of Ca^{2+} for excitation-contraction and stimulus-secretion coupling processes, it was extremely difficult to determine whether the effects of variation of Ca^{2+} levels were attributable to effects at a receptor-linked Ca^{2+} site or whether they arose from effects at other Ca^{2+} binding loci. The successful isolation of ACh-receptors from electric organs of electric fish has provided the opportunity for direct studies of the interaction of Ca^{2+} with the ACh-receptor, and the effect of receptor activators and inhibitors on Ca^{2+} mobilization.

THE ACH-RECEPTOR OF *TORPEDO* ELECTRIC ORGANS

Electric organs of *Torpedo* are tissues extremely rich in cholinergic synapses, whose pharmacology is similar to that of mammalian skeletal muscles. Their content of ACh-receptors may reach 1-2 nmoles of receptor sites per g of tissue, which is 4 to 5 orders of magnitude its density in mammalian skeletal muscles[3]. The ACh-receptor of these tissues is identified by its specific binding of radiolabeled cholinergic ligands, such as [^{3}H]ACh[4] or neurotoxins from elapid venoms (α-bungarotoxin[5] or cobra neurotoxin[6]). Purified receptor preparations are obtained from detergent extracts of electric organ membranes by affinity chromatography[5-7], with a specific binding reaching up to 10 and 12 nmoles of ACh or α-bungarotoxin per mg

---

[*]Present address: Institut für Virologie, Universität Giessen, 63 Giessen, Frankfurterstr. 107, Germany.

protein.

The ACh-receptor of *Torpedo* electric organ is a glycoprotein containing 0.9-2 mole % hexosamines and other sugars[5,6]. Its content of tryptophan (1.4-2.4 mole %) imparts strong fluorescence to the receptor molecule[7]. In 0.1% Triton, the receptor exists as an oligomer of 330,000 daltons, but apparently aggregates into higher molecular weight species at low detergent concentrations[8]. The protomer carrying a single ACh binding site is about 80,000 daltons, and the major subunit obtained by sodium dodecyl sulfate (SDS) electrophoresis is about 40,000 daltons. There is accumulating evidence which suggests that the ACh-receptor molecule carries sites that bind agonists (e.g. ACh or carbamylcholine) which are separate from those that bind antagonists (e.g. α-bungarotoxin or d-tubocurarine[3,9-11]).

THE ACH-RECEPTOR AS A Ca^{2+} BINDING PROTEIN

In 1974, we discovered that the purified ACh-receptor preparations contained large amounts of bound Ca^{2+}, as determined by atomic absorption[12]. The amount of bound Ca^{2+} was dependent upon the availability of Ca^{2+} during purification of the receptor (Fig. 1). Even in the presence of 1 mM EDTA in all solutions used, the receptor molecule still had 15 moles of Ca^{2+} bound per mole of ACh-binding site, amounting to 0.6% of the molecular weight of the receptor. It led to the suggestion that this bound Ca^{2+} might be sequestered or tightly bound. Chang and Neumann[13] confirmed these findings and obtained 18 moles of bound Ca^{2+} per mole of ACh-binding site.

In order to determine whether bound Ca^{2+} plays a role in the regulatory mechanisms resulting from transmitter-receptor interactions, and whether binding of ACh to its receptor affects the Ca^{2+}-receptor interactions, we have utilized the fluorescent lanthanide terbium (Tb^{3+}) as a probe[14]. Binding of Tb^{3+} to the ACh-receptor is accompanied by a fluorescence enhancement of 10^4 (λ excitation 295 nm, λ emission 546 nm) (Fig. 2). The ACh-receptor binds 10 moles of Tb^{3+} per mole of α-bungarotoxin or ACh binding site as detected by changes in fluorescence which accompany binding and by neutron activation analysis[14]. In presence of 8 mM Ca^{2+}, two types of Tb^{3+}-binding sites are revealed, both with dissociation constants (for

118

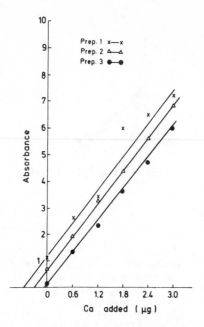

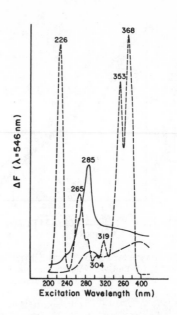

Fig. 1. The relationship between absorbance and Ca^{2+} added to three differently prepared *Torpedo* ACh-receptors. The intercept of each line with the x-axis is the amount of Ca^{2+} in each preparation. Prep. 1 is purified in Krebs original Ringer phosphate solution containing 0.67 mM Ca^{2+}. Prep. 2 is purified in a solution of 0.2 M NaCl and 5 mM Na_2HPO_4. Prep. 3 is purified in the same solution as that used for prep. 2, except that 1 mM EDTA is added (from Eldefrawi *et al.*[12]).

Fig. 2. Fluorescence excitation spectra for emission at 546 nm at 20°C, pH 6.5. (----), 10^{-2} M $TbCl_3$ in H_2O; (— —) ACh-receptor (0.7 μM α-bungarotoxin binding sites); (———) ACh-receptor in presence of only 10^{-4} M Tb^{3+}. The purified ACh-receptor is dialyzed extensively at 4°C against 10 mM Pipes buffer, pH 6.5 (adjusted with Tris), 0.03% Triton X-100 and 10 μM $TbCl_3$ until free of Ca^{2+} as determined by atomic absorption. Tb^{3+} is then removed by extensive dialysis in same Pipes buffer in absence of any $TbCl_3$ (from Rübsamen *et al.*[14]).

Tb^{3+}) in the 18-25 μM range (Fig. 3). About 60% of these sites also bind Ca^{2+} with an apparent K_D of 1 mM. Thus, for each mole of ACh-binding site , there are 6 moles of sites that bind both Tb^{3+} and Ca^{2+}, 4 moles of sites that bind only Tb^{3+} (Fig. 3) and a minimum of 9 moles of Ca^{2+} binding sites (15 moles of Ca^{2+} sites detected by atomic absorption[12] minus the 6 moles of Tb^{3+}-Ca^{2+} binding sites).

THE EFFECT OF DRUG-RECEPTOR INTERACTION ON BOUND Tb^{3+} AND Ca^{2+}

The Tb^{3+} bound to the purified ACh-receptor (at a Tb^{3+} concentration less than K_{Tb} so as to maximize the signal change) is displaced by increasing ACh concentra-

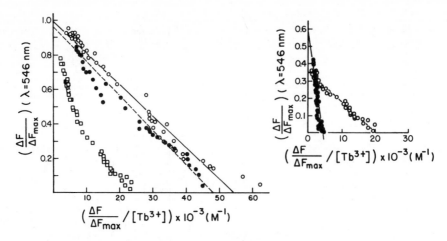

Fig. 3. Scatchard plot of the binding of Tb^{3+} to ACh-receptor in presence and absence of 8 mM $CaCl_2$. ΔF is the observed fluorescence intensity (λEM = 546 nm-535 nm) in arbitrary units at a given concentration of Tb^{3+} and ΔF_{max} the maximum observed fluorescence intensity when all ion binding sites are occupied. o and o, in absence of $CaCl_2$ (o, μ = 7 mM; •, μ = 100 mM); □, in presence of 8 mM $CaCl_2$, μ = 55 mM. Excitation at 295 nm. The slope of the line gives a value for K_{Tb} of 18 ± 0.5 µM. <u>Inset</u>: A replot of the Tb^{3+}-binding data obtained in presence of 8 mM $CaCl_2$ on the basis of two types of binding sites. o, binding site not affected by 8 mM $CaCl_2$ (K_{Tb} = 18 ± 0.05 µM); •, binding affected by 8 mM $CaCl_2$ (K_{Tb} = 150 ± 11 µM) (from Rübsamen *et al.*[14]).

tions (Fig. 4)[15]. Quantitative analysis of these data gives a K_D of ACh from the receptor of 0.7 µM (Fig. 5a). A K_i value of 0.3 µM is obtained for Tb^{3+} through its inhibition of [^{3}H]ACh measured by equilibrium dialysis in same buffer (Fig. 5b). In presence of 8 mM Ca^{2+} (a concentration at which Ca^{2+} displaces 60% of bound Tb^{3+}), ACh fails to displace any Tb^{3+}. Thus, it is suggested that ACh displaces Tb^{3+} from the Ca^{2+}-binding sites. Other receptor activators, such as carbamylcholine and decamethonium also cause release of bound Tb^{3+} similar to the effect of ACh[14]. The K_D of 10 ± 4 µM for Tb^{3+}-receptor complex, obtained by its displacement with ACh, is in agreement with the K_D of 18-25 µM determined by direct fluorescent titration (Table 1). In contrast to activators, two inhibitors of neural transmission, <u>d</u>-tubocurarine and α-bungarotoxin, used at up to 100 times their K_D value, do not induce the receptor to release Tb^{3+} (Fig. 4). Chang and Neumann[13], using murexide as an optical indicator to determine changes in Ca^{2+} concentration, confirmed the displacement of Ca^{2+} from the ACh-receptor by agonists. However, they calculated 2-3 Ca^{2+} ions bound per ACh-binding site as compared to

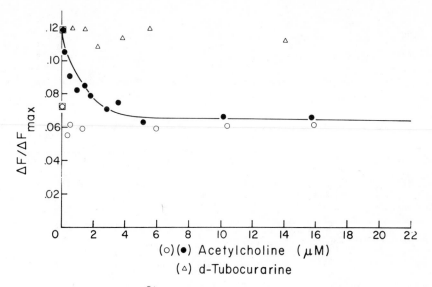

(o)(●) Acetylcholine (μM)

(△) d-Tubocurarine

Fig. 4. Displacement of Tb^{3+} from the purified ACh-receptor (0.6 μM α-bungarotoxin binding sites) by ACh (in 2 mM Pipes/Tris buffer, pH 6.5, 0.03% Triton X-100, 3.3 μM $TbCl_3$, 20°C). ●, titration with ACh in absence of $CaCl_2$; o, titration with ACh in presence of 6 mM $CaCl_2$; △ titration with d-tubocurarine; ◙ , receptor-Tb^{3+} complex before addition of ligand; ◻ , receptor-Tb^{3+} complex in presence of 6 mM $CaCl_2$ before addition of ligand (from Rübsamen et al.[15]).

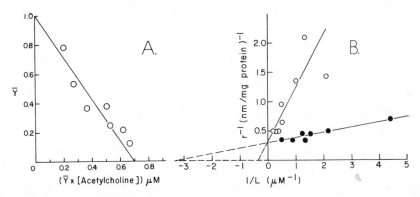

Fig. 5. Competition of ACh with Tb^{3+}. A. Increase in fluorescence signal due to displacement of Tb^{3+} from the receptor by ACh is plotted according to a linear equation: $\bar{Y} = \Delta F'_\alpha/\Delta F_\alpha = 1-\Delta F'_\alpha/\Delta F_\alpha([L]/K_D)$, where [L] is the molar concentration of ACh and K_D the receptor-ACh dissociation constant. Subtracting the value for ΔF obtained at saturating concentrations of ACh from the observed values obtained in absence and presence of ACh gives the values of ΔF_α and $\Delta F'_\alpha$, respectively. The reciprocal slope of the graph gives K_D = 0.7 μM. B. Binding of [^3H]ACh to the ACh-receptor (in 2 mM Pipes/Tris buffer 0.1% Triton X-100, pH 6.5) measured by equilibrium dialysis for 16 h at 4°C in absence of Tb^{3+} (●) and in presence of 50 μM Tb^{3+} (o). r, nmoles ACh bound/mg protein; L, ligand concentration; $K_{D(Tb)}$ = 10 ± 4 μM; $K_{D(ACh)}$ = 0.3 ± 0.03 μM (from Rübsamen et al.[15]).

TABLE 1

Dissociation constants of inorganic ions from the ACh-receptor, its
subunit IV and peptides determined by fluorescence measurements of Tb^{3+}
binding in absence and presence of other inorganic ions (from Rübsamen et al.[15])

Preparation	K_D (mM)				
	Tb^{3+*}	Ca^{2+}	Mg^{2+}	Na^+	K^+
ACh-receptor	0.018–0.025	1	2	60	130
Subunit IV	19	4			
ACh-receptor peptides**	22	3.9			

*Percent of Tb^{3+}-binding sites interacting with competing ion is 60% in
case of Ca^{2+} and Mg^{2+} on the intact ACh-receptor and 100% in all other
cases.

**Peptides of molecular weight of 8000 or less, produced by exposure to
trypsin and chymotrypsin.

our calculation of 3-6 Ca^{2+} ions per ACh-binding site[15]. They also reported that
a receptor inhibitor such as α-bungarotoxin did not cause release of Ca^{2+} from the
receptor; but that subsequent to receptor activation, α-bungarotoxin caused reuptake of up to 6 Ca^{2+} ions.

Ca^{2+}-BINDING AND THE SUBUNIT STRUCTURE OF THE ACH-RECEPTOR

The ACh-receptor of electric organs is made up of two major subunits plus one or
more minor ones[5,6,16], depending on purity and denaturation. Of the two major
subunits, the one with a molecular weight of about 40,000 carries binding sites for
the agonist-site-directed affinity label [³H]p-(N-maleimido)-α-benzyltrimethylam-
monium iodide and the α-neurotoxins[6]. It is also the same subunit that binds Tb^{3+}
and Ca^{2+} (Fig. 6).

Two experimental findings indicate that Ca^{2+} binding sites are determined by
structural features of the polypeptide rather than by the three-dimensional arrangement of the intact receptor. First, the Tb^{3+} binding sites are preserved and

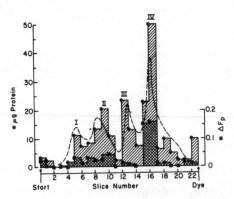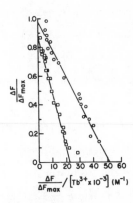

Fig. 6. The subunits of the ACh-receptor isolated by SDS acrylamide slab electrophoresis after denaturation of receptor in SDS and β-mercaptoethanol. The Tb^{3+}
fluorescence/mg protein, ΔF_P (▨) and protein concentration (▧) were corrected
by subtracting values obtained from a blank gel run in parallel. Superimposed on
the protein profile, as obtained by Lowry analysis of the isolated fractions, is
the densitometer trace (----) of a slice cut at the side of the gel and stained for
protein. Inset: Scatchard plot of the binding of Tb^{3+} to the isolated subunit IV
in absence (o) and presence (□) of 4.3 mM Ca^{2+}, at 20°C, pH 6.5, 2 mM Pipes/Tris
Buffer, 0.03% Triton X-100. K_{Tb} = 19 ± 1 μM. Assuming competitive inhibition
between Tb^{3+} and Ca^{2+}, K_{Ca} is calculated to be 4 ± 0.7 mM (from Rübsamen et al.[16]).

their affinity is unchanged when the receptor is dissociated into subunits by

SDS[16]. Second, when the receptor is degraded by trypsin and chymotrypsin to pep-

tides of molecular weight of 8000 or less, these small size peptides contain all

the Tb^{3+}-binding sites. It is interesting to note that while in the intact recep-

tor only about 60% of the Tb^{3+}-binding sites interact with Ca^{2+}, all the Tb^{3+}-

binding sites detected in the subunit and the receptor peptides interact with Ca^{2+}.

Presumably, the Tb^{3+}-binding sites, which do not interact with Ca^{2+} depend on the

three-dimensional structure of the protein or are located on other subunits. In-

vestigations are in progress to determine whether or not the *Torpedo* ACh-receptor

contains γ-carboxyglutamic acid, a Ca^{2+} binding amino acid present in prothrombin[17].

Ca^{2+} MOBILIZATION AND RECEPTOR REGULATED CONDUCTANCE

A minimum interpretation of the data is that Ca^{2+} regulates receptor-mediated

conductance changes by competing for activator-binding sites. Other explanations

would be that Ca^{2+} acts as a physical barrier closing ion channels[18], while its dis-

sociation by ACh opens them or that Ca^{2+} stabilizes a receptor conformation, which

does not permit transfer of inorganic ions through the membrane. This would be analogous to the model described by Chang and Triggle for the role of Ca^{2+} in activation of ACh-receptors of smooth muscle[19].

However, our recent studies on the binding of $[^3H]$perhydrohistrionicotoxin, which inhibits receptor mediated ion conductances without inhibiting ACh-binding to its receptor, and does not bind to the purified receptor protein, suggest that the functional receptor-ion conductance unit consists of two proteins[20]: One, the ACh-receptor, which carries the drug recognition sites. The other, the ion conductance modulator, which has high affinity for histrionicotoxins and separates from the receptor protein upon solubilization by detergents and removal of receptor by affinity chromatography or immunoprecipitation. Thus, while our data suggest a role for Ca^{2+} in ACh-receptor activation, an additional role may involve binding of Ca^{2+} to the ion conductance modulator.

Desensitization of the ACh-receptor is caused by exposure to a high concentration of ACh or by prolonged exposure to a low concentration. The presence of high external Ca^{2+} leads to faster desensitization[1]. However, incubation of the purified receptor protein with desensitizing doses of ACh still causes release of Ca^{2+} instead of reuptake (Fig. 4). It is possible that during desensitization Ca^{2+} binds to the ion conductance modulator instead of the ACh-receptor protein. This can be tested when the ion conductance modulator protein, which is presently undergoing characterization and isolation, is purified.

In conclusion, the evidence at hand is suggestive of a Ca^{2+} role in the ACh-receptor induced permeability changes. Complete understanding of the role of transmitter induced Ca^{2+} mobilization must await further clarification of receptor-ion permeability coupling.

ACKNOWLEDGMENTS

This work was supported in part by National Institutes of Health grant GM 04842 and National Science Foundation grants 76-21683 and BMS 72-01908. Helga Rubsamen was recipient of a Max Kade Postdoctoral Research Grant.

REFERENCES

1. Takeuchi, N. (1963) *J. Physiol.* <u>167</u>, 141-155.

2. Nastuk, W. L. and Parsons, R. L. (1970) *J. Gen. Physiol.* <u>56</u>, 218-248.

3. Eldefrawi, M. E. and Eldefrawi, A. T. (1977) In *Receptors and Recognition* <u>3</u> (Cuatrecasas, P. and Greaves, M. F., eds.) Chapman and Hall, London (in press).

4. Eldefrawi, M. E., Britten, A. G. and Eldefrawi, A. T. (1971) *Science* <u>173</u>, 338-340.

5. Michaelson, D., Vandlen, R., Bode, J., Moody, T., Schmidt, J. and Raftery, M. A. (1974) *Arch. Biochem. Biophys.* <u>165</u>, 796-804.

6. Meunier, J.-C., Sealock, R., Olsen, R. and Changeux, J.-P. (1974) *Eur. J. Biochem.* <u>45</u>, 371-394.

7. Eldefrawi, M. E., Eldefrawi, A. T. and Wilson, D. B. (1975) *Biochemistry* <u>14</u>, 4304-4310.

8. Edelstein, S. J., Beyer, W. B., Eldefrawi, A. T. and Eldefrawi, M. E. (1975) *J. Biol. Chem.* <u>250</u>, 6101-6106.

9. Fu, J.-J., Donner, D. B., Moore, D. E. and Hess, G. P. (1977) *Biochemistry* <u>16</u>, 678-684.

10. Bulger, J. E., Fu, J.-J. L., Hindy, E. F., Silberstein, R. L. and Hess, G. P. (1977) *Biochemistry* <u>16</u>, 684-692.

11. Eldefrawi, M. E. (1974) In *"Peripheral Nervous System"* (Hubbard, J. J., ed.) Plenum Publ. Corp., N. Y., 181-200.

12. Eldefrawi, M. E., Eldefrawi, A. T., Penfield, L. A., O'Brien, R. and Van Campen, D. (1975) *Life Sci.* <u>16</u>, 925.

13. Chang, H. W. and Neumann, E. (1976) *Proc. Nat. Acad. Sci. U.S.A.* <u>73</u>, 3364.

14. Rübsamen, H., Hess, G. P., Eldefrawi, A. T. and Eldefrawi, M. E. (1976) *Biochem. Biophys. Res. Comm.* <u>68</u>, 56-63.

15. Rübsamen, H., Eldefrawi, A. T., Eldefrawi, M. E. and Hess, G. P., *Biochemistry* (in press).

16. Rübsamen, H., Montgomery, M., Hess, G. P., Eldefrawi, A. T. and Eldefrawi, M. E. (1976) *Biochem. Biophys. Res. Comm.* 70, 1020-1027.

17. Magnusson, S., Sottrup-Jensen, L., Petersen, T. E., Morris, H. R. and Dell, A. (1974) *FEBS Letters* 44, 189-193.

18. Nachmansohn, D. and Neumann, E. (1975) *Chemical and Molecular Basis of Nerve Activity*. Academic Press, New York.

19. Chang, K. J. and Triggle, D. J. (1973) *J. Theor. Biol.* 40, 125-154.

20. Eldefrawi, A. T., Eldefrawi, M. E., Albuquerque, E. X., Oliveira, A. C., Mansour, N., Adler, M., Daly, J. W., Brown, G. B., Burgermeister, W. and Witkop, B. (1977) *Proc. Nat. Acad. Sci. U.S.A.* 74, 2172-2176.

Calcium-Binding Proteins in Electroplax

Steven R. Childers, Ari Sitaramayya, Joan C. Egrie, James A. Campbell and
Frank L. Siegel
Departments of Pediatrics and Physiological Chemistry
University of Wisconsin Center for Health Sciences
Madison, Wisconsin 53706

INTRODUCTION

At the first International Symposium on Calcium-Binding Proteins we described
the isolation and purification of two calcium-binding proteins from pig brain[1].
The impetus for this research derived from physiological studies which suggested
that the actions of calcium on the release of neurotransmitters required the
participation of an intracellular calcium receptor[2]. It now appears unlikely
that either of the two proteins which we have described is involved directly in
transmitter release - this role is more probably fulfilled by the contractile
proteins troponin, actin and myosin which have recently been found in brain[3,4].
Calcium participates in neural events other than transmitter release, however,
and one of the two soluble calcium-binding proteins in brain has been shown to
activate brain cyclic nucleotide phosphodiesterase[5] and adenylate cyclase[6]. This
finding is of particular interest since cyclic nucleotides have been implicated
in the mediation of slow physiological actions of neurotransmitters in several
neuronal systems[7]. This protein, which we initially named CBP II, is now
designated as activator protein, modulator protein or calcium-dependent regulator
(CDR), the designation which we will use in this report. Cheung had shown that
CDR is present in virtually all mammalian tissues; brain contains by far the
highest levels of this protein. The estimation of CDR levels is based upon the
ability of boiled tissue extracts to activate brain CDR-deficient cyclic nucleotide phosphodiesterase, and it is thus conceivable that CDR activity in different
tissues derives from different, tissue-specific activators. That this is
probably not the case that was demonstrated in a subsequent study from this
laboratory which demonstrated the chemical identity of CDR isolated from brain
and from adrenal medulla[8], another rich source of this protein. Similar data
also confirmed the chemical identity of bovine testicular and brain CDR.

Within the nervous system CDR is found in all brain regions, although the
levels in grey matter are generally higher than in white matter[9]. This protein
is present in both neuronal and glial cell bodies, but the highest levels in
brain are found in synaptosomal cytosol[9], suggesting a possible role in events
related to neurotransmission. To determine the functions of CDR in neural events
of brain is difficult due to the regional differences in neurotransmitters and
the cellular heterogeneity of this tissue. To circumvent these difficulties we
have chosen to study the problem in simpler neural systems such as adrenal

127

medulla[8,10,11] and electroplax, a modified neuromuscular junction receiving only cholinergic innervation. This report describes recent studies of CDR in electroplax.

   Isolation of CDR from Electroplax - We have investigated several aspects of CDR function in electroplax from two marine organisms, the electric eel Electrophorus electricus and the electric ray Torpedo californica. CDR was isolated from extracts of eel electroplax prepared by homogenizing the tissue in two volumes of Tris-HCl, pH 7.4 in a Waring Blendor[12]. The resulting homogenate was centrifuged at 18,000 x g for 30 min and the supernatant from this step was brought to 48 per cent saturation with respect to ammonium sulfate. After removal of the protein precipitate by centrifugation the supernatant was brought to saturation with ammonium sulfate. The protein precipitate was collected by centrifugation and dialyzed against 50 mM Tris-HCl, pH 7.4. At this point the extract is quite viscous due to the high mucin content of electroplax. The mucins interfere with further purification steps and are partially removed by a brief boiling procedure. The boiling step removes not only the mucins, but much of the non-CDR protein, and thus serves to affect an early purification of CDR. The supernatant fraction from the boiling step is subjected to chromatography on ECTEOLA-cellulose and the proteins eluting between 0.1 M NaCl in 50 mM Tris-HCl, pH 7.4 and 0.2 M NaCl in the same buffer are taken for further purification. This acidic protein fraction is dialyzed to remove salt and is fractionated by gel filtration in the presence of ^{45}Ca. The resulting profile of radioactivity indicates the presence of a single peak of protein-bound calcium (Fig. 1).

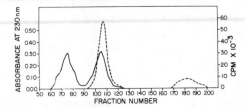

Fig. 1. Gel filtration on Sephadex G-100 of the acidic protein fraction from ECTEOLA-cellulose. This sample was incubated with 25 μCi of ^{45}CaCl$_2$ for 30 min and applied to a column (2.5 x 100 cm) of Sephadex G-100. The column was eluted with 50 mM Tris-HCl, pH 7.4. Fraction size was 2.5 ml; the column was monitored by measurement of absorbance at 230 nm, and radioactivity was determined on 0.1-ml aliquots in 10 ml Bray's scintillator (_____) absorbance; (----) radioactivity.

This fraction corresponds to a single species of protein, as judged by its electrophoretic purity on 15% polyacrylamide gels. The purified protein was found to bind calcium with $K_D = 2.1 \times 10^{-5}$, as compared to a $K_D = 1.7 \times 10^{-5}$ which we previously reported for mammalian CDR[8].

Comparison of the Electroplax Protein and Mammalian CDR - The chromatographic, electrophoretic and calcium-binding properties of the electroplax calcium-binding protein suggested that this protein is similar to or identical to mammalian CDR. This identity has been established by co-electrophoresis and comparisons of the amino acid composition (Table I) and peptide maps (Fig. 2) of the two proteins. We conclude that the two proteins are very similar or identical - proof of identity must await sequence studies of the two proteins.

Table I

Amino Acid Composition of CDR from Electroplax and Pig Brain

Amino acid	Mol per 14 000 g protein	
	Electroplax	Pig brain
Aspartic acid	15.6	16.7
Threonine	8.4	8.5
Serine	3.5	4.1
Proline	2.6	3.2
Glutamic acid	19.0	21.6
Glycine	7.9	8.5
Alanine	7.5	8.3
Valine	4.5	4.9
Methionine	2.9	3.5
Isoleucine	6.7	6.1
Leucine	10.0	7.4
Tyrosine	0.9	1.5
Phenylalanine	6.0	6.2
Lysine	6.0	6.1
Histidine	1.1	1.1
Arginine	3.9	4.4
Cysteine	N.D.	N.D.
Tryptophan	N.D.	N.D.

N.D., not detected.

Is Electroplax CDR a Parvalbumin? Electroplax is embryologically derived from skeletal muscle and since fish muscle contains a low molecular weight calcium-binding protein known as parvalbumin it was necessary to determine if electroplax CDR is in fact a parvalbumin. To resolve this question we have isolated the parvalbumin from skeletal muscle of Electrophorus and compared its chemical and physical properties to those of CDR isolated from electroplax[13]. Eel skeletal muscle was homogenized in a Waring Blendor for 2 min with four volumes of 50 mM Tris-HCl, pH 7.4 and the soluble proteins which precipitated between 70 and 100 per cent ammonium sulfate were collected and dialyzed against 50 mM Tris. This protein fraction was incubated with ^{45}Ca and applied to a column of Sephadex G-75. One peak of protein bound radioactivity was found (Fig. 3) and the proteins in this peak were fractionated by ion-exchange chromatography on QAE-Sephadex. Three protein components were resolved by this procedure; equilibrium dialysis revealed that only the second peak has significant calcium-binding activity. Polyacrylamide gel electrophoresis revealed a single protein component in this active fraction. The molecular weight of this muscle calcium-binding protein was estimated to be 11,000 daltons by SDS gel electrophoresis.

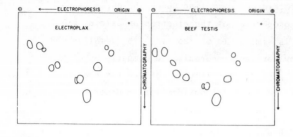

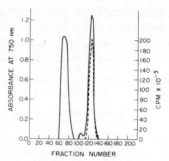

Fig. 2. Tracings of peptide maps of trypsin peptides from electroplax and bovine testis calcium-binding proteins. Descending chromatography was carried out in n-butantol/ acetic acid/water (4:1:5, v/v) for 19 h, while electrophoresis was carried out at pH 3.7 for 45 min at 3000 V.

Fig. 3. Gel filtration of soluble eel muscle proteins on Sephadex G-75. Proteins precipitating between 70 and 100% $(NH_4)_2SO_4$ were incubated with 25 μCi of $^{45}CaCl_2$ and applied to a column of Sephadex G-75 which was eluted with 50 mM Tris HCl, pH 7.4. (———) Lowry protein absorbance at 750 nm; (----) radioactivity.

CDR has a molecular weight of 15,000 daltons, and the two proteins have clearly different electrophoretic mobility. By a variety of other criteria we have concluded that the muscle protein is a typical parvalbumin and is non-identical to electroplax CDR. The major differences between the two proteins include amino acid composition (parvalbumin contains no tyrosine), UV spectra (the 260:280 ratio for parvalbumin is 4.78 and for CDR it is 0.88) and the K_D for calcium binding (1×10^{-7} for parvalbumin and 2.1×10^{-5} for CDR). To confirm the non-identity of the two proteins we prepared an antiserum to eel parvalbumin and demonstrated that there is no cross-reactivity with CDR[13]. Quantitative immunoprecipitation assays were also done and the level of parvalbumin in eel skeletal muscle was found to be 0.1 mg/mg protein, whereas in electroplax and a variety of other tissues the level of parvalbumin was less than 0.4 μg/mg protein, the lower limit of sensitivity of the assay.

Tissue Distribution of CDR in Electrophorus - Polyacrylamide gel electrophoresis of the soluble proteins from several eel tissues revealed that a protein band with electrophoretic mobility of CDR was a major protein component only of electroplax, although a faint band was visible in extracts of brain, liver and spleen (Fig. 4). No bands of similar mobility were seen in extracts of kidney, heart or skeletal muscle. In order to quantitate the levels of CDR it was first necessary to determine the ability of eel CDR to activate mammalian PDE, as this activation is the basis of the CDR assay.

Biological Activity of Electrophorus CDR - The relative ability of eel CDR and mammalian CDR to activate CDR-deficient cyclic nucleotide phosphodiesterase (PDE) prepared from pig brain was compared. Electroplax CDR was found to have

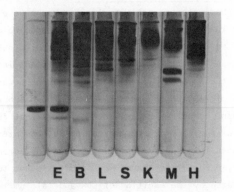

Fig. 4. Polyacrylamide gel electrophoresis of electric
eel tissues. Each tissue was homogenized and precipitated
by saturated $(NH_4)_2SO_4$. Each sample was applied to 15%
polyacrylamide gels and run at pH 8.7; the anode is at
the bottom. The gel on the left is purified electroplax
calcium-binding protein; from left to right, the tissue
gels are: E, electroplax, B, brain; L, liver, S, spleen;
K, kidney; M, skeletal muscle; H, heart.

activity equal to the protein from mammalian tissues, validating the enzymatic
assay of CDR in electric fish and confirming further the contention that eel CDR
is chemically identical to the mammalian protein. The results of this quantitative assay are given below in Table II.

Table II

CDR Levels in Electrophorus Tissues and Torpedo Electroplax

Tissue	μg CDR/mg protein
Electrophorus:	
Electroplax	23.9
Brain	3.54
Muscle	2.67
Liver	1.78
Kidney	0.85
Spleen	0.79
Heart	0.65
Torpedo electroplax	7.48
Pig Brain	2.00

The finding that CDR levels in eel electroplax are greater than ten times those
of mammalian brain, the richest source previously known, and the observation that
CDR is the major soluble protein in electroplax suggests that this protein may
underwrite a specialized function of the cholinergic synapse. The first candidate
for this role was regulation of cyclic nucleotide phosphodiesterase.

Electroplax Cyclic Nucleotide Phosphodiesterase - Greater than 80% of the cyclic nucleotide phosphodiesterase activity in Electrophorus electroplax was found to be located in the soluble fraction; neither the soluble nor particulate activity was inhibited by EGTA. With 5 μM cyclic AMP as the substrate, EGTA at concentrations as high as 1 mM failed to produce significant inhibition of enzyme activity; under the same conditions porcine brain PDE was inhibited 60% at EGTA concentrations of 0.05 mM or greater. Brain PDE has been shown to be calcium and CDR dependent[5] and this experiment indicates that CDR does not regulate the electroplax enzyme. A word of caution in the interpretation of such experiments has been introduced by similar studies of adrenal medullary PDE[10], however. We noted that this enzyme was also resistant to inhibition by EGTA and was not activated by the addition of CDR and calcium. Subsequent to our report of these observations, Uzunov et al demonstrated that polyacrylamide gel electrophoresis resolves adrenal medullary PDE into one principal activity with a shoulder which is activated by exogenous calcium and CDR[14]. More recently we have cleanly resolved this adrenal activity into three components by gel filtration and two components by density gradient centrifugation[15]. One activity was shown to be CDR dependent. This experience tells us that failure to observe inhibition of PDE activity by EGTA, especially in crude preparations, is not in itself an adequate demonstration of the independence of that activity of calcium and CDR. We therefore sought to determine if more than one species of PDE is present in electroplax, and if a minor activity is CDR-dependent. An aqueous extract of eel electroplax was subjected to density gradient centrifugation and the PDE activity was determined on the fractions collected following centrifugation. Two peaks of cyclic AMP PDE were detected[15]; each of these yielded a single activity upon recentrifugation. Neither peak was inhibited by EGTA nor activated by the addition of calcium and CDR or by cyclic GMP, which does activate the major adrenal medullary PDE[11]. We therefore conclude that in electroplax of Electrophorus CDR does not function to activate cyclic nucleotide phosphodiesterase. Removal of endogenous CDR from electroplax PDE by DEAE cellulose chromatography does not confer CDR sensitivity upon the enzyme; this observation tends to confirm our conclusion that electroplax PDE is CDR - independent. Preliminary experiments indicate that neither adenylate cyclase nor acetylcholinesterase of electroplax are activated by CDR, and we turned our attention to experiments designed to determine the nature of interactions between CDR and other electroplax proteins.

CDR Binding Protein in Electroplax - Soluble proteins of electroplax which precipitated at between 50 and 100% ammonium sulfate were fractionated by gel filtration on Sephadex G-100 in the presence and in the absence of EGTA. In the absence of EGTA, CDR eluted with the major protein peak at the void volume of the column (Fig. 5). In the presence of EGTA CDR eluted within the included volume of the column, indicating a calcium-dependent association of CDR with one or more high molecular weight electroplax proteins. To explore this association we prepared an affinity column by coupling CDR to Sepharose 4B. The CDR-free electroplax protein fraction excluded by Sephadex G-100 in the presence of EGTA was loaded on the CDR column. Proteins not specifically bound were eluted with 50 mM Tris-HCl, pH 7.4 containing 5 µM $CaCl_2$, and proteins which bound to the column were eluted with 50 mM Tris-HCl, pH 7.4 containing 100 µM EGTA. CDR binding activity was assayed by the ability of a given fraction to inhibit the CDR activation of brain CDR-deficient PDE at 1 mM cAMP. This inhibition assay depends upon the absence of PDE activity in electroplax at substrate concentrations of 1 mM; thus no PDE activity is contributed by the electroplax proteins, only inhibition of the activation of the standard brain PDE. The results of affinity chromotography (Fig. 6) indicate that the protein which inhibits CDR activation of PDE also binds to the CDR affinity column. Efforts are presently underway to purify this protein; hopefully this will provide us with further insights as to possible functions of CDR in electroplax.

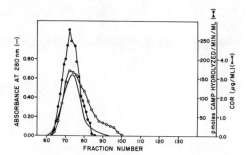

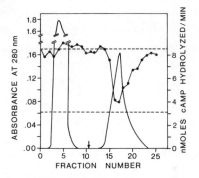

Fig. 5. Gel filtration of electroplax cyclic AMP phosphodiesterase on Sephadex G-100 in the absence of EGTA. The sample was applied to a column (2.5 x 100 cm) of Sephadex G-100 which had been equilibrated with 50 mM Tris-HCl, pH 7.4, and then eluted with the same buffer. (____) absorbance at 280 nm; (●), phosphodiesterase; (o) CDR, µg/ml.

Fig. 6. Fractionation of electroplax proteins on the CDR-affinity column. An aliquot (5 mg) of CDR-deficient protein was applied to the column in 50 mM Tris-HCl, pH 7.4, containing 50 µM $CaCl_2$. The first peak was eluted, then the column was washed with 50 mM Tris-HCl, pH 7.4, containing 100 µM EGTA. Aliquots of each fraction were added to tubes containing 25 mM Tris-HCl, and 50 ng CDR, and tubes were assayed with porcine phosphodiesterase for competition. Dotted lines represent stimulated and nonstimulated levels of phosphodiesterase. (____) absorbance; (●), phosphodiesterase activity.

133

REFERENCES

1. F.L. Siegel, J.C. Brooks, S.R. Childers and J.A. Campbell (1974) in Proc. Int. Symp. Calcium-Binding Proteins (Ed. W. Drabikowski, H. Strzelecka-Gotaszewska and E. Carafoli) Elsevier, Amsterdam pp 721-738.

2. B. Katz and R. Miledi (1968) J. Physiol. 195, 481-492.

3. R. Fine, W. Lehman, J. Head and A. Blizt (1975) Nature 258, 260-262.

4. S. Berl, S. Puszkin and W.J. Nicklas (1973) Science 179, 441-446.

5. Y.M. Lin and W. Y. Cheung (1974) J. Biol. Chem. 249, 4943-4954.

6. C.O. Brostrom, Y.C. Huang, B. McL. Breckenridge and D. J. Wolff (1975) Proc. Nat. Acad. Sci. U.S. 72, 64-68.

7. P. Greengard (1976) Nature 260, 101-108.

8. J. C. Brooks and F.L. Siegel (1973) J. Biol. Chem. 248, 4189-4193.

9. J.C. Egrie, J.A. Campbell, A. Flangas and F.L. Siegel, J. Neurochem., In Press.

10. J.C. Egrie and F.L. Siegel (1975) Biochem. Biophys. Res. Commun. 67, 662-669.

11. J.C. Egrie and F.L. Siegel, Biochim. Biophys. Acta, In Press.

12. S.R. Childers and F.L. Siegel (1975) Biochim. Biophys. Acta 405, 99-108.

13. S.R. Childers and F.L. Siegel (1976) Biochim. Biophys. Acta 439, 316-325.

14. P. Uzunov, M.E. Gregy, A. Revuela and E. Costa (1976) Biochem. Biophys. Res. Commun. 70, 132-138.

15. A. Sitaramayya and F.L. Siegel, In Preparation.

Acknowledgment

This research was supported by Public Health Service grant NS 11652.

SARCOPLASMIC RETICULUM

Kinetics of Passive Ca^{2+} Transport by Skeletal Sarcoplasmic Reticulum

Duncan H. Haynes and Vincent C.K. Chiu

Department of Pharmacology
University of Miami School of Medicine
Miami, Fla. 33152

INTRODUCTION

The sarcoplasmic reticulum (SR) has three major functions: Ca^{2+} uptake, storage, and release. The simplicity of its functions and its composition[7] make it an its composition (see contribution of Dr. D.H. MacLennan, this volume) make it an excellent subject for study as a transport system. The present contribution deals with the best understood functions of the SR: Ca^{2+} uptake and Ca^{2+} storage. In vitro studies of SR have given us the general outline of the Ca^{2+} uptake mechanism. Two Ca^{2+} are taken up per ATP split[1,2] and the SR pump function is able to reduce the external Ca^{2+} to submicromolar concentrations. The uptake is sufficiently rapid to account for the rate of muscle relaxation.[3,4] A model of the Ca^{2+}-Mg^{2+}-ATPase function as a carrier has been proposed.[5] The model is based on the dependence of the degree of phosphorylation of the ATPase upon Ca^{2+} and Mg^{2+} concentrations in the presence of gamma-labelled ATP. It predicts that the active transport process involves the obligatory exchange of Ca^{2+} for Mg^{2+} or 2 K^{+} (or other available cations). On the other hand, most published studies of the Ca^{2+} transport reaction have measured the cotransport (and coprecipitation) of Ca^{2+} and oxalate. Although oxalate plays no "physiological" role in the Ca^{2+} uptake function of the SR, the fact that it is cotransported indicates that the transport of other anions might be a step in the uptake mechanism.

Figure (1) illustrates four simple testable models for the active transport process. In models I and III the transport is an obligatory exchange diffusion reaction and the reaction is not influenced by anions. Models II and IV predict Ca^{2+} and anion (A^{-}) cotransport reactions, with no direct involvement of Mg^{2+} or K^{+}. Models III and IV describe the Ca^{2+}-ATPase as an electrogenic pump; "energization" produces a membrane potential which drives electrophoretic cation or anion transport in order to achieve net Ca^{2+} movement. Models I and III predict no membrane potential involvement in the reaction. We are attempting to determine which of these models most adequately describes the function of the ATPase in the active uptake of Ca^{2+} by the SR. We have begun by studying the passive permeability of the organelle.

The problems discussed above are related to the problem of Ca^{2+} storage by the SR. The purification and characterization of a Ca^{2+} (and Mg^{2+}) binding protein "calsequestrin" from SR by MacLennan[6,7] has offered important information about the storage function of the SR. MacLennan proposed that this class of proteins is

137

responsible for complexing the transported Ca^{2+}, thereby lowering the gradient against which it must be transported.[6,7] In a recent study,[8] we have given kinetic evidence that the calsequestrin-class of Ca^{2+} binding is located in the inner aqueous space of the SR. In this study we used spectroscopic indicators of free Ca^{2+} and Mg^{2+} together with the stopped-flow rapid mixing technique. This method allows us to measure reactions with half-times $(t_{\frac{1}{2}})$ of 3 msec or more, offering distinct advantages over conventional methods such as millipore filtration. Two classes of binding site were observed in this study: One with low capacity and high affinity (ca. 35 nmole/mg protein Ca^{2+} binding capacity; K_d = 17.5 μM) and one with a high capacity and low affinity (ca. 820 nmole/mg protein; K_d = 1.9 mM). The low affinity binding site has the characteristics of the calsequestrin class of acidic binding proteins. It showed no Ca^{2+} over Mg^{2+} specificity in its binding reactions and the $t_{\frac{1}{2}}$ for the binding of externally-added Ca^{2+} was about 2 sec. The slow kinetics of binding constitute evidence that the binding sites are located behind the SR membrane in the internal aqueous space. This supports MacLennan's suggestion that they have a storage function.[6,7]

The binding of added Ca^{2+} to the high affinity site is faster than 3 msec, the mixing time of the stopped-flow apparatus. The kinetics of this reaction could not be resolved. We suggested that the high affinity site was the Ca^{2+}-ATPase and presented evidence , based on the comparison of rate data,[8] that the passive Ca^{2+} permeability of the SR was ATPase-mediated. These studies were carried out using SR prepared by differential centrifugation. Meissner[9] has shown that the SR

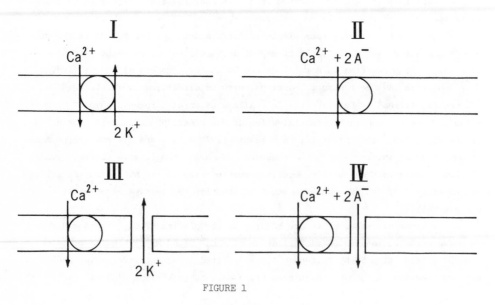

FIGURE 1

138

can be separated further by sucrose density gradient techniques to obtain a low density fraction which is rich in the ATPase (90 protein percent) and a high density fraction which is rich in the calsequestrin class of acidic binding proteins (50 protein percent). We have applied his procedure in slightly modified form and have found that the ATPase-rich fraction has no detectable low affinity binding but contains 22 nmole/mg of high affinity Ca^{2+} binding sites (V. Chiu and D.Haynes to be published). This corresponds to 2.2 Ca^{2+} binding sites per ATPase molecule, assuming the total protein to be ATPase and calculating with a molecular weight of 102,000 daltons.[7] The behavior of the low affinity sitewith the slow binding kinetics is therefore referred to the heavy SR vesicle fraction.

Obviously the light, ATPase-rich fraction is the most ideal preparation for studying the kinetics of the ATPase in unperturbed form. However, the lack of calsequestrin in the aqueous interior renders it difficult if not impossible to study passive transport using indicators of Ca^{2+} in the aqueous phase. We chose therefore to use the binding of the fluorescent probe 1-anilino-8-naphthalenesulfonate (ANS⁻) to the inner surface of the membrane as an indicator of the Ca^{2+} concentration in the inner aqueous phase. ANS⁻ shows no fluorescence in water but has strong fluorescence when bound in hydrophobic environments[10] such as the space between polar head groups in phospholipid membranes.[11] Our studies have shown that the binding to phospholipid membranes increases with increasing concentrations of monovalent (M^+) or divalent (M^{2+}) cations in the medium. The increase is primarily due to an increase in binding, with salt-dependent changes in quantum yield making only minor contributions. The binding equilibrium is coupled to the cation concentration through the electrostatic surface potential of the membrane. Increasing the concentration of M^+ or M^{2+} provides increased shielding of the existing negative charges in the membrane and of the negative charge induced by ANS⁻ binding. This allows for greater ANS⁻ binding. We have modelled this behavior in terms of the Gouy-Chapman theory of the electric double layer.[12] What is critical for our present application is that ANS⁻ fluorescence arising from the inside (or outside) surface will, under the proper choice of conditions, report the cation concentrations in the inside (or outside) medium.[12,13] The enhancement effects of divalent cations occur at much lower concentrations than do the effects of monovalent cations. This is due to the larger screening contributions of the former. ANS⁻ binding to the inside and outside surfaces of phospholipid vesicles can be differentiated on the basis of their kinetics in stopped-flow experiments.[13] Binding to the outside surface is rapid ($t_{1/2}$ = 100 μsec) while binding to the inside surface is slow ($t_{1/2}$ between 5 and 100 sec) due to the rate limitation imposed by the membrane. Valinomycin in the presence of K^+ can speed up this process.

MATERIALS AND METHODS

The materials are described in previous publications.[11,13] ATPase-rich SR was prepared by a density gradient method. Briefly, 40 gm of rabbit back muscle were homogenized in 400 ml of buffer (20 mM histidine, pH 7.2, 0.25 M sucrose) for 90 sec in a Waring Blendor. Myofibrils, mitochondria, nuclei and cell debris are removed by centrifugation (20 min at 10,000 g). The supernate is pelleted at 100,000 g. The pellet is washed, resuspended and centrifuged twice. Then the resuspended pellet is subjected to sucrose gradient centrifugation using a Beckman type 35 rotor. Two distinct bands are observed with densities corresponding to 29% and 39% sucrose. The fractions between 28% and 31% sucrose are rich in ATPase (ca. 90% of the total protein is ATPase) and were used for permeability studies.

Rapid mixing experiments with ANS$^-$ were performed with an Aminco-Morrow Stopped-Flow Apparatus (Cat. No. 4-8409). The apparatus mixes equal volumes of two solutions. The excitation monochromator was set at 368 nm with a Schott GG420 cutoff filter placed in front of the photomultiplier. In the cases where multiple kinetic processes were observed, the $t_{\frac{1}{2}}$ and amplitude values were obtained by a graphical peeling technique.

RESULTS AND DISCUSSION

Figure (2) is an oscillograph trace showing the time course of the fluorescence increase observed when the ATPase-rich SR fraction is mixed rapidly with ANS$^-$. An instantaneous increase is observed, followed by a single exponential kinetic phase with a $t_{\frac{1}{2}}$ of 8 sec. The first process corresponds to binding to the outside surface of the membrane and the second process corresponds to the movement of ANS$^-$ across the membrane. There are several features of the reaction which are similar

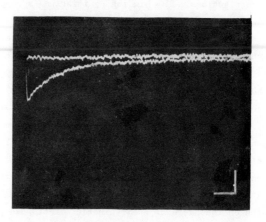

Fig. 2. Stopped-flow trace showing permeation of ANS$^-$ through ATPase-rich SR vesicle membranes. The vertical axis represents fluorescence (0.5 V/cm). The horizontal axis represents time (5 sec/cm, two consecutive sweeps). The markers represent one cm. Syringe A contained 0.2 mg/ml SR protein, zero Ca^{2+} (buffered with 2.4 x 10^{-4}M EGTA) and buffer. Syringe B contained 3 x 10^{-5}M ANS$^-$ and buffer. The buffer solution contained 10 mM histidine, pH 7.2, 10 mM KCl and 0.6 M sucrose.

to the ANS$^-$ transport reaction in dimyristoyl phosphatidylcholine (PC) vesicles: The $t_{1/2}$ values are similar (8 sec and 15 sec, respectively) and the ratios of the amplitude of the slow phase (A_i) to that of the instantaneous phase are similar (0.43 and 0.63, respectively). In dimyristoyl PC vesicles ANS$^-$ is accompanied by K^+ across the membrane.[13] For low ANS$^-$ concentrations this movement does not affect the internal K^+ concentration. In both membrane systems the rate of the slow phase is increased by the addition of valinomycin. The amplitudes of both phases (in both membrane systems) increase when the monovalent or divalent cation concentrations in the A and B syringes are increased. We will make use of this property to determine the rate of K^+ and Ca^{2+} equilibration across the membrane.

Figure (3) shows the result of an experiment in the mixing configuration (SR) vs. (ANS$^-$ + Ca^{2+}). In this case, three kinetic phases are observed: (a) the instantaneous phase ($t_{1/2} \leq 3$ msec), (b) the slow phase ($t_{1/2} = 8$ sec) and (c) a still slower phase ($t_{1/2} = 120$ sec). The 120 sec phase arises from the increase in ANS$^-$ binding to the inside surface which results from movement of Ca^{2+} into the aqueous interior. This conclusion is supported by the observation that the 120 sec phase is not seen in the absence of divalent cations, that its amplitude increases with increasing M^{2+} concentration and that its $t_{1/2}$ is decreased by the addition of the Ca^{2+} ionophore X537A. When the experiment of Figure (3) is repeated using dimyristoyl PC vesicles, a third phase is also observed with a $t_{1/2}$ value of ca. 120 sec. It is thus unlikely that the reaction observed results from the coupled permeation of Ca^{2+} and Cl^- via an intrinsic permeability system (Case II or IV in the Introduction). We take the difference between the $1/t_{1/2}$ values of the SR and PC membrane systems to represent the upper limit of the rate contribution of such a transport system.

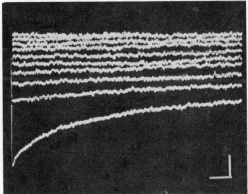

Fig. 3. Stopped-flow trace showing the influence of externally-added Ca^{2+} on the kinetics of ANS$^-$ permeation through ATPase-rich SR vesicle membranes. The vertical axis represents fluorescence (1.0 V/cm). The horizontal axis represents time (5 sec/cm with repetitive sweeps). The markers represent one cm. Syringe A contained 0.2 mg/ml SR protein and buffer. Syringe B contained 3×10^{-5} M ANS$^-$, 15 mM $CaCl_2$ and buffer. The buffer was as in Figure (2). T = 23°C.

Table I shows the dependence of the $t_{1/2}$ and amplitude values on the added Ca^{2+}
concentration. The $t_{1/2}$ values are essentially invariant with Ca^{2+} concentration.
This is what is expected for uncatalyzed copermeation of cations with ANS^- through
phospholipid regions of the membrane (cf. ref. 13). The amplitudes of all three
phases are observed to increase with increasing Ca^{2+} concentration. An increase
in amplitude of a kinetic phase represents an increase in the Ca^{2+} concentration
responsible for that phase. Thus the increase in ANS^- binding to the outside (A_o)
is readily expected. The amplitude of the third phase ($A_{i,b}$) represents equilibration of Ca^{2+} across the membrane via an ANS^--induced non-specific permeability. It
is also expected to increase with increasing Ca^{2+} concentration. However the amplitude of the second phase ($A_{i,a}$) which results from copermeation of K^+ and ANS^- by
a non-specific mechanism (cf. ref. 13) is not expected to be a function of the Ca^{2+}
concentration. Control experiments with dimyristoyl PC vesicles fail to show a
Ca^{2+} effect on $A_{i,a}$. We are left to conclude that the increase in $A_{i,a}$ in the SR
experiments results from Ca^{2+} movement into the aqueous interior via an intrinsic
permeability system (the Ca^{2+}-ATPase).[*] Before describing further experiments
which we designed to test this conclusion we must consider an important property of
exchange diffusion systems - the equilibria which they set up.

If Ca^{2+} movement is influencing the amplitude of the second kinetic phase, then
this movement must occur with a $t_{1/2}$ of 8 sec or less. The persistance of the slower
Ca^{2+}-dependent process would thus indicate that at the end of the second phase, the
Ca^{2+} concentration in the aqueous interior ($[Ca^{2+}]_{in}$) is still lower than the Ca^{2+}

---

[*] If this amplitude were due to a "leak", then we would expect the rates of both
phases to be increased substantially.

TABLE 1

DEPENDENCE OF $t_{1/2}$ AND AMPLITUDE OF ANS^- PERMEATION ON ADDED CALCIUM[*]

$[Ca^{2+}]_{added}$	A_o (volts)	$A_{i,a}$ (volts)	$t_{1/2,a}$ (sec)	$A_{i,b}$ (volts)	$t_{1/2,b}$ (sec)
0.0[**]	3.2	1.0	8	0.0	–
2.0×10^{-5} M	4.3	1.6	8	0.3	120
2.0×10^{-4} M	5.8	1.5	8	0.6	120
2.0×10^{-3} M	13.3	2.0	8	1.1	120
8.0×10^{-3} M	14.2	2.3	8	2.4	120
3.0×10^{-2} M	19.6	2.6	8	3.2	120
6.0×10^{-2} M	21.8	2.8	8	3.5	120

---

[*] Experimental conditions are given in Figure (3). The results ($t_{1/2}$ and relative
amplitudes) do not depend on the ANS^- concentration.
[**] 2×10^{-4} M EGTA added

concentration outside ($[Ca^{2+}]_{out}$). This gives rise to the question of why the postulated intrinsic permeability system would give less than "perfect" equilibration of the Ca^{2+} gradient. The answer to this is readily apparent if the permeability system is considered to act as an obligatory Ca^{2+}/K^+ exchange diffusion carrier. Influx of Ca^{2+} will result in depletion of K^+ from the aqueous interior until the two ionic species come to equilibrium. A simple calculation given below shows that $[Ca^{2+}]_{in} < [Ca^{2+}]_{out}$ at the equilibrium selected by the permeability system.

The reaction for a Ca^{2+}/K^+ passive exchange diffusion system would be

$$Ca^{2+}_{out} + 2 K^+_{in} \xleftarrow{\quad K_{eq} \quad}\rightarrow Ca^{2+}_{in} + 2 K^+_{out} \tag{1}$$

where K_{eq}, the equilibrium constant for the reaction, is equal to one. For the case where the SR is preequilibrated with 10 mM KCl ($[K^+]_{in} = [K^+]_{out} = 10$ mM*) and the SR is mixed with $CaCl_2$ to a final concentration of 15 mM, the final equilibrium concentration of Ca^{2+} inside (= x) can be calculated from

$$([K^+]_{in,o} - 2x)^2/[K^+]^2_{out} = x/[Ca^{2+}]_{out} \tag{2}$$

where the subscript o represents the initial condition. Solution of the quadratic equation gives $[Ca^{2+}]_{in} = 2.83$ mM. This value is substantially lower than the "true" equilibrium value (15mM) which is reached after completion of the third process. Thus only a fraction of the equilibration occurs in the second phase and a substantial amplitude is permitted for the third kinetic process. The exchange-diffusion models I and III are consistent with this argument and the data of Table 1. Models II and IV are excluded by the same reasoning.

The above reasoning indicates that the Ca^{2+}/K^+ exchange diffusion permeability system under study here must have a $t_{\frac{1}{2}}$ of 8 sec or less for Ca^{2+}/K^+ equilibration. Table 2 gives the results of a series of experiments in which Ca^{2+} was mixed with SR in the presence of ANS^-, valinomycin and K^+. The ionophore serves to speed up the rate of ANS^- equilibration across the membrane** such that it is an adequate indicator of the internal Ca^{2+} concentration during the course of the (ATPase) carrier-mediated Ca^{2+}/K^+ equilibration across the membrane. Comparison of Expt.(6) with the experiment of Figure (2) demonstrates that valinomycin addition has reduced the $t_{\frac{1}{2}}$ for ANS^- equilibration across the membrane to 24 msec. Comparison of Expts.

---

* The absence of low affinity non-specific Ca^{2+} binding in this fraction (V. Chiu and D.Haynes, to be published) renders this assumption quite plausible.
** It can be shown that electrophoretic permeability of valinomycin-K^+ would have no effect on Models I and III. Cotransport of K^+ and ANS^- via valinomycin can be shown to make only a small perturbation of the internal ion concentrations since ANS^- is present at a concentration of 3.3×10^{-5}M (1/100 th of that of the other ions). Cotransport of KCl by valinomycin is probably slow (cf. ref. 13). The upper limit of its contribution can be estimated from the small increase in $1/t_{\frac{1}{2}}$ of the third process upon valinomycin addition.

TABLE 2

KINETICS OF Ca^{2+} MOVEMENT INDICATED BY ANS^-

Expt.	Mixing Configuration	$t_{\frac{1}{2},a}$	$A_{i,a}$	$t_{\frac{1}{2},b}$	$A_{i,b}$
1	(SR) vs. (30 mM Ca^{2+})	8 sec	1.0	100 sec	5.0
2	(SR + $\frac{1}{2}$val)** vs. (30 mM Ca^{2+})	45 msec	0.9	120 sec	5.7
3	(SR + val) vs. (30 mM Ca^{2+})	47 msec	1.0	120 sec	6.0
4	(SR + val) vs. (60 mM Ca^{2+})	21 msec	1.1	70 sec	6.5
5	(SR + val, no K^+) vs. (30 mM Ca^{2+}, no K^+)	--	--	55 sec	5.0
6	(SR + 15 mM Ca^{2+} +val, no ANS) (15 mM Ca^{2+} + 2x ANS***)	24 msec	6.0	--	--

* In all cases, both reservoirs contained 3.3×10^{-5}M ANS^-, 0.6 M sucrose, 5 mM histidine buffer, pH 7.2 and 10 mM KCl unless otherwise indicated. Val denotes 1.4×10^{-5}M valinomycin. T = 23°C. The maximal experimental uncertainty is $\pm$ 10% for A values and 15% for $t_{\frac{1}{2}}$ values.
** 7×10^{-6}M valinomycin
*** 6.6×10^{-5}M ANS^-

(1), (2) and (3) indicates that the ANS^- responds to the Ca^{2+}/K^+ exchange diffusion process registering a $t_{\frac{1}{2}}$ of 47 msec. The control experiment (Expt. (5)) in which K^+ has been omitted gives no amplitude attributable to an exchange diffusion process. This is the expected behavior derived from the exchange diffusion model. This experiment also proves that the 47 msec reaction is not an artifact due to the presence of valinomycin in the membrane. Comparison of Expts. (2), (3) and (6) indicates that the measured $t_{\frac{1}{2}}$ values must be characteristic of the kinetics of the exchange diffusion system. The rate of the ANS^- transport reaction catalyzed by valinomycin is not limiting. Comparison of Expts. (3) and (4) indicates that the rate of Ca^{2+} equilibration increases with increasing Ca^{2+} concentration. This aspect of the reaction will be considered in greater detail elsewhere. In summary, our experiments show that in the absence of ATP, Ca^{2+} is freely permeable across the SR membrane via an exchange diffusion system (Model I or III) which equilibrates Ca^{2+} and K^+ with a half-time of ca. 47 msec. The half-time for Ca^{2+} equilibration by permeability mechanisms involving anions is at least four orders of magnitude slower ($t_{\frac{1}{2}} \geq 120$ sec).

It is of interest to compare the rate of passive Ca^{2+} transport measured in the ATPase-rich fraction with that previously measured in the calsequestrin-rich fraction.[8] Such a comparison can be made if the $t_{\frac{1}{2}}$ value in the present experiments can be converted into numbers of Ca^{2+} crossing the membrane per mg protein per second. In the experiment of Figure (5) the calculated $[Ca^{2+}]_{in}$ at the end of the 50 msec phase is 2.83 mM. Taking the internal water volume to be 5×10^{-6} 1/mg protein[14] we calculate a Ca^{2+} movement of 14.2 nmole/mg protein. Since the high affinity Ca^{2+} binding capacity due to the ATPase is 22 nmole/mg (V. Chiu and D. Haynes, to be published) less than a single turnover of the enzyme is necessary to equilibrate the inner aqueous phase. The initial rate of passive Ca^{2+} transport can be calculated from the $t_{\frac{1}{2}}$ (=47 msec) using the exponential relationship using

$$\text{Initial rate} = (Ca^{2+} \text{ moved}) \times 0.69/t_{\frac{1}{2}} \tag{3}$$

This gives us an initial rate of 208 nmole/mg/sec. This is about 3.2 times as fast as the initial rate of the active uptake reaction measured by Inesi and Scarpa[4] in unfractionated SR and is about 6.1 times as fast as the rate of the passive transport that we observed in unfractionated SR.[8] We can calculate the turnover number of the ATPase in the ATPase-rich fraction assuming its composition to be 100% and using 1×10^5 daltons as its molecular weight. To make this calculation for the calsequestrin-rich SR vesicles in the unfractionated SR preparation we must divide the rate by the fraction of the total protein which is in the calsequestrin-rich fraction (ca. 0.2) and divide by the fraction of ATPase in that band (0.5). We obtain values of 20.8 and 34.0 for the turnover numbers of the ATPase in the ATPase-rich and in the calsequestrin-rich SR vesicles, respectively. Further experimentation will be conducted to determine whether this minor difference results from experimental errors and the approximations that we have used in our calculations or whether there are true differences in the behavior of the ATPase in the two types of vesicle. For future studies we plan to determine the role of exchange diffusion in the active Ca^{2+} uptake mechanism.

ACKNOWLEDGEMENTS

We wish to thank Mr. Donald Mouring for his excellent technical assistance. This work was supported by NIH grants 1 P01 HL 16117 and R01AM20086. Dr. Chiu is supported by a post-doctoral fellowship from the Florida Heart Association.

REFERENCES

1. Hasselbach, W., and Makinose, M. Biochem. Z. 339:94-111 (1963)

2. Weber, A., Herz, R., and Reiss, I. Biochem. Z. 345:329-369 (1966)

3. Ohnishi, T. and Ebashi, S. J. Biochem. (Tokyo) 55:599-603 (1964)

4. Inesi, G. and Scarpa,. A. Biochemistry 11:356-359 (1972)

5. Kanazawa, T., Yamada, S., Yamamoto, T. and Tonomura, Y. J. Biochem. (Tokyo) 70:95-123 (1971)

6. MacLennan, D.H. and Wong, P.T.S. Proc. Nat. Acad. Sci. (U.S.A.) 68:1231-1235 (1971)

7. MacLennan, D.H. and Holland, P.C. Ann. Rev. Biophys. Bioeng. 4:377-404 (1975)

8. Chiu, V.C.K. and Haynes, D.H. Biophys. J. 18:3-22 (1977)

9. Meissner, G. Biochim. Biophys. Acta. 389:51-68 (1975)

10. Weber, G. and Laurence, D.J.R. Biochem. J. 56:xxxi (1954)

11. Haynes, D.H. and Staerk, H. J. Membrane Biol. 17:313-340 (1974)

12. Haynes, D.H. J. Membrane Biol. 33:63-108 (1974)

13. Haynes, D.H. and Simkowitz, P. J. Membrane Biol. 33:63-108 (1977)

14. Duggan, P.F. and Martonosi, A. J. Gen. Physiol. 56:147-167 (1970)

Calcium Efflux
from Sarcoplasmic Reticulum Vesicles*

Arnold M. Katz[†], Doris I. Repke
Gary Fudyma[†] and Munekazu Shigekawa[†]
Department of Medicine
Mount Sinai School of Medicine
New York, New York, 10029
USA

INTRODUCTION

Calcium movements across the membrane of the sarcoplasmic reticulum (SR) of muscle serve two important functions in skeletal[1] and cardiac[2] muscular function. Calcium efflux from this intracellular structure, by providing calcium for binding to troponin, represents the final step in the process linking excitation at the sarcolemma to the initiation of the contractile process. Calcium influx into the SR lowers cytosolic Ca^{2+} concentration to levels sufficiently low to cause this cation to become dissociated from troponin, thereby effecting relaxation. These two intracellular calcium fluxes reflect the operation of an uphill, active, transport process and the passive movement of this ion, respectively.

Calcium uptake into the SR is mediated by an ATP-dependent calcium pump that utilizes the energy available from the hydrolysis of one mole of ATP to move two moles of Ca^{2+} against an electrochemical gradient[1,3]. Details regarding the mechanism and kinetics of this active transport process are presented elsewhere in this volume. The present report describes our recent work on the second of the calcium fluxes in muscle, the release of calcium from the SR.

MATERIALS AND METHODS

Sarcoplasmic reticulum vesicles were prepared from the heavy microsomal fraction of rabbit "white" skeletal muscle as described previously[4]. The calcium

*Supported by Research Grants HL-13191, HL-18801 and AA-00316 from the National Institutes of Health, and a Grant-in-Aid from the New York Heart Association.

†Present Address: Department of Medicine, University of Connecticut Health Center, Farmington, Connecticut, 06032 USA.

content of SR vesicles was measured after ATP or acetyl-P supported Ca uptake by counting the filtrates obtained after reaction mixtures containing [45]Ca-labelled calcium were filtered through Millipore filters mounted in Swinney adapters[4]. Rates of calcium uptake[§], release, influx and efflux were calculated as described in Reference 5. Unless otherwise specified, reactions were carried out in 120 mM KCl, 40 mM histidine buffer (pH 6.8) and 5 mM MgATP at 25° C. Where acetyl-P instead of ATP was substrate, $MgCl_2$ concentration was 5 mM and temperature 37° C. In studies of the Mg^{2+}-dependence of Ca efflux, various concentrations of $MgCl_2$ were added to ATP (final ATP concentration was 5 mM) and pH adjusted to 6.8 prior to use. Protein concentrations were generally between 10 and 20 µg/ml.

Concentrations of Mg^{2+}, and of Ca^{2+} inside (Ca_i) and outside (Ca_o) the SR vesicles were calculated as described previously[4,6]. Ca_o was varied by changing the amounts of $CaCl_2$ added at the start of the Ca uptake reaction. Ca_i was calculated from the solubility products of Ca oxalate or Ca phosphate[4] in experiments where these anions stabilized internal Ca^{2+} through the formation of insoluble precipitates within the vesicles.

RESULTS

Effects of Ca_i and Ca_o on calcium efflux and calcium permeability: Ca efflux rates were measured at or shortly after the Ca content of the SR vesicles reached an initial peak following initiation of Ca uptake reactions. Under these conditions, Ca efflux rate increases with increasing Ca_o in the range between 0.1 and 3.3 µM[5,7]. Increasing Ca_i in the range between 4 and 750 µM, however, does not cause a proportional increase in Ca efflux rate in spite of the fact that Ca_i represents the "driving force" for Ca efflux.

A calcium permeability coefficient for the SR membranes can be calculated from the equation:

$$\text{Ca Permeability Coefficient} = \frac{\text{Ca Efflux Rate}}{Ca_i}$$

[§] Ca uptake and Ca release refer to the net rate of gain or loss of Ca by the vesicles, Ca influx and Ca efflux refer to the rates of unidirectional Ca fluxes into or out of the vesicles.

According to this equation, the failure of Ca efflux rate to increase in proportion to increasing Ca_i indicates that Ca permeability falls at higher Ca_i.

The dependence of Ca permeability on Ca_i and Ca_o can, as a first approximation, be depicted as illustrated in Figure 1, where the Ca permeability coefficient is plotted against the ratio Ca_i/Ca_o. Calcium permeability is seen to vary almost 1000-fold, decreasing as this ratio increases from 1 to 3000.

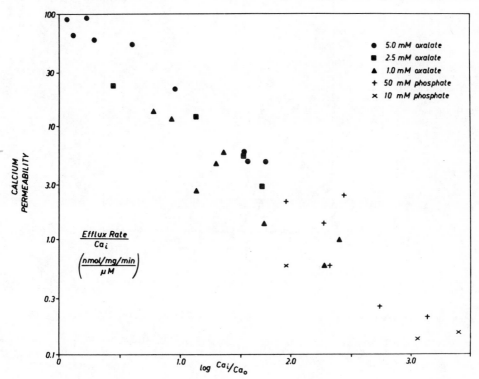

Fig. 1. Relationship between the Ca permeability coefficient and the ratio Ca_i/Ca_o. Each symbol represents data from a reaction mixture containing a different Ca precipitating agents, as shown in the figure. Reproduced from Katz et al.[7]

Kinetics of the dependence of Ca efflux on Ca_o: The mechanism underlying the relationship between Ca permeability and the Ca^{2+} concentrations inside and outside the vesicles was examined by measuring Ca efflux rates at constant Ca_i over a wide range of Ca_o. The ability of increasing Ca_o to promote Ca efflux exhibits saturation kinetics[6]. In the experiment shown in Figure 2, where Ca_i was maintained at ~150 μM by the use of 50 mM phosphate as Ca-precipitating agent, V_{max}

was ~400 nmol/mg·min and K_{Ca} (Ca_O where Ca efflux velocity was half-maximal) was 0.8 µM. Values for K_{Ca} averaged approximately 1 µM in a series of experiments

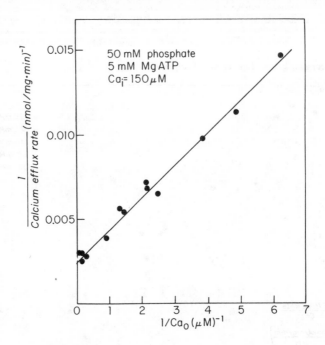

Fig. 2. Dependence of 1/Ca efflux velocity on $1/Ca_O$. Reproduced from Katz <u>et al</u>.[6]

such as that shown in Figure 2, and did not change significantly when Ca_i was increased 5-fold by reducing the phosphate concentration to 10 mM, or when Ca_i was reduced approximately 40-fold by the use of 5 mM oxalate as the Ca-precipitating agent[6].

<u>Effects of Mg^{2+} and Na^+ on Ca efflux</u>: Variation of Mg^{2+} concentration in the medium outside the SR vesicles significantly influences Ca efflux. Ca efflux at Ca_O = 0.2 µM and Ca_i = 50 µM decreased approximately 60% when Mg^{2+} was increased from 0.1 to 1.1 mM[6].

Substitution of 120 mM NaCl for the 120 mM KCl in the reaction mixture was without significant effect on either Ca efflux velocity or the Ca permeability coefficient[6].

<u>Effects of replacement of ATP with acetyl-P on Ca efflux</u>: When 2-10 mM acetyl-P instead of ATP was the energy donor during the initial Ca uptake reaction, Ca efflux following the achievement of maximal Ca content was markedly inhibited, and K_{Ca} for Ca efflux increased approximately ten-fold[6].

DISCUSSION

The stimulation of Ca efflux from SR vesicles by increasing Ca_o and the phenomenon of Ca-triggered Ca release described previously in "skinned" skeletal and cardiac muscle fibers[8,9] may reflect the operation of a single mechanism. Increasing Mg^{2+} inhibits both the ability of increasing Ca_o to promote Ca efflux from SR vesicles and the tension response to a transient increase in Ca^{2+} in the medium surrounding a "skinned" fiber, suggesting that changes in Ca permeability such as those described in the present report may play a role in controlling Ca release from the SR of intact muscle. The absence of detectable effects of replacing K^+ with Na^+ does not support the hypothesis that the SR membrane contains a permeability control mechanism similar to that reported to effect a Na-Ca exchange in the sarcolemma of various tissues[10,11].

The relationship between Ca efflux and the Ca^{2+} concentrations inside and outside the SR vesicles shown in Figure 1 appears not to reflect a direct effect of the Ca^{2+} concentration gradient across this membrane upon Ca permeability. The ability of increasing Ca_o to promote Ca efflux by a mechanism that exhibits saturation kinetics (Figure 2) suggests, instead, that Ca^{2+} binding to a site on the external surface of the membranes increases Ca permeability. Lack of significant effects of changing Ca_i on the apparent Ca^{2+} affinity of this putative external Ca^{2+}-binding site suggests that Ca^{2+}-sensitive sites at the internal and external surfaces of the vesicles interact independently with Ca^{2+}. These findings thus do not support the view that Ca efflux is mediated by a carrier with a single Ca^{2+}-binding site that moves between the inner and outer surface of the membrane.

The marked decrease in Ca efflux seen when acetyl-P is substituted for ATP suggests that the nucleotide can modulate Ca efflux, but the mechanism responsible for this proposed role of ATP remains unknown.

A number of similarities observed between factors that influence Ca efflux from the SR and those that influence Ca influx via the Ca pump are listed in Table 1. These similarities suggest that the Ca pump can also modulate Ca efflux.

TABLE 1

SIMILARITIES BETWEEN FACTORS CONTROLLING

CA EFFLUX AND THE CA PUMP IN SR VESICLES.

Variable	Ca Efflux	Ca Pumping
Increasing Ca_o	Increased	Increased
Increasing Ca_i	Decreased	Decreased
Replacement of ATP with acetyl-P	Decreased	Decreased
Increasing Mg^{2+} in mM range	Decreased	Decreased
K_{Ca} in 5 mM MgATP	~1 µM	~1 µM
K_{Ca} in acetyl-P	~10 µM	~10 µM

The hypothesis that the mechanism responsible for Ca pumping by the SR when muscle relaxes participates also in the control of the Ca release which initiates contraction is further supported by the finding that incorporation of the ATPase protein of the SR into phospholipid vesicles increases their Ca permeability[12,13] and that this protein can exhibit properties similar to those of a Ca^{2+} ionophore[14].

A role in controlling Ca efflux for one or more of the minor proteins found in SR preparations is suggested by the finding of Repke et al.[15] that reconstituted SR vesicles that lack most of their calsequestrin show little inhibition of their capacity to pump and to store calcium, whereas their ability to release this cation is markedly inhibited. A reduced rate of calcium release in the face of a virtually normal rate of calcium uptake can account for the high Ca capacity of these reconstituted vesicles to store Ca oxalate[15]. A similar conclusion is suggested by the data of Meissner[16] who found that light skeletal muscle microsomes,

152

which are virtually devoid of the minor proteins, release Ca more slowly and have a much higher Ca oxalate storage capacity than heavy microsomes which contain these minor proteins. It is possible, therefore, that calcium release from the SR involves an interaction between the ATPase protein and calsequestrin.

SUMMARY

1. Ca efflux from skeletal muscle SR vesicles increases with increasing Ca_o.

2. Ca permeability coefficients decrease with increasing Ca_i.

3. Ca permeability coefficients, as a first approximation, are inversely proportional to the ratio Ca_i/Ca_o, but this correlation does not appear to reflect an effect of the Ca^{2+} concentration gradient, per se, on Ca permeability.

4. The ability of increasing Ca_o to promote Ca efflux exhibits saturation kinetics, K_{Ca} being ~1 μM and V_{max} ~400 nmol/mg·min in 5 mM MgATP and 50 mM phosphate.

5. The K_{Ca} for Ca efflux does not change in parallel to variations in Ca_i, indicating that Ca^{2+}-sensitive sites at the inside and outside of the SR membrane interact independently with Ca^{2+}.

6. Replacement of 120 mM KCl with 120 mM NaCl does not significantly modify Ca efflux, indicating that this permeability control mechanism differs from that reported to mediate Na-Ca exchange by the sarcolemma.

7. Like "Ca-triggered Ca release" in skinned skeletal muscle fibers, Ca efflux from SR vesicles is inhibited by increasing Mg^{2+} in the millimolar range.

8. Substitution of acetyl-P for ATP in the reaction mixture reduces maximal Ca efflux velocity, and causes an approximately ten-fold increase in K_{Ca}.

9. A number of similarities between factors influencing Ca influx and those known to modify the Ca pump are consistent with the view that the Ca pump can also modulate Ca efflux.

10. Calsequestrin and other minor proteins found in these SR preparations may play a role in the control of Ca efflux.

REFERENCES

1. Ebashi, S. (1976) Ann. Rev. Physiol. 38: 293-313.

2. Katz, A.M. (1977) Physiology of the Heart. Raven Press, New York, pp. 137-159.

3. Hasselbach, W. (1964) Prog. Biophys. Mol. Biol. 14: 167-222.

4. Katz, A.M., et al. (1977) J. Biol. Chem. 252: 1938-1949.

5. Katz, A.M., et al. (1977) J. Biol. Chem. 252: 1950-1956.

6. Katz, A.M., et al. (1977) J. Biol. Chem. 252: In press.

7. Katz, A.M., et al. (1976) FEBS Letters 67: 207-208.

8. Endo, M. (1977) Physiol. Rev. 57: 71-108.

9. Fabiato, A. and Fabiato, A. (1977) Circulation Res. 40: 119-129.

10. Reuter, H. (1974) Circulation Res. 34: 599-605.

11. Blaustein, M.P. (1976) Federation Proc. 35: 2574-2578.

12. Jilka, R.L., et al. (1975) J. Biol. Chem. 250: 7501-7510.

13. Jilka, R.L. and Martonosi, A. (1977) Biochim. Biophys. Acta 466: 57-67.

14. Shamoo, A.E. and MacLennan, D.H. (1974) Proc. Nat. Acad. Sci. USA 71: 3522-3526.

15. Repke, D.I., et al. (1976) J. Biol. Chem. 251: 3169-3175.

16. Meissner, G. (1975) Biochim. Biophys. Acta 389: 51-68.

Proposal for a Mechanism of Ca^{2+} Transport

Efraim Racker
Section of Biochemistry, Molecular & Cell Biology, Wing Hall
Cornell University, Ithaca, New York 14853

> *Choose your hypothesis;*
> *I have chosen mine.*
>
> *from Thomas Henry Huxley*
> *"On a Piece of Chalk"*

INTRODUCTION

In recent years it has become evident that an ATP-driven proton pump plays a key role in the generation of ATP during oxidative phosphorylation in mitochondria[1]. Our studies on the mitochondrial proton-translocating ATPase revealed that it has a complex structure with water-soluble and water-insoluble components. It contains a water-soluble ATPase (F_1) with five subunits - two water-soluble coupling factors (F_6 and OSCP) which are required for the attachment of F_1 to the membrane, one water-soluble coupling factor (F_2) which is probably required to seal the pump against proton leaks and at least one or two water-insoluble proteins that serve as transmembranous proton channel[2].

In contrast to this complex structure, the purified Ca^{2+}ATPase from sarcoplasmic reticulum contains one polypeptide chain of about 100,000 molecular weight[3] and a second polypeptide which has been identified as a proteolipid[4]. This purified ATPase preparation has been incorporated into liposomes and shown to catalyze ATP-dependent Ca^{2+} transport[5]. Both sarcoplasmic reticulum vesicles[6,7] and reconstituted Ca pump vesicles[8] can be loaded with Ca^{2+}. On addition of P_i, ADP and EGTA, efflux of Ca^{2+} from the vesicles gives rise to the formation of ATP. It thus resembles the process of H$^+$-driven ATP generation in mitochondria, chloroplasts and bacteria.

The formulation of the mechanism of this process presented in this paper is based on three experimental facts. The first is that the isolated Ca^{2+} ATPase catalyzes the formation of ATP from P_i and ADP although it cannot generate a Ca^{2+} gradient[9]. The second fact is that the addition of Mg^{2+} to the enzyme results in heat release of about 40 kcal per mole of protein[10]. The third fact is that after reconstitution an ATPase preparation with a high content of proteolipid, pumps Ca^{2+} more rapidly and more efficiently than an ATPase with little proteolipid[11].

Before describing the new hypothesis I want to discuss the interpretation and significance of these three experimental observations.

The formation of phosphoenzyme and of ATP by isolated Ca^{2+} ATPase

Among the appealing features of the Ca^{2+} ATPase that distinguishes it from the proton ATPase is that chemical intermediates can readily be demonstrated. In the

155

presence of ATP and Ca^{2+} a phosphoenzyme ($E_1{\sim}P$) is formed[12,13] with an aspartyl residue of the protein as acceptor of the phosphoryl group[14,15]. In the reverse direction phosphoenzyme ($E_2{\sim}P$) is formed in the presence of P_i and Mg^{2+} by sarcoplasmic reticulum vesicles[16] and also by purified Ca^{2+} ATPase[9].

It was furthermore shown that $E_2{\sim}P$ is a high-energy intermediate capable of generating ATP by donating the phosphoryl group to ADP[9]. Similar findings were recorded with Na^+K^+ ATPase[17]. The thermodynamic dilemma posed by these observations did not emerge clearly, until it became apparent[9] that a) the ATP was not enzyme-bound but free in solution and b) the process could be repeated provided that after each cycle the enzyme was reprecipitated in the presence of EDTA (Table I).

TABLE I

ATP FORMATION BY ISOLATED Ca^{2+} ATPase

	Step I $E_2{\sim}P$ formation	Step II ATP formation
First cycle	1.71	0.93
Second cycle	2.53	1.76

Experimental conditions were as described[9]. Values are expressed in nmoles/mg protein.

There are some features of this process that warrant discussion. First it should be pointed out that it is different from ATP formation by intact sarcoplasmic reticulum vesicles during the reversal of ATP-dependent Ca^{2+} transport. In this system the first step is the loading of the vesicles by Ca^{2+} followed by the simultaneous addition of P_i, ADP and EGTA. In contrast, in the process catalyzed by the purified enzyme, the first step of $E_2{\sim}P$ formation in the presence of Mg^{2+} and P_i must be performed in the absence of Ca^{2+}, since traces of this cation inhibit the reaction. In the second step ADP must be added simultaneously with Ca^{2+} or once again $E_2{\sim}P$ is hydrolyzed. No gradient of Ca^{2+} can play a role in this sequence of events since without reconstitution the purified enzyme cannot transport Ca^{2+} and establish a gradient.

A second point of interest is a comparison of $E_1{\sim}P$ and $E_2{\sim}P$. Although both appear to be chemically aspartyl-P enzymes, they have different properties, particularly with respect to stability in the presence of Ca^{2+}, ADP, etc. Thus, it appears that conformational states of the enzyme determine the access of water or other phosphoryl acceptors to the active site. An interesting difference in the accessibility to Mg^{2+} was recently reported[18]. Many differences between $E_1{\sim}P$ and $E_2{\sim}P$ have also been recorded with the Na^+K^+ ATPase[19] which will not be reviewed here.

156

A third point that has been debated both in the case of the Na^+K^+ ATPase and the Ca^{2+} ATPase is the question whether the aspartyl phosphate enzyme is indeed a true intermediate or an artifact that occurs during protein denaturation. Indeed, many attempts in our and other laboratories to trap such an intermediate with a nucleophilic reactant without protein denaturation, have thus far been unsuccessful. It is therefore quite possible that the true intermediate is not a covalent protein derivative, but an activated phosphoenzyme complex of the type recently discussed by Jencks[20]. This possibility is of particular interest in the case of the proton ATPase where a chemical intermediate has not been demonstrated.

Whatever the nature of the true intermediate, the key point established by the experiments with the isolated Ca^{2+} ATPase is that $E_2 \sim P$ formed from P_i is capable of generating ATP in solution.

What is the source of energy that drives the net formation of ATP from P_i and ADP with stoichiometric amounts of enzyme?

Calorimetric measurements with Na^+K^+ ATPase and Ca^{2+} ATPase

We have approached this thermodynamic dilemma by measuring heat changes that occur when these enzymes interact with Mg^{2+}, P_i and other reactants[10,21]. Since these data will be presented later during this symposium, I shall only summarize the key observations. When Mg^{2+} was added either to Na^+K^+ ATPase from electric eel or to Ca^{2+} ATPase from sarcoplasmic reticulum, a large release of heat (40 to 60 kcal/mole of protein) was observed. Artifacts due to aggregation or due to interaction with phospholipids, were ruled out. Moreover the equilibrium constant for the Mg^{2+} enzyme interaction, determined by calorimetry, was very close to that determined by $E_2 \sim P$ formation (Table II). The agreement for the P_i values was not quite as good, probably due to the fact that $E_2 \sim P$ formation was measured in seconds while heat measurements required hours for heat equilibration. Similar findings were made with the Ca^{2+} ATPase and some of the discrepancies noted were attributed to the limited stability of the enzyme during heat equilibration. We are now attempting to improve these measurements with more stable enzyme preparations.

TABLE II

EQUILIBRIUM CONSTANTS FOR THE INTERACTION OF Mg^{2+} AND P_i

WITH Na^+K^+ ATPase

Reaction	Calorimetric	$E_2 \sim P$ formation
$E + Mg^{2+} \rightleftarrows E\ Mg^{2+}$	0.8 mM	1.1 mM
$E + P_i \rightleftarrows E\ P_i$	3.0 mM	0.67 mM

Experimental conditions were as described[21].

Among the most interesting findings was the observation shown in Table III. When Mg^{2+} was added to the enzyme in the presence of P_i under conditions of $E_2 \sim P$ formation, the heat changes diminished with increasing P_i concentrations. This finding is consistent with the idea that the conformational changes of the enzyme induced by Mg^{2+} can be used to drive $E_2 \sim P$ formation. In addition to calorimetry, other methods were used to demonstrate major conformational changes of the enzyme induced by Mg^{2+} [21].

TABLE III

ENTHALPHY CHANGES ASSOCIATED WITH THE INTERACTION OF Mg^{2+} WITH Na^+K^+ ATPase
IN THE PRESENCE OF VARYING AMOUNTS OF P_i

P_i (mM)	Q (kcal/mol enzyme)
0	- 42
2	- 30
5	- 15
10	- 6
15	- 3.3

Experimental conditions were as described[21].

The role of the proteolipid

Because of space limitation, I shall only briefly summarize our views on the role of the proteolipid which was discussed elsewhere[2].

The basic observation which prompted us to propose a biological role for the proteolipid was that ATPase preparations that had the same ATPase activity but a high proteolipid content pumped Ca^{2+} rapidly and efficiently (as determined by Ca^{2+}/ATP ratios), whereas ATPase preparations with low proteolipid content pumped poorly[11]. Unfortunately neither we nor other investigators have as yet succeeded in a complete separation of the proteolipid from the ATPase without loss of enzyme activity. Preparations that are low in proteolipid, when reconstituted with a heated extract of a preparation that is high in proteolipid, were stimulated 2 to 3 fold in the rate of transport and in efficiency[11].

SDS-acrylamide gel patterns of R_{3a} ATPase (low transport activity), R_{3e} ATPase (high transport activity) and of a heated extract of R_{3e}, are shown in Fig. 1, and the transport activities are summarized in Table IV. We noted that excess proteolipid added during reconstitution resulted in a decrease rather than increase in transport and efficiency. This was reminiscent of our experiments with the hydrophobic protein of the mitochondrial ATPase which also inhibited oxidative phosphorylation when added in excess[22]. Addition of the proteolipid to protein-free liposomes that were loaded with ^{45}Ca resulted in release of counts[11] suggesting that the proteolipid acts as an ionophore.

158

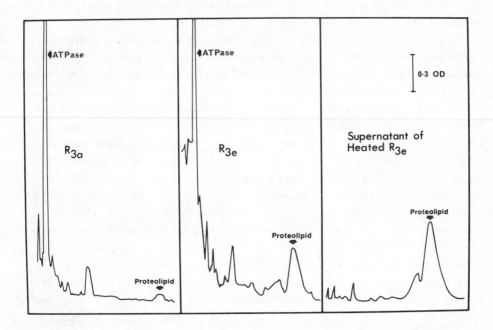

Fig. 1. Polyacrylamine gel scans of Ca^{2+} ATPase fractions and of heated proteolipid. Experimental conditions were as described[11].

TABLE IV

STIMULATION OF Ca^{2+} TRANSPORT BY HEAT-STABLE EXTRACT OF Ca^{2+} ATPase

ATPase preparations	ATPase µmoles/min/mg	Ca^{2+} uptake	ATPase (reconstituted)	Ca^{2+}/ATP ratio
		nmoles/min/mg protein		
R_{3a}	5.6	50	160	0.31
R_{3e}	5.0	826	490	1.7
R_{3a} + heated R_{3e} (4.5 µg)	–	370	510	0.72
R_{3a} + heated R_{3e} (15 µg)	–	120	1010	0.12

Experimental conditions were as described[11].

These are the experimental facts. What comes next is conjecture, mainly based on analogies with the proton pump of mitochondria. Since we know that F_1 is not intramembranous (for evidence see reference 2), a transmembranous channel for the protons must exist. This channel was isolated as a hydrophobic protein fraction which contained 2-3 protein bands[23]. One of these bands is the DCCD-reactive polypeptide which was identified as a proteolipid[24]. The role of this proteolipid as part of the proton channel is now widely accepted.

The exact position of the Ca^{2+} ATPase in the membrane is still controversial, but it appears[25] that the spheres seen in electromicrographs protruding from the membrane belong to the ATPase and may in fact contain the active site of the enzyme as in the case of F_1. Obviously, there is need for communication between this site and the inner compartment of the vesicles. Is this communication achieved by a proteolipid as in the case of the mitochondrial ATPase or by another segment of the ATPase as proposed by Shamoo and MacLennan[26]? These authors have shown that a tryptic digest of the Ca^{2+} ATPase acts as an electrogenic Ca^{2+} ionophore. This raises the interesting question whether the Ca^{2+} pump of sarcoplasmic reticulum is electrogenic? Our attempts to demonstrate electrogenicity with sensitive dyes and by the methods of Grinius et al[27] have been constantly negative. I will be the last one to make an issue based on negative experiments. Has anyone shown that the Ca^{2+} pump is electrogenic?

In any case, the hypothesis formulated below does not rest on the role of the proteolipid as a channel. If a segment of the ATPase turns out to be responsible for the delivery of Ca^{2+} to the inner compartment, so be it. Until this has been shown, I shall include the proteolipid as part of the pump - if for no other reasons - for the sake of the analogy with the proton pump.

The hypothesis

The hypothesis begins with the formulation of the events that take place with the isolated ATPase. As shown in Fig. 2 the addition of Mg^{2+} either induces a major conformational change in the protein or shifts the equilibrium of two stages of the enzyme to the right to a form which accepts the phosphate and yields $E_2 \sim P$. In the second step Ca^{2+} induces a conformational change in the protein $(E_1 \sim P)$ in which the phosphoryl group is accessible to ADP thereby permitting ATP formation. If Ca^{2+} is present without ADP, the active site is accessible to water and the phosphoryl group is discharged.

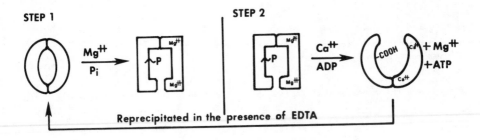

Fig. 2. Proposed mechanism of ATP formation by isolated Ca^{2+} ATPase.

The events shown in Fig. 3 are visualized to take place during ATP formation in Ca^{2+} loaded vesicles. In the presence of Mg^{2+} and P_i, phosphoenzyme is formed as in the case of the isolated enzyme. But within the membrane the access to the cation site from inside the vesicles is closed off until the formation of $E_2{\sim}P$ induces a conformational change that opens the channel. Now Ca^{2+} has access to the active site, displaces Mg^{2+} into the medium and allows formation of ATP. With the discharge of the phosphoryl group the enzyme returns to closed channel state. Now the Ca^{2+} pressure is eliminated, Mg^{2+} returns to the cation site and Ca^{2+} is pushed into the medium. This cycle of events is probably also facilitated by corresponding changes in the affinity of the enzyme to Ca^{2+} and Mg^{2+}. The key feature of this hypothesis is the cyclic attachment and detachment of Mg^{2+}, with the Ca^{2+} gradient achieving what was accomplished by reprecipitation of the enzyme in the presence of EDTA in the experiment shown in Fig. 2. The temporal sequence of this latter experiment is replaced by the closing and opening of the channel by conformational adaptation of the enzyme.

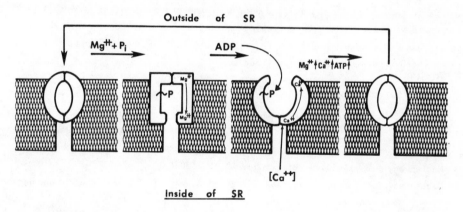

ATP FORMATION BY Ca⁺⁺ PUMP

Fig. 3. Proposed mechanism for ATP formation during reversal of Ca^{2+} transport in sarcoplasmic reticulum.

Finally, how is ATP-driven Ca^{2+} translocation achieved? With Ca^{2+} and ATP on the outside, the first step is the formation of $E_1{\sim}P$ and the placement of Ca^{2+} in a strategic position close to the channel. The conformation changes shown in Fig. 4 that accompany the formation and hydrolysis of phosphoenzyme, result in the opening and closing of the channel required for the unidirectional displacement of Ca^{2+} from one side of the membrane to the other. Electrical neutrality is achieved, most likely, by the counter movement of K^+ or of another cations.

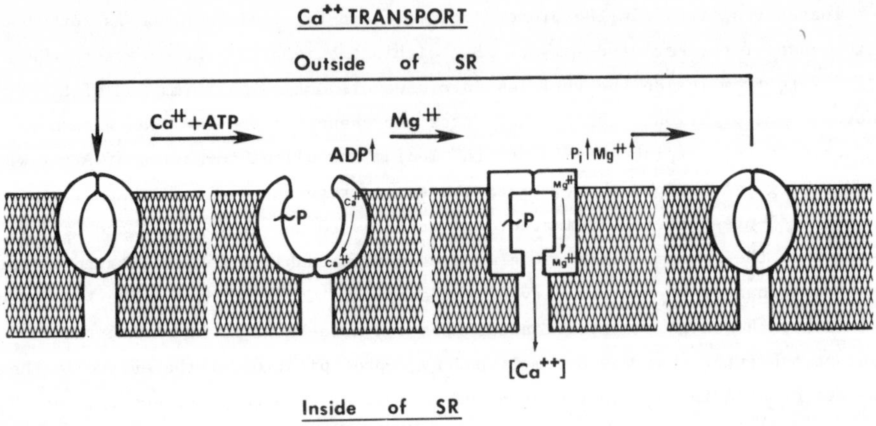

Fig. 4. Proposed mechanism for ATP-driven Ca^{2+} transport in sarcoplasmic reticulum.

I cannot conclude without pointing out that the scheme shown in Fig. 3 is easily converted into a new hypothesis of oxidative phosphorylation by substituting H^+ for Ca^{2+}. Accordingly, the major function of the proton gradient generated by the respiratory chain is the cyclic displacement of Mg^{2+} ions. One can without much strain associate such movements with attachments and detachments of adenine nucleotides which take place during the process.

REFERENCES

1. Mitchell, P. (1966) Biol. Rev. Cambridge Phil. Soc. 41, 445-502.

2. Racker, E. (1976) "A New Look at Mechanisms in Bioenergetics," Academic Press, New York.

3. MacLennan, D.H. (1970) J. Biol. Chem. 245, 4508-4518.

4. MacLennan, D.H., Yip, C.C., Iles, G.H. and Seaman, P. (1972) Cold Spring Harbor Symp., Quant. Biol. 37, 469-478.

5. Racker, E. (1972) J. Biol. Chem. 247, 8198-8220.

6. Hasselbach, W. (1972) in H.H. Weber, ed., Molecular Bioenergetics and Macromolecular Biochemistry, Springer-Verlag, pp. 149-171.

7. Panet, R. and Selinger, Z. (1972) Biochim. Biophys. Acta, 255, 34-42.

8. Knowles, A.F. and Racker, E. (1975) J. Biol. Chem. 250, 3538-3544.

9. Knowles, A.F. and Racker, E. (1975) J. Biol. Chem. 250, 1949-1951.

10. Epstein, M., Kuriki, Y., Biltonen, R. and Racker, E. (1977) Int'l Symp. on Calcium Binding Proteins, Elsevier, North-Holland, Amsterdam, in press.

11. Racker, E. and Eytan, E. (1975) J. Biol. Chem. 250, 7533-7534.

12. Yamamoto, T. and Tonomura, Y. (1967) J. Biochem. Tokyo, 62, 55a.

13. Martonosi, A. (1967) Biochem. Biophys. Res. Commun. <u>29</u>, 753-757

14. Bastide, F., Meissner, G., Fleischer, S. and Post, R.L. (1973) J. Biol. Chem. <u>248</u>, 8385-8391.

15. Degani, C. and Boyer, P.D. (1973) J. Biol. Chem. <u>248</u>, 8222-8226.

16. Masuda, H. and de Meis, L. (1973) Biochemistry, <u>12</u>, 4581-4585.

17. Taniguchi, K. and Post, R.L. (1975) J. Biol. Chem. <u>250</u>, 3010-3018.

18. Garrahan, P.J., Rega, A.F. and Alonso, G.L. (1976) Biochim. Biophys. Acta, <u>448</u>, 121-132.

19. Askari, A., ed. (1974) "Properties and Functions of $(Na^+ + K^+)$-Activated Adenosine-triphosphatase," Ann. N.Y. Acad. Sci., Vol. 242, pp. 6-2741.

20. Jencks, W.P. (1975) Adv. in Enzym. <u>43</u>, 219-410.

21. Kuriki, Y., Halsey, J., Biltonen, R. and Racker, E. (1976) Biochemistry, <u>15</u>, 4956-4961.

22. Racker, E. and Kandrach, A. (1973) J. Biol. Chem. <u>248</u>, 5841-5847.

23. Serrano, R., Kanner, B.I. and Racker, E. (1976) J. Biol. Chem. <u>251</u>, 2453-2461.

24. Cattell, K.J., Knight, I.G., Lindop, C.R. and Beechey, R.B. (1970) Biochem. J. <u>125</u>, 169-177.

25. MacLennan, D.H. and Holland, P.C. (1975) Ann. Rev. Biophys. and Bioengineering, <u>4</u>, 377-404.

26. Shamoo, A.E. and MacLennan, D.H. (1974) Proc. Nat. Acad. Sci. <u>71</u>, 3522-3526.

27. Grinius, L.L., Jasaitis, A.A., Kadziauskas, J.P., Liberman, E.A., Skulachev, V.P., Topali, V.P., Tsofina, L.M. and Vladimirova, M.A. (1970) Biochim. Biophys. Acta, <u>216</u>, 1.

Structural Studies on the Ca^{++} Transporting ATPase
of Sarcoplasmic Reticulum

N.M. Green, G. Allen, G.M. Hebdon and D.A. Thorley-Lawson

National Institute for Medical Research, London, N.W.7. 1AA, England

The Ca transporting ATPase of sarcoplasmic reticulum is a large, predominantly polar protein which performs a tightly coupled translocation of two Ca^{++} ions across the membrane for each mole of ATP hydrolysed. These characteristics suggests that its orientation is fixed and that the flow of Ca^{++} is controlled by a gated pore. In order to provide a firmer basis for speculations on mechanism we have started work on its primary structure, which should eventually enable one to determine the nature of the gate and the pore, if indeed they exist.

Topography of the ATPase

Electron microscopy of negatively stained preparations shows that at least half the molecule projects from the membrane [1,2] while freeze fractured preparations show intramembranous particles [2]. We will first consider the relation of these gross features to proteolytic fragments of several kinds.

Mild trypsin digestion cleaves the molecule into fragments A and B of molecular weight 60,000 and 55,000 respectively [2,3,4]. The larger, fragment A is cleaved more slowly to A$_1$ (molecular weight, 33,000) and A$_2$ (molecular weight 24,000). The active site aspartyl phosphate is located on A$_1$ [4,5]. Although labelling with lactoperoxidase suggested that fragment A was less buried in the membrane than was fragment B, attempts to release A$_1$ or A$_2$ selectively from the membrane failed. It was possible to separate the fragments only in the presence of sodium dodecyl sulphate [6]. It was therefore likely that each fragment had a water soluble and a lipid soluble segment. This was consistent with the absence of significant difference between their hydrophobicities as reflected by their amino acid compositions.

In order to differentiate the aqueous and membrane buried regions of the molecule we have carried out more extensive proteolysis of acid treated sarcoplasmic reticulum, from which most of the minor proteins had been removed by extraction with EDTA. The acid treatment did not disrupt the membrane and smooth vesicles could be sedimented from the digest after proteolysis even when 80% of the peptide material had been removed. The results in Table 1 show that a number of different extensive digestions left about the same proportion of peptide material in the membrane. The membrane retained a high proportion of the tryptophan as measured by ultraviolet absorption, but

Table 1

Extensive proteolysis of sarcoplasmic reticulum

Sarcoplasmic reticulum (20mg/ml) was extracted twice with EDTA (2mM in 50mM tris-HCl pH 8.4 [7]. Before digestion with pepsin (1/50) or cleavage with CNBr it was carboxymethylated with [^{14}C]-iodoacetate (5mM), washed and suspended in 0.1M HCl. Before digestion with trypsin (1/50) it was suspended in 0.1M HCl, centrifuged and resuspended in tris to adjust pH to 8.4. It was then carboxymethylated concurrently with tryptic digestion.

| | Percent remaining in membrane | | |
Proteolysis by:-	Lowry	E_{280}	CM-cysteine
Trypsin	30	61	9
CNBr and trypsin	-	60	12
Trypsin and S. aureus protease	24	53	5
Pepsin	25	43	6

only a very low proportion of the reactive cysteine residues, which had been carboxymethylated. Similar results were obtained using preparations of the ATPase that had been purified by solution in Triton X-100 and chromatography on DEAE-cellulose and reconstituted with egg lecithin [8]. With this preparation only 15% of original Lowry colour remained with the membrane after digestion with pepsin followed by trypsin. The peptides were further characterized by gel electrophoresis on 12% acrylamide gels in SDS-urea (8M), which showed two bands corresponding to molecular weights of approximately 3000 and 6000. Amino acid analysis of the mixture showed a high proportion (37%) of valine, leucine, isoleucine and tryptophan. When the vesicles were dissolved in chloroform/methanol or in acetone the peptide material remained in the organic solvent together with the lipid. The peptides could be delipidated by gel filtration in cholate. They remained aggregated and they emerged at the void volume of Sephadex G-100 even in 6M guanidinium chloride/2% sodium cholate. Further chacterization of this part of the molecule requires suitable solvent systems for separating the peptides.

The location and reactivity of the cysteine residues

Most of our structural work so far has been concentrated on the more tractable water soluble regions of the molecule and in particular upon the cysteine containing peptides, since these were readily labelled by carboxymethylation and since there was prior evidence that they were involved in the hydrolytic site [9].

ATPase, purified by the method of MacLennan [10] with the addition of dithiothreitol, contained 26 residues of cysteine or cystine per 115,000 g [10]. Only 20 of these could be labelled directly with [^{14}C]-N-ethylmaleimide

Table 2

Distribution of thiol and disulphide residues among tryptic fragments of the ATPase [11].

The thiol distribution was determined by labelling the cleaved protein with [^{14}C]-N-ethylmaleimide in SDS followed by separation by gel electrophoresis, slicing and counting. The distribution of disulphides was determined similarly after treatment with unlabelled N-ethylmaleimide, reduction and treatment with [^{14}C]-N-ethylmaleimide.

| Fragment | M.W x 10^{-3} | Cysteine residues | | | |
		reactive	buried	-S-S-	Total
ATPase	115	15	5	6	26
A	60	11	2	0	13
B	55	5	2	6	13
A$_1$	33	9	1	0	10
A$_2$	24	2	1	0	3

in SDS, but a further six residues were labelled after reduction with dithiothreitol in SDS. The distribution of cysteine residues among the major tryptic fragments is shown in Table 2. The largest concentration of reactive cysteines was on fragment A$_1$, which also carried the active site. There were also two reactive thiols on A$_2$ and five on B. Since almost all the reactive cysteines were removed by extensive proteolysis (Table 1) each of the fragments was partially accessible to proteases. Conversely each fragment carried at least one buried thiol group, which was consistent with the presence of a membrane buried segment on each. The disulphide bonds were located exclusively on fragment B.

The kinetics of the reaction of these thiols with an excess of DTNB [5,5' dithiobis-(2-nitrobenzoate)] showed that four to five of the twenty cysteines were unreactive and that the reaction of the remainder could be described by two rate constants. A large class of thirteen reacted slowly and a small class of two reacted about ten times faster, (fig. 1). The enzyme lost activity with a rate constant which was about the same as that of the slow class. The rates were considerably decreased by substrate (Fig. 1) but there was no increase in the number of unreactive thiols. It was of particular interest that all but two of the large slow class were equally protected and that the new first order rate constant was again close to the rate constant for inactivation. The protection of such a large number of thiols and their reaction as a single class implied either that there was some conformational constraint on their reactivity which was increased when ATP was bound or that there was a clustering of thiols in the region of the hydrolytic site.

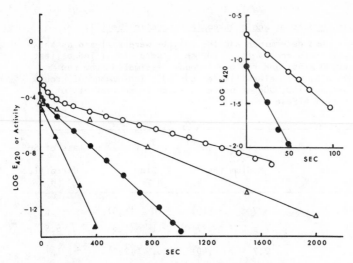

Fig. 1. Reaction of purified ATPase with DTNB in the presence and absence of substrate [11]. In the absence of substrate, (● ▲) the reaction mixtures (1.0ml) contained ATPase vesicles (0.3mg) in tris-HCl buffer (0.05M, pH 8.4) and DTNB (0.5mM). In the presence of substrate (O △) ATP (6mM) and Ca^{++} (0.02mM) were included. The reaction was followed at 420nm (25°C). ATPase activities were determined in a coupled enzyme system [12] ▲. Inset:- first order plot of the faster reacting thiol groups after substracting the contribution of the main set.

We have further evidence that both of these factors are relevant.

Evidence for dependence of reactivity on conformation

Similar results were obtained using either vesicles regenerated from purified ATPase (Fig. 1) or whole sarcoplasmic reticulum (Table 3). The main difference between the two preparations was the presence of two extra buried thiols in the latter. Although Ca^{++} and ATP together exerted a marked protective effect on the thiols, Ca^{++} by itself in low concentration had almost no effect on their reactivity. A possible corollary of this is the observation of Meissner (13) that reaction of the thiols with N-ethylmaleimide, which abolished binding of ATP, left the binding of Ca^{++} relatively unchanged.

Other conformational changes increased the reactivity of the thiol groups. Solution of the vesicles in deoxycholate or SDS (Table 3) brought about an increase in reactivity of all thiols, including the buried ones, and split the slow class into several more reactive sub-classes (see also Murphy [14]). The effects of low concentrations of deoxycholate were reversible. Irreversible inactivation of the ATPase at pH 4 also increased the reactivity (Fig.2), but did not affect the buried groups. The most interesting activation was that caused by thiol reagents themselves. Both mercurials and disulphides such as DTNB inactivated the ATPase irreversibly. Regeneration of the thiols

Table 3

Reactivity of thiol groups of whole sarcoplasmic reticulum [11]

The reaction rates were determined and the results were analysed as shown in Fig. 1. Groups with rate constants 100 sec^{-1} were classified as fast. Reaction ceased after 30 mins. and residual groups, estimated from the extent of reaction in SDS, were classified as buried. The number of groups in each class is given per 150,000g of protein, which is equivalent to approximately one mole of ATPase.

Additions to reaction	Rate constants (sec^{-1} x 10^3) for each class of SH						
	Fast		Medium fast		Slow		Unreactive
	n	k	n	k	n	k	n
0.2mM EGTA	0		3.7	(19)	10	(1.9)	5.5
0.02mM Ca^{++}	0		2.5	(27)	11	(1.8)	5.5
0.2% deoxycholate	3.5		2.7	(40)	6.5	(2.2)	1
					5.5	(5.2)	
1% deoxycholate	11.4	(>100)	4.0	(23)	3.5	(3.2)	0
1.0% sodium dodecyl sulphate	6.9	(>100)	7.9	(26)	4.2	(4.6)	0

with dithiothreitol gave an inactive product which showed a spectrum of rate constants, not unlike the acid inactivated ATPase (Fig. 2). It would be interesting to have further evidence to show whether these increases in reactivity of thiols are necessarily associated with a decreased ATPase activity, for, although detergent-solublized ATPase is still active, the activity is low unless the detergent is diluted out in the assay system [15]. The similarity of the rate constants for reaction of the slow class of thiol groups and for inactivation suggests that blocking of a critical thiol of this class is responsible for inactivation. The simplest way of rationalizing the existence of a large uniformly reacting class of thiols, all but two protected by Ca^{++}/ATP, and the irreversibility of the inactivation accompanied by increased reactivity of the thiols, is to postulate that these results are all consequences of a conformational change following the reaction of a critical thiol. However, this simple hypothesis is inadequate since the rate constant for inactivation was not identical with that for the slow class of thiols and the ratio of the rate constants varied under different conditions. In the presence of Ca^{++}/ATP at low DTNB concentration (0.2mM) it was 0.6, while in the absence of Ca^{++}/ATP at high DTNB concentration (2mM) it was 3.1.

Clustering of thiols in tertiary and in primary structure

Although it is possible that the thiols are widely scattered in the tertiary structure of the ATPase and that the conformational changes which

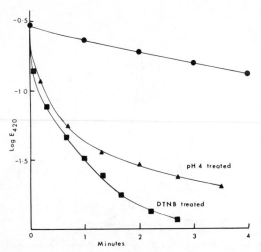

Fig. 2. Reactivity of thiol groups of sarcoplasmic reticulum before (●)
and after inactivation at pH 4 (▲) or by DTNB (■). The DTNB-inactivated
sample was reduced by dithiothreitol (5mM) and the reaction products were
removed on a column of Dowex 2 in 50mM Tris-HCl, pH 8.4. The subsequent
reaction with DTNB (1mM) was performed as described for Fig. 1.

affect their reactivity are also widespread, there is evidence for clustering
of thiols which suggests that the changes may be localized. In the first
place, when the thiols were titrated with stoicheiometric amounts of DTNB
it was observed that two moles of thionitrobenzoate were liberated for each
mole of DTNB that was added, almost up to the end point. This shows that six
or seven disulphide bonds were formed and that most of the reactive thiols
were within reach of a potential partner. Strong evidence for clustering
comes from our work on the primary structure of the ATPase. This work
commenced with an examination of tryptic peptides derived from delipidated
reduced and carboxymethylated ATPase, containing 25-27 [^{14}C]-carboxymethyl
groups per 115,000g. These peptides were first fractionated on a Sephadex
G-50 column with the interesting result shown in Fig. 3. About half the
peptide material emerged in the void volume together with about 90% of the
280nm absorbance and 20% of the radioactivity. Most of the radioactivity
was in smaller peptides which contained a few tyrosine residues and no tryptophan. Sixteen different cysteine-containing sequences were identified in
peptides purified from this region of the column. Three more sequences were
identified in chymotryptic digests of the aggregated peptides from the void
volume material. There was no evidence for any duplicated sequences so that
five or six cysteines remain to be identified, unless the molecular weight
happens to be lower than 115,000 and there are fewer than 26 residues per mole.
    Further digests have been made with thermolysin, chymotrypsin, pepsin and

169

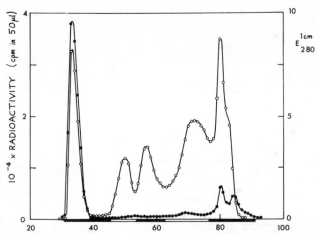

Fig. 3. Gel filtration of tryptic peptides from reduced and carboxymethylated ATPase. The ATPase was reduced and carboxymethylated in the presence of guanidinium chloride and sodium taurodeoxycholate. After dialysis and delipidation by gel filtration in SDS the protein was digested with trypsin and fractionated on a column of Sephadex G-50 (superfine) in 0.05M ammonium bicarbonate.

cyanogen bromide, but no further cysteine sequences have been identified. However, in all cases a proportion of the radioactivity remained in the void volume along with most of the 280nm absorbance and this material has not been fully analysed. The aggregated peptides from this region are hydrophobic and, in view of this and their high tryptophan content, it is likely that they include almost all the membrane buried peptides resulting from the digestion of the acid denatured intact membranes (Table 1). We have so far made little progress with sequencing this part of the molecule, but we have located nearly six hundred amino acids in ten runs of 20-120 residues from the more soluble regions.

The peptides that we have sequenced include the amino and carboxyl terminal sequences and seventy residues around the active site aspartyl phosphate (Table 4). The blocked amino terminus is derived from fragment B [5], the carboxyl terminus therefore belongs to either A_1 or A_2. We have previously located the active site on A_1 [4]. Apart from these we do not yet know which of these peptides derive from which of the major tryptic fragments of the molecule. The active site peptide contains three cysteine residues within a sequence of twenty residues around the aspartyl phosphate. Two other peptides contain five cysteines in short stretches so it would not be unreasonable to suggest that many of the cysteine residues protected by ATP are close to one another in the tertiary structure.

170

Table 4

Active site [16]	-Thr	Leu	Gly	Cys	Thr	Ser	Val	Ile	Cys	Ser
	Asp	Lys	Thr	Gly	Thr	Leu	Thr	Thr	Asn	Gln
	Val	(Cys	Met	Ser	Lys)-					
Cysteine clusters	-Ala	Thr	Ile	Cys	Ala	Leu	Cys	Asx	Asx	Ser-
	-Arg	Glu	Ala	Cys	Arg	Arg	Ala	Cys	Cys	Phe-
N.terminus	Ac	Met	Glu	Ala	Ala	His	Ser	Lys	Ser	Thr-
C-terminus	-Ile	Ala	Arg	Asn	Tyr	Leu	Glu	Gly-COOH		

Conclusions

The results described here relate to a variety of different aspects of the ATPase molecule. It is not yet possible to fit them all into a unified picture of a functioning Ca^{++} pump. However, some tentative conclusions emerge which enable one to frame more specific questions for the future.

Electron microscopy shows that both extra-membranous projections and intra-membranous particles are located in or on the cytoplasmic half of the bilayer, that is on the side to which the ribosomes were originally attached. We assume that the protein penetrates to the intra-cisternal space, since the calcium is delivered there, but there is no direct evidence for this and the fraction of the peptide chain exposed on this face is likely to be small [4]. The aminoterminal 33 residues, which belong to fragment B are fairly polar and are probably located on the cytoplasmic side. The shorter C-terminal sequence may also be on this side, since it would be the last to leave the ribosome. The 20-30% of the peptide chain that is left in the membrane after extensive proteolysis probably traverses the bilayer several times. It is likely that all three major tryptic fragments $(A_1, A_2$ and B) contribute to this region of the molecule, forming a channel for the passage of Ca^{++} and Mg^{++}. This region is quite distinct from the cysteine-rich region around the hydrolytic site, since most of the reactive cysteines can be stripped from the membrane by proteolysis (Table 1). Although many of the reactive cysteines are located on fragment A_1 there are also several on B which have similar kinetic properties, so that regions of both fragments are affected by binding of ATP.

In the context of this meeting, this blurred picture of the structure of the ATPase unfortunately resembles a performance of Hamlet without the Prince of Denmark since we have little information about the nature of the Ca^{++} binding site. We have not yet seen any sign of a sequence of amino acids resembling the binding site of troponin-like proteins, with their alternating residues of aspartic acid [17]. There is, however, a sequence containing four glutamyl residues and an aspartyl residue within a total of eight residues. In spite of the controlling effect of Ca^{++} on the hydrolysis of ATP it appears

171

that the Ca^{++} site is distinct from the hydrolytic site and is coupled to it indirectly. This is consistent with the results of Stewart et al [5] and with the suggestion that the Ca^{++}-independent, basal ATPase activity of native sarcoplasmic reticulum is a function of the ATPase protein itself [18,19]. Granted this identity, the insensitivity of the basal ATPase to blocking of thiols [9], which we have confirmed, is evidence that a thiol is not essential for the hydrolytic step per se.

Acknowledgements

We thank Mr. E.J. Toms, Mrs. J. North and Mr. B. Trinnaman for skilled technical assistance.

References

1. Hardwicke, P.M.D. and Green, N.M., 1974, Eur. J. Biochem. 42, 183-193.
2. Stewart, P.S. and MacLennan, D.H., 1974, J.B.C. 249, 985-993.
3. Migala, A., Agostini, B. and Hasselbach, W., 1973, Z. Naturforsch, 28, 178-182.
4. Thorley-Lawson, D.A. and Green, N.M., 1973, Eur. J. Biochem. 40, 403-413.
5. Stewart, P.S., MacLennan, D.H. and Shamoo, A.E., 1976, J. Biol. Chem. 251, 712-719.
6. Thorley-Lawson, D.A. and Green, N.M., 1975, Eur. J. Biochem. 59, 193-200.
7. Duggan, P.F. and Martonosi, A., 1970, J. Gen. Physiol. 56, 147-167.
8. Green, N.M. 1975, in Calcium Transport in Contraction and Secretion (E. Carafoli, F. Clementi, W. Drabikowski and A. Margreth, Eds.) p. 339-348, North Holland, Amsterdam, Oxford.
9. Hasselbach, W. and Seraydarian, K. 1966, Biochem. Zeit. 345, 159-172.
10. MacLennan, D.H. 1970, J. Biol. Chem. 245, 4508-4518.
11. Thorley-Lawson, D.A. and Green, N.M. 1977, Biochem. J. In press.
12. Neet, K.E. and Green, N.M. 1977, Arch. Biochem. Biophys. 178, 588-597.
13. Meissner, G. 1973, Biochim. Biophys. Acta, 298, 906-926.
14. Murphy, A.J. 1976, Biochemistry, 15, 4492-4496.
15. Rizzolo, L.J. le Maire, M., Reynolds, J.A. and Tanford, C. 1976, Biochemistry, 15, 3433-3437.
16. Allen, G. and Green, N.M. 1976, FEBS Lett. 63, 188-192.
17. Tufty, R.M. and Kretsinger, R.H. 1975, Science, 187, 167-169.
18. Froelich, J.P. and Taylor, E.W. 1976, J. Biol. Chem. 251, 2307-2315.
19. Inesi, G., Cohen, J.A. and Coan, C.R. 1976, Biochemistry, 15, 5293-5298.

Ca^{2+} Ionophore from Ca^{2+} + Mg^{2+} ATPase

Adil E. Shamoo and Jonathan J. Abramson
Department of Radiation Biology and Biophysics
University of Rochester School of Medicine and Dentistry
Rochester, New York 14642

INTRODUCTION

A great deal of knowledge about the sarcoplasmic reticulum (SR) and the function of proteins contained in this vesicular membrane have been presented in the last few years. By hydrolysis of ATP, the sarcoplasmic reticulum can form a 1000-3000 fold gradient of Ca^{2+}, lowering the Ca^{2+} concentration to micromolar levels (1). This depletion of Ca^{2+} leads to muscle relaxation. Hydrolysis of 1 mole of ATP results in the uptake of 2 moles of Ca^{2+} (2).

The protein responsible for the active uptake of Ca^{2+} into the SR is the main constituent of the SR (i.e., Ca^{2+} + Mg^{2+} dependent ATPase). Phosphorylation of the ATPase is Ca^{2+} dependent, and dephosphorylation is Mg^{2+} dependent (3). Dephosphorylation of the enzyme is inhibited by Ca^{2+} ions which compete with Mg^{2+} (4), and also by the absence of phospholipid (5). In addition, this pump is reversible (6). With high concentrations of Ca^{2+} inside SR vesicles and low Ca^{2+} concentrations outside, the addition of ADP to the outside results in the decrease in the amount of phosphorylated intermediate and results in the synthesis of ATP.

The main protein constituent of the sarcoplasmic reticulum, Ca^{2+} + Mg^{2+} ATPase has been purified by MacLennan (7) using small concentrations of deoxycholate followed by ammonium acetate fractionation. The ATPase molecule has a molecular weight of 102,000 and an activity ranging from 28-35 μmole/min. mg. The protein was later purified by Ikemoto (8) by solubilizing it in Triton X-100 followed by Sephadex chromatography, and by Warren (9) who treated the SR with 10% DOC and centrifuged the solubilized proteins into a sucrose gradient. Warren obtained a fraction containing the purified Ca^{2+} + Mg^{2+} ATPase with less than .3% of the added DOC present.

Racker was able to reconstitute the isolated Ca^{2+} + Mg^{2+} ATPase into vesicles and obtain ATP dependent loading (10,11). These experiments were done by a cholate dialysis technique or by sonication technique. Both experiments were done in the presence of a calcium precipitable anion, either potassium phosphate or potassium oxalate. Racker's group also shows that Ca^{2+} + Mg^{2+} ATPase activity could be reactivated after delipidating the isolated enzyme by gel filtration followed by replacing the lipids with phosphatidylcholine (12). ATP dependent loading was also reconstituted, after delipidation when phosphatidylethanolamine was added as the only lipid. Warren, *et. al.* (9) showed that after removal of 98% of the endogenous lipid and its replacement with dioleoyl lecithin, the Ca^{2+} pump

173

could be restored after the addition of excess sarcoplasmic reticulum lipids. The reconstituted system accumulates Ca^{2+} at a rate and a level comparable to the native sarcoplasmic reticulum. Warren's experiment could be done in the absence of oxalate.

These results strongly suggest that $Ca^{2+} + Mg^{2+}$ ATPase acts as both an ATPase, and as the sole carrier of Ca^{2+}. The reconstitution experiments of Racker and Warren show that the isolated $Ca^{2+} + Mg^{2+}$ ATPase molecule in the appropriate lipid environment hydrolyzes ATP and translocates Ca^{2+}. The protein contains both the site of ATP hydrolysis and the ion translocating or ionophoric site.

Using black lipid membrane (BLM) conductance as an assay for ionophoric activity, our laboratory, in collaboration with Dr. MacLennan's laboratory, showed that $Ca^{2+} + Mg^{2+}$ ATPase from sarcoplasmic reticulum is a Ca^{2+} dependent (requiring Ca^{2+}) and selective (relatively high permeability to Ca^{2+}) ionophore (13). Calsequestrin, the high affinity Ca^{2+} binding protein and the proteolipid from SR all failed to act as ionophores. Conductance increases were seen after (a) $Ca^{2+} + Mg^{2+}$ ATPase was sonicated into the lipid forming solution, (b) the ATPase was mildly succinylated to increase its water solubility and, (c) the protein underwent mild digestion by trypsin. The relative conductance, $G_D{}^{2+}/G_{Ca}{}^{2+}$, and the relative permeability $P_D{}^{2+}/P_{Ca}{}^{2+}$, where D is a divalent cation, followed the same sequence

$$Ba^{2+} > Ca^{2+} > Sr^{2+} > Mg^{2+} > Mn^{2+}.$$

We have shown that the ionophoric activity of the intact enzyme is inhibited by Zn^{2+}, Mn^{2+}, La^{3+} and Hg^{2+} (14). Ruthenium red, which inhibits ATPase activity in SR (15) inhibits ionophoric activity of the Ca^{2+} ionophore (16). Mercuric chloride inhibits ionophoric activity of the intact enzyme when assayed in a BLM, while methylmercury has no effect on the ionophoric activity. This is especially interesting when the effect of these two compounds is studied in the sarcoplasmic reticulum (14). With methylmercury, we see a congruence of the dose-response curves for the inhibition of Ca^{2+} transport and $Ca^{2+} + Mg^{2+}$ ATPase activity. In the case of mercuric chloride, the active transport of Ca^{2+} ions is more susceptible to inhibition than ATPase activity. This is consistent with the preferential inhibition of ionophoric activity by mercuric chloride and the inhibition of ATPase activity by CH_3HgCl. This data suggests that there are two sites on the intact enzyme, one site at which ATP is hydrolyzed and one which is the ionophoric site.

In order to further elucidate the molecular mechanism of Ca^{2+} transport, we have subjected the sarcoplasmic reticulum to controlled tryptic digestion and isolated the peptides, resulting from digestion by SDS column chromatography (17) and SDS-preparative gel electrophoresis (18). The 102,000 dalton $Ca^{2+} + Mg^{2+}$ ATPase is initially cleaved into a 55,000 dalton piece and a 45,000 dalton piece.

174

Further tryptic digestion leads to the cleavage of the 55,000 dalton fragment into a 30,000 dalton fragment and a 20,000 dalton fragment. The properties of these fragments are outlined in Figure 1.

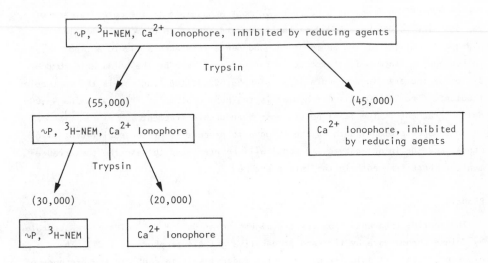

$Ca^{2+} + Mg^{2+}$ ATPase (102,000)

FIGURE 1

Schematic diagram of $Ca^{2+} + Mg^{2+}$ ATPase
tryptic fragments with their related functions

The intact $Ca^{2+} + Mg^{2+}$ ATPase, the 55,000, 45,000, and 20,000 dalton fragments all act as Ca^{2+} dependent and selective ionophores. When the active site of ATP hydrolysis is labelled with $[\gamma-^{32}P]ATP$, after tryptic digestion of the ATPase molecules, the intact enzyme, and the 55,000 and 30,000 dalton tryptic fragments are all found to be phosphorylated ($\sim$P) (17). The sulfhydryl group at the site of ATP hydrolysis, when labelled with N-ethyl $[2-^3H]$ maleimide (^{3}H-NEM) is also found in the intact enzyme and the 55,000 and 30,000 dalton fragments (17). The absence of ^{3}H-NEM binding to the 20,000 dalton fragment is expected due to the absence of cysteine in the 20,000 dalton fragment.

When sarcoplasmic reticulum is tryptically digested for various amounts of time with various doses of trypsin, and ATPase activity, and calcium uptake are measured, we make an interesting discovery (19,20). ATPase activity is constant throughout this experiment. Ca^{2+} uptake is undiminished by the cleavage of the intact enzyme into 45,000 and 55,000 dalton fragments. Uptake is, however, abolished when the 55,000 dalton fragment is cleaved into 20,000 and 30,000 dalton

175

fragments. ATP hydrolysis and calcium uptake is uncoupled by the tryptic cleavage of the 55,000 dalton fragment into two fragments. The 20,000 dalton fragment contains the ionophoric activity and the 30,000 dalton fragment contains the site of ATP hydrolysis.

The use of antibodies against the various fragments leads to a better understanding of the arrangement of these tryptic fragments in the SR membrane (17). The 55,000 dalton fragment appears to be most accessible to the outer surface of the SR. The 20,000 and 45,000 dalton fragments appear to be more buried in the hydrophobic milieu of the SR membrane, and hence less accessible to antibodies raised against them. This picture is further confirmed by the amino acid composition of the tryptic fragments (17). The 55,000 dalton fragment is the most polar fragment. The 20,000 dalton fragment is the next most polar fragment. The 45,000 dalton fragment is the least polar, most hydrophobic fragment.

This paper presents data on the ionophoric properties of these different fragments, and the intact enzyme. A model will be presented to describe the arrangement of these fragments in the intact enzyme.

RESULTS

The peptide fragments from tryptic digestion of $Ca^{2+} + Mg^{2+}$ ATPase are separated by column chromatography (17) and preparative gel electrophoresis (18). The digested protein is solubilized by sodium dodecyl sulfate (SDS) in the presence of β-mercaptoethanol and separate fractions containing the 20,000 dalton fragment, the 30,000 dalton fragment, and a fraction containing a mixture of the 45,000 and the 55,000 dalton fragments are eluted from a Bio-Gel A 1.5 m column. The 20,000 dalton fragment is then concentrated and passed through a Bio-Gel P-100 column. The 45,000 dalton fragment is separated from the 55,000 dalton fragment by preparative gel electrophoresis (23). The sample from the column is concentrated and applied to a 5% Weber and Osborn gel. Using a modified preparative gel electrophoresis setup similar to one previously described (18), we can elute off fractions containing "pure" 45,000 dalton fragment and "pure" 55,000 dalton fragment.

Attempts have been made to separate the tryptic fragments of $Ca^{2+} + Mg^{2+}$ ATPase without the use of SDS. Thorley-Lawson and Green (21) have attempted to dissociate the tryptic fragments using Triton X-100. The association between these fragments is very strong. Such reagents as Guanidine-HCl (2 M, 4 M and 6 M), NaBr (3.5 M), Urea (2 M, 4 M, 6 M and 8 M), and Propionic acid (.1 M, .5 M, and 1 M) all failed to dissociate the fragments (21). In our laboratory we have also attempted to dissociate the fragments using Urea (1 M, 3 M and 6 M) and NaSCN (1 M, 3 M and 6 M) (unpublished data). The enzyme seems to dissociate from the membrane before the fragments separate from each other. The SDS column chromatography and preparative gel electrophoresis methods described in the last

176

paragraph are the only known ways of separating the tryptic fragments from each other. In order to remove the SDS bound to the fragments, the samples are dialyzed against 8 M Urea for seven (7) days followed by water for three (3) days. It is these purified samples that we study in an artificial black lipid membrane.

Table 1 shows the permeability ratio $P_{Ca^{2+}}/P_{D^{2+}}$, where D^{2+} is the divalent cation on the side of the BLM opposite Ca^{2+}, of the tryptic fragments of Ca^{2+} + Mg^{2+} ATPase. All measurements were made across an oxidized cholesterol membrane in the presence of 5 mM $CaCl_2$, 5 mM HEPES (N-2-Hydroxyethylpiperazine-N'-2-Ethane-sulfonic Acid), pH = 7.3. Notice that the 55,000 dalton fragment and the intact enzyme show the same selectivity. The 20,000 dalton fragment is less selective among divalent cations than the intact enzyme. The 45,000 dalton fragment shows similar selectivity to the 20,000 dalton fragment, but shows less cation/anion selectivity. The intact enzyme and each of the fragments show the same ion dependence sequence.

TABLE 1

RELATIVE PERMEABILITY TO DIVALENT CATION $[(P(Ca^{2+})/P(D^{2+})]$
(assayed in an oxidized cholesterol membrane)

	Intact Ca^{2+} ATPase	45,000 dalton fragment	55,000 dalton fragment	20,000 dalton fragment
Ba^{2+}	0.55	1.03	0.50	0.69
Sr^{2+}	1.49	1.19	1.50	1.10
Mg^{2+}	1.89	1.29	1.80	1.34
Mn^{2+}	2.04	1.50	2.00	1.39
$P_{Ca^{2+}}/P_{Cl^-}$	4.30	1.76	4.30	2.30

Phosphatidylcholine:cholesterol bilayers (5 mg:mg) provide an environment more similar to the sarcoplasmic reticulum membrane (22). The 20,000 dalton fragment in a phosphatidylcholine:cholesterol membrane is more selective among divalent cations (20,24) than we report here in an oxidized cholesterol membrane. We also note that $P(Mg^{2+})$ is greater than $P(Mn^{2+})$ in an oxidized cholesterol membrane while $P(Mg^{2+})$ is less than $P(Mn^{2+})$ in a PC:cholesterol membrane. The relative permeability to Mg^{2+} has decreased in the more native lipid environment of PC:cholesterol. The concentration needed to see an increase in conductance after a given amount of time is approximately the same (5×10^{-9} M) for any of the peptides listed in

177

Table I. We have also incorporated the 20,000 dalton fragment into a phosphatidyl-choline vesicle, equilibrated it with ^{45}Ca, placed it into buffered medium without Ca^{2+}, and measured ^{45}Ca efflux. The efflux measured is much larger than we see when the 30,000 dalton fragment is incorporated into a vesicle, or when pure phospholipid vesicles are used as a control (unpublished data).

We have also studied the effect of reducing agents on the various fragments (23). The fragment is incubated at room temperature for 30 minutes at various concentrations of either dithiothreitol (DTT) or β-mercaptoethanol. Both reducing agents inhibited the ionophoric activity of the 45,000 dalton fragment. Probably because of its smaller size, β-mercaptoethanol was the more effective inhibitor of transport in the BLM. When the succinylated enzyme was incubated in β-mercaptoethanol, its ionophoric activity was also found to be inhibited. The 55,000 dalton fragment was not inhibited by equal concentrations of β-mercapto-ethanol.

CONCLUSIONS

We have shown that Ca^{2+} + Mg^{2+} ATPase, the primary pump for calcium in the sarcoplasmic reticulum acts as an ionophore which shows dependency and selectivity toward Ca^{2+}. We offer Figure 2 as a model to describe the relative orientation of the tryptic fragments of Ca^{2+} + Mg^{2+} ATPase in the SR membrane. Tryptic digestion initially cleaves a lysine or arginine at point 1, leaving the 45,000 dalton fragment and the 55,000 dalton fragment. Further digestion cleaves a similar bond at point 2. The 55,000 dalton fragment is cleaved into 20,000 and 30,000 dalton fragments. It is the cleavage of this second bond that uncouples Ca^{2+} uptake from ATP hydrolysis.

The energy for Ca^{2+} transport is provided by the hydrolysis of ATP at the 30,000 dalton fragment. Calcium is transported through, or mediated by the 20,000 dalton fragment and the 45,000 dalton fragment. The data presented here strongly suggest that the 55,000 dalton fragment and the 45,000 dalton fragment are in series. Addition of β-mercaptoethanol in large doses inhibits the ionophoric activity of the 45,000 dalton fragment and the intact enzyme, but does not affect the ionophoric activity of the 55,000 dalton fragment. This inhibition is probably due to the reduction of a disulfide bond, essential for transport, located in the 45,000 dalton fragment. Transport through the ATPase molecule requires intact function of both the 45,000 and 55,000 dalton fragments. Cleavage of the intact enzyme into the 45,000 and 55,000 dalton fragments does not disrupt transport through the ATPase. The inhibition of one of the fragments causes loss of transport activity through the whole enzyme.

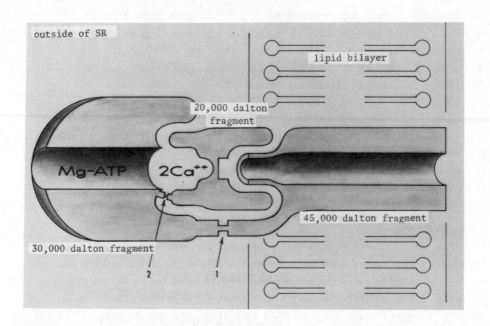

FIGURE 2

Model of Ca^{2+} + Mg^{2+} ATPase in the
Sarcoplasmic Reticulum Membrane

REFERENCES

1. Hasselbach, W. and Makinose, M. (1962) Biochem. Biophys. Res. Comm. 7: 132-136.

2. Hasselbach, W. and Makinose, M. (1963) Biochem. Zeitschrift 339:94-111.

3. Inesi, G., et al. (1970) Arch. Biochem. Biophys. 138:285-294.

4. Yamada, S. and Tonomura, Y. (1972) J. Biochem. 72:417-425.

5. Martonosi, A., et al. (1971) Arch. Biochem. Biophys. 144:529-540.

6. Kanazawa, T., et al. (1970). J. Biochem. 68:593-595.

7. MacLennan, D. H. (1970). J. Biol. Chem. 245:4508-4518.

8. Ikemoto, N., et al. (1971) Biochem. Biophys. Res. Comm. 44:1510-1517.

9. Warren, G. B., et al. (1974) Proc. Nat. Acad. Sci. USA 71:622-626.

10. Racker, E. (1972) J. Biol. Chem. 247:8198-8200.

11. Racker, E. and Eytan, E. (1973) Biochem. Biophys. Res. Comm. 55:174-178.

12. Knowles, A., et al. (1976) J. Biol. Chem. 251:5161-5165.

13. Shamoo, A. E. and MacLennan, D.H. (1974) Proc. Nat. Acad. Sci. USA 71:

3522-3526.

14. Shamoo, A. E. and MacLennan, D. H. (1975) J. Memb. Biol. 25:65-74.

15. Vale, M. G. P. and Carvalho, A. P. (1973) Biochem. Biophys. Acta 325:29-37.

16. Shamoo, A. E., et al. (1975) J. Biol. Chem. 250:8289-8291.

17. Stewart, P. S., et al. (1976) J. Biol. Chem. 251:712-719.

18. Ryan, T. E., et al. (1976) Analyt. Biochem. 72:359-365.

19. Scott, T. L. and Shamoo, A. E. (1977) Biophys. J. 17:185a.

20. Shamoo, A. E., et al. (1977) J. Supramol. Struct., in press.

21. Thorley-Lawson, D. A. and Green, N. M. (1975) Eur. J. Biochem. 59:193-200.

22. MacLennan, D. H., et al. (1971) J. Biol. Chem. 246:2702-2710.

23. Abramson, J. and Shamoo, A. E. (1977) Biophys. J. 17:185a.

24. Shamoo, A. E. (1977) Biophys. J. 17:184a.

ACKNOWLEDGEMENTS

This paper is based on work performed under contract with the U.S. Energy Research and Development Administration at the University of Rochester Biomedical and Environmental Research Project and has been assigned Report No. UR-3490-1151. The paper is also supported in part by NIH Grant 1 RO1 AM18892; Center Grant ESO-1247; Program Project Grant ES-10248 from NIEHS; the Muscular Dystrophy Assn. of America and the Upjohn Company.

Jonathan J. Abramson is a fellow of the Muscular Dystrophy Association of America.

Adil E. Shamoo is an Established Investigator of the American Heart Association.

Freeze-Fracture and Enzymatic Studies
of Ca^{2+} Transport ATPase in Adipocyte Endoplasmic Reticulum

David E. Bruns, Betty Black, Jay M. McDonald and Leonard Jarett
Division of Laboratory Medicine, Departments of Medicine and Pathology
Washington University School of Medicine and Barnes Hospital,
St. Louis, Missouri 63110

The endoplasmic reticulum appears to play an important role in the regulation and control of intracellular calcium metabolism. Previous studies of calcium regulation from our laboratory, using highly enriched mitochondrial, endoplasmic reticulum (ER) and plasma membrane fractions from the rat adipocyte, have identified an active Ca^{2+} transport system in the ER[1] which has a Km of ~ 1 μM Ca^{2+} and is modulated by insulin[2].

The properties of the Ca^{2+} transport system in ER showed marked qualitative similarities to those of sacroplasmic reticulum (SR) but the transport rates were more than an order of magnitude less than in SR. Similarly, the number of high affinity (passive) binding sites in E.R. was 2% of that of SR[3]. In skeletal muscle SR, high affinity Ca^{2+} binding can be entirely accounted for by the binding of 2 Ca^{2+} per Ca^{2+}-transport ATPase[4]. If a similar relationship exists for adipocyte ER, then the different transport rates in ER and SR could simply reflect a quantitative difference in the concentration of a similar Ca^{2+} transport ATPase in the two organelles.

The present studies are the first in a series of investigations into the mechanism of calcium transport in ER and were undertaken to quantitate (a) the calcium-stimulated MgATPase of ER and (b) the density of intramembranous particles which in SR appear to represent interruptions of the lipid bilayer by the Ca^{2+} transport proteins. The findings suggest that Ca^{2+} is transported in ER by Ca^{2+}-stimulated ATPase which can be visualized by freeze fracture electron microscopic techniques.

METHODS

Fat cells were isolated by collagenase digestion from rat epididymal fat pads, homogenized in 0.25M sucrose with 10 mM Tris/HCl, pH 7.4 at 4°C, and fractionated

as previously described[1].   ATPase activities were measured in the presence and
absence of 2.5 or 5 mM oxalate using $(\gamma^{32}P)$-ATP, 0.5 mM ATP, 1 mM $MgCl_2$, 0.1 M KCl,
pH 6.7 at $37^{O}C$, and 10-15 µg microsomal protein in a total volume of 500 µl;
200 µM EGTA or 10 µM $CaCl_2$ was included to measure "basal" or "total" ATPase activity, respectively. The Ca^{2+}-stimulated ATPase was calculated as the difference
between the "total" and "basal" activities. Ca^{2+} concentrations above 10 µM did
not result in measurably higher activities. Calcium transport in ER was measured
using the 20,000 xg supernatant fraction as previously described[1]. Incubations
were at 37° using the ATPase assay buffers containing 2.5 or 5 mM oxalate and
10 µM $CaCl_2$ with $^{45}CaCl_2$. Ca^{2+} uptake was terminated after 1 or 2 min by Millipore filtration. Intramembranous particles were visualized in ER visicles isolated as a microsomal fraction. Freeze-fracture replicas were made and particle
densities determined as described for SR by Tillack et al[5].

RESULTS AND DISCUSSION

The calcium-stimulated ATPase in 7 microsomal preparations averaged 23.6 $\pm$
3.6 (SE) nmol/mg microsomal protein/min at 37° with 10 µM $CaCl_2$. Essentially
identical results were obtained using $CaCl_2$ and EGTA in a molar ratio of 0.9 to
achieve a similar free Ca^{2+} concentration. When 5 mM oxalate was added to the
buffers, Ca^{2+} stimulated ATPase decreased to 11.5 $\pm$ 2.6 nmol/mg/min (n=6). Calcium
transport averaged 11.6 $\pm$ 1.5 nmol/mg/min (n=6) under the same conditions in the
presence of oxalate. These values suggest a stoichiometry of 1 Ca^{2+} transported
per ATP hydrolyzed. This ratio should be considered as a lower limit since only
net calcium accumulation was measured under these assay conditions[1]. Both calcium
transport and the calcium stimulation of ATP-ase activity were markedly inhibited
by 50 µM p-chloromercuribenzene sulfonate.

Freeze-fracture electron microscopy revealed 75$\overset{o}{A}$ particles (Plate 1) which were
more numerous on the concave than convex fracture faces. The total particle density was 80-390/um^2. These values are near the particle density of 300/um^2 predicted from the number of high affinity binding sites in ER (0.28 nmol/mg protein),
assuming the binding of 2 Ca^{2+} per ATPase and estimating the surface area of the
vesicles to be similar to that of isolated SR vesicles (2.9 x 10^{14} um^2/g protein[6]).

182

All of the measured properties of the Ca^{2+} transport system in ER are approximately 1-10% of the values for adult rabbit skeletal muscle SR. However, the ER system appears quantitatively similar to that in chick embryo skeletal SR as judged by CaATPase activities, Ca^{2+} transport rates, and particle densities[5]. These data are consistent with the concept that the physiologic functions of the adipocyte do not require as rapid a Ca^{2+} transport system as adult skeletal muscle. The calcium transport system in rat adipocyte ER has the capability to alter cytosol calcium concentrations under physiological conditions. Modulation of this system would lead to profound effects on subcellular calcium distribution and alter the activities of intracellular calcium-sensitive enzymes. Further investigation is needed into the mechanisms of calcium transport in ER and the modulation of the system by hormones and other agents.

This work was supported by USPHS research grants AM11892 and RR05389 and a grant from the Juvenile Diabetes Foundation.

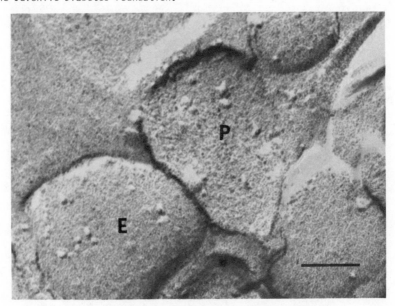

Plate 1. Freeze fracture replicas of endoplasmic reticulum vesicles.
Lines indicate 0.1 μ. Concave (P) and convex (E) fracture faces
contain 75Å intramembranous particles.

REFERENCES

1. Bruns, D.E., McDonald, J.M., and Jarett, L. (1976) J. Biol. Chem. 251: 7191-7197.

2. McDonald, J.M., Bruns, D.E., and Jarett, L. (1976) Endocrinology 90 (Suppl), 90 (abstr).

3. Bruns, D.E., McDonald, J.M., and Jarett, L. (1977) J. Biol. Chem. 252: 927-932.

4. MacLennan, D.H., and Holland, P.C. (1975) Annu. Rev. Biophys. Bioeng. 5:377-404.

5. Tillack, T.W., Boland, R., and Martonosi, A. (1974) J. Biol. Chem. 249:624-633.

6. Scales, D. and Inesi, G. (1976) Biophys. J. 16:735-751.

Identification of Two Intrinsic Proteins Uniquely Associated with the Terminal Cisternae of the Sarcoplasmic Reticulum

Kevin P. Campbell and Adil E. Shamoo

Department of Radiation Biology and Biophysics
University of Rochester School of Medicine and Dentistry
Rochester, New York 14642

Introduction

The membranes directly involved in excitation-contraction coupling in skeletal muscle are the transverse tubular membrane and the junctional sarcoplasmic reticulum membrane[1]. It is generally accepted that depolarization at the transverse tubular membrane initiates the release of calcium from the terminal cisternae of the sarcoplasmic reticulum[2]. There have been recent studies toward the understanding of the electrophysiological properties of the membranes involved in excitation-contraction coupling[3]. However, little is known about the membrane proteins or their role in excitation-contraction coupling. In this study, we have shown that sarcoplasmic reticulum vesicles derived from the terminal cisternae contain two intrinsic membrane proteins which are probably unique to the terminal cisternae.

Materials and Methods

A combination of differential and isopycnic zonal ultracentrifugation was used to isolate light and heavy sarcoplasmic reticulum vesicles (LSR, HSR). LSR and HSR were obtained from the 30-32.5% (w/w) and 39-42.5% (w/w) region of the sucrose gradient, respectively. Our method is similar to that of Meissner[4] and Caswell[5]. Deoxycholate (DOC) treatment (5-10 mg protein/ml, 0.1 mg DOC/mg protein, 1 M NaCl and 1 mM DTT) was done essentially according to the method of MacLennan[6]. Triton treatment was carried out using the method of Ikemoto[7]. Electron microscopy was performed by Dr. Franzini-Armstrong[8]. Gel electrophoresis profiles were obtained by the method of Swank and Munkres[9] (SDS-Urea PAGE) and Laemmli[10] (SDS-PAGE). The procedure of Fairbanks[11] was used for glycoprotein staining. SR calcium oxalate loading and sucrose gradient separation was carried out according to the method of Levitsky[12]. Cytochrome (a+a3) was assayed using the dual wavelength method of Levitsky[12].

Results

In thin section electron microscopy LSR appear as empty vesicles whereas the HSR appear as vesicles filled with electron dense material, similar to that seen in the terminal cisternae of the sarcoplasmic reticulum. A further description of LSR and HSR will appear in a paper in preparation[8].

SDS-PAGE shows that the LSR (Fig. 1) contains essentially one protein having a MW of approx. 102,000 ($Ca^{2+}+Mg^{2+}$-ATPase). Five main protein bands are consistently observed on SDS-PAGE of HSR (Fig. 1) having MWs of approx. 102,000 ($Ca^{2+}+Mg^{2+}$-ATPase), 64,000 (calsequestrin), 55,000 (high affinity calcium binding protein), 34,000 and 30,000. SDS-Urea PAGE gives qualitatively the same protein composition except that the SR proteolipid is visible in gels of both LSR and HSR. The results of staining the gels of HSR with Schiff's reagent indicate that the two intrinsic proteins have little or no carbohydrate.

Deoxycholate, at low ratios of detergent to protein, has been used by Mac-Lennan[6] to selectively solubilize the extrinsic membrane proteins from SR vesicles. DOC treatment of the LSR results in the further purification of $Ca^{2+}+Mg^{2+}$-ATPase. DOC treatment of HSR results first, in the extraction of calsequestrin and high affinity calcium binding protein into the soluble fraction; and second, in the enrichment of the 30,000 and 34,000 dalton peptides in the insoluble fraction

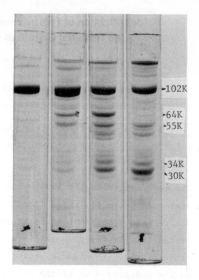

1 2 3 4

Figure 1. Laemmli SDS-PAGE of light (1), intermediate (2), heavy SR vesicles (3) and of the insoluble fraction of DOC-treated HSR (4) showing the enrichment of the 30,000 and 34,000 dalton proteins.

(Fig. 1). Triton treatment of the HSR also results in the enrichment of these two proteins in the insoluble fraction. Using additional DOC followed by ammonium acetate fractionation yields two fractions from HSR both have the ATPase peptide but one is enriched in the 30,000 dalton peptide and the other in the 34,000 dalton peptide.

Mitochondrial contamination as measured by cytochrome ($a+a_3$) heme content is less than 1% in LSR and less than 3% in HSR. Calcium oxalate-loaded HSR, separated from mitochondrial contamination, has been shown to contain the 30,000 and 34,000 dalton proteins.

Discussion

Isopycnic zonal ultracentrifugation has been used to isolate light and heavy sarcoplasmic reticulum vesicles from rabbit skeletal muscle. LSR and HSR differ in their protein composition as seen by SDS-PAGE and morphology as seen by thin section electron microscopy. The biochemical and morphological data indicate that most of the LSR and HSR are derived from the longitudinal and terminal regions of the SR, respectively. DOC treatment of LSR and HSR reveals that the HSR contains two intrinsic proteins which are unique to the HSR.

Since the two intrinsic proteins account for a small percentage of the HSR protein, experiments were conducted to rule out the possibility that they might arise from contamination. The level of cytochrome $(a+a_3)$ indicates that the level of mitochondrial contamination in the HSR is quite low. In order to further rule out mitochondrial contamination, we actively loaded HSR with calcium oxalate in the presence of azide to inhibit mitochondrial calcium accumulation. We were then able to separate the loaded HSR from mitochondrial fragments using a step sucrose gradient. The isolated calcium oxalate loaded HSR was then shown to contain both intrinsic proteins.

Since our preparation does contain some transverse tubular contamination, mostly attached to the HSR in the form of dyads, we have not ruled out that one or both of these proteins are located in the T-tubule membrane and/or the junction between the T-system and the terminal cisternae.

We are presently carrying out further experiments on the localization of these proteins and the relationship of these proteins to the mechanism of calcium release.

Acknowledgements

This paper is based on work supported by US-ERDA contract and supported in part by NIH, MDA and Upjohn Company, Report # UR-3490-1108. Adil E. Shamoo is an Established Investigator of the American Heart Association.

References

1. Franzini-Armstrong, C. (1975) Fed. Proc. 34: 1382-1389.
2. Ebashi, S. and Endo, M. (1968) Progr. Biophys. Mol. Biol. 18: 123-183.
3. Fuchs, F. (1974) Annual Rev. Physiol. 36: 461-502.
4. Meissner, G. (1975) Biochim. Biophys. Acta 389: 51-68.
5. Caswell, A. H., et al. (1976) Arch. Biochem. Biophys. 176: 417-430.
6. MacLennan, D. H. (1970) J. Biol. Chem. 245: 4508-4518.
7. Ikemoto, N. (1975) J. Biol. Chem. 250: 7219.
8. Campbell, K. P., Franzini-Armstrong, C. and Shamoo, A. E. (manuscript in preparation).
9. Swank, R. T. and Munkres, K. D. (1971) Anal. Biochem. 39: 462-477.
10. Laemmli, U.K. (1970) Nature 227: 680-685.
11. Fairbanks, G., et al. (1971) Biochem. 10: 2606-2617.
12. Levitsky, D. O., et al. (1976) Biochem. Biophys. Acta 443: 468-484.

Characteristics and Ionophoric Activity
of Cardiac Sarcoplasmic Reticulum Calcium ATPase

M. Chiesi, G.F. Prestipino, E. Wanke, and
E. Carafoli, Laboratory of Biochemistry,
Swiss Federal Institute of Technology (ETH)
Universitätstrasse 16, 8092 Zurich, Switzer-
land, and Laboratory of Cybernetics and
Biophysics of CNR, Camogli, Italy.

The function, and the protein composition, of sarcoplasmic reticulum
(SR) membranes may vary with the type of muscle. Therefore, we have
carried out a comparative study of some characteristics of microsomes
isolated from dog skeletal muscle and from dog heart muscle. The SR
preparation of skeletal muscle has been obtained essentially by the
method of Suko et al. (1976) and that of heart muscle by the zonal
centrifugation procedure of Affolter et al. (1976). The protein composition of the two preparations has been examined with the aid of
SDS-polyacrylamide gel electrophoresis.

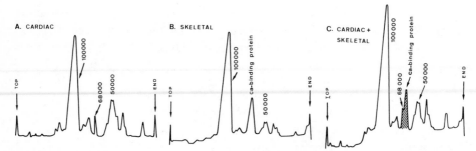

Figure 1. A comparison between the protein components of dog heart
and skeletal muscle microsomal membranes. SDS polyacrylamide gels,
containing 8 % and 4 % acrylamide in the separating and stacking gel
respectively, were prepared and run accordingly to Laemmli (1970)
with the small modifications described by Affolter et al. (1976). The
figures represent the densitometry tracings of Coomassie Blue-Stained
gels.

Figure 1 shows that the main membrane component (the ATPase) of both
kind of vesicles has a molecular weight of approximately 100'000, but
the number of protein components present in heart SR is higher than
in skeletal muscle. The so-called "calsequestrin", which is normally present
in skeletal muscle SR, however, is absent in heart SR.

The Ca^{2+}-related enzymatic activities of the two preparations are summarized in Table 1. The most striking difference between heart and skeletal muscle SR is the reduced energy-dependent binding of Ca^{2+} and the reduced rate of active uptake in the former. The reduced binding capacity in heart SR could possibly be due to the absence of the calsequestrin.

	ATPase activity (μmoles P_i/mg.min.)		Ca^{2+}-binding (-oxalate)	Ca^{2+}-uptake (+ oxalate)	
	$- Ca^{2+}$	$+ Ca^{2+}$	(nmoles/ mg/.5min.)	(nmoles/ mg/min.)	(nmoles/ mg/20min.)
Dog cardiac vesicles	0.7	1.4	25-30	200-250	2000-3000
Dog skeletal vesicles	0.1	1.5	100-120	700-1000	3000-5000

Table 1. A comparison between the ATPase activity and Ca^{2+} uptake capacity of dog heart and skeletal muscle SR. Ca^{2+}-binding (-oxalate), Ca^{2+} uptake (+ oxalate) and ATPase activity have been determined respectively with the millipore filtration technique and by the P_i liberation technique as described by Affolter et al. (1976).

The heart ATPase has been extracted in soluble form and its ionophoric characteristics have been studied. Several extraction methods have been tested and it has been found that the gradual solubilization with alkylating agents has been the most successful. The vesicles are normally incubated at 10° C in a medium containing 0.5 M sucrose and 0.02 M Tris-Cl, pH 8.0, at a concentration of 2 mg protein/ ml. Citraconic acid anhydride is added slowly (reagent to protein ratio = 5 : 1 in weight) while the pH is maintained at 7.8 - 8.0 with 0.2 M NaOH. At the end of the reaction (normally about 15 minutes) the medium is centrifuged at 160'000 g/25 minutes. The pellet is carefully resuspended with the aid of a Pasteur pipette, taking care to remove only its loosely packed portion

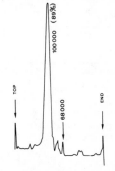

Figure 2. Densitometry tracing of SDS-polyacrylamide gel electrophoresis of the alkylated cardiac ATPase preparation.

and washed once in the sucrose-Tris medium. After a second centrifugation, the ATPase can be fully solubilized by a second treatment
with citraconic anhydride (3 : 1 in weight). The SDS polyacrylamide
gel electrophoresis (Figure 2) shows that the ATPase had been enriched by this procedure to about 88 - 90 % with respect to the starting vesicular material. It is noteworthy that the treatment of skeletal muscle SR with the same reagent under milder conditions (2 : 1 in
weight) is sufficient to remove completely the Ca^{2+}-binding protein
and to "enrich" the ATPase to more than 95 % (not shown). This may be
due to differences in chemical properties between the skeletal muscle
Ca^{2+}-binding protein and the cardiac 68'000 MW protein, which seems
to be very hydrophobic, and is still present as a mayor contaminant
of the cardiac ATPase preparation obtained by alkylation (see Figure
2).

The alkylated cardiac ATPase, which has now lost the ATPase activity
as is the case for the succinylated skeletal muscle preparation of
Shamoo et al. (1974) , has been incorporated in planar lipid bilayers
(black films). The conductance measurements were performed essentially as indicated by Shamoo et al. (1974). As can be seen from Table 2

	mg/ml amount of protein required to induce a 10x increase in conductance	cationic-anionic selectivity $P_{Ca^{2+}}/P_{Cl^-}$	cationic/cationic selectivity $P_{Ca^{2+}}/PMe$					
			Me=Ba²⁺	Ca²⁺	Sr²⁺	Mg²⁺	Na⁺	K⁺
cardiac ATPase	2.10^{-4}	3-4	0.78	1.0	1.9	1.9	0.9	0.9
skeletal ATPase	2.10^{-5}	4-5	0.6	1.0	1.5	1.9	1.1	1.3

Table 2. Ionophoric characteristics of cardiac SR ATPase. Electrical
conductance measurements and ionic permeability ratios were determined as described by Shamoo et al. (1974). The cardiac ATPase preparation was used for ionic permeability determination at a final concentration of 5.10^{-4} mg/ml.

the ionophoric properties at the two ATPases are very similar. The
electrical conductance increase induced by the alkylated cardiac
ATPase was detectable only at Ca^{2+} concentrations higher than 2 - 3
mM and when oxidized cholesterol membranes were used, and was strongly

inhibited by Zn^{2+} and Na^+. In symmetric conditions, monovalent cations were completely uneffective (Ba^{2+}, Ca^{2+}, Sr^{2+}>Mg^{2+}>> K^+, Na^+, Zn^{2+}=O). Noise fluctuation experiments were also carried out using an experimental set up similar to that described by Wanke et al. (1974). Both heart and skeletal muscle ATPase proteins had a similar behaviour. The cation and lipid requirement were exactly the same as those described for the conductance measurement. However, no specific kinetic behaviour of the two ATPase "channels" could be observed, even when purified, and enzymatically active, tryptic fragments of skeletal muscle SR ATPase were used, or when the medium was supplemented with $MgCl_2$ and ATP, in addition to $CaCl_2$. The power spectra of the current fluctuations invariably gave a typical $1/f$ curve. An explanation for this may be either a loss of the hypothetical gating mechanism, or a very high leakage rate when the protein is incorporated in the membrane (it must be stressed that oxidized cholesterol membranes are very rigid).

The qualitative similarity between the two SR ATPases isolated from the two types of muscle has already been observed [cf. for example, Suko et al. (1976)] with respect to MW and characteristics of phosphoenzyme intermediate formation. It can now be further extended to their ionophoric behaviour. The functional differences between microsomal preparations of (fast) skeletal and heart muscle may be related to the presence of additional regulatory proteins and/or to the absence of the Ca^{2+}-binding protein in the latter.

Acknowledgement
Parts of the original research described have been carried out with the financial assistance of the Swiss Nationalfonds (Grants No. 3.597.0-73 and 3.597.075).

References
Affolter, H., Chiesi, M., Dabrowska, R., and Carafoli, E. (1976) Eur. J. Biochem. 67, 389-396.
Laemmli, U.K. (1970) Nature 229, 680-685.
Shamoo, A.E., and MacLennan, D.H. (1974) Proc. Nat. Acad. Sci. USA 71 3522-3526.
Suko, J., and Hasselbach, W. (1976) Eur. J. Biochem. 64, 123-130.
Wanke, E., DeFelice, L.J., and Conti, F. (1974) Pflügers Arch. 347, 63.

Calorimetric Studies of the Interaction
of Ionic Ligands with Ca^{2+} ATPase from Sarcoplasmic Reticulum

M. Epstein, Y. Kuriki, E. Racker
Cornell University
Ithaca, New York

and

R. Biltonen
University of Virginia
Charlottesville, Virginia

It was shown previously (Kuriki et.al., Biochemistry $\underline{15}$: 4956, 1976) that the interaction of (Na$^+$K$^+$)ATPase with Mg^{2+} or P$_i$ results in large heat changes (>40 kcal/mole of enzyme). We observed similar heat changes with the (Ca^{2+}) ATPase from SR. The dependency on ion concentrations of this heat release was similar to that observed in kinetic measurements of phosphoenzyme formation. Based on these studies we propose a scheme in which an equilibrium between two enzyme conformations is shifted by interaction with Mg^{2+} yielding a low enthalpy form of the enzyme which becomes phosphorylated.

If P$_i$ is added first, the apparent heat of Mg^{2+} binding decreases without a significant change in the observed affinity of the enzyme for Mg^{2+}. The same phenomenon is observed with P$_i$ if Mg^{2+} is added first. However, the binding of Ca^{2+} and of the substrate analog AMPPNP are independent of each other and their heat effects are additive. Our findings allow us to estimate the heat changes associated with the hydrolysis of the covalent phosphoenzyme intermediate.

We propose that the conformational changes induced by ligand binding drive the formation of phosphoenzyme and eventually of ATP in the isolated system described by Knowles and Racker (J. Biol. Chem. $\underline{250}$, 1949, 1975). The conformations of the enzyme with their different affinities for the ionic ligands seem to play a role in ion transport.

Calcium Movements between Myofibrils, Parvalbumins and Sarcoplasmic Reticulum in Muscle

J.M. Gillis and Ch. Gerday

Laboratoire de Physiologie Générale, Université de Louvain
B-1200 Bruxelles and Laboratoire de Biochimie musculaire,
Université de Liège, B-4000 Liège, Belgium

Parvalbumins (PA) are soluble, low molecular weight proteins which form a large fraction (up to 25 %) of the muscle myogen in lower vertebrates (1). Though less abundant, they have also been found in muscles from mammals (2). PA have two binding sites for Ca, the affinity of which is similar to that of troponin C (K_d = 0.2 - 0.4 x 10^{-6}M) (3). Though the physiological role of PA is still unclear, the latter characteristic indicates a possible involvement in the Ca movements in muscle. To check this hypothesis, we have investigated the following points.

I. THE MAGNITUDE AND THE KINETICS OF THE Ca TRANSFER FROM PA TO THE SARCOPLASMIC RETICULUM (SR)

a. Native PA (containing 2Ca/PA) were mixed with fragmented SR (both from frog) in a medium containing MgATP and oxalate (2 mM). The relative concentrations of PA and SR was adjusted from 0.5 to 2.2 (weight/weight) to cover the *in vivo* situation where PA concentration is around 3-5 g/kg wet weight of muscle, and reticular proteins around 5-10 g/kg. The extent of removal was studied after separation of PA and SR by high speed centrifugation. PA were quantitatively recovered in the supernatant and analysed for Ca. After an incubation of 10 min, at 20°C, 95 to 100 % of the Ca originaly bound to the PA had been accumulated by the SR. The effect is independent of Mg^{2+} (from 10 µM to 1.0 mM) and is not affected by pH variations ranging from 6.9 to 7.4. This results show that the Ca bound to PA is exchangeable, therefore, in a resting muscle, PA are probably in a Ca-free form; moreover they show that like the native PA (2), Ca-free PA do not bind to the SR either.

b. The kinetics of this transfer has been studied using ^{45}Ca labelled PA. Native PA were first decalcified as above, then recalcified with an excess of $CaCl_2$ containing ^{45}Ca (0.7 C/mole). PA were then concentrated and washed by ultrafiltration on a molecular membrane (AMICON UM2) until the Ca content was reduced to 2-2.5 Ca/PA. Labelled PA were mixed with the SR in the same medium as above, and separated after given intervals by rapid filtration on Millipore filters (0.22 µm, the filters were precoated with serum albumin to saturate protein binding sites : in this case PA retention was usually less than 1 %, and that of SR around 90 %). The kinetics of the Ca accumulation by the SR is illustrated at the Fig. 1, for a case where the PA/SR ratio was 1.1.

For the sake of comparison, we also studied the rate of accumulation of free Ca^{2+} in solution. (In both experiments, the total Ca concentration was nearly the same : 32 μM for Ca-PA and 37 μM for Ca^{2+}). The results of Fig. 1 are expressed

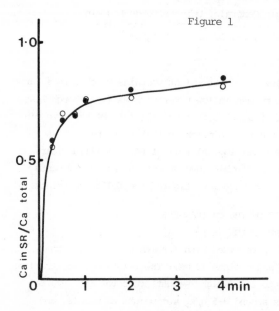

Figure 1

as the percentage of the total Ca present at the start of the experiment, which has been accumulated after a given time of incubation. Two facts are clearly illustrated (i) the kinetics of Ca removal from PA is as fast as from a mere $CaCl_2$ solution. This indicates that the rate of dissociation of the Ca-PA complex is not the limiting factor (at least after 15 sec); (ii) PA are already half decalcified after 15 sec (the shorter time investigated) indicating that the first of the two bound Ca is very rapidly exchangeable.

II. EFFECT OF PARVALBUMINS ON THE ATPase ACTIVITY OF MYOFIBRILS

Besides their elevated and specific affinity for Ca, parvalbumins and troponin C have also structural analogies (4) suggesting that they might be functionnally interchangeable. This hypothesis has been ruled out : PA neither confer Ca sensitivity (5) nor bind to myofibrils (6). Our demonstration that Ca-free PA could be obtained by a cellular system (SR) opens another possibility : that PA affect the Ca^{2+} level of the medium and this way controls contraction.

We have checked this idea by studying the effect of Ca-free PA and the ATPase activity of washed frog myofibrils (estimated by the H^+ production). A typical experiment is illustrated at the Fig. 2. The enzymic activity is started by addition of ATP (in the presence of 7 μM Ca^{2+}); then Ca-free PA are added in increasing quantities to get a complete inhibition of the ATPase activity. This relaxing effect of Ca-free PA is due to their binding of Ca : (i) the amount of PA needed corresponds to the amount of Ca in the medium (the amount actually needed was slightly higher than the one expected from a 2Ca/PA ratio, but in the presence of millimolar concentrations of Mg^{2+}, a competition of Mg vs Ca is to be expected); (ii) the activity can be fully recovered by addition of an excess of Ca; (iii) the inhibition is the same as that produced by the Ca-chelating agent EGTA (5 mM);

194

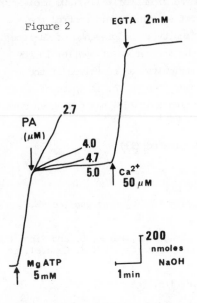

Figure 2

(iv) native PA (2Ca/PA) have no effect (not shown).

Effect of modified PA. We have taken advantage of this effect of Ca-free-PA on the ATPase activity of myofibrils to study if specific antibodies of PA could prevent relaxation by affecting the Ca binding properties (done in collaboration with Dr. Cécile Gosselin-Rey). In a first type of experiments, purified anti-PA were mixed with Ca-activated myofibrils (they produced no effect by themselves). Then Ca-free PA were added. We found that the amount of PA needed to inhibit the myofi-brillar ATPase activity was the same as in the absence of antibodies. In a second type of experiments, Ca-free PA and anti-PA were first mixed; then the complex was added to relaxed myofibrils; thirdly, the amount of Ca needed to activate the ATP-ase was determined. In control experiments, Ca-free PA alone were added, or no PA at all. We found that the amount of Ca needed was, as expected, larger in the presence of Ca-free PA, but the anti-PA had no effect. Thus the Ca binding proper-ties of parvalbumins are not modified by reaction with specific antibodies.

The ability of Ca-free PA to control contraction has also been demonstrated in more intact systems (7) : Ca-free PA applied to skinned fibres prevent caffein induced contractions, probably by trapping the Ca released from the SR by the drug.

CONCLUSIONS : Our results and others show that there is an affinity gradient for Ca of the type : myofibrils $\rightarrow$ parvalbumins $\rightarrow$ sarcoplasmic reticulum. PA can then play the role of a "SHUTTLE MECHANISM" for Ca between myofibrils and the SR.

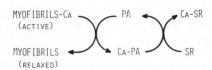

In this context it is noteworthy that Ca accumulation by the SR is not fast enough to account for the rate of relaxation of a living muscle (8, 9) and that it is proposed that a fast binding step preceeds the transport into the SR (9). Parvalbumins might play that role : the amount of

PA is large enough to fix all the Ca to be removed from myofibrils (200 nmoles/g) at least in fast muscles of lower vertebrates; if so, they could control the kinetics of relaxation. This view raises a difficulty : the presence of Ca-free PA in a resting muscle would strongly prevent the activation of myofibrils, by trapping the Ca released by excitation. This difficulty can be solved if the kinetics of the Ca binding to troponin C is faster than to PA, in spite of similar binding constants. Indirect evidence supporting this idea has just been obtained by Pechère et al. (10).

Some of these results have already been communicated (11).

REFERENCES

(1) Gosselin-Rey, C. (1974) in *Symposium on Calcium binding Proteins*, ed. W. Drabikowski et al., Amsterdam: Elsevier.

(2) Pechère, J.F., Demaille, J., Capony, J.P. Dutruge, E., Baron, C. and Pina, C. (1975) in *Calcium transport in secretion and contraction*, ed. E. Carafoli et al., Amsterdam: Elsevier.

(3) Benzonana, G., Capony, J.P. and Pechère, J.F. (1972) *Biochim. biophys. Acta*, 278, 40.

(4) Collins, J.H. (1976) *Symp. Soc. exp. Biol.* XXX, 303.

(5) Demaille, J., Dutruge, E., Eisenberg, E., Capony, J.P. and Pechère, J.F. (1974) *FEBS letters* 42, 173.

(6) Hitchcock, S.E. and Kendrick-Jones, J. (1975) in *Calcium transport in secretion and contraction*, ed.: E. Carafoli et al., Amsterdam: Elsevier.

(7) Fisher, E.M., Becker, J.V., Blum, H.E., Byers, B.,Heizmann, C., Kerrick, G.W., Lehky, P., Malencik, D.A. and Pocinwong, S. (1976) in *Molecular basis of Motility*, ed. L.M.G. Heilmeyer et al., Berlin: Springer Verlag.

(8) Ebashi, S. (1976) *Ann. Rev. Physiol.* 38, 293.

(9) Hasselbach, W. (1976) in *Molecular basis of Motility*, ed. L.M.G. Heilmeyer et al., Berlin: Springer Verlag.

(10)Pechère, J.F., Derancourt, J. and Maiech, J. (1977), *FEBS letters* 75, 111.

(11)Gerday, C. and Gillis, J.M. (1976) *J. Physiol.* 258, 96P.

We acknowledge the excellent assistance of Mrs Claire Vercruysse-Meganck.

Sarcoplasmic Calcium Pump and Phosphatase Activity

Madoka Makinose

Max-Planck-Institut für medizinische Forschung

Abteilung Physiologie, Heidelberg

B.R.Deutschland

Phloridzin, an inhibitor of Na-dependent glucose transport[1] and of Na-K-transport ATPase[2], affects the activity of the sarcoplasmic calcium pump. As shown in Fig. 1, the rate of calcium uptake and that of calcium activated ATP splitting by the isolated membrane of the sarcoplasmic reticulum (SR) are lowered to 50% of the control value by 0.1 to 0.2 mM Phloridzin. This inhibition is reversible, seems to be unspecific (non competitive) and independent of the calcium precipitating agent present in the reaction mixture. Phloridzin inhibits also the activity of p-nitro phenyl phosphate (pNPP) splitting of the SR membrane (Fig. 2) in contrast to the phosphatase activity of the Na-K-transport enzyme[3]. Activity of ATP-ADP exchange reaction (Fig. 3) and the formation of the phosphoprotein during the active accumulation of calcium are unaffected by Phloridzin. These results indicate that Phloridzin inhibits the reaction step of dephosphorylation of the activated intermediate formed during the active calcium accumulation. This type

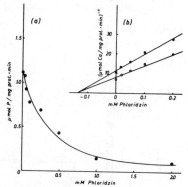

Fig. 1. Inhibition of the SR calcium pump by Phloridzin.

2PEP+PK, 0.1Ca. a) 5ATP,7Mg,5 oxalate. b) 5 or 0.5ATP, 5Mg and 5 PO$_4$. Ordinates: a) Rate of ATP splitting. b) Reciprocal of the rate of calcium transport.

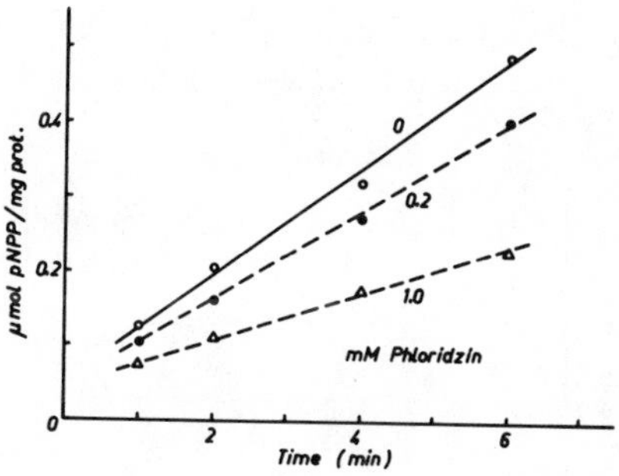

Fig. 2. Inhibition of the SR pNPPase by Phloridzin.

5 pNPP, 5 Mg, 1 Ca-EGTA and 20% DMF.

The experiments presented were carried out at 20°C, μ = 0.1 and pH 7 (20mM histidine). Essential components in each assay were given in the legends (numbers represent the concentration of the substance in mM).

197

of inhibition has been observed with prenylamine or chlorpromazine on the SR calcium pump.

Dimethyl formamide (DMF), an aprotonic solvent, modifies the SR calcium pump in another manner. The rate of the transport ATP splitting, that of the calcium transport and of the ATP-ADP exchange are reduced lower than 50% in the presence of 20% (v/v) DMF. Dimethyl sulfoxide, another aprotonic solvent, shows some similar effects[4]. DMF does not affect the niveau of the phosphoprotein formed during the active calcium accumulation. Obviously, in the presence of DMF, the rate of dephosphorylation of the phosphoprotein is inhibited and, additionally, the affinity of ADP to the enzyme as phosphate acceptor is reduced. Contrary to the Na-K-transport system[5], DMF activates strongly the phosphatase activity of the SR membrane together with the coupled transport activity for the calcium ions (Figg. 4 and 5). Since the activation of pNPPase activity by DMF is observed with phospholipase A treated leaky SR membrane (NNV), this activation is not a result of the shift of ionic activities inside and outside of the SR vesicles. Assuming that the same calcium transport mechanism of the SR membrane is activated by pNPP or acetyl phosphate (AcP) as by natural nucleoside triphosphates, the activation of phosphatase and the coupled calcium uptake should be the result of enhancement of a reaction step other than the step of the dephosphorylation; perhaps the binding of pNPP or AcP to the transport enzyme is enhanced.

In the presence of DMF, the rate of the calcium accumulation in an AcP containing mixture is further activated by addition of ADP in relatively low concentrations (Fig. 5). Since this ADP activation can be observed in the presence of glucose and hexokinase (HK), mobilization of occult acetate kinase activity by DMF is hardly to be assumed. Under usual conditions, ADP prevents the active

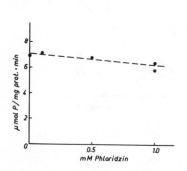

Fig. 3. Effect of Phloridzin on the rate of ATP-ADP exchange.

5 ATP, 2 ADP, 7 Mg, 5 PO$_4$ and 1 Ca-EGTA.

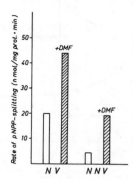

Fig. 4. Activation of the SR phosphatase activity by DMF addition.

5 pNPP, 5 Mg and 0.1 Ca.
NV : Native membrane. NNV : Membrane treated with phospholipase A.
+DMF : Dimethyl formamide (20%, v/v).

198

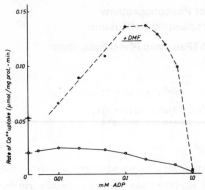

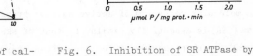

Fig. 5. Activation of the rate of calcium uptake by ADP in the acetyl phosphate containing assay.

4 AcP, 7 Mg, 5 PO_4, 0.1 Ca and 100 glucose + HK.

Fig. 6. Inhibition of SR ATPase by Isoptin.

5 ATP, 7 Mg, 10 PO_4 and 0.2 Ca.

calcium accumulation as a potent phosphate acceptor removing energy rich bound phosphate from the activated intermediate. The activation effect of ADP shown in Fig. 6 may, therefore, reflect some unknown properties of ADP (or ATP) in the function of the SR calcium pump. ADP may sustain the conformation of the transport enzyme favorable for the active calcium accumulation. IDP cannot substitute the effect of ADP. In the mixtures containing pNPP or ITP, ADP shows no activating effect of this kind.

Taking into account many structural and functional common features between the calcium and Na-K transport systems, the opposite effects of Phloridzin and DMF on the phosphatase activity of both systems comes to one's attention. Several facts indicate that the lipid components of the system play here essential roles. For example, as shown in Fig. 6, 1mM Isoptin inhibits the ATPase activity of the native SR vesicles (NV) about 30% and that of NNV nearly completely. On the activity of delipidated and reactivated (by addition of oleate) SR ATPase, Isoptin shows no inhibition. Thus, the effectiveness of modifying agents can be predominated by the lipid components associating to the transport enzyme molecule.

REFERENCES

1. Goldner, A.M., Schultz, S.G. and Curran, P.F. (1969) J.Gen.Physiol. 53, 362.
2. Wheeler, K.P. and Whittam, R. (1964) Biochem.J. 93, 349.
3. Robinson, J.D. (1969) Mol.Pharmacol. 5, 584.
4. The, R. and Hasselbach, W. (1977) Eur.J.Biochem. 74, 611.
5. Kaniike, K., Erdmann, E. and Schoner,W. (1974) Biochim.Biophys.Acta 352, 275.

Significance of Two Classes of Phosphoproteins in the Function of Cardiac Sarcoplasmic Reticulum: Phosphorylation of Ca^{2+}-Dependent ATPase and Phospholamban*

Michihiko Tada, Fumio Ohmori, Naokazu Kinoshita, and Hiroshi Abe

The First Department of Medicine, Osaka University
Medical School, Fukushima-Ku, Osaka 553, Japan

INTRODUCTION

The molecular mechanism of active calcium transport by cardiac sarcoplasmic reticulum (CSR) is essentially similar to that of skeletal muscle sarcoplasmic reticulum (SSR), in that the 100,000-dalton ATPase enzyme (E) within the membrane is assumed to serve as an energy transducer as well as a translocator of Ca^{2+} (1). The amount of acyl phosphoprotein intermediate (EP) formed during the ATPase reaction of CSR was about one fourth that of SSR ($2,3$), in accord with the findings that the content of E in CSR was lower ($4,5$). CSR was also shown to possess a control mechanism that is not seen in SSR (fast-contracting muscle). Thus, Tada et $al.$ ($6-8$) and Kirchberger et $al.$ ($9-11$) indicated that the rates of both calcium uptake and Ca^{2+}-dependent ATPase of CSR could be regulated by phosphorylation and dephosphorylation of a protein of 22,000 daltons in CSR, catalyzed by adenosine 3':5'-monophosphate (cAMP)-dependent protein kinase (A kinase) and protein phosphatase, respectively. This phosphoprotein possessed stability characteristics of a phosphoester in which the phosphate was incorporated largely into serine (9). Based on these observations, Tada and co-workers proposed a possible mechanism in which the 22,000-dalton protein, referred to as $phospholamban$ (7), serves as a modulator of Ca^{2+}-dependent ATPase of CSR ($7,12,13$). The present paper reports further observation on the chemical and enzymatic properties of the two classes of phosphoproteins and their interactions.

EXPERIMENTAL PROCEDURES

CSR was prepared as described previously (9), and the ATPase enzyme of this membrane was partially purified by treatment with deoxycholate (0.1 mg/mg CSR) by the method of MacLennan (14). CSR proteins were fractionated by gel filtration on a column (2.5 X 90 cm) of Ultrogel AcA 34 (LKB-Produktor) in 0.5% sodium dodecyl sulfate (SDS), 1% β-mercaptoethanol, 3 mM NaN$_3$, and 10 mM sodium phosphate buffer (pH 7.0). Purity of fractions was determined by SDS-polyacrylamide gel electrophoresis. Microsomal phosphorylation catalyzed by A kinase was determined as described previously ($7,9$). For the determination of EP, CSR (0.1 mg/ml) was incubated with

*This work was supported by research grants from the Ministry of Education, Science and Culture, Japan, and from the Japan Heart Foundation, and by a Grant-In-Aid from the Muscular Dystrophy Association of America.

10 μM [γ-^{32}P]ATP (0.1 mCi/μmole) in 2 mM MgCl$_2$, 100 mM KCl, 40 mM histidine–HCl (pH 6.8), various concentrations of Ca^{2+} (Ca-EGTA buffer), and 5 mM NaN$_3$ at 0° or 15°.

RESULTS AND DISCUSSION

When CSR was kept on ice after preparation, the amount of the 22,000–dalton phosphoprotein formed by incubation with A kinase remained unchanged, whereas the amount of *EP* decreased precipitously, becoming less than half within 24 hours after preparation. The decrease could be prevented to some extent by freezing the sample, but not by lyophilization. The 100,000–dalton ATPase of CSR accounted for 25 to 30% of total protein when estimated by measuring the density profile of stained proteins after SDS-polyacrylamide gel electrophoresis. Attempts at puri- fying the ATPase of CSR with full enzyme activity through deoxycholate-treatment was not always successful, mainly because this enzyme was much less stable than that of SSR. Column chromatography on Ultrogel in SDS gave purification of the 100,000–dalton protein of CSR to near homogeneity. Although the identification of this protein as the ATPase through the activity measurement was not possible, it was quite likely that it represented the ATPase, since the immuno-precipitation test with anti-ATPase (SSR) antibodies was positive, and its amino acid composition was quite similar to that of SSR. Thus, polar amino acids accounted for a rather large fraction (about 45%) of the total amino acids, and glutamic and aspartic acids accounted for about 22% of the total, outnumbering other polar amino acids. Contents of histidine and cysteine were low. The 22,000–dalton protein phospho- lamban accounted for about 3% of the total protein of CSR. Gel filtration on Ultrogel in SDS did not clearly separate this protein from neighboring components. Purification by preparative SDS-polyacrylamide gel electrophoresis is being at- tempted. Treatment of CSR with deoxycholate, a procedure known to extract extrin- sic proteins of SR, did not significantly lower the amount of phospholamban phos- phorylation. When the treatment with deoxycholate was carried out after CSR was phosphorylated by A kinase and [γ-^{32}P]ATP, virtually all of ^{32}P label remained associated with the 22,000–dalton protein of the pellet after centrifugation. CSR phosphorylated by A kinase and unlabeled ATP was passed through a column of Dowex 1 X 8, and was subjected to assay for the formation of *EP* at various concentra- tions of Ca^{2+}. The steady-state levels of *EP* of CSR previously phosphorylated by A kinase were not significantly different from those of the control CSR (Table I).

The present findings indicated that CSR exhibited two types of phosphorylation, one occurring at 100,000–dalton ATPase (acyl phosphoprotein) and the other at the

Table I	Ca^{2+}	Control CSR	Phosphorylated CSR
Effects of Phospholamban Phosphorylation on the	*M*	*μmoles EP / g of CSR*	
Steady-state Levels of EP	5 X 10^{-8}	0.038	0.040
in CSR at 0°.	1 X 10^{-6}	0.311	0.308

22,000-dalton protein (phosphoester phosphoprotein). Formation of the latter phosphoprotein did not result in appreciable alteration in the steady-state levels of the former phosphoprotein. Chemical and enzymatic analyses showed that the ATP-ase enzyme is almost analogous to that of SSR, in accord with earlier observations (2-5). Comparison between the 22,000-dalton protein and other possible substrates for A kinase such as troponin I and histone should be made when purification of this protein is completed. Table II summarizes several properties of the two proteins of CSR, that were obtained from the previous and present studies. In order to substantiate the previous proposal that the 22,000-dalton protein serves as a modulator of Ca^{2+}-dependent ATPase, it remains to be seen whether any of the elementary steps of the ATPase reaction (1) is altered by the phosphorylation of the 22,000-dalton protein. A report on further attempts at clarifying this contention should be published elsewhere (15).

Table II

Comparison of the Two Phosphoproteins Formed in the Cardiac Sarcoplasmic Reticulum

	Phospholamban	ATPase
1) Molecular Weight	22,000	100,000
2) P-amino acid	Serine	Aspartic acid
3) Nature of phosphoprotein	Phosphoester (stable in hot acid and in hydroxylamine)	Acyl phosphate (labile in hydroxylamine)
4) Amount of phosphorylation	1.0—1.4 nmoles/mg	~1.2 nmoles/mg
5) Dependency of phosphorylation on Ca^{2+}	Independent of Ca^{2+}	Dependent on Ca^{2+}
6) Sensitivity to trypsin	Labile to trypsin	Relatively resistant to trypsin
7) Physiological function	Substrate of cyclic AMP-dependent protein kinase Modulator of Ca^{2+}-transportt ATPase	Transports Ca^{2+}, which is directly coupled to the contraction-relaxation cycle

Acknowledgment--
We thank Dr.M Ooe for performing the amino acid analyses, and Drs.Yuji Tonomura, and T. Yamamoto for the critical discussion during the course of this study.

REFERENCES

1. Tada, M., Yamamoto, T., and Tonomura, Y. (1978) *Physiol. Rev.* in press
2. Pang, D.C., and Briggs, F.N. (1973) *Biochemistry* 12, 4905-4911
3. Shigekawa, M., Finegan, J.M., and Katz, A.M. (1976) *J. Biol. Chem.* 251, 6894
4. Suko, J., and Hasselbach, W. (1976) *Eur. J. Biochem.* 64, 123-130
5. Affolter, H., Chiesi, M., Dabrowska, R., and Carafoli, E. (1976) *Eur. J Biochem.* 67, 389-396
6. Tada, M., Kirchberger, M.A., Repke, D.I., and Katz, A.M. (1974) *J. Biol. Chem.* 249, 6174-6180
7. Tada, M., Kirchberger, M.A., and Katz, A.M. (1975) *J. Biol. Chem.* 250, 2640-2647
8. Tada, M., Kirchberger, M.A., and Li, H.-C. (1975) *J. Cycl. Nuc. Res.* 1, 329-338
9. Kirchberger, M.A., Tada, M., and Katz, A.M. (1974) *J. Bio. Chem.* 249, 6166-6173
10. Kirchberger, M.A., and Chu, G. (1976) *Biochim. Biophys. Acta* 419, 559-562
11. Kirchberger, M.A., and Raffo, A. (1977) *J. Cycl. Nuc. Res.* 3, 45-53
12. Katz, A.M., Tada, M., and Kirchberger, M.A. (1975) *Advan. Cycl. Nuc. Res.* 5, 453-472
13. Tada, M., and Kirchberger, M.A. (1975) *Acta Cardiologica* 30, 231-237
14. MacLennan, D.H. (1970) *J. Biol. Chem.* 245, 4508-4518
15. Tada, M., Ohmori, F., Kinoshita, N., and Abe, H. *Advan. Cycl. Nuc. Res.* 9, in press

Proteins Associated with Purified Sarcoplasmic Reticulum Vesicles in Ovine Nutritional Muscular Dystrophy

Martha J. Tripp, Philip D. Whanger, Paul H. Weswig and John A. Schmitz
Oregon State University
Corvallis, Oregon 97331

INTRODUCTION

Some skeletal muscles exhibit more calcification than others during onset of selenium and vitamin E responsive muscular dystrophy in lambs. Lambs developing dystrophy do so within three to nine weeks after birth. In moderate cases, the lambs can recover without treatment, but the semitendinosus and biceps femoris muscles may be calcified while semimembranosus muscle is damaged little, if at all. In severe cases, all skeletal muscles are calcified and the lambs die. Heart muscle may also be involved[1,2]. In severe dystrophies, Ca-uptake activity of semitendinosus muscle is reduced 10-fold, but total adenosine triphosphatase (ATPase) activity is only slightly reduced[3].

METHODS AND MATERIALS

Sarcoplasmic reticulum (SR) vesicles were prepared at 0-4°C by homogenizing semitendinosus, semimembranosus or biceps femoris muscle tissues in 10% (w/w) sucrose containing 0.05M phosphate, pH 6.3, and 10^{-5}M phenylmethylsulfonylfluoride (PMSF), centrifuging at 10,000 xg (max) for 20 min, filtration of the supernatant through glass wool to remove coagulated fats and again centrifuging the filtrate at 160,000 xg (max) for 90 min. This pellet was rinsed and suspended in the homogenizing buffer, layered on top of linear sucrose gradients (typically 25-45% with a 50% cushion) and centrifuged at 85,000 xg (average) for 12 to 15 hours. All sucrose solutions contained 0.05M phosphate, pH 6.3, and 10^{-5}M PMSF. Sucrose concentrations were confirmed on a Bausch and Lomb "ABBE-3L" refractometer. Ca^{++}-uptake assays were performed by the millipore filter technique[4] in media containing oxalate; adenosine triphosphatase (ATPase) activity was measured through analysis of adenosine diphosphate (ADP)[5] produced during hydrolysis of adenosine triphosphate (ATP). Ca^{++}-ATPase was determined by difference of ATPase activities in media containing Ca^{++} or free of Ca^{++}. The most active fractions from the first sucrose gradient were applied to second linear sucrose gradients. Fractions rebanded at the same density and were then assayed by sodium dodecylsulfate polyacrylamide gel electrophoresis (SDS PAGE)[6,7].

RESULTS

Each muscle had a characteristic sedimentation profile on the gradient and a distinct distribution of Ca^{++}-uptake and ATPase activities within the profile.

The authors wish to thank Mr. Richard J. Cushman for his technical assistance. This work was supported by N.I.H. Grant No. NS07413 and P.H.S. Fellowship No. 5 F32AM05441-02 to the senior author.

Figure 1 illustrates these features for semitendinosus muscle from a normal lamb. The vesicles typically localized in two 280 nm absorption peaks which centered at 27-30% and 34-38% sucrose. Ca^{++}-uptake and ATPase activities were demonstrable in both peaks. Semimembranosus muscle profiles (data not shown) had 280 nm absorption peaks centering at 27%, 31%, 34% and 37% sucrose. Ca^{++}-uptake and ATPase activities localized in two of these peaks, at 31% and 37% sucrose. Biceps femoris (data not shown) had 3 280 nm absorption peaks centering at 27%, 35% and 40% sucrose. Only the peak sedimenting at 40% sucrose had Ca^{++}-uptake and ATPase activities. Vesicles isolated from semimembranosus and biceps femoris lost Ca^{++}-uptake activity rapidly although total ATPase activity remained constant. Research is continuing to find conditions which will stabilize these vesicles.

With onset of dystrophy, both sedimentation profiles and enzymatic activities were changed. Figure 2 describes data from the semitendinosus muscle of a moderately dystrophic lamb which had been showing clinical signs for 17 days. The sedimentation profile showed a slight increase in amount of the more dense vesicles. However, the Ca^{++}-uptake vesicles were localized only in the less dense region of the gradient. Ca^{++}-uptake activity was about 1/4 of normal (3) at 66 nmoles $Ca^{++} \cdot min^{-1} \cdot mg^{-1}$ and Ca-ATPase was also low 18 nmoles $ADP \cdot min^{-1} \cdot mg^{-1}$) but there was a large quantity of Ca-independent ATPase activity (47 nmoles $ADP \cdot min^{-1} \cdot mg^{-1}$). Following treatment with vitamin E (a single 750 I.U. i.m. injection per 11 kg lamb 10 days earlier) there was an increased quantity of less dense vesicles (at 30% sucrose) in the sedimentation profile. Ca^{++}-uptake activity (112 nmoles $Ca^{++} \cdot min^{-1} \cdot mg^{-1}$) had not yet recovered to normal, Ca^{++}-ATPase was also low at 54 nmoles $ADP \cdot min^{-1} \cdot mg^{-1}$ and quantities of Ca^{++}-independent ATPase were still present (31 nmoles $ADP \cdot min^{-1} \cdot mg^{-1}$) (Figure 3). All Ca^{++}-uptake was confined to the less dense vesicles. Treatment with selenium (a single 1 mg Se i.m. injection as Na_2SeO_3 per 11 kg lamb 10 days earlier) resulted in nearly complete Ca^{++} activity recovery (Ca^{++}-uptake activity = 182 nmoles $Ca^{++} \cdot min^{-1} \cdot mg^{-1}$, Ca^{++}-ATPase activity = 40 nmoles $ADP \cdot min^{-1} \cdot mg^{-1}$ and resulted in an appearance of the more dense vesicles having Ca^{++}-uptake activity (Figure 4). Ca^{++}-independent ATPase was also 40 nmoles $ADP \cdot min^{-1} \cdot mg^{-1}$.

As might be expected, the number of protein bands (SDS PAGE) associated with a particular fraction varies with its position in the gradient and degree of apparent purification. There are 8 protein bands associated with the more dense Ca^{++}-uptake vesicles of semitendinosus muscle and 16 bands with the less dense vesicles.

CONCLUSIONS

1. Under the isolation conditions listed, each muscle will produce SR vesicles having densities characteristic of the muscle.

2. Ca^{++}-uptake vesicles are confined to discrete density regions characteristic of the muscle of origin and the dystrophic status of the animal.

3. With onset of dystrophy, only the less dense Ca^{++}-uptake vesicles appear in the sedimentation profile.

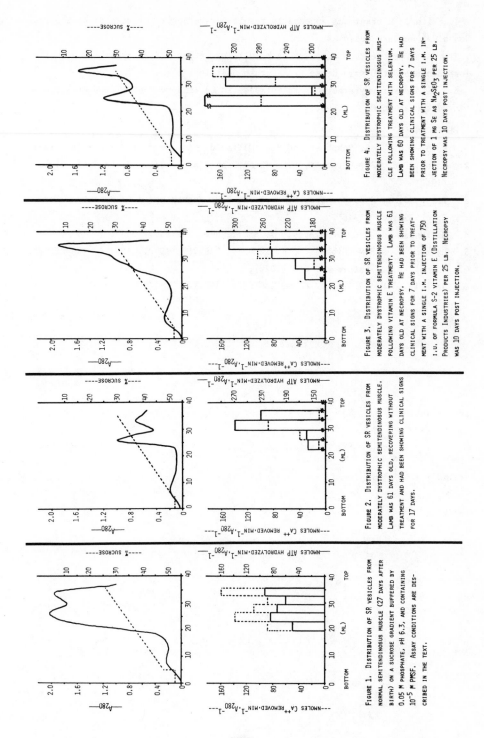

FIGURE 1. DISTRIBUTION OF SR VESICLES FROM NORMAL SEMITENDINOSUS MUSCLE (27 DAYS AFTER BIRTH) ON A SUCROSE GRADIENT BUFFERED BY 0.05 M PHOSPHATE, PH 6.3, AND CONTAINING 10⁻⁵ M PMSF. ASSAY CONDITIONS ARE DESCRIBED IN THE TEXT.

FIGURE 2. DISTRIBUTION OF SR VESICLES FROM MODERATELY DYSTROPHIC SEMITENDINOSUS MUSCLE. LAMB WAS 61 DAYS OLD, RECOVERING WITHOUT TREATMENT AND HAD BEEN SHOWING CLINICAL SIGNS FOR 17 DAYS.

FIGURE 3. DISTRIBUTION OF SR VESICLES FROM MODERATELY DYSTROPHIC SEMITENDINOSUS MUSCLE FOLLOWING VITAMIN E TREATMENT. LAMB WAS 61 DAYS OLD AT NECROPSY. HE HAD BEEN SHOWING CLINICAL SIGNS FOR 7 DAYS PRIOR TO TREATMENT WITH A SINGLE I.M. INJECTION OF 750 I.U. OF FORMULA S-2 VITAMIN E (DISTILLATION PRODUCTS INDUSTRIES) PER 25 LB. NECROPSY WAS 10 DAYS POST INJECTION.

FIGURE 4. DISTRIBUTION OF SR VESICLES FROM MODERATELY DYSTROPHIC SEMITENDINOSUS MUSCLE FOLLOWING TREATMENT WITH SELENIUM. LAMB WAS 60 DAYS OLD AT NECROPSY. HE HAD BEEN SHOWING CLINICAL SIGNS FOR 7 DAYS PRIOR TO TREATMENT WITH A SINGLE I.M. INJECTION OF 1 MG SE AS Na₂SeO₃ PER 25 LB. NECROPSY WAS 10 DAYS POST INJECTION.

4. Vitamin E treatment appears to result in synthesis (or repair) of the less dense Ca^{++}-uptake vesicles.

5. Selenium treatment appears to result in synthesis (or repair) of both the less dense and the more dense Ca^{++}-uptake vesicles.

6. With dystrophy, there is an increase in Ca^{++}-independent ATPase with concomitant decrease in Ca^{++}-ATPase.

REFERENCES

1. Muth, O.H., J.R. Schubert and J.E. Oldfield. Am. J. Vet. Res. 22:466 (1961).

2. Hogue, D.E., J.F. Proctor, R.G. Warner and J.K. Loosli. J. An. Sci. 21:25 (1962)

3. Tripp, J.M., P.D. Whanger, R.S. Black, P.H. Weswig and J.A. Schmitz. Fed. Proc. 35:1664 (1976).

4. Martonosi, A. and R. Feretos. J. Biol. Chem. 239:648 (1964).

5. Latzko, E. and M. Gibbs. Meth. in Enzy. 24B:267 (1972). Ed. A. San Pietro.

6. Laemmi, U.K. Nature 227:680 (1970).

7. Neville, D.M. J. Biol. Chem. 246:6328 (1971).

Changes in Localization of Trinitrophenyl Peptides of the Ca^{2+}-Dependent ATPase of Sarcoplasmic Reticulum at Its Different Enzymic States

Taibo Yamamoto

Dept. of Biology, Faculty of Science, Osaka University

Toyonaka, Osaka 560, JAPAN

INTRODUCTION

In the preceding study ([1]), we modified exposed lysine residues of the Ca^{2+}-dependent ATPase of sarcoplasmic reticulum (SR) with an impermeant reagent, trinitrobenzenesulfonate (TBS) without loss in its activity. The number of modified residues reached the saturated level after about 10 min reaction at pH 8.0 and 0°. The maximum number varied with varing enzymatic states of the ATPase. When an enzymatic state was changed to another one during the modification, the final number of modified residues was larger than those obtained by the reaction of SR with TBS in either one of the two states. The results could be readily explained by assuming that the ATPase molecule rotates within the membrane during the ATP hydrolysis ([1]). To understand more detailed mechanism of functional movement of the Ca^{2+}-dependent ATPase, we examined changes in the distribution of exposed lysine residues among the tryptic subfragments of the SRATPase with change in its enzymatic state.

EXPERIMENTAL

SR vesicles were incubated with 0.5 mM TBS at pH 8.0 and 0° for 15 min in presence of 0.1 M Tris-HCl and 50 mM KCl plus:

2 mM EGTA and 5 mM MgCl$_2$ to obtain the enzymatic state "^{Mg}E",

2 mM EGTA, 5 mM MgCl$_2$, and 5 mM ATP to obtain the enzymatic state "$^{Mg}E_{ATP}$",

or 10 mM CaCl$_2$ and 5 mM ATP to obtain the state of the phosphorylated intermediate "$E_{\sim p}$".

After terminating the reaction by adding 0.5 M Tris-maleate at pH 6.0, the trinitrophenylated (TNP) SR was washed exhaustively by centrifugation. TNP-SR thus obtained were treated with trypsin (SR protein : trypsin = 20 : 1) at pH 7.0 and 32 - 35° for 25 min in the presence of 0.5 M sucrose. The suspension was centrifuged at 150,000 x g for 60 min. The pellet of the SR vesicles was suspended, and washed 2 times with "Buffer A" of Stewart et al. ([2]). The washed TNP-SR was then treated with DOC to remove extrinsic proteins, as described by MacLennan ([3]). The trypsinized TNP-SR was dissolved in 1 % SDS, and applied to tops of 20 columns of the 10 % polyacrylamide gel. Electrophoresis was performed in a buffer of 0.1 % SDS and 50 mM NaP$_i$ at pH 7.0. One of the gel columns was stained with Coomassie blue to estimate mobilities of protein bands, and other gel columns were sliced into 3 mm fractions at the corresponding positions of protein bands to those in the stained gel. After eluting the protein from the pool of the gel slices with 1 % SDS, the concentrations of TNP and protein were determined, respectively, by

207

measuring the absorbance at 340 nm and by the method of Lowry et al.

RESULTS

The electrophoretic profile of the tryptic digest of SR protein consisted of three major bands of subfragments I, II, and III, which migrated with mobilities corresponding to molecular weights of about 50,000, 32,000, and 22,000. The mobility of a minor component between bands I and II corresponded to that of calsequestrin with molecular weight of 44,000 ($\underline{3}$). These results are in agreement with those reported by other workers ($\underline{2},\underline{4}$).

Table I is representative for the measurements of the amount of TNP incorporated into the three tryptic subfragments of the Ca^{2+}-dependent ATPase of SR which had been reacted with TBS at the different enzymatic states of ^{Mg}E, $^{Mg}E_{ATP}$, and $E_{\sim p}$. The sum of amounts of TNP incorporated into the three subfragments of the ATPase were about 2.3, 0.8 - 0.9, and 4 moles per mole at enzymatic state ^{Mg}E, $^{Mg}E_{ATP}$, and $E_{\sim p}$, respectively. These values were almost similar to those reported previously ($\underline{1}$). The amount of TNP incorporated into subfragment I, which is assumed to represent a hydrophobic core of the enzyme ($\underline{2}$) decreased from 1.1 - 1.4

Table I. Incorporation of TNP to tryptic subfragments of SR ATPase.

State	Exp.	Subfragment	Protein (mg/ml)	TNP (nmole/ml)	TNP/subfragment (nmole/mg)	TNP/subfragment (mole/mole)	TNP/ATPase (mole/mole)
^{Mg}E	1	I	0.075	2.1	28	1.40	
		II	0.052	0.6	16	0.50	
		III	0.033	0.5	16	0.35	2.23
	2	I	0.13	2.9	22.3	1.12	
		II	0.10	1.9	19.0	0.50	
		III	0.07	2.1	30.0	0.66	2.28
$^{Mg}E_{ATP}$	1	I	0.394	4.0	10	0.50	
		II	0.270	2.5	9	0.28	
		III	0.175	0.0	0	0.0	0.78
	2	I	0.25	2.3	9.2	0.46	
		II	0.13	1.4	10.7	0.34	
		III	0.10	0.6	6.0	0.13	0.93
$E_{\sim p}$	1	I	0.096	2.6	27	1.35	
		II	0.044	1.9	43	1.35	
		III	0.064	3.9	61	1.34	4.04
	2	I	0.27	7.4	27	1.35	
		II	0.25	10.5	42	1.32	
		III	0.14	8.4	60	1.32	3.99

to 0.5 mole per mole with change in the enzymatic state from ^{Mg}E to $^{Mg}E_{ATP}$ and increased again to 1.4 mole per mole at the enzymatic state E_p. The amount of TNP incorporated into subfragment II, which contains phosphorylation site ($\underline{2},\underline{4}$), unaltered by changing the state ^{Mg}E to $^{Mg}E_{ATP}$, while increased from 0.5 to 1.3 - 1.4 mole per mole with changing ^{Mg}E to $E_{\sim p}$. When SR was reacted with TBS at the enzymatic state ^{Mg}E, 0.4 - 0.7 mole of TNP were incorporated into subfragment III, which has a Ca^{2+}-selective ionophore activity ($\underline{2}$). The amount of bound TNP decreased to nearly zero with changing ^{Mg}E into $^{Mg}E_{ATP}$, and increased to 1.3 mole

per mole at the enzymatic state of $E_{\sim p}$.

For simplicity, for the numbers of exposed lysine residues the observed values in Table I were approximated to either of 0, 0.5, 1.0 or 1.5 mole per mole of subfragment, and are summarized as follows:

State	Total	I	II	III
Mg_E	2.0	1.0	0.5	0.5
$Mg_{E_{ATP}}$	1.0	0.5	0.5	0
$E_{\sim p}$	3.5	1.5	1.0	1.0

A simple model that would account for many facts reported here is depicted in Fig. 1. In this model, the ATPase molecule consisting of three subfragments I, II, and III, is assumed to be capable of rotating within the SR membrane, as already assumed in our previous papers (1). This model assumes that lysine residues locating on the boundary between the outer surface and lipid bilayer of the membrane (◖) give the value of 0.5 as the number of modified residue. This model also accomodates our previous and other biochemical studies: such as, changes in the number of the exposed lysine residues of the ATPase with changing its enzymatic state during the modification (1), or movement of the Ca^{2+} site to an internal orientation with formation of E_p, and locations of phosphorylation and cation sites at various enzymatic states (5).

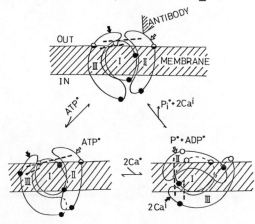

Fig. 1. Diagrammatic model for change in the amount of exposed lysine residues with change in the enzymatic state. Exposed and buried unreactive lysine residues are represented by ○ and ●, respectively, while the symbol ◖ indicates a lysine residue located at the outer boundary which is assumed to give the value of 0.5 as the number of modified residue. The arrows ⇓, ⬇, and ⇓ represent the sites for ATP binding or phosphorylation, Ca^{2+} binding, and antigenic sites, respectively.

REFERENCES

1. Yamamoto, T. & Tonomura, Y. (1976) J. Biochem. 79, 693-707
2. Stewart, P.S., MacLennan, D.H., & Shamoo, A.E. (1976) J. Biol. Chem. 251, 712-719
3. MacLennan, D.H. (1970) J. Biol. Chem. 245, 4508-4518
4. Tholley-Lawson, D.A. & Green, M. (1975) Eur. J. Biochem. 59, 193-200
5. Tada, M., Yamamoto, T., & Tonomura, Y. (1977) Physiol. Rev. in press

CALCIUM-BINDING PROTEINS
IN MUSCLE

The Significance of Parvalbumins among Muscular Calciproteins

J.-F. Pechère

Centre de Recherches de Biochimie Macromoléculaire du CNRS
Route de Mende, B.P. 505I, 34033 Montpellier,France

Our understanding of muscular parvalbumins has progressed significantly in recent times. After years of frequently fruitless efforts, a rather coherent picture of their significance is now emer-
ging in spite of the remaining, or the new coming up, of several
problems. A survey will thus be given here of our present knowledge
on the functional role and the evolutionary pattern of these proteins, two aspects which, although dependent on very different experimental approaches, are actually closely linked and constitute
the key to the relation of the parvalbumins to the other muscular
calciproteins.

The function of muscular parvalbumins

The physiological role of parvalbumins has remained obscure for
a long time. The discovery of their specific and very strong calcium binding properties [1,2] led to the inference that they could be
connected with the crucial role of this ion in the regulation of
muscular activity, but no direct evidence for it was available. In
order to solve this problem, a systematic search was started[3,4] and
the following results can be recalled :

- Parvalbumins are typically muscular proteins. Among all the
tissues examined, only brain, beside muscle, has been found to contain some. They are thus not essential to an ubiquitous major metabolic pathway, aerobic or anaerobic (cf also ref.5).

- However they are not present in all types of muscle. They are
thus not part of the contractile machinery itself. Their association
only with fast, nerve-impulse activated, skeletal muscle, suggests,
instead, a connection with the regulation of this type of muscle.

- Parvalbumins are genuine low-molecular weight sarcoplasmic
proteins, present in the 2-0.05 m$\underline{M}$ range and in adequate quantity to
bind all the Ca^{++} of the sarcoplasm. They furthermore do not interact
with any other sarcoplasmic macromolecule, either in the presence or
absence of Ca^{++}. They thus seem to constitute an autonomous regula-

tory system, which can also be shown to be independent of any phosphorylation process[6].

These results in fact only stressed the apparently paradoxical situation existing in living muscle. The high affinity of parvalbumins for Ca^{++} ($K_d \simeq 10^{-7}$ $\underline{M}$), about an order of magnitude higher than that of troponin-C (TN-C) ($K_d \simeq 10^{-6}$ $\underline{M}$), made it indeed very difficult to understand how Ca^{++} from the sarcoplasmic reticulum would ever be able to cross, during the activation process, the barrier of parvalbumins present in skeletal muscle around the myofibrils at a concentration frequently higher than that of TN-C ($\underline{ca}$ 0.1 m$\underline{M}$). Physico-chemical studies[7,8], however, suggested that the binding of Ca^{++} by parvalbumins could be pH-dependent in a way limiting their actual intervention to alter the acidification resulting from the hydrolysis of ATP during contraction. A hypothetical scheme was proposed on this basis[4] depicting the possible involvement of parvalbumins in

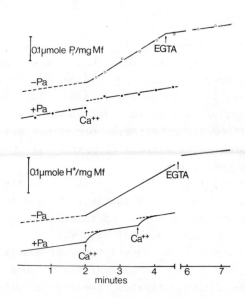

Fig. 1 - Splitting of ATP by myofibrils initially in the Ca^{++}-free state and subsequently activated by the addition of $CaCl_2$. The production of hydrogen ions (———) or of P_i (o, ●), appearing in a 0.79/1.00 ratio, is reported in function of time, either in absence (- Pa) or in presence (+ Pa) of Ca^{++}-free parvalbumin. In absence of Ca^{++}-free parvalbumin, the levels of ATPase activity corresponding to the relaxed state were measured at the end of the experiments by addition of EGTA (final concentration 2 mM) and are reported as dotted lines at the start of the experiments. Reproduced by permission from North Holland.

the movement of Ca^{++} ions during a contraction cycle in fast skeletal muscle, at the relaxation phase.

Two points were confirmed subsequently in relation to this scheme. The first is that the sarcoplasmic reticulum is indeed able to recover all the Ca^{++} from Ca^{++}-loaded parvalbumins[9]. The other is that Ca^{++}-free parvalbumins are well capable to compete with the Ca^{++}-activation of myofibrillar ATPase[9,10]. In contrast, experiments designed to test the proposal of a modulation of this capacity by pH[11] lead to the conclusion that this cellular parameter can not be invoked to understand the non-interference of parvalbumins with the activation process.

When going over the preceding experiments, however, it came to mind that they all shared a characteristic which might not prevail in vivo : they only considered the situation at equilibrium. The competition with the myofibrillar ATPase was therefore reexamined in a system which, by the order of addition of the reactants and by its pace, did not encompass this limit and could thus be closer to reality. The striking observation was then made (Fig.1) that the blocking of the myofibrillar ATPase by Ca^{++}-free parvalbumins occurs only after this activity has taken place for a short time, suggesting that kinetic effects are crucial to our appreciation of the

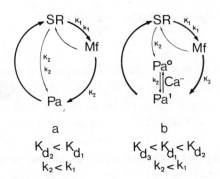

a

$K_{d_2} < K_{d_1}$
$k_2 < k_1$

b

$K_{d_3} < K_{d_1} < K_{d_2}$
$k_2 < k_1$

Fig. 2 - Scheme depicting the possible involvement of parvalbumins in the movement of Ca^{++}-ions during a contraction cycle in fast skeletal vertebrate muscle. The width of the arrows symbolically reflects the magnitude of the Ca^{++}-fluxes resulting from the overall kinetics of the different steps considered. (a) Scheme involving one parvalbumin state. (b) Scheme involving two parvalbumin states whose interconversion could be Ca^{++}-dependent. The original version of the last scheme has been modified slightly to take subsequent results (see below) into account. Reproduced by permission from North Holland.

actual process. This situation can be represented by a scheme[11], analogous to the one proposed earlier[4], but where the widths of the arrows now refer to relative Ca^{++}-fluxes associated with magnitudes of kinetic constants rather than of affinities at equilibrium (Fig.2).

Further insight into the operation of such a cycle involves a detailed knowledge of the Ca^{++}-binding by muscular parvalbumins at both the kinetic and equilibrium levels. As the detection of kinetically important steps presupposes the identification of the different molecular species which are possibly involved and which can be traced experimentally, a closer study of the equilibrium state has been undertaken first. At present, experiments are conducted, among others, in which UV difference spectroscopy is used to follow the conformational changes concomittant with the removal of Ca^{++} by EGTA, starting from the Ca^{++}-loaded state. The last point is in contrast to earlier studies[2] where some of the results obtained were suspected to be of questionable significance. Essentially similar results, nevertheless, were obtained presently under comparable conditions (T = 25°C, pH 7.55, $\Gamma/2$ = 0.25 incl. 2 m$\underline{M}$ Mg^{++}) confirming the recognition of two binding sites with $K_d \simeq 10^{-7}$ $\underline{M}$. In contrast, in the absence of Mg^{++}, a sequential mechanism

$$\text{Pa} \underset{K_1}{\rightleftharpoons} \text{PaCa} \underset{K_2}{\rightleftharpoons} \text{PaCa}_2$$

involving two different and slightly cooperative sites (calculated Hill coefficient = 1.05), with K_{d2} = 2.10^{-7} $\underline{M}$ and K_{d1} = 5.10^{-9} $\underline{M}$, appears to be the only one compatible with the observations. The experiments of Donato and Martin[8], of Moews and Kretsinger[12] and of Nelson et al.[13] - all conducted in the absence of Mg^{++} - clearly permit to associate K_2 with the EF site and K_1 with the CD one. The mutual dependence of the two sites will result from the fact that the more difficult departure of the CD Ca^{++}-which has to be considered as a semi-structural element- is seen to be conditioned by the easier departure of the EF one. Preliminary data further suggest that the process characterized by K_2 is slightly pH-dependent above pH 7.6 while that characterized by K_1 is ionic strength-dependent and is in fact better delineated by a two-step mechanism

$$Pa^0 \underset{K_0}{\rightleftharpoons} Pa^1 \underset{K_1}{\rightleftharpoons} Pa^1Ca$$

with $K_{d1} (1 + K_{do}) = 1.75 \cdot 10^{-8}$ $\underline{M}$.

When the two sets of experiments are compared, they lead to the conclusion that, in a way exactly comparable to what has been found for TN-C[14], one of the sites, with lower affinity (EF), is Ca^{++}-specific while the Ca^{++}-affinity of the other (CD) is in some way dependent on the Mg^{++} concentration. The nature of this dependence cannot be ascertained at the present time. The earlier observations[2] suggesting that a slight cooperativity exists between the two sites, even in the presence of Mg^{++}, would favour the belief that the sequential mechanism is in fact also operating under these conditions. In this case, Mg^{++} ions could exert their influence by modulating the $Pa^0 \rightleftharpoons Pa^1$ equilibrium, e.g. by favouring the $Pa^0 \rightarrow Pa^1$ conversion through a Pa-Mg^{++} complex of conformation close to that of Pa^1. This situation could have important consequences from the kinetic point of view, as the $Pa^0 \rightarrow Pa^1$ transformation will take place at the early stage of the activation of the myofibrils by Ca^{++} (as depicted on scheme b of Fig.2), and is probably rate-limiting. Conformational states analogous to Pa^0, and with lower affinity for Ca^{++}, might be the ones which, on the other hand, are involved in the interactive discharge of Ca^{++}-loaded parvalbumins on the SR-receptor. The whole system would thus appear, in fact, to be very similar to that considered by Homig and Reddy[15] when their observations on TN-C lead them to suggest that several conformational states of this protein, with varying affinities for Ca^{++}, are involved in its function, according to the progress of its interactions with TN-I and TN-I/TN-T.

In their present form, at any rate, the above results already imply that the performance of fast skeletal vertebrate muscle is not the outcome of a simple back-and-forth movement of Ca^{++} ions between the sarcoplasmic reticulum and the myofibrils but that it has to be associated with the transport of these ions through a one-way cycle in which parvalbumins play an essential role, similar in many respects to that of TN-C. It also establishes definitely the existence, the identity and the localization of the "soluble relaxing factor" which has been conjectured at times[16] to play a part in the regula-

lation of the action of calcium in muscle. The relaxing action of
parvalbumins might be useful in fast contracting muscle as a compen-
sation for the kinetic insufficiencies of the Ca^{++}-recapture by the
sarcoplasmic reticulum[17]. It is not impossible that the small amount
of parvalbumin which has been detected in brain tissue[3], where TN-C
has also been found[18], is in fact associated, in a way comparable to
what exists in muscle, with the requirements of a rapid transient
activation process. A similar role, on the other hand, can tentatively be ascribed, to the recently described calciprotein which is
found in crayfish tail muscle[19], devoid of parvalbumin, but thin-
filament regulated[20].

It thus appears that all these various processes are governed by
a subtle interplay of Ca^{++} ions with at least two -and perhaps
more- calciproteins.

Evolutionary situation of muscular parvalbumins

The study of the evolutionary relationships between these several
calciproteins constitutes another interesting way of extending our
understanding of this remarkable regulatory conformation. The amino
acid sequences of thirteen parvalbumins are now available as well as
that of nine other muscular calciproteins, such as myosin light chains (LC) and TN-Cs. Fig.3 depicts the tentative dendrogram which has
been obtained from these in a current study aiming at an extension
of previous work on a smaller number of sequences[21].

The following conclusions can be drawn from this maximum parsimony dendrogram in spite of its few unresolved aspects :

- the parvalbumins are closer related to TN-Cs and A-LCs than to
EDTA-and DTNB-LCs, and they seem to have evolved before the TN-C/
A-LC divergence. It should be noted, however, that the last outcome
is particularly dependent on the alignement adopted for the different sequences. This partially explains that a different topology
has been obtained in independent studies[22]. Only further investigation will permit a clarification of this important point. At any rate
the wide distribution all over the dendrogram of the several light
chains considered suggests that this last type of protein is the
basic element of the complete family, where TN-Cs and parvalbumins
appear rather as particular side branches. This view is supported
by considering the evolution of the Ca^{++}-binding capacity of the
various proteins. All the light chains have either one or no Ca^{++}-
binding sites; the TN-Cs from animals which have diverged earlier

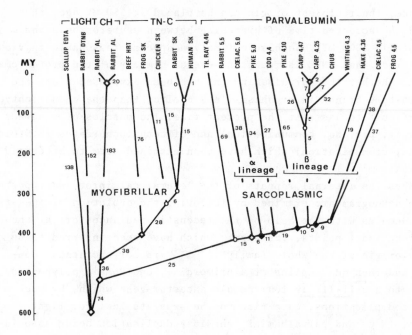

Fig. 3 - Phylogeny of parvalbumins and other homologous muscular calciproteins as determined by the maximum parsimony method. Circles refer to species divergence, which only can be dated paleontologically. Squares refer to gene duplications; a black square designates a gene duplication postulated to account for a deviation of the topology from that expected on biological grounds. The number along the branches represent nucleotide replacements.

also have a smaller number of Ca^{++}-binding sites (I in lobster tail, 2 in crayfish tail) but this number has increased progressively (2 in dogfish and varan; 3 in hake and python) to four (in rabbit). It is conceivable that it is in parallel with this evolution of TN-C that the light chains have progressively lost their original Ca^{++}-binding capacity.

- The evolution of parvalbumins -as well as that of the myosin light chains and of the TN-Cs-has occurred at varying rates, being faster at the beginning. The data of Fig.3 indeed yield an average of 17.5 NR% (nucleotide replacements per hundred codons/100 MY) in the case of parvalbumins since the time of divergence of the chondrichtyes and the bony fishes (425 MY), but only an average of 8.2 NR% since the time of divergence of the bony fishes and the amphibians (400 MY). If the highest of these rates is admitted to have been prevalent during the time separating the parvalbumins'divergence from the TN-C/A-LC line and the chondrichtyes/bony fishes divergence, their closest origin can be placed at 550 MY ago, thus around the accepted time of branching between deuterostomia and protostomia (600 MY). Such an estimation appears reasonable in view of the fact that TN-C is found in several members of the protostome branch. Whether the parvalbumin gene has been expressed along the deuterostome branch in organisms less evolved than the chondrichtyes is difficult to asses at the present time. The muscles of agnathes (C.Gerday, personal communication) and of cephalochordates (E.Stein, personal communication) which have been examined seem to be devoid of parvalbumin.

- When the internal repeats of the several calciproteins reported on the dendogram are examined by following the evolution of the structural domains with the help of the reconstituted ancestors at the different nodal points, the events which have been inferred to be at the origin of the whole family[21,23,24] are corroborated : doubling and then quadrupling of a primordial, one-domain polypeptide gene into a I-II-III-IV four domain ancestor gene which, by successive duplications, gave rise to the separate loci for light chains, TN-Cs and parvalbumins, the last duplication being associated with a partial gene deletion yielding proteins of the type II-III-IV. Data are not yet sufficient for deciding if the primordial polypeptide was Ca^{++}-binding. Recent investigations indicate that Ca^{++}-ions are not bound to the 76-108 fragment of carp 4.47 parvalbumin with a $K_d < 10^{-2}$ M (cf also refs 25,26). In contrast, the 1-75 fragment of the same protein does bind Ca^{++} in the absence of Mg^{++} with a $K_d = 0.9.10^{-4}$ M. Such a situation may simply reflect the inclusion, in the last fragment, of the CD site with a higher affinity (see above). Alternatively, it may represent the improvement of the Ca^{++}-binding capability of the two-domain ancestral polypeptide with respect to its one-domain prototype, thus providing it, perhaps, with some selective advantage.

Acknowledgements

The present contribution is the outcome of the author's close association with several individuals : Dr J.-P.Capony, Dr J.-P. van Eerd, Dr J.Jauregui-Adell, Prof. M.Goodman, Prof. J.Demaille, Mrs C.Ferraz, Mr J.Derancourt and Mr J.Haiech. Their kind collaboration is gratefully acknowledged, as well as that of Miss M.Sevegner for the typographical work.

References

1 Pechère, J.-F., Capony, J.-P. & Rydèn, L. (1971) Eur.J.Biochem., 23, 421-428.

2 Benzonana, G., Capony, J.-P. & Pechère, J.-F. (1972) Biochim. Biophys.Acta, 278, 110-116.

3 Baron, G., Demaille, J. & Dutruge, E. (1975) FEBS Letters, 56, 156-160.

4 Pechère, J.-F., Demaille, J. Dutruge, E., Capony, J.-P., Baron, G. & Pina, C. (1975) in Calcium Transport in Contraction and Secretion, E.Carafoli, F. Clementi, W. Drabikowski & E. Margreth eds, North Holland Publ.Cy, Amsterdam, pp 459-468.

5 Pechère, J.-F. & Focant, B. (1965) Biochem.J., 96, 113-118.

6 Demaille, J., Dutruge, E., Baron, G., Pechère, J.-F. and Fischer, E.H. (1975) Biochem.Biophys.Res.Comm., 67, 1034-1040.

7 Burstein, E.A., Permyakow, E.A., Emelyanenko, V.I., Bushueva, T.L. & Pechère, J.-F. (1975) Biochim.Biophys.Acta, 400, 1-16.

8 Donato, H.Jr. & Martin, R.B. (1974) Biochem., 13, 4575-4579.

9 Gerday, C. & Gillis, J.-M. (1976) J.Physiol., 258, 96-97P.

10 Fischer, E.H., Becker, J.U., Blum, H.E., Byers, B., Heizman, C., Kerrick, G.W., Lehky, P., Malencik, D.A. & Pocinwong, S. (1976) in Molecular basis of Motility, L.M.G. Heilmeyer, J.-C. Rüegg & Th. Wieland, eds, Springer Verlag, Berlin, pp 137-153.

11 Pechère, J.-F., Derancourt, J. & Haiech, J. (1977) FEBS Letters, 75, 111-114.

12 Moews, P.C. & Kretsinger, R.H. (1975) J.Mol.Biol., 91, 201-225.

13 Nelson, D.J., Opella, S.J. & Jardetzky, O. (1976) Biochem., 15, 5552-

14 Potter, J.D. & Gergely, J. (1975) J.Biol.Chem., 250, 4628-4633.

15 Honig, C.R. & Reddy, Y.S. (1975) Am.J.Physiol., 228, 172-178.

16 Briggs,F.N. (1965) Fed.Proc., 24, 208 and (1975) ibid., 34,540.

17 Ebashi, S. (1976) Ann.Rev.Physiol., 38, 293-313.

18 Fine, R., Lehman, W., Head, J. & Blitz, A. (1975) Nature (London) 258, 260-262.

19 Cox, J.A., Wnuk, W. & Stein, E.A. (1976) Biochem., 15, 2613-2618.

20 Lehman, W. (1976) Internat. Rev.Cytol., 44, 55-92.

21 Goodman, M. & Pechère, J.-F. (1977) J.Mol.Evol., 9, 131-158.

22 Barker, W.C. & Dayhoff, M.O. (1976) in Atlas of Protein Sequence and Structure, Supp. 2, 245-252.

23 Weeds, A.G. & McLachlan, A.D. (1974) Nature (London), 252, 646-649.

24 Collins, J.H. (1976) Symp.Soc.Exptl Biol., 30, 303-334.

25 Coffee, C.J. & Solano, C. (1976) Biochim.Biophys.Acta, 453,67-80.

26 Maximov, E.E. & Mitin, Y.V. (1976) Studia Biophys., 60, 149-156.

Structure and Function of Myosin Light Chains

Alan Weeds, Paul Wagner, Ross Jakes and John Kendrick-Jones
MRC Laboratory of Molecular Biology,
Hills Road, Cambridge CB2 2QH
England

STRUCTURAL RELATIONSHIPS OF MYOSIN LIGHT CHAINS

All myosins studied to date have a similar subunit structure composed of two heavy chains and two pairs of light chains. These light chains are non-covalently associated with the globular "heads" which form the cross-bridges between the overlapping filaments. The presence of two distinct classes of light chains in vertebrate striated muscle was first demonstrated by chemical means[1,2]. These light chains could also be distinguished functionally, since the 19,000 mol. wt. component was largely dissociated from myosin by reaction with 5, 5' dithiobis-(2-nitrobenzoic acid)(DTNB) in the presence of EDTA, without loss of enzymatic activity[2,3]. Hence this light chain, termed the DTNB light chain, is not required for ATPase activity. More stringent conditions were required for dissociation of the second class of light chains, which resulted in total loss of enzymatic activity, e.g. at alkaline pH, and these we called the Alkali light chains[2]. Rabbit fast-twitch muscle myosin contains two Alkali light chains of 21,000 and 17,000 mol. wt. which are identical in amino acid sequence over their C-terminal 141 residues but differ significantly in their N-terminal sequences[5]. Thus although they are structurally very similar, they are genotypically different. These chains occur in unequal and non-integral ratios[6-8], and the relative amounts change during development[9], suggesting the presence of myosin isoenzymes within a single histochemically homogeneous muscle type.

Chemically homologous light chains have been identified in other striated muscle myosins. Both cardiac and slow-twitch muscle myosins contain chemically homologous Alkali light chains of molecular weight about 21,000[10,11], and the 19,000 mol. wt. light chain of cardiac myosin shows sequence homologies with the DTNB light chain[12]. Whilst the presence of certain sulphydryl sequences has provided a valuable marker for the identification of Alkali light chains, thiol residues do not appear to have been conserved in the different 19,000 mol. wt. chains. However, the different 19,000 mol. wt. light chains show other features in common which suggest structural and functional similarities. The DTNB light chain is phosphorylated at a single serine residue[13,14], as are also the 19,000 mol. wt. light chains of cardiac and slow-twitch muscle[15]. Phosphorylation has been demonstrated for myosin light chains from both smooth muscle[14,16] and platelets[16], but there is no evidence that the Alkali light chains are phosphorylated.

Studies on molluscan myosins have also revealed the presence of two distinct classes of light chains[17], one of which may be removed by EDTA with consequent loss of calcium regulation of the actomyosin ATPase activity[18]. Calcium regulation is restored when the light chain is reassociated, showing that the EDTA light chains are directly involved in regulation of contraction in molluscan muscles. As in the case of vertebrate muscle myosins, removal of the second class of light chains cannot be achieved without loss of ATPase activity. A functional connection between the EDTA light chains of scallop myosin and the DTNB light chains of rabbit myosin was indicated by the observation that a DTNB light chain can replace one of the EDTA light chains in restoring calcium regulation to desensitized scallop actomyosin[19]. Similar observations have been made for some of the other phosphorylatable 19,000 mol. wt. light chains[20], which suggest a common structure and function for these light chains, all of which may be defined as "regulatory". More recent observations have shown that these different light chains do not respond in an identical manner in the resensitization experiments, indicating changes in their regulatory role during evolution.

An important contribution to our understanding of the relationship between these various light chains has been the determination of their amino acid sequences. Although only a few sequences are complete, the data reveal interesting features in common between these proteins as well as extensive homologies with two calcium binding proteins from muscle, troponin C and the calcium binding parvalbumin. The significance of these comparisons is considerably enhanced by a knowledge of the three-dimensional structure of carp calcium binding parvalbumin[21], which provides a structural basis from which the homologies may be evaluated. Parvalbumin contains two calcium binding sites, and internal sequence repeats suggest gene duplication of a primitive calcium binding unit which Kretsinger has termed the "EF-hand"[21] (see also this volume). Each calcium site lies in a pocket between two helices and the structure is stabilized by the packing of apolar residues present within these helices into a hydrophobic "core". Thus comparison of the sequence and structure reveals a clear pattern for the distribution of apolar residues[21-24] (Fig. 1). Conservation of this pattern of apolar residues in troponin C[25,26] and the different light chains (Alkali[23,27], DTNB[28], and EDTA[29]) suggests that these structural features have been preserved during the evolution of these proteins. Thus a family of proteins has evolved in muscle descended from an ancestral calcium binding protein, whose structural features have been conserved but whose functions appear to have diverged as the calcium binding sites have been lost. It is therefore important to examine the sequences in greater detail, particularly those parts related to the calcium binding sites of parvalbumin and troponin C, to search for information which may relate these structures to current knowledge of their biological roles.

223

The calcium ions in parvalbumin are ligated to oxygen atoms donated by six different amino acids, four of these residues being acidic. Sequence comparisons between troponin C and parvalbumin show a similar distribution of oxygen ligating residues in each of the four sites in troponin C (Fig. 2) (see ref. 30 for discussion). In analysing the sequences of the light chains, it is relevant to ask whether the residues in the ligating positions within the potential calcium binding sites have been conserved? Changes in these positions may be significant in loss of calcium binding, though it must be remembered that the conformations of the different light chains are undoubtedly influenced by their interaction with the heavy chains of myosin.

FUNCTION OF MYOSIN LIGHT CHAINS

EDTA Regulatory Light Chain

Molluscan myosin contains two high affinity calcium binding sites and requires the presence of specific regulatory light chains for calcium regulation of its interaction with actin[17,18]. Selective removal of one of the two moles of this EDTA light chain from scallop myosin results in complete loss of calcium regulation and the loss of one calcium binding site. The isolated EDTA light chain, however, does not bind calcium, but it readily recombines with the "desensitized" myosin and both calcium regulation and complete calcium binding are restored[19,31]. These results suggest that calcium binding is dependent on the correct stereochemical configuration of groups which may originate in both light and heavy chains. Alternatively, if calcium ligation depends solely on residues in the light chains, then the conformation of the light chain is altered by interaction with the heavy chains.

Structural homology between the EDTA light chain and parvalbumin is suggested on the basis of the distribution of apolar residues as stated above, and the presence of four potential calcium binding sites is predicted. Close examination of the sequences in these regions[29] shows that in only one case (site 1) is there an acidic residue in the -Z co-ordinating position (Fig. 2). This site is also the only one to contain four acidic residues among the six potential oxygen ligating groups. Furthermore, the other three sites show amino acid substitutions and deletions which suggest structural distortion of these sites. Both sites 2 and 4 have deletions and proline residues are present in potential calcium ligating positions, while site 3 has lost the important glycine residue (Fig. 2). While the amino acid residues in site 1 appear to conform to the requirements of an "EF-hand", aspartic acid is found in the -Z co-ordinating position in place of glutamic acid, which is present in all four sites of troponin C and in both parvalbumin sites. Nevertheless, it is reasonable to propose that this single

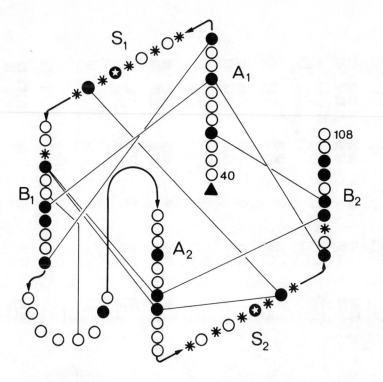

Fig. 1. Schematic representation of the structure of the two calcium sites of
carp calcium binding parvalbumin. Symbols:- A1, B1 etc. are the helical segments
and S1, S2 denote the calcium binding sites. Circles represent amino acid
residues, solid circles being used for buried apolar groups while the lines
between these mark the main van der Waals' interactions. $*$, residues ligated
to calcium ions; ⬤ , glycine residues in the calcium sites; ▲ , the start
of helix A1 is at residue 40 in the sequence. (See ref. 23 for further details.)

site in the EDTA light chain is the high affinity calcium binding site of
scallop myosin[29] ($K_b = 10^6$ M^{-1}), though calcium binding depends on association of
both light and heavy chains.

These conclusions, together with other experiments[29] suggest that the EDTA
light chains of molluscan myosin may inhibit actomyosin interaction sterically
in the absence of calcium ions, while in their presence the conformation of the
light chain is altered to expose sites on the myosin "heads" which interact with
actin and facilitate contraction. This simple model, based on steric inhibition
of interaction sites, is analogous to models of calcium regulation on the thin
filaments of striated muscle mediated by the movement of tropomyosin and troponin.

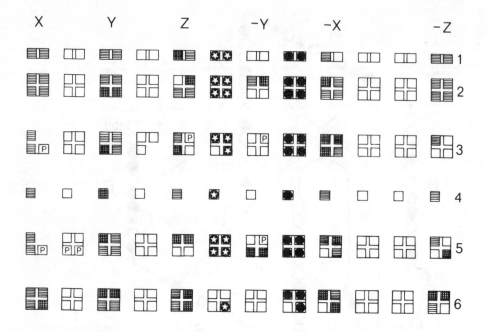

Fig. 2. Residues occupying calcium ligating positions in parvalbumin and troponin C, with equivalent residues in various myosin light chains. X,Y,Z etc. refer to the octahedral vertices of the calcium ligating positions[24] (those positions denoted ✳ in Fig. 1). Symbols:- ●, Ile or exceptionally other large apolar residues; ⊛, glycine; ▤ Asp or Glu and ▦, Asn, Gln, Ser or Thr in calcium ligating positions; P, denotes the presence of proline. The order of the sites is numbered from top left (site 1) to bottom right (site 4). A deletion is indicated by the absence of a small square in a tetrad. 1) calcium binding parvalbumin; 2) troponin C; 3) scallop EDTA light chain; 4) site 1 of gizzard regulatory light chain; 5) DTNB light chain; 6) Alkali light chain. N.B. The oxygen ligating atom in the -Y co-ordinating position is the carbonyl oxygen of the peptide bond.

Smooth Muscle Myosin Regulatory Light Chains

Only the N-terminal portion of the amino acid sequence of gizzard myosin regulatory light chain has been determined, but this region contains the potential calcium binding site by analogy with the sequence of the EDTA light chain[14,29] (Fig. 2). The distribution of apolar groups in the "helices" surrounding this potential site is conserved and the site contains four acidic amino acids together with the glycine and isoleucine residues characteristic of the "EF-hand" (Fig. 2). It is significant that the gizzard light chain will functionally replace the EDTA light chain in restoring both calcium regulation and calcium binding to desensitized scallop myosin, showing that it fully substitutes for the scallop regulatory light chain.

Bremel has shown that calcium regulation of gizzard actomyosin does not depend on troponin but is linked to myosin[32], though the precise nature of this system remains controversial. There is general agreement that actomyosin interaction in smooth muscle depends on phosphorylation of the myosin regulatory light chains, and the activity of the light chain kinase is calcium dependent[16,33,34]. Where there is less agreement concerns the presence of a direct calcium ON/OFF switch mediated by the light chains in a manner analogous to molluscan myosin. Actin activation of the ATPase activity of vas deferens myosin requires phosphorylation of the regulatory light chain but calcium regulation has been shown to be independent of phosphorylation, thus suggesting the presence of a direct switch[33]. This has not been confirmed however for other smooth muscles[34]. Nevertheless calcium ions appear to control two processes:- 1) the activity of the light chain kinase which thereby promotes actomyosin interaction, and 2) a direct calcium mediated regulatory response. Thus certain parallels appear to exist between the regulation of smooth muscle and the more primitive molluscan muscle, but evolution of a calcium sensitive phosphorylation mechanism indicates modification of the overall regulatory process.

Sequence analysis has located the site of phosphorylation to a serine residue located on the N-terminal side of the potential calcium binding site[14]. Although there is a serine residue in the equivalent position in the EDTA light chain, it is not phosphorylated. The molecular weight of the EDTA light chain is 17,000, while that of the gizzard regulatory light chain is 20,000, suggesting that this additional sequence, located mainly in the N-terminal region of the gizzard light chain, may serve as a recognition site for the specific kinase[29]. This hypothesis is supported by recent observations that proteolytic fragments of gizzard light chain which retain the essential serine residue cannot be phosphorylated[14].

DTNB Regulatory Light Chain

Calcium control in vertebrate striated muscle is exercised through the tropomyosin-troponin system associated with the actin filaments and there is no evidence that a functional myosin-linked regulatory system is present. Nevertheless, the demonstration that both forms of regulation occur in certain worms and insects[35] has led to the suggestion that dual regulation may have considerable advantages in precision and may therefore be more widespread than has been supposed, particularly if the mode of myosin regulation has been further modulated during subsequent evolution. Thus, following the emergence of actin-linked regulation as the dominant calcium control mechanism, myosin linked regulatory systems would be free to mutate as part of a "silent gene pool" without being

227

lethal[36]. Any selective advantage gained during this silent phase of mutation would subsequently be expressed as a refinement of the regulatory process or in some other functional role.

At present there is no direct evidence that the DTNB light chains are involved in calcium regulation, but they bind calcium ions although this binding is weak when magnesium is present[37]. Only one of the four potential calcium binding sites derived from sequence comparisons with parvalbumin and troponin C has sufficient oxygen ligating residues to fulfill Kretsinger's criteria, and this site (site 1, Fig. 2) has only three acidic residues instead of four[30]. Experimental evidence suggests that this site is a general divalent cation site[38,39]; it differs from the calcium binding site of molluscan myosin in that magnesium and manganese ions will bind competitively to this site though they do not compete for the high affinity calcium sites of molluscan myosin[38]. In this respect it appears to be analogous to the two $Ca^{2+}.Mg^{2+}$ sites of troponin C whose functions may be to stabilize the structure rather than mediate calcium regulation[26,40].

The main evidence supporting a possible regulatory role for the DTNB light chains is that they will restore calcium regulation to desensitized molluscan myofibrils or actomyosin[19]. More recent experiments have shown, however, that they do not resensitize purified scallop myosin in the absence of actin, nor do they restore calcium binding[31]. These results indicate loss of regulatory function as compared to the EDTA light chains and gizzard regulatory light chains, though the site responsible for attachment to scallop myosin heavy chains has been preserved. Whether the DTNB light chain provides more than a passive support in the restoration of calcium regulation to molluscan actomyosin remains unclear: the residual EDTA light chain may be sufficient to mediate calcium control as long as the site for the second EDTA light chain is occupied[31].

The presence of a phosphorylation site on the DTNB light chain in a position identical to that found for the gizzard regulatory light chain suggests a further evolutionary relationship between these two proteins[14]. The role of this phosphorylation has not been established, though recent studies on the phosphorylation of the homologous cardiac 19,000 mol. wt. light chain indicate a possible involvement in actin-myosin interaction[41]. Other experiments also suggest that the DTNB light chains influence actin-myosin interaction[42]. But in the absence of any conclusive experiments, the role of the DTNB light chains remains an enigma. They may be located near the subfragment-1/subfragment-2 "hinge" region to influence the orientation of the cross-bridges[38,39], but since they bind magnesium as well as calcium, this does not support a calcium regulatory mechanism. Another possibility is that they regulate the level of calcium by divalent cation buffering, in a manner similar to that reported

recently by Pechère *et al.*[43] for the parvalbumins (C. Bagshaw, personal communication). Future electron paramagnetic resonance studies may help resolve this
confusion.

Vertebrate Striated Myosin Alkali Light Chain

The evolution of myosin light chains away from a direct calcium regulatory
role is completed when we analyse the Alkali light chains. They do not bind
calcium ions, nor do they restore calcium regulation to desensitized molluscan
actomyosin. Sequence analysis shows that the essential features characteristic
of parvalbumin have been preserved[23,24] though only one of the four potential
calcium binding sites has retained four acidic groups among its six ligating
residues (site 3, Fig. 2) and this site appears to be distorted since it lacks
both the glycine and isoleucine residues characteristic of an "EF-hand" and
contains bulky apolar residues which might block the site.

The role of the Alkali light chains has been controversial for many years, but
although their presence is required for ATPase activity there is no evidence that
they participate directly in the catalytic cleavage of ATP by myosin alone[44].
Recent experiments have shown that they influence the actin-activated ATPase,
indicating a possible role in cross-bridge interaction[45]. The precise nature of
this involvement has not been established but hybridization experiments, in which
light chains from one muscle source are complexed with heavy chains from a
different source, suggest that the light chains may help determine the maximum
activity of the actin-activated ATPase, and hence be implicated in controlling
the muscle's maximum shortening velocity. In this respect they may be regarded
as "catalytic" in their function, but further evidence will be required to confirm
this terminology.

The class of light chains to which the Alkali light chains belong are termed
"essential" since they cannot be dissociated without loss of ATPase activity.
This observation indicates that they are much more tightly associated to the
heavy chains than the regulatory light chains but it makes analysis of their
function more difficult to test. Further experiments will be necessary to
establish their specific roles.

CONCLUSIONS

It is clear from amino acid sequence comparisons that a family of muscle
proteins exists, structurally related to the calcium binding parvalbumins and
evolved from a primitive calcium binding ancestral protein by gene duplication
and mutation. This evolution has produced functional diversity in the myosin
light chains. While the molluscan EDTA light chains have retained a calcium

229

regulatory role they do not bind calcium in the isolated state. Gizzard myosin light chains also appear to be involved in calcium regulation, but the response is more complex, depending in addition on phosphorylation of the light chain catalysed by a calcium-dependent light chain kinase. In vertebrate striated muscle myosin, the regulatory DTNB light chains bind divalent cations in a manner which may stabilize the myosin cross-bridge structure and orientation, but their specific function remains uncertain. Finally the Alkali light chains have no metal dependent function and they bind to the heavy chains more tightly than the regulatory light chains. Current evidence suggests they are involved in acto-myosin interaction. While sequence homologies and structural relationships inferred from them provide evidence for an evolutionary connection between these different light chains and other calcium binding proteins in muscle, their functional diversity suggests that considerable modification of the three-dimensional structures has occurred which may largely be due to their interactions with myosin heavy chains and other proteins in the contractile system. Thus confirmation that a true structural relationship has been preserved will depend ultimately on determination of the three-dimensional structures of these different myosins.

REFERENCES

1. Weeds, A.G. (1969) Nature (London) 223, 1362-1364.
2. Weeds, A.G. and Lowey, S. (1971) J. Mol. Biol. 61, 701-725.
3. Gazith, J., Himmelfarb, S. and Harrington, W.F. (1970) J. Biol. Chem. 245, 15-22.
4. Kominz, D.R., Carroll, W.R., Smith, E.N. and Mitchell, E.R. (1959) Arch. Biochem. Biophys. 79, 191-199.
5. Frank, G. and Weeds, A.G. (1974) Eur. J. Biochem. 44, 317-334.
6. Lowey, S. and Risby, D. (1971) Nature (London) 234, 81-85.
7. Sarkar, S. (1972) Cold Spring Harbor Symp. Quant. Biol. 37, 14-17.
8. Weeds, A.G., Hall, R. and Spurway, N.C. (1975) FEBS Lett. 49, 320-324.
9. Pelloni-Müller, G., Ermini, M. and Jenny, E. (1976) FEBS Lett. 70, 113-117.
10. Weeds, A.G. (1975) FEBS Lett. 59, 203-208.
11. Weeds, A.G. (1976) Eur. J. Biochem. 66, 157-173.
12. Léger, J.J. and Elzinga, M. (1977) Biochem. Biophys. Res. Comm. 74, 1390-1396.
13. Perrie, W.T., Smillie, L.B. and Perry, S.V. (1973) Biochem. J. 135, 151-164.
14. Jakes, R., Northrop, F. and Kendrick-Jones, J. (1976) FEBS Lett. 70, 229-234.
15. Frearson, N. and Perry, S.V. (1975) Biochem. J. 151, 99-107.
16. Adelstein, R.S., Chacko, S., Barylko, B. and Scordilis, S.P. (1976) Contractile Systems in Non-Muscle Tissues, eds. S.V. Perry *et al.*,

Elsevier/North Holland Biomedical Press, 153-163.

17. Kendrick-Jones, J., Szentkirályi, E.M. and Szent-Györgyi, A.G. (1972) Cold Spring Harbor Symp. Quant. Biol. 37, 47-53.

18. Szent-Györgyi, A.G., Szentkirályi, E.M. and Kendrick-Jones, J. (1973) J. Mol. Biol. 74, 179-203.

19. Kendrick-Jones, J. (1974) Nature (London) 249, 631-634.

20. Kendrick-Jones, J. (1975) 26th Mosbach Colloquium on 'Molecular Basis of Motility', eds. Heilmeyer *et al*., Springer-Verlag, 122-136.

21. Kretsinger, R.H. and Nockolds, C.E. (1973) J. Biol. Chem. 248, 3313-3326.

22. Pechère, J.-F., Capony, J.-P. and Demaille, J. (1973) Systematic Zoology 22, 533-548.

23. Weeds, A.G. and McLachlan, A.D. (1974) Nature (London) 252, 646-649.

24. Tufty, R.M. and Kretsinger, R.H. (1975) Science 187, 167-169.

25. Collins, J.H., Potter, J.D., Horn, M.J., Wilshire, G. and Jackman, N. (1973) FEBS Lett. 36, 268-272.

26. Kretsinger, R.H. and Barry, C.D. (1975) Biochim. Biophys. Acta 405, 40-52.

27. Collins, J.H. (1974) Biochem. Biophys. Res. Comm. 58, 301-308.

28. Collins, J.H. (1976) Nature (London) 259, 699-700.

29. Kendrick-Jones, J. and Jakes, R. (1977) International Symp. Myocardial Failure (Tegernsee, Munich, June 1976) in press.

30. Collins, J.H. (1976) Symp. Soc. Experimental Biol. XXX, 303-334.

31. Kendrick-Jones, J., Szentkirályi, E.M. and Szent-Györgyi, A.G. (1976) J. Mol. Biol. 104, 747-775.

32. Bremel, R.D. (1974) Nature (London) 252, 405-407.

33. Chacko, S., Conti, M.A. and Adelstein, R.S. (1977) Proc. Natl. Acad. Sci. 74, 129-133.

34. Sobieszek, A. (1977) Eur. J. Biochem. 73, 477-483.

35. Lehman, W. and Szent-Györgyi, A.G. (1975) J. Gen. Physiol. 66, 1-30.

36. Hartley, B.S. (1974) Symp. Soc. Gen. Microbiology XXIV, 151-182.

37. Werber, M.M., Gaffin, S.L. and Oplatka, A. (1972) J. Mechanochem. Cell Motility 1, 91-96.

38. Bagshaw, C.R. (1977) Biochemistry 16, 59-67.

39. Weeds, A.G. and Pope, B. (1977) J. Mol. Biol. 111, 129-157.

40. Potter, J.D., Seidel, J.C., Leavis, P., Lehrer, S.S. and Gergely, J. (1976) J. Biol. Chem. 251, 7551-7556.

41. Frearson, N., Solaro, R.J. and Perry, S.V. (1976) Nature (London) 264, 801-802.

42. Margossian, S.S., Lowey, S. and Barshop, B. (1975) Nature (London) 258, 163-166.

43. Pechère, J.-F., Derancourt, J. and Haiech, J. (1977) FEBS Lett. 75, 111-114.

44. Taylor, R.S. and Weeds, A.G. (1977) FEBS Lett. 75, 55-60.

45. Wagner, P.D. and Weeds, A.G. (1977) J. Mol. Biol. 109, 455-473.

Aminoacid Sequence of Frog Skeletal Troponin C

Jean-Paul van Eerd[+]
Pharmakologisches Institut der Universität Zürich
Gloriastrasse 32, 8006 Zürich, Switzerland

and

Jean-Paul Capony and Jean-François Pechère
Centre de Recherches de Biochimie Macromoléculaire du CNRS
B.P. 5051, 34033 Montpellier, France

INTRODUCTION

Troponin is a protein located on the thin filaments of muscle. It plays an important role in the regulation of muscular contraction. Troponin is composed of three non-identical subunits. One subunit, troponin C, reversibly binds calcium ions. Muscular contraction occurs when calcium ions are bound to troponin C. When Ca^{++} is removed from troponin C relaxation of the muscle occurs (for a review see Ebashi, 1974)[1].

The primary structure of rabbit skeletal troponin C has been determined by Collins et al.[2,3]. Collins et al.[2] showed that troponin C is homologous to parvalbumins, a group of low molecular weight Ca^{++}-binding proteins from the sarcoplasm[4]. Because the three-dimensional structure of a parvalbumin from the sarcoplasm of carp is known[5], Collins et al. and Weeds and McLachlan[6] tentatively localized four Ca^{++}-binding sites in troponin C. The aminoacid sequence of bovine cardiac troponin C has been determined by van Eerd and Takahashi[7,8]. It is homologous to rabbit skeletal troponin C. It has been suggested by van Eerd and Takahashi that there are only three Ca^{++}-binding sites in bovine cardiac troponin C[7].

There is about 35% difference between the aminoacid sequences of bovine cardiac troponin C and rabbit skeletal troponin C. Because the mutation rate of parvalbumins is low, van Eerd and Takahashi[7] postulated that the difference in aminoacid sequence between rabbit skeletal troponin C and bovine cardiac troponin C, is due to the difference in tissue rather than to the difference in species. The best way to check this postulate would have been to determine the aminoacid sequence of either rabbit cardiac troponin C or bovine skeletal troponin C. It was decided however to determine the amino-

---

[+]Present address: Biochemisch Laboratorium, Rijksuniversiteit te Groningen, Zernikelaan, Groningen, The Netherlands.

232

acid sequence of frog skeletal troponin C because most physiological experiments and X-ray diffraction experiments to study the contractile process have been performed with frog muscle. The determination of the primary structure of frog skeletal troponin C should answer the question whether the aminoacid sequence difference between bovine cardiac troponin C and rabbit skeletal troponin C is due to species difference or due to tissue difference, and also provide additional data for phylogenetic studies[9]. Moreover, the determination of the aminoacid sequence of frog skeletal troponin C should give an indication whether there is a large difference in troponin C from frog muscle and rabbit muscle, used to study muscular contraction in a physiological system and in a biochemical system respectively.

Recently the aminoacid sequences of human skeletal troponin C[10] and chicken skeletal troponin C[11] have been published.

In this symposium the nearly complete aminoacid sequence of frog skeletal troponin C is presented. It is compared with the known aminoacid sequences of other troponins C. The difference in aminoacid sequence between skeletal and cardiac muscle troponin C is considerably larger than the differences between the skeletal troponins C, which means that the difference in aminoacid sequence between cardiac troponin C and skeletal troponins C is largely due to the difference in tissue. The difference in aminoacid sequence between frog and rabbit skeletal troponin C is not large and there is probably no significant difference in functional properties. Therefore, as far as troponin C is concerned, there are no major differences between frog muscle (physiological system) and rabbit muscle (biochemical system).

MATERIALS AND METHODS

Material. Frog (Rana temporaria) skeletal troponin C was prepared according to the method of Greaser and Gergely[12].

Cleavage methods. CNBr cleavage: Native frog skeletal troponin C was cleaved by CNBr according to the method of Gross and Witkop[13]. Tryptic digestion: Performic acid-oxidized troponin C was digested with trypsin. A tryptic digest was also made from phthalylated[14] troponin C.

Purification methods. Initial purification of the CNBr peptides and the tryptic peptides was performed by chromatography on Sephadex G-50 fine (400 x 0.9 cm). Further purification of the large peptides

(> 15 residues) was on DEAE-Sephadex or DEAE-cellulose. The small peptides (< 15 residues) were purified using a Technicon peptide analyser or by high-voltage paper electrophoresis.

Aminoacid analysis. Peptides were hydrolysed by 6 N HCl at 110°C in evacuated and sealed glass tubes. Aminoacid analysis was performed on a Beckman Multichrom aminoacid analyser using a single column analysis system.

Aminoacid sequence determination. CNBr peptides were sequenced by manual Edman degradation. PTH-aminoacids were identified by High Performance Liquid Chromatography (HPLC) using the method of Frank and Strubert[15]. Tryptic peptides were sequenced by the dansyl-Edman method. The amide-containing residues were located by high-voltage paper electrophoresis according to Offord[16]. The largest tryptic peptide (residues 48-84) was sequenced by automatic Edman degradation in a Socosi Sequencer using hake parvalbumin as a carrier[17].

RESULTS AND DISCUSSION

The aminoacid sequence of frog skeletal troponin C has almost been completed. Experiments are still continuing to determine the aminoacid sequence of the N-terminal CNBr peptide and to complete one tryptic peptide. The tentative sequence of frog skeletal troponin C is shown in Fig. 1, where the sequence is compared with the known sequences of other troponins C. Frog skeletal troponin C consists of a single polypeptide chain of 162 aminoacid residues. There are three extra aminoacids at the N-terminal as compared with rabbit skeletal troponin C. There is a single Cys residue at position 101, and a single Tyr residue at position 112. The N-terminus of frog skeletal troponin C is blocked but the nature of this blockage has not yet been investigated. It is probably acetylated like most muscular proteins.

The sequence of the parts in brackets (Fig. 1) is arranged assuming maximum homology. Wilkinson[11] was unable to cleave the peptide chain behind the Met residue in the N-terminal part of chicken troponin C with CNBr. Since a Met-Thr bond is known to be difficult to cleave with CNBr[18], there is probably a Met-Thr sequence in the first four aminoacid residues of chicken troponin C. When we place this tentative Met-Thr sequence as residues 3 and 4 there can be homology with the Thr residue of the N-terminal part of rabbit skeletal troponin C and human skeletal troponin C.

```
                                             10
        rabbit skeletal      Ac(-.-.-.-)A E - - - Y - -
        human skeletal       X(-.-.-)- A E - - - Y - -
        chicken skeletal     X(S.-.M.-)- - - A E - - A - - -
        frog skeletal        X(P.A.E.T.D.Q.Q)M D A R S F L S
        bovine cardiac       Ac M D - I Y K A - V E Q - T


       20              30                40              50
- - - - - - - - - - - - - - - A - - - - - - - V - - - - - - - - - - -
- - - - - - - - - - - - - - - A - - - - - - - V - - - - - - - - - - -
- - - - - - - - - - - - - - - A - - - - - - - - - - - - - - - - - - -
E E M I A E F K A A F D M F D  T D G G G D I S T K E L G T V M R M L G Q
- - Q K N - - - - - - - I - V L G A E D - C - - - - - - K - - - - - -


            60              70              80              90
- - - - - - - - - - - - - (-.-.-.-.-.-.-) - - - - - - - - - - - - - - - E - - K
- - - - - - - - - - - - - - - - - - - - - - - - - - - - - - - - - - - E - - K
N - - - - - - - - - - - - - - - - - - - - - - - - - - - - - - - - - - E - - K
T P T K E E L D A I I E E V D E D G S G T I D F E E F L V M M V R Q M K Q D A E
N - - P - - - Q E M - D - - - - - - - - - V - - D - - - - - - - C - - D - S K


            100             110             120
- - - - - - - - - - - - - - - R - - - - - - - A - - - A - - F - A - - -
- - - - - - - - - - - - - - - R - - - - - - - P - - - A - - F - A - - -
- - - - - - - - D - - - - - - - - - - F - - I - - - - - - - A T - -
G K S E E E L A E C F R I(F.D)K N A D G Y I D S E E L G E I L R S S G E
- - - - - - - S D L - - M - - - - - - - - - - L - - - K I M - Q A T - -


   130             140             150             160
H V - - - - - - S - - - - - - - - - - - - R - - - - - - - - - - - - - -
H V - - - - - - S - - - - - - - - - - - - R - - - - - - - - - - - - - -
H V - E - D - - D - - - - S - - - - - - R - - - - - - - - - - - - - -
S I T D E E I E E L M K D G D K N N D G K I D F D E F L K M M E G V Q
T - - E D D - - - - - - - - - - - - - R - - Y - - - - E F - K - - E
```

Fig. 1. The aminoacid sequence of frog skeletal troponin C. The
sequence is compared with the known sequences of other troponins C.
The underlined regions indicate the likely calcium-binding sites.
Explanation of the one-letter code for aminoacids: A stands for Ala,
C for Cys, D for Asp, E for Glu, F for Phe, G for Gly, H for His,
I for Ile, K for Lys, L for Leu, M for Met, N for Asn, P for Pro,
Q for Gln, R for Arg, S for Ser, T for Thr, V for Val, X for blocked
N-terminus, and Y for Tyr. Only differences with the aminoacid
sequence of frog skeletal troponin C are indicated for other tropo-
nins C. The exact sequence of the parts in brackets is unknown.

Homology with other troponins C. In Fig. 2 is shown the difference
matrix for the number of aminoacid replacements between the 5 troponin C sequences. The difference matrix for the minimum number of
nucleotide replacements is shown in Fig. 3. There is a comparable
number of aminoacid replacements between rabbit and chicken troponin
C (15), between rabbit and frog troponin C (16) and between chicken
and frog troponin C (21). The number of nucleotide replacements

235

involving two bases is similar (3,4) for these three pairs of
skeletal troponins C. The number of aminoacid replacements between
the skeletal troponins C and bovine cardiac troponin C is much larger.

| Aminoacid Replacements | | | | | | Nucleotide Replacements | | | | |
	RS	HS	CS	FS	BH		RS	HS	CS	FS	BH
RS	0	1	15	16	55	RS	0	1	18	20	75
HS	1	0	15	16	55	HS	1	0	18	20	75
CS	15	15	0	21	49	CS	18	18	0	24	65
FS	16	16	21	0	54	FS	20	20	24	0	72
BH	55	55	49	54	0	BH	75	75	65	72	0

Fig. 2   Fig. 3

Fig. 2. Difference matrix indicating the number of aminoacid replacements between the known sequences of troponin C. RS stands for rabbit
skeletal, HS for human skeletal, CH for chicken skeletal, FS for frog
skeletal and BH for bovine heart.

Fig. 3. Difference matrix indicating the number of nucleotide replacements between the known sequences of troponin C. For an explanation of the abbreviations see legend Fig. 2.

Cardiac and skeletal troponin C. From the matrices in Figs. 2 and
3 it is clear that the 4 skeletal troponins C are much closer related to one another than to bovine cardiac troponin C. Therefore the
difference observed between bovine cardiac troponin C and the
skeletal troponins C cannot be accounted for by the difference in
species but must be largely the result of difference in tissue as
postulated before by van Eerd and Takahashi[7].

Calcium-binding sites. Collins et al.[2] assigned 4 calcium-binding
sites to rabbit skeletal troponin C, because of homology with the
calcium-binding sites of parvalbumins. Van Eerd and Takahashi[7]
suggested that in bovine cardiac troponin C the region between
residues 30 and 41 has lost its ability to bind Ca^{++} because of the
insertion of an aminoacid between positions 30 and 31, and because
of the large number of aminoacid replacements in this region.
Frog skeletal troponin C, when compared with rabbit skeletal
troponin C contains some aminoacid replacements in the 4 calcium-
binding regions, but the replacements are conservative so that no
drastic changes in those regions can be expected. Therefore frog
skeletal troponin C probably contains 4 calcium-binding sites like

the other skeletal troponins C.

The single Tyr residue of frog skeletal troponin C is located in
the third calcium-binding site. Therefore frog skeletal troponin C
should be very suitable to study calcium-induced conformational
changes in troponin C by observations of its intrinsic tyrosine
fluorescence[19].

Variable regions in troponin C. When we look at Fig. 1 we see
that most aminoacid replacements between the skeletal troponins C
are located in two regions. The first region is the N-terminal part
of the protein. The second region is situated between the third and
the fourth calcium-binding site. If we assume that the troponin C
molecule is composed of four homologous domains each consisting of
a helix - calcium-binding site - helix, we see that the variable
region at the N-terminal is situated just before the first domain. The
second variable region is located in the second helix of the third
domain, the loop region between the third and fourth domain, and the
first helix of the fourth one. Particularly noteworthy is position
115. The aminoacid at position 115 is different for all the troponins
C listed in Fig. 1. It is homologous with the aminoacid at position
60 of carp parvalbumin[2]. When we look at the three-dimensional
structure of carp parvalbumin[5] we see that the aminoacid corresponding to position 115 in troponin C is located at a sharp bend. It
forms the end of a β-pleated sheet at the beginning of a helix, at
right angle with the pleated sheet structure. Though residue 115 is
located in the third calcium-binding site it does not form a ligand
for the binding of the calcium ion.

Extrapolation of physiological and biochemical data on muscle.
Most physiological experiments on muscle have been performed with
frog muscle while most biochemical experiments on muscle have been
performed on rabbit muscle. It may be asked to what extent the
results obtained with frog muscle can be extrapolated to rabbit
muscle and vice versa.

There are 16 aminoacid replacements and 3 additional aminoacids
at the N-terminal in frog skeletal troponin C as compared with
rabbit skeletal troponin C. However from what has been said before,
there are apparently no large differences in structure and calcium-
binding properties. Therefore as far as troponin C is concerned it
seems reasonable to extrapolate experimental results from frog
muscle to rabbit muscle.

Acknowledgements. Part of this work was performed during the tenure of a short-term EMBO fellowship by JPvE. This study was supported by a grant from the Swiss National Foundation (Grant Nr. 3.2130.73). We thank Dr. H. Rochat from Marseille for the performance of a run on the automatic sequencer and Mrs. C. Ferraz for her skilful technical assistance.

REFERENCES

1. Ebashi, S. (1974) Essays in Biochemistry, 10, 1-36.
2. Collins, J.H., Potter, J.D., Horn, M.J., Wilshire, G. and Jackman, N. (1973) FEBS Lett., 36, 268-272.
3. Collins, J.H. (1974) Biochem. Biophys. Res. Commun., 58, 301-308.
4. Pechère, J.F., Capony, J.P. and Ryden, L. (1971) Eur. J. Biochem., 23, 421-428.
5. Kretsinger, R.H. and Nockholds, C.E. (1973) J. Biol. Chem., 248, 3313-3326.
6. Weeds, A.G. and McLachlan, A.D. (1974) Nature, 252, 646-649.
7. van Eerd, J.P. and Takahashi, K. (1975) Biochem. Biophys. Res. Commun., 64, 122-127.
8. van Eerd, J.P. and Takahashi, K. (1976) Biochemistry, 15, 1171-1180.
9. Goodman, M. and Pechère, J.F. (1977) J. Mol. Evol., 9, 131-158.
10. Romero-Herrera, A.E., Castillo, O. and Lehmann, H. (1976) J. Mol. Evol., 8, 251-270.
11. Wilkinson, J.M. (1976) FEBS Lett., 70, 254-256.
12. Greaser, M. and Gergely, J. (1971) J. Biol. Chem., 246, 4226-4233.
13. Gross, E. and Witkop, B. (1962) J. Biol. Chem., 237, 1856-1860.
14. Bertrand, R., Pantel, P. and Pechère, J.F. (1974) Biochimie, 56, 515-522.
15. Frank, G. and Strubert, W. (1973) Chromatographia, 6, 522-524.
16. Offord, R.E. (1966) Nature, 211, 591-593.
17. Rochat, M., Bechis, G., Kopeyan, C., Gregoire, J. and Van Rietschoten, J. (1976) FEBS Lett., 64, 404-408.
18. Schroeder, W.A., Shelton, J.B. and Shelton, J.R. (1969) Arch. Biochem. Biophys., 130, 551-557.
19. van Eerd, J.P. and Kawasaki, Y. (1972) Biochem. Biophys. Res. Commun., 47, 859-865.

Calcium-Binding Proteins: Relationship of Binding, Structure, Conformation and Biological Function

James D. Potter*, J. David Johnson, John R. Dedman, William E. Schreiber, Frederic Mandel, Richard L. Jackson and Anthony R. Means

Departments of Cell Biophysics and Cell Biology, Baylor College of Medicine
Houston, Texas 77030

SUMMARY

The calcium binding parameters of cardiac TnC (C-TnC), paravalbumin (MCBP) and testis phosphodiesterase Ca^{2+}-dependent regulatory protein (CDR) have been determined. Previous studies have shown that skeletal TnC (S-TnC) contains two high affinity Ca^{2+}-Mg^{2+} sites and two lower affinity Ca^{2+}-specific sites. We show that C-TnC, in contrast, lacks a Ca^{2+}-specific site. Since it has been predicted from sequence studies that site I (res. 29-40) of C-TnC does not bind Ca^{2+} while the other three sites are preserved, this would mean that site I of S-TnC (res. 27-38) is the location of one of the Ca^{2+}-specific sites. MCBP contains two high affinity Ca^{2+}-Mg^{2+} sites and CDR contains four Ca^{2+}-specific sites. We have partially determined the sequence of CDR and compared it to the known sequences of the above proteins. From these studies, it has been possible to classify some of the known Ca^{2+} binding site sequences as being either Ca^{2+}-Mg^{2+} or Ca^{2+}-specific. Of the nine Ca^{2+}-specific site sequences studied all contained glycine between the Y and Z Ca^{2+} coordination sites. We also show that TnC sites I and II are Ca^{2+}-specific and that sites III and IV are Ca^{2+}-Mg^{2+}.

The fluorescence of a Dansyl probe attached to S-TnC undergoes a large enhancement when Ca^{2+} binds to either of the two Ca^{2+}-specific sites. The kinetics of this Ca^{2+}-induced conformational change has been measured by stopped-flow spectrofluorometry. The results of these studies indicate that the conformational change induced by Ca^{2+} binding to the Ca^{2+}-specific sites occurs very rapidly. These results are consistent with the idea that the Ca^{2+}-specific sites are the sites which regulate muscle contraction. Dansyl labelled TnC has also been used for fluorescence depolarization measurements. The results of these studies show that TnC has a dimeter of $\sim37\text{Å}$ and the overall size does not change when 0, 2 or 4 Ca^{2+} are bound. Previous C.D. measurements of TnC suggested that Ca^{2+} binding to the Ca^{2+}-Mg^{2+} sites produces all of the α-helix change. However, our recent measurements show that about 35% of this change is actually produced by Ca^{2+} binding to the Ca^{2+}-specific regulatory sites.

CDR, MCBP and TnC all activate c-AMP phosphodiesterase (PDE) in the presence of Ca^{2+}. The Ca^{2+} dependence of this activation and of changes in CDR tyrosine fluorescence and C.D. have been compared. The changes in tyrosine fluorescence and C.D. appear to be induced by Ca^{2+} binding at any of the four Ca^{2+}-specific sites and the activation of PDE involves some type of cooperative process. PDE activation by MCBP is the first demonstration of a biological effect of this protein and these results may shed some light on the heretofore unknown biological role of MCBP.

Calcium Binding

To date, the most widely studied and characterized Ca^{2+} binding protein has been troponin-C (TnC) from rabbit skeletal muscle. The involvement of TnC in muscle regulation (1,2), its Ca^{2+} and Mg^{2+} binding properties (3), and effects of Ca^{2+} and Mg^{2+} on its conformation (4-7) have been carefully documented. The complete amino acid sequence of TnC from rabbit (8) and chicken (9) skeletal muscle (S-TnC) and from bovine cardiac muscle (C-TnC) (10) have also been determined.

Inquiries should be directed to Dr. James Potter, Dept. of Pharmacology and Cell Biophysics, University of Cincinnati College of Medicine, Cincinnati, Ohio 45267.

239

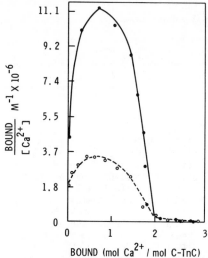

Figure 1. Scatchard plot of Ca^{2+} binding to bovine C-TnC. Ca^{2+} binding was carried out on pure C-TnC using the method previously described for S-TnC (3). Conditions: 0.1M KCl, 20mM imidazole, pH 7.0, 0.1mM EGTA. ●, no $MgCl_2$; O, 2mM $MgCl_2$. A calculated (3) amount of $CaCl_2$ was added to achieve the $[Ca^{2+}]$ indicated on the abscissa.

Several investigators (11-13) have noted the homology of TnC to paravalbumin (MCBP) and, based on this homology, have predicted the location of the four Ca^{2+} binding sites in the sequence of TnC. In addition, Kretsinger and Barry (14) have constructed a model of the predicted three dimensional structure of TnC. Ca^{2+} binding measurements (3) have established that rabbit S-TnC contains two high affinity Ca^{2+} binding sites that bind Mg^{2+} competitively (Ca^{2+}-Mg^{2+} sites) and two lower affinity sites that are specific for Ca^{2+} (Ca^{2+}-specific sites). By studying the Ca^{2+} binding properties and comparing the sequences of bovine C-TnC, carp MCBP"B" (15) and rat testis phosphodiesterase Ca^{2+}-dependent regulatory protein (CDR), it has been possible to identify the predicted Ca^{2+} binding site sequences in S-TnC as being either of the Ca^{2+}-Mg^{2+} or of the Ca^{2+}-specific type.

The Ca^{2+} binding results for C-TnC are presented in Figure 1. C-TnC contains two high affinity sites (K_{Ca} = 1.2 x 10^7 M^{-1}) whose affinity is lowered in the presence of 2 mM $MgCl_2$ (K_{Ca}' = 3.6 x 10^6 M^{-1}). The calculated K_{Mg} for these two sites is 1.2 x 10^3 M^{-1}. Thus, C-TnC contains two high affinity Ca^{2+}-Mg^{2+} sites similar to the two found in rabbit S-TnC (3), with the exception that the Scatchard plot of Ca^{2+} binding to C-TnC exhibits positive cooperativity. If one assumes that there is no heterogeneity between the two Ca^{2+}-Mg^{2+} sites in C-TnC, then the minimum degree of cooperativity (16) between the sites is 30, i.e., binding to either of the two unoccupied sites would increase the affinity of the other site by a factor of 30. It should be pointed out that although the Scatchard plot of Ca^{2+} binding to the Ca^{2+}-Mg^{2+} sites of S-TnC is linear, this does not prove that the binding is not cooperative since a linear Scatchard plot can be due to positive or no cooperativity (16).

C-TnC contains only one Ca^{2+}-specific site (K_{Ca} = 2 x 10^4 M^{-1}) in contrast to the two (3) found in S-TnC (K_{Ca} = 2 x 10^5 M^{-1}). van Eerd and Takashi (10) have predicted that C-TnC would bind only three Ca^{2+} since in Ca^{2+} site I (res. 29-40) the X and Y ASP residues, whose side chain carboxyls are required for Ca^{2+} coordination (15), are replaced by LEU and ALA respectively (Fig. 4). Thus, site

I in C-TnC is a defective Ca^{2+}-specific site. Since the other three C-TnC site sequences (C-TnC II-IV, Fig. 4) are essentially identical to those in S-TnC (S-TnC II - IV, Fig. 4), this would mean that region I (Fig. 4) of S-TnC (res. 29-40) is the location of one of its Ca^{2+}-specific sites.

Evidence presented below suggests that sites II in S-TnC and C-TnC (Fig. 4) are also Ca^{2+}-specific sites and it is not clear why the C-TnC site has a 10-fold lower affinity than its counterpart in S-TnC. Since sites I and II are close to each other in the proposed structure of TnC (14) it is possible that an altered conformation in region I of C-TnC may affect the conformation and Ca^{2+} affinity of site II.

The Ca^{2+} binding results for MCBP "B" are shown in Fig. 2. MCBP "B" contains two high affinity sites (K_{Ca} = 2.5 x 10^8 M^{-1}) whose affinity is lowered to 7.9 x 10^6 M^{-1} in the presence of 3 mM Mg^{2+}. The calculated K_{Mg} is 1.1 x 10^4 M^{-1}.

Again, the linear Scatchard plot does not reveal any information regarding cooperativity. It should also be mentioned that if these binding data are correct, it would not be possible to selectively remove one of the Ca^{2+} from MCBP (17). MCBP, therefore, contains two high affinity Ca^{2+}-Mg^{2+} sites whose affinity is the same as the two Ca^{2+}-Mg^{2+} sites in native unfractionated Tn (3). The lower affinity of the sites in TnC may be due to incomplete renaturation of TnC after the 6M urea purification step (7). The original affinity, however, can be restored by combining TnC with TnI (3). It is generally agreed (for review see 18) that MCBP evolved from a TnC-like ancestor with the C-terminal portion conserved. This would mean that the two Ca^{2+} sites (EF and CD, Fig. 4) remaining in MCBP (15) would correspond to sites III and IV of TnC (Fig. 4). Indeed, sequence comparisons show the C-terminal portions of TnC and MCBP to be quite similar (18). Since the two MCBP sites are Ca^{2+}-Mg^{2+} sites, this would mean that sites III and IV in S-TnC (and C-TnC) would also be Ca^{2+}-Mg^{2+} sites. Other studies (6) also suggest that site III in TnC is a Ca^{2+}-Mg^{2+} site, leaving sites I and II of TnC to be Ca^{2+}-specific sites and confirming the above conclusions regarding site I in C-TnC.

The Ca^{2+} binding properties of CDR from different tissues have been studied by many laboratories, yet there is no clear agreement about the number and affinity of Ca^{2+} binding sites. Part of the disagreement may stem from the fact that metal buffers were not employed to accurately regulate the free Ca^{2+} concentration. The technique (3) used in the present work overcomes this problem through the use of an EGTA-Ca^{2+} buffer system in combination with equilibrium dialysis. The Ca^{2+} binding results for rat testis CDR are shown in Fig. 3. CDR contains four Ca^{2+} sites with an affinity of 4.2 x 10^5 M^{-1} and Mg^{2+} does not affect their affinity (data not shown). Therefore, CDR contains four Ca^{2+}-specific sites with the same affinity as the Ca^{2+}-specific sites of S-TnC (3).

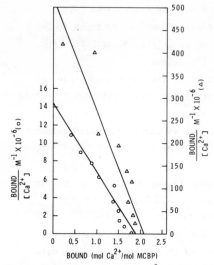

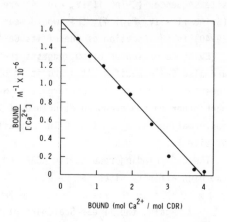

Fig.2. Scatchard plot of Ca^{2+} binding to MCBP"B". See Fig.1 for procedure. Conditions: △, 0.1M KCl, 20mM imidazole pH 7.5, 0.1mM EDTA; ○, 0.1M KCl, 20mM imidazole pH 7.0, 0.1mM EGTA, 3mM MgCl$_2$.

Fig.3. Scatchard plot of Ca^{2+} binding to CDR. See Fig.1 for procedure. Conditions: 0.1M KCl, 20mM imidazole pH 6.5, 0.1mM EGTA.

	X	Y	Z			-Y		-X			-Z	
STnC-I	ASP	ALA	ASP	GLY	GLY	GLY	ASP	ILE	SER	VAL	LYS	GLU #
STnC-II	ASP	GLU	ASP	GLY	SER	GLY	THR	ILE	ASP	PHE	GLU	GLU #
CDR-I	ASP	LYS	ASP	GLY	ASP	GLY	THR	ILE	THR	THR	LYS	GLU #
CDR-II	ASX	ASX	ASX	GLY	ALA	GLY	THR	ILE	ASX	PHE	PRO	GLX #
CDR-III	ASX	LYS	ASX	GLY	ASP	GLY	TYR	ILE	SER	ALA	GLU	ALA #
CTnC-I	LEU	GLY	ALA	GLU	ASP	GLY	CYS	ILE	SER	THR	LYS	GLU , #
CTnC-II	ASP	GLU	ASP	GLY	SER	GLY	THR	VAL	ASP	PHE	LYS	ASP #
STnC-III	ASP	ARG	ASN	ALA	ASP	GLY	TYR	ILE	ASP	ALA	GLU	GLÛ *
STnC-IV	ASP	LYS	ASN	ASN	ASP	GLY	ARG	ILE	ASP	PHE	ASP	GLU *
DLC	ASP	GLN	ASN	ARG	ASN	GLY	ILE	ILE	ASP	LYS	GLU	ASP *
CTnC-III	ASP	LYS	ASN	ALA	ASP	GLY	TYR	ILE	LEU	ALA	GLU	GLU *
CTnC-IV	ASP	LYS	ASN	ASN	ASP	GLY	ARG	ILE	ASP	TYR	ASP	GLU *
MCBP-EF	ASP	GLN	ASP	LYS	SER	GLY	PHE	ILE	GLU	GLU	ASP	GLU *
MCBP-CD	ASP	SER	ASP	GLY	ASP	GLY	LYS	ILE	GLY	VAL	ASP	GLU *

Fig.4. Comparison of Ca^{2+} binding site sequences of different Ca^{2+} binding proteins. S-TnC I-IV, rabbit S-TnC (18); DLC, DTNB light chain of rabbit fast skeletal muscle myosin (19); C-TnC I-IV, bovine C-TnC (10); CDR I-III, rat testis PDE Ca^{2+} dependent regulatory protein. # = Ca^{2+}-specific site, * = Ca^{2+}-Mg^{2+} site. X, Y, Z, -X, -Y, -Z are the residues involved in Ca^{2+} coordination.

242

The Ca^{2+} site sequences of S-TnC and C-TnC, the rat testis CDR Ca^{2+}-specific site sequences and the single Ca^{2+}-Mg^{2+} site sequence of the DTNB light chain are compared in Fig. 4. This comparison is interesting since we can now assign them as being Ca^{2+}-Mg^{2+} or Ca^{2+}-specific and look for clues which determine the nature of these sites. The most obvious finding is that all of the Ca^{2+}-specific site sequences contain GLY between the Y and Z coordination sites, with the exception of C-TnC-I which contains a GLU in this position. It should be remembered, however, that C-TnC-I does not bind Ca^{2+} due to two other amino acid replacements (see above). In addition, the two Ca^{2+}-specific sites in chicken S-TnC (9) contain GLY at that position. Kretsinger (2) has suggested that the GLY in this position might account for the lower affinity of the Ca^{2+}-specific sites and the data presented here support this hypothesis; also 6 of the 7 Ca^{2+}-Mg^{2+} sites studied here (Fig. 4) as well as those of chicken S-TnC (9) do not contain a GLY at this position. It seems clear from these comparisons that a GLY at this position may be an important determinant of whether a site is Ca^{2+}-specific.

Conformational Studies

Recent studies (3) on the effect of Ca^{2+} and Mg^{2+} on myofibrillar ATPase suggest that the Ca^{2+}-specific sites, but not the Ca^{2+}-Mg^{2+} sites of S-TnC, are involved in the regulation of muscle contraction, although it is difficult to unequivocally exclude the involvement of the latter. In any event, Ca^{2+} must be bound to the Ca^{2+}-specific sites before the ATPase is switched on (3,21) making binding to these sites a key step in the regulation of muscle contraction. Much is known about the conformational changes which occur when Ca^{2+} binds to the Ca^{2+}-Mg^{2+} sites (4-7), but little is known about the conformational changes associated with Ca^{2+} binding to the Ca^{2+}-specific regulatory sites (6). Recent measurements (7) of the enthalpy of Ca^{2+} binding to S-TnC suggest that some change in protein conformation is occurring upon binding of Ca^{2+} to these sites. We report here additional evidence that conformational changes occur when Ca^{2+} binds to the Ca^{2+}-specific sites.

S-TnC has been labelled with dansylaziridine (DANZ). The reactivity of DANZ to TnC when all four of its Ca^{2+} sites are filled is twice as high as when two or no Ca^{2+} are bound. Thus, the reactivity toward this label is greatly enhanced by Ca^{2+} binding at the Ca^{2+}-specific sites. The labelling appears to be specific since only one mole of DANZ reacts per mole of TnC. Fig. 5 shows that the fluorescence of TnC_{DANZ} is enhanced approximately two fold when Ca^{2+} binds to the Ca^{2+}-specific sites and this is accompanied by a 10 nm blue shift (see inset). In addition, the Ca^{2+} titration is essentially the same in the absence of Mg^{2+}. Labelling of TnC with DANZ does not seriously perturb the structure of Ca^{2+} induced structural changes in TnC, since equilibrium dialysis revealed that TnC_{DANZ} still binds four moles of Ca^{2+} per mole and the Ca^{2+} induced C.D. changes

243

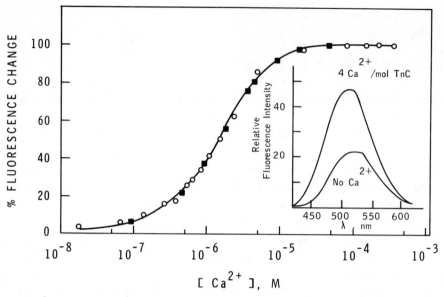

Fig.5. Ca^{2+}-dependence of TnC_{DANZ} fluorescence. Conditions: $3 \times 10^{-6}M$ TnC_{DANZ}, 2mM EGTA, 2.4mM $MgCl_2$, 90mM KCl, 10mM phosphate (K^+) buffer pH 7.0, 23°. A calculated (6) amount of $CaCl_2$ was added to achieve the $[Ca^{2+}]$ shown on the abscissa and the pH was maintained at 7.0 with KOH. o, TnC_{DANZ} fluorescence, (excitation=340nm and emission=510nm). ■, theoretical fluorescence change produced by Ca^{2+} binding to either of the two Ca^{2+}-specific sites. Insert: fluorescence spectra of TnC_{DANZ}.

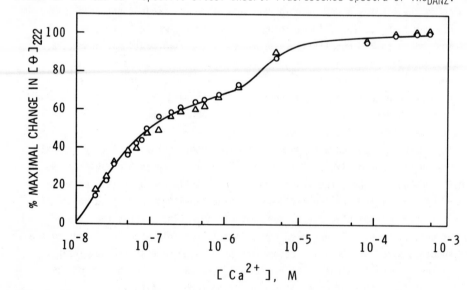

Fig. 6. Circular dichroism of TnC and TnC_{DANZ}. Conditions: 0.2 mg/ml protein, 2mM EGTA, 90mM KCl, 10mM phosphate (K^+) buffer pH 7.0, 23°. $\Delta[\theta]_{222}$ = 6,396 and 5,848 for TnC (o) and TnC_{DANZ} (△) respectively. Ca^{2+} titrations were performed as in Fig. 5.

are the same for TnC$_{DANZ}$ as for native TnC (see Fig. 6). The K$_{Ca}$ for this change calculated from the midpoint of the transition (Fig. 5) is 4 x 10^5 M^{-1}. A theoretical curve assuming that these changes arise from Ca^{2+} binding to <u>either</u> of two Ca^{2+}-specific sites of TnC is shown in Fig. 5 and suggests that the change in fluorescence arises from a conformational change around the bound fluorophore occurring with Ca^{2+} binding to <u>either</u> of the two Ca^{2+}-specific regulatory sites. This is the first report of a conformational probe which <u>only</u> detects Ca^{2+} binding to these sites.

It is now well established that Ca^{2+} binding to the Ca^{2+}-specific sites is the key step in the regulation of muscle contraction (3,21). It would be of great interest to know the rate at which the Ca^{2+}-specific site induced conformational changes occur. To investigate this, we have used stopped-flow spectrofluorometry to measure the rate of change of TnC$_{DANZ}$ fluorescence. The instrument (Durrum) used for these measurements has a mixing time of approximately 2 msec. In these experiments, we studied the rate of change of TnC$_{DANZ}$ fluorescence on going from the state where both Ca^{2+}-Mg^{2+} sites are filled (but not the two Ca^{2+}-specific sites) to the state where all four Ca^{2+} sites are filled. Under these conditions, the fluorescence was enhanced approximately two fold as in Fig. 5 and was due only to Ca^{2+} binding at the Ca^{2+}-specific sites. The rate of this change was so fast that it was complete within the mixing time of the instrument. Since muscle contraction occurs very rapidly, one would expect the Ca^{2+}-specific regulatory site induced conformational changes to be rapid since it is these conformational changes which presumably (3,21) initiate the series of events (1) that allow myosin and actin to interact. Previous workers have examined the rate of conformational changes induced by Ca^{2+} binding to the Ca^{2+}-Mg^{2+} sites (22). These workers used an instrument that had a longer mixing time (20 msec) than ours and found a rate constant of 13.7 sec^{-1} which was preceded by a faster reaction. This rate is considerably slower than those induced by Ca^{2+} binding at the Ca^{2+}-specific sites reported here.

Fluorescence depolarization spectroscopy has proven useful in determining the rates of rotation of macromolecules labelled with fluorescent dyes (23). If a protein dye complex is spherical (globular), then a plot of -Log anisotropy (A) versus time should be linear and the rotational correlation time ϕ of the protein may be evaluated from the slope. ϕ can then be used to determine the volume of the hydrated sphere (protein). Plots of -Log A versus time for TnC$_{DANZ}$ with 0, 2 and 4 Ca^{2+} bound per mole TnC$_{DANZ}$ are shown in Fig. 7. ϕ for each of these plots was found to be 7.3 nsec, suggesting in each case that TnC$_{DANZ}$ rotates as a globular protein of $\sim$37Å in diameter. Thus, the Ca^{2+} induced conformational changes reported by other techniques apparently do not alter the overall size or shape of TnC.

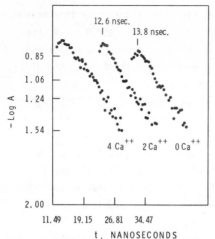

12.6 nsec.

13.8 nsec.

4 Ca^{++} 2 Ca^{++} 0 Ca^{++}

−Log A

0.85
1.06
1.24
1.54

2.00

11.49   19.15   26.81   34.47

t, NANOSECONDS

Fig. 7. Fluorescence depolarization of TnC$_{DANZ}$. The -Log of the anisotropy is plotted as a function of time. Fluorescence decay curves of the parallel and perpendicular components were collected for 2000 seconds each with a modified Ortec nanosecond time resolved single photon counting spectrofluorometer described elsewhere (24). Excitation was through a Corning CS 7-51 filter and emission was monitored at 510 nm. Conditions: 3×10^{-6} M TnC$_{DANZ}$, 2 mM EGTA, 90 mM KCl, 10 mM phosphate (K$^+$) buffer, pH 7.0, 23°. 0 Ca^{2+} (bound/mol TnC$_{DANZ}$) = no added Ca^{2+}; 2 Ca^{2+} (bound/mol TnC$_{DANZ}$) = 1.4×10^{-7} M [Ca^{2+}]; 4 Ca^{2+} (bound/mol TnC$_{DANZ}$) = 4×10^{-4} M [Ca^{2+}]. The 0 Ca^{2+} and 2 Ca^{2+} decay curves have been shifted in time for comparison.

There have been several conflicting reports on the Ca^{2+} induced changes in ellipticity or percent α-helix in TnC (4,5). We have reexamined the Ca^{2+} dependence of this change using an EGTA-Ca^{2+} buffer system and the percent change in [θ]$_{222}$ as a function of [Ca^{2+}] is shown in Fig. 6. We find that $\sim$35% of the change in ellipticity occurs with Ca^{2+} binding to the Ca^{2+}-specific sites. It is clear from Fig. 6 that the ellipticity changes are biphasic and may be separated into two distinct changes. The K$_{Ca}$ estimated from the midpoints of each of these transitions are 2.7×10^7 M^{-1} and 3.3×10^5 M^{-1} representing binding to the Ca^{2+}-Mg^{2+} and Ca^{2+}-specific sites, respectively. The total change in α-helix is from $\sim$34% to $\sim$50% in agreement with Kawasaki and van Eerd (4). However, the Ca^{2+} dependence of this change (Fig. 6) is quite different from theirs. Thus, changes in C.D. as well as in TnC$_{DANZ}$ fluorescence occur upon Ca^{2+} binding to the Ca^{2+}-specific regulatory sites.

C-AMP Phosphodiesterase Activation by CDR, TnC and MCBP

Much attention has been focused lately on the Ca^{2+}-dependent regulation of c-AMP phosphodiesterase (PDE) and of some adenylate cyclases. Cheung (25,26) and Kakiuchi et al (27) have reported that a soluble, low molecular weight thermostable protein can activate PDE in brain. Similar proteins have been found in almost every tissue tested (28-30) and the activation of PDE by these proteins requires micromolar levels of calcium (30-32). Thus, the protein has been referred to as a calcium-dependent regulatory protein, or CDR (33). The activation is not limited, however, to phosphodiesterases since Brostrom et al (33) and Lynch et al (34) have demonstrated that solubilized brain adenylate cyclase can also be regulated by CDR. Lin et al (35) and Teo et al (36) initially purified CDR from bovine brain and heart (respectively) and found it to be acidic calcium-binding protein with a molecular weight of approximately 15,000 to 20,000. Wang and coworkers (37,38) first noted the structural similarity

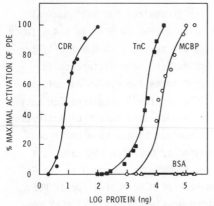

Fig. 8. Activation of PDE by S-TnC, CDR and MCBP. The percent maximal activation of PDE is plotted as a function of protein concentration. The assays were carried out as previously described (40).

between CDR and TnC. Based on preliminary structural studies of bovine brain CDR, other workers (39) have noted similar sequences between CDR and TnC, suggesting a homologous origin of the protein. We have previously described the biological similarities between these proteins; rat testis CDR and rabbit S-TnC will substitute for one another in their respective biological systems (40). We report here the partial sequence of rat testis CDR as well as the relationship of CDR Ca^{2+} binding and conformational changes to PDE activation. We also report the first demonstration of PDE activation by MCBP and the relationship of MCBP Ca^{2+} binding to this activation.

A comparison of CDR, TnC and MCBP "B" activation of PDE is shown in Fig. 8. The amount of testis CDR required for half-maximal stimulation of rat brain PDE in the presence of Ca^{2+} is 8 ng. The same amount of PDE requires 5 and 11 μg of TnC (data from (40)) and MCBP respectively for half-maximal stimulation. Bovine serum albumin (BSA), also an acidic protein, has no effect on the activity of PDE at concentrations up to 200 μg per assay. The lack of activation by BSA demonstrates that the stimulatory nature of MCBP is not a non-specific ionic or acidic effect. Not only does MCBP activate PDE, but this activation is Ca^{2+} dependent. The Ca^{2+}-dependence of CDR activation of PDE is shown in Fig. 9 and of MCBP activation in Fig. 10.

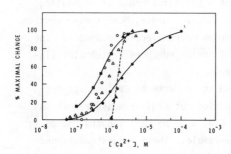

Fig. 9. Comparison of CDR Ca^{2+} binding, circular dichroism, tyrosine fluorescence and PDE activation. The Ca^{2+} binding data (●) are from Fig. 3. The C.D. (O) and tyrosine fluorescence (△) measurements were carried out using the same conditions; 0.2 mg/ml protein, 2 mM EGTA, 90 mM KCl and 10 mM phosphate (K^+) buffer, pH 7.0. Ca^{2+} titrations were performed as described in the legend to Fig. 5. The Ca^{2+} dependence of PDE activation was carried out (▲) as previously described (40). (■) Theoretical fluorescence or C.D. Change produced by Ca^{2+} binding to any of the four CDR Ca^{2+}-specific sites.

247

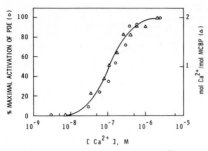

Fig. 10. Comparison of MCBP Ca^{2+} binding and MCBP activation of PDE. The Ca^{2+} binding data are from Fig. 3. (3 mM MgCl$_2$). The Ca^{2+} dependence of PDE activation was carried out as previously described (40).

The [Ca^{2+}] required for half-maximal activation of PDE by MCBP is 1.4 x 10^{-7}M in contrast to 1.2 x 10^{-6} M [Ca^{2+}] for CDR activation. The difference in slopes of these two curves as well as the different [Ca^{2+}] required for activation demonstrates that MCBP activation of PDE can not be due to a small contamination by CDR. In Fig. 10, Ca^{2+} binding to MCBP is compared to the Ca^{2+}-dependence of MCBP activation of PDE. As shown previously (Fig. 2) MCBP contains two Ca^{2+}-Mg^{2+} sites and it appears from Fig. 10 that binding to a specific one of these sites is responsible for the activation.

The function of the MCBP's is still not entirely clear. The fact that they bind Ca^{2+} with a high affinity has led many to hypothesize that they function as a Ca^{2+} trap or transport protein. Our data demonstrate that homogenous MCBP "B" purified from carp muscle can regulate rat brain phosphodiesterase in a Ca^{2+}-dependent manner. The amount necessary for maximal activation was considerably higher than the amount of native CDR required. The difference in the activation of the brain enzyme by the two proteins may be due to species and/or tissue differences. Since troponin C can also activate phosphodiesterase (40), the results regarding MCBP suggest that these homologous proteins have a similar enzyme binding site even though MCBP is about half the size of CDR and TnC. Noteworthy is the fact that parvalbumins of fish muscle are found in concentrations approximately 1000-fold greater than those employed in the phosphodiesterase assays. Should parvalbumin prove not to be a phosphodiesterase regulator in vivo, the present results provide supportive evidence of a biological role in regulating other Ca^{2+}-dependent processes which occur in muscle. Hence, MCBP's may function, in the intact cell, as subunits of a more complex macromolecular assembly. The finding that the parvalbumins are also present in the advanced vertebrates (including mammals) suggests a universal function.

Several recent papers (41,42) have examined the effects of Ca^{2+} on various physical properties of brain CDR. We have carried out similar measurements on testis CDR and the results are summarized in Fig. 9. The Ca^{2+} induced change in C.D. ([θ]$_{222}$) is from 14,696 to 18,196 deg cm^2/dmole. These values correspond to a change in α-helix from 45% to 54%. This is the first report where the free Ca^{2+} concentration was carefully controlled through the use of an EGTA-Ca^{2+} buffer system. Changes in C.D. have been noted previously (41,42), but the range

248

of $[Ca^{2+}]$ needed to produce these changes was not accurately determined. When Ca^{2+} binds to CDR there is also a 2.5 fold enhancement of tyrosine fluorescence (Fig. 9). The shapes of the C.D. and tyrosine fluorescence curves are very similar and the midpoint of the transitions occur around 7×10^{-7} M $[Ca^{2+}]$. These curves are shifted to the left of the Ca^{2+} binding curve and this may be explained by the theoretical curve shown in Fig. 9 which represents the theoretical change of either the C.D. or tyrosine fluorescence if Ca^{2+} binding to any of the four CDR Ca^{2+}-specific sites can produce the change. Although the curves are not exactly superimposable, they seem to fit this model reasonably well. The CDR activation curve is also shown in Fig. 9 and it is clear that most of the physical changes are nearly complete before any activation of PDE is observed. In addition, the activation curve is much steeper than the C.D. or fluorescence curves, indicating a cooperative activation mechanism.

It appears from these findings that the binding of Ca^{2+} to any of the four Ca^{2+}-specific CDR sites can produce these conformational changes and that these changes are essentially complete before activation occurs. This suggests that more than one Ca^{2+} must be bound to CDR for activation to occur. The activation step may involve the initial binding of CDR to PDE followed by a cooperative step, e.g., the affinity of the remaining CDR Ca^{2+} sites for Ca^{2+} is increased by the interaction with PDE and binding of the subsequent calciums brings about the activation. Another possibility would be that more than one CDR molecule must be bound to PDE before activation occurs and binding of the first CDR to PDE may facilitate the binding of the second (positive cooperativity). There are, of course, many other possibilities and further knowledge of the exact mechanism of activation will require knowing more about the CDR-PDE interaction.

ACKNOWLEDGMENTS

The research was supported in part by a grant from the Muscular Dystrophy Association, by the American Heart Association (75-818) and by HEW Grant HD-07503. JDP and RLJ are Established Investigators of the American Heart Association, JDJ is a Fellow of the Muscular Dystrophy Association, JRD has a NIH Postdoctoral Fellowship HD-01925-03. ARM is a NIH Research Career Development Awardee.

1. Potter, J.D. and Gergely, J. (1974) Biochemistry 13: 2697-2703.
2. Hitchcock, S.E. (1975) Eur. J. Biochem. 52: 255-263.
3. Potter, J. D. and Gergely, J. (1975) J. Biol. Chem. 250: 4628-4633.
4. Kawasaki, Y. and vanEerd, J.P. (1972) Biochem.Biophys.Res.Commun.49: 898-905.
5. Murray, A. and Kay, C.M. (1972) Biochemistry 11: 2622-2627.
6. Potter, J.D., Seidel, J.C., Leavis, P., Lehrer, S.S. and Gergely, J. (1976) J. Biol. Chem. 251: 7551-7556.
7. Potter, J. D., Hsu, F.J. and Pownall, H.J. (1977) J.Biol.Chem.252:2452-2454.
8. Collins, J.H., Greaser, M.L., Potter, J.D. and Horn, M.J. (1977) J. Biol. Chem., In Press.
9. Wilkinson, J. M. (1976) FEBS Letters 70: 254-256.
10. vanEerd, J.P. and Takahashi, K. (1976) Biochemistry 15: 1171-1180.
11. Collins, J.H., Potter, J.D., Horn, M.J., Wilshire, G. and Jackman, N. (1973) FEBS Letters 36: 268-272.
12. Weeds, A.G. and MacLachlan, A. D. (1974) Nature 252: 646-649.
13. Tufty, R.M. and Kretsinger, R.H. (1975) Science 187: 167-169.
14. Kretsinger, R.H. and Barry, C.D. (1975) Biochim.Biophys.Acta 405: 40-52.
15. Kretsinger, R.H. and Nockolds, C.E. (1973) J. Biol. Chem. 248: 3313-3326.
16. Mandel, F. (1977) Biophysical J. 17: 279a.
17. Donato, H. and Martin, R.B. (1974) Biochemistry 13: 4575-4579.
18. Collins, J.H. (1976) in: Calcium in Biological Systems (Soc. Exptl.Biol.Symp. 30, C.J. Duncan, ed). The University Press, Cambridge, pp. 303-334.
19. Collins, J.H. (1976) Nature 259: 699-700.
20. Kretsinger, R.H. (1974) in: Perspectives in Membrane Biology (S. Estrada-O. and C. Gitler, eds), Academic Press, New York, pp. 229-262.
21. Bremel, R.D. and Weber, A. (1972) Nature New Biol. 238: 97-101.
22. Iio, T., Mihashi, K. and Kondo, H. (1976) J. Biochem. 79: 689-691.
23. Yguerabide, J. (1973) in: Fluorescent Techniques in Cell Biology (A. A. Thaer and M. Sernitz, eds). Springer-Verlag, New York, pp. 311-331.
24. Avouris, J.K. and El-Bayoumi, A. (1974) Chem. Phys. Lett. 26: 373-376.
25. Cheung, W.Y. (1970) Biochem.Biophys.Res.Commun. 33: 533-538.
26. Cheung, W.Y. (1971) J. Biol. Chem. 246: 2859-2869.
27. Kakiuchi, S., Yamazaki, R. and Nakajima, H. (1970) Proc.Jap.Acad.46: 587-592.
28. Waisman, D., Stevens, F.D. and Wang, J.H.(1975) Biochem.Biophys.Res.Commun. 65: 975-982.
29. Smoake, J.A., Song, S.Y. and Cheung, W.Y.(1974) Biochim.Biophys.Acta 341: 402-411.
30. Kakiuchi, S., Yamazaki, R., Teshima, R., Uenishi, K. and Miyamoto, E. (1975) Biochem. J. 146: 109-120.
31. Teo, T. S. and Wang, J. H. (1973) J. Biol. Chem. 248: 5950-5955.
32. Wolff, D.J. and Brostrom, C.O. (1974) Arch.Biochem.Biophys. 163: 349-358.
33. Brostrom, C.O., Huang, Y.C., Breckenridge, B. M. and Wolff, D. J. (1975) Proc. Nat. Acad. Sci. USA 72: 64-68.
34. Lynch, T. J., Tallant, E. A. and Cheung, W. Y. (1976) Biochem. Biophys. Res. Commun. 68: 616-624.
35. Lin, Y. M., Liu, Y.P. and Cheung, W.Y. (1974) J.Biol. Chem. 249: 4943-4954.
36. Teo, T.S., Wang, T.H. and Wang, J.H. (1973) J.Biol.Chem. 248: 588-595.
37. Wang, J.H., Teo, T.S., Ho, H.C. and Stevens, F.C. (1974) Adv. Cyc. Nucleotide Res. 5: 179-194.
38. Stevens, F.C., Walsh, M., Ho, H.C., Teo, T.S. and Wang, J.H. (1976) J. Biol. Chem. 251: 4495-4500.
39. Watterson, D.M., Harrelson, W.G.,Jr., Keller, P.M., Sharief, F. and Vanaman, T.C. (1976) J. Biol. Chem. 251: 4501-4513.
40. Dedman, J.R., Potter, J.D. and Means, A.R. (1977) J.Biol.Chem. 252:2437-2440
41. Liu, Y.P. and Cheung, W.Y. (1976) J.Biol.Chem. 251: 4193-4198.
42. Klee, C.B. (1977) Biochemistry 16: 1017-1024.

The Role of Ca^{2+} and Myosin Phosphorylation in Regulating Actomyosin in Smooth Muscle and Non-Muscle Cells

Robert S. Adelstein, Samuel Chacko, Stylianos P. Scordilis, Barbara Barylko,
Mary Anne Conti and John A. Trotter
Section on Molecular Cardiology,
National Heart, Lung, and Blood Institute,
Bethesda, Maryland   USA   20014

SUMMARY

1) Phosphorylation of the 20,000 dalton light chain of platelet, proliferative myoblast and vas deferens myosin results in an increased actin-activated
myosin ATPase activity. Dephosphorylation decreases this activity.

2) The kinase catalyzing the phosphorylation of platelet and proliferative
myoblast myosin is not dependent on Ca^{2+} for its activity. The kinase catalyzing
the phosphorylation of vas deferens myosin is dependent on Ca^{2+}.

3) The actin-activated myosin ATPase activity of phosphorylated vas deferens
myosin is inhibited by EGTA in the absence of kinase and phosphatase. The actin-
activated myosin ATPase activity of phosphorylated platelet and proliferative
myoblast myosin is not altered when Ca^{2+} is removed.

4) A model for the manner by which Ca^{2+} may regulate vas deferens smooth
muscle actin-activated myosin ATPase activity, is presented.

INTRODUCTION

The contractile proteins actin and myosin are present in many, if not all
eukaryotic cells. Moreover, many of the structural and biological properties
first characterized for these muscle proteins are also found in their analogous
non-muscle proteins.

Myosin from both muscle and non-muscle sources is composed of six polypeptide
chains. Two of the chains can be classified as heavy chains (molecular weight
200,000) and four as light chains (molecular weight 15-25,000). In each molecule
of myosin the light chains are probably present as two identical pairs (see Fig. 1).

Ca^{2+} plays an important role in initiating muscle contraction as well as in
regulating numerous cellular functions in non-muscle cells. In muscle cells there
are two well-described mechanisms by which Ca^{2+} regulates actin-myosin interaction.
First, in skeletal and cardiac muscle Ca^{2+} ions initiate the contractile event by
binding to the regulatory protein troponin-C.[1] This removes the inhibition of the
actin-myosin interaction imposed by the complex of troponin-tropomyosin in the absence of the divalent cation. When assayed in vitro in the absence of the regulatory proteins troponin and tropomyosin, actin-activation of the myosin ATPase
activity occurs in the presence and absence of Ca^{2+}. Addition of troponin-

251

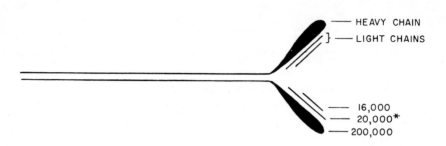

<table>
<tr><td></td><td>—— HEAVY CHAIN</td></tr>
<tr><td></td><td>} —— LIGHT CHAINS</td></tr>
<tr><td></td><td>—— 16,000</td></tr>
<tr><td></td><td>—— 20,000*</td></tr>
<tr><td></td><td>—— 200,000</td></tr>
</table>

Fig. 1. Schematic representation of the platelet myosin molecule. *Indicates the phosphorylated light chain. (From Adelstein et al.[22] after Lowey et al.[23]).

tropomyosin to the incubation mixture confers "Ca^{2+} sensitivity"; that is, the activation of myosin ATPase by actin is reduced as the Ca^{2+} concentration is lowered.

A second mechanism for the regulation of actin-myosin interaction has been described by Szent-Gyorgyi and his colleagues for the contractile proteins of scallops.[2] In this case Ca^{2+} regulation is mediated by a particular light chain of myosin. When a full complement of myosin light chains is present the scallop myosin ATPase activity can only be activated by actin in the presence of Ca^{2+}. Removal of 50% of this "regulatory" light chain of scallop myosin results in a loss of "Ca^{2+} sensitivity" that is, actin-activation can now occur in the presence of both Ca^{2+} and EGTA. Ca^{2+} sensitivity can be restored by readdition of the scallop regulatory light chains, or by substitution of the 20,000 dalton light chain of smooth muscle or cardiac muscle, or the 19,000 dalton light chain of skeletal muscle myosin in place of the scallop light chain.[3]

In this paper we shall describe a different type of regulatory mechanism that controls actin-myosin interaction in smooth muscle and non-muscle cells. This regulation results from the reversible phosphorylation of the 20,000 dalton light chain of myosin which is catalyzed by the specific enzymes myosin light chain kinase and phosphatase.[4] We shall review evidence showing that: a) phosphorylation of vas deferens,[5] platelet[6] and proliferative myoblast myosin[7] results in an increase in the actin-activated myosin ATPase activity. Dephosphorylation of vas deferens and platelet myosin decreases this activity; b) Ca^{2+} is required by the smooth muscle kinase for activity, but is not required by the myosin light chain kinase isolated from non-muscle cells, c) phosphorylated vas deferens myosin which has been purified to remove the kinase and phosphatase, has a higher actin-activated ATPase activity in the presence of Ca^{2+} than EGTA. Finally, we shall present a scheme indicating how Ca^{2+} may work to regulate the actin-activation of vas deferens myosin in smooth muscle cells.

METHODS

All procedures were carried out at 4°C unless otherwise noted. Deionized water was used throughout. Human blood platelets, guinea pig vas deferens and Yaffe L-5 810 cloned proliferative myoblasts were used as sources of myosin and the phosphorylating systems.

The details of the preparation and purification of the proteins, as well as the techniques used for phosphorylating myosin and assaying myosin kinase and ATPase activities have previously been published; for platelets, see Adelstein and Conti[6], for proliferative myoblasts, Scordilis and Adelstein[7] and for vas deferens smooth muscle, Chacko, et al.[5]

The partially-purified platelet phosphatase was prepared from an alcohol-ether powder[8] of platelets as described by Barylko et al[9] (see also 10).

Phosphatase activity was assayed by using an isolated fraction[11] of ^{32}P-labelled platelet, rabbit skeletal muscle or smooth muscle light chains as a substrate. The incubation mixture consisted of 4 mM $MgCl_2$, 0.1 mM $CaCl_2$, 10 mM Imidazole-HCl (pH 7.3) and 20 mM KCl. The assay was carried out at room temperature with enough phosphatase to allow for a 50% decrease in ^{32}P cpm in approximately 20-30 min. The extent of dephosphorylation was determined using a Millipore sampling manifold with Millipore HA filters.[12]

Incorporation, release, and the exact location of ^{32}P, as well as the purity of proteins, was monitored by 1% SDS-7 1/2 or 10% polyacrylamide gel electrophoresis according to Fairbanks et al.[13] In some cases diallyltartardiamide (Biorad) was substituted for bis-acrylamide in the same molar ratio to the acryl-amide[14] to maximize the recovery of radioactivity. Gels were stained with 0.03% Coomassie brilliant blue, scanned at 584 nm using a Gilford Model 2520 gel scanner and the radioactivity eluted as previously described.[5,7]

RESULTS

Fig. 2 is a time course of ^{32}P incorporation into the 20,000 dalton myosin light chain of a) guinea pig vas deferens smooth muscle myosin, b) rat proliferative myoblast myosin. Both experiments were carried out with ammonium sulfate fractions (30-70% for the vas deferens and 35-55% for the myoblast) and γ-labelled AT^{32}P. Whereas the phosphorylation of the vas deferens myosin is dependent on the presence of Ca^{2+} (2a), phosphorylation of myoblast myosin proceeds in the presence of Ca^{2+} or EGTA (2b). The latter finding is similar to that for platelet myosin phosphorylation.[12]

The difference in these two phosphorylating systems is due to the presence of a Ca^{2+} independent kinase in the case of proliferative myoblasts and a Ca^{2+} dependent enzyme in the case of vas deferens smooth muscle. Table 1 illustrates

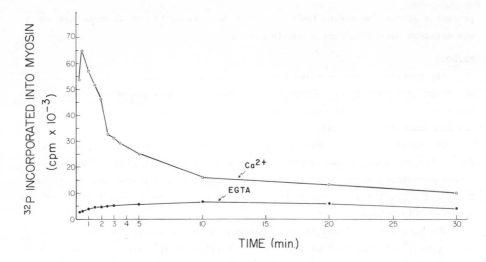

Fig. 2a. Time course for the phosphorylation of vas deferens myosin. The 35–70% ammonium sulfate fraction was incubated in the presence of 0.1 mM ATP and either 0.1 mM $CaCl_2$ or 2 mM EGTA. Aliquots were removed at the indicated times and incorporation of ^{32}P was determined. Maximum incorporation in this preparation was 0.4 mol of P_i/mol of 20,000–dalton light chain (from Chacko et al.[5]).

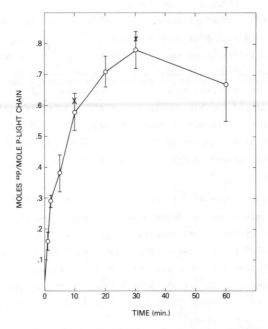

Fig. 2b. Time course for the phosphorylation of proliferative myoblast myosin. The 35–55% ammonium sulfate fraction was incubated with either 0.2 mM $CaCl_2$ (O) or 2 mM EGTA (x) in the presence of 0.1 mM ATP. Aliquots were removed at the designated times and ^{32}P incorporation determined as outlined previously[7] (from Scordilis and Adelstein,[7] modified for inclusion of EGTA).

254

Table 1

Effect of Calcium on the Activity of the Myosin Light Chain Kinase

Kinase	Substrate	Moles ^{32}P Incorporated per Mole P-Light Chain	
		+ Ca^{2+}	− Ca^{2+}
Proliferative Myoblast	Myoblast Myosin	0.75	0.81
	Skeletal Myosin Light Chains	0.91	0.86
Rat Thigh Muscle	Skeletal Myosin Light Chains	0.93	0.01
	Platelet Myosin Light Chains[1]	0.84	0.01

Data is from reference 7 except [1]unpublished observation S.P. Scordilis.

Table 2

Myosin Phosphorylation and the Actin-Activated ATPase Activity

Myosin Source	Phosphorylated		Dephosphorylated		Unphosphorylated
	ATPase	P_i	ATPase	P_i	ATPase
Platelets	.21	.8−.9	.05	.3	.03
Proliferative Myoblasts	.18	.8			.02
Vas Deferens	.10	.4−.6	.01	<.05	.01

ATPase: μMoles P_i/mg myosin/min at 37°

$\underline{P_i}$ Moles P_i bound/mole of 20,000 dalton light chain

that in addition to phosphorylating its own myosin, the Ca^{2+} independent kinase
from cultured myoblasts can phosphorylate the isolated myosin light chains from
skeletal muscle myosin. This phosphorylation, in contrast to that catalyzed by
the rat skeletal muscle kinase, proceeds in the presence or absence of Ca^{2+}. The
table also shows that Ca^{2+} dependence is a property of the kinase and not of the
light chains, since the skeletal muscle kinase retains its Ca^{2+} requirement when
platelet light chains are used for a substrate.

Myosin light chain phosphorylation has been shown to markedly increase the
actin-activation of myosin ATPase activity in three systems studied to date: a)
platelets,[6] b) smooth muscle (vas deferens,[5] chicken gizzard[15,16] and pig stomach,[20]) and c) cultured myoblasts.[7] Table 2 summarizes data for phosphorylated
and non-phosphorylated myosin from each of these systems relating the extent of
phosphorylation to the actin-activated myosin ATPase activity.

Dephosphorylation of previously phosphorylated platelet myosin was accomplished by incubating the myosin in the presence of a partially purified platelet
myosin phosphatase.[10] Although dephosphorylation of platelet and vas deferens
myosin lowered the actin-activated ATPase activity and released radioactive ^{32}P
from the 20,000 dalton light chain, it had no effect on the K^+-EDTA and Ca^{2+} activated ATPase activities measured at high (0.5 M) KCl concentration.[5,10] This
finding as well as the preservation of the original pattern of myosin heavy and
light chains on SDS-polyacrylamide gel electrophoresis, indicated that dephosphorylation did not denature or degrade the myosin molecule.

Dephosphorylation of smooth muscle myosin was accomplished by extending the
incubation period to 20 minutes (see Figure 2a). This resulted in dephosphorylation of the 20,000 dalton light chain by the endogenous light chain phosphatase,
after the ATP had been depleted. Unphosphorylated myosin refers to myosin isolated in the unphosphorylated state and not subsequently phosphorylated (Table 2).

Table 3 shows the effect of Ca^{2+} and EGTA on the actin-activated ATPase of
purified phosphorylated smooth muscle myosin isolated from guinea pig vas deferens,
as well as phosphorylated human platelet and rat proliferative myoblast myosin.
The myosin in each case was purified from the ammonium sulfate fraction by chromatography on Sepharose 4B, which separates actomyosin and myosin from the endogenous kinase, phosphatase and other proteins. In contrast to the phosphorylated myosins from non-muscle cells the guinea pig vas deferens myosin shows a
higher actin-activated ATPase activity in the presence of Ca^{2+}. Thus removing
Ca^{2+} has an inhibitory effect on the actin-activated myosin ATPase activity, although the amount of inhibition observed (40-68%) is less than when smooth muscle
myosin is dephosphorylated (Table 2). Complete dephosphorylation of vas deferens
myosin results in a 90% suppression of the actin-activated myosin ATPase activity.

TABLE 3

The Effect of Ca^{2+} on the Actin-Activated ATPase Activity of Phosphorylated Myosin

Myosin Source	ATPase Activity (μmoles P_i/mg myosin/min at 37°C	
	Ca^{2+}	EGTA
Vas Deferens[+] - guinea pig	.042	.025
	.085	.028
	.105	.034
Platelets - human	.18	.18
Proliferative myoblasts - rat (cultured)	.18	.18

[+]Individual determinations

DISCUSSION

Figure 3 is a scheme summarizing how phosphorylation may act to regulate the actin-activated myosin ATPase activity of non-muscle and smooth muscle cells. Equating the physiological process of contraction and relaxation with the bio-chemical findings of increased and "no change" in the actin-activated ATPase activity of myosin must be regarded as unproven until the proper physiological experiments are performed. Moreover, it should be appreciated that in non-muscle cells there is no evidence that directly correlates actin-myosin interaction with any well defined cellular process.

Part A of Fig. 3 is meant to emphasize the following points: a) the kinase responsible for phosphorylating the 20,000 dalton light chains has been identified in the non-muscle cells and smooth muscle cells listed, b) in non-muscle cells, the kinase is equally active in the presence or absence of Ca^{2+} whereas in smooth muscle cells the kinase is dependent on Ca^{2+} for activity, c) in the presence of MgATP and the kinase, myosin is phosphorylated when the γ-phosphate is transferred to the 20,000 dalton light chain to yield P-Myosin and MgADP. d) P-Myosin can be activated by actin as indicated by the "↑ ATPase" activity. e) Dephosphorylation of P-Myosin is catalyzed by a phosphatase which has been identified in the non-muscle and smooth muscle cells outlined above with the exception of astrocytes and HeLa cells. It has been purified from skeletal muscle by Morgan et al[18] and partially purified from platelets by Barylko et al.[9] f) Dephosphorylation results in hydrolysis of the phosphate ester bond and the production of (dephosphorylated) "Myosin" which cannot be activated by actin as indicated by "no Δ in ATPase" (no change in ATPase) activity on the addition of actin. Data from both proliferative myoblasts[7] as well as vas deferens smooth muscle[5] myosin shows that phosphorylation is a pre-requisite for actin-activation.

257

Inspection of the diagram reveals that two different but not mutually exclusive mechanisms, both involving Ca^{2+}, can regulate actin-myosin interaction in vas deferens smooth muscle cells (A and B). The first of these (A) involves the reversible phosphorylation of the 20,000 dalton light chain of myosin. Actin-myosin interaction in both schemes is initiated by raising the free Ca^{2+} concentration to 10^{-7} - 10^{-6} M.[17] At this concentration of Ca^{2+} the kinase is active and phosphorylation of myosin in the presence of MgATP results. Contraction would then be correlated with the ability of actin to activate the phosphorylated form of myosin.

Relaxation is initiated by lowering the Ca^{2+} concentration to less than 10^{-7} M.[17] Under these circumstances, as outlined in A, the Ca^{2+} dependent kinase would be inhibited and the active phosphatase would dephosphorylate myosin. Actin would no longer be capable of interacting with myosin as indicated by the lack of actin-activation of the ATPase activity of dephosphorylated myosin. Again it should be emphasized that information is not yet available on the state of myosin phosphorylation in contracted and relaxed smooth muscles.

An additional mechanism (Fig. 3B) for regulation of vas deferens smooth muscle is suggested by the data presented in Table 3. The finding that EGTA inhibits the actin-activated ATPase activity of vas deferens phosphorylated myosin, in the absence of kinase and phosphatase, is evidence for a second form of control. This regulation of the actin-activated ATPase activity of smooth muscle myosin differs from the first form of regulation illustrated in A, in that suppression of the ATPase activity occurs without dephosphorylation of myosin. In this case a decrease in the Ca^{2+} concentration would result in an inhibition of the ATPase activity independent of dephosphorylation. This form of myosin mediated regulation by Ca^{2+} is similar to the mollusc system described previously by Szent-Gyorgyi et al.[2] However, whereas in smooth muscle, phosphorylation of myosin is necessary for actin-activation, there is no evidence to date that phosphorylation plays a role in the molluscan system.

Fig. 3. This model relates the in vitro findings of the effect of myosin phosphorylation on actin-activation with the in vivo process of contraction and relaxation. Although phosphorylation is necessary for actin-activation of myosin ATPase activity in smooth muscle and proliferative myoblasts in vitro, it has not been shown to play any role in vivo. The state of myosin phosphorylation in contracted and relaxed smooth muscle is unknown. Moreover, the terms "Contraction" and "Relaxation" applied to non-muscle cells do not at present relate to any well defined physiological function. A and B are described in the text. [1]Small and Sobieszek (20); [2]Sobieszek (17); [3]Scordilis et al. (24); [4,5]Trotter, J.A., unpublished observation.

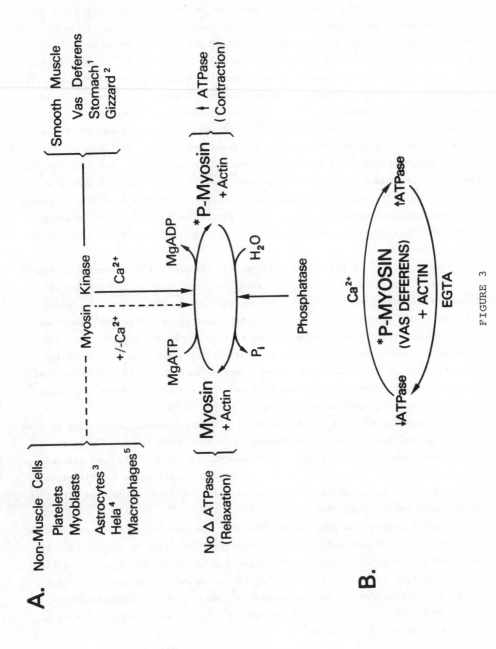

A. Non-Muscle Cells
   Platelets
   Myoblasts
   Astrocytes[3]
   Hela[4]
   Macrophages[5]

Smooth Muscle
   Vas Deferens[1]
   Stomach
   Gizzard[2]

Myosin Kinase

$+/-Ca^{2+}$

Ca^{2+}

MgADP

MgATP

*P-Myosin
+ Actin

$\uparrow$ ATPase
(Contraction)

H_2O

P_i

Phosphatase

Myosin
+ Actin

No Δ ATPase
(Relaxation)

B.

Ca^{2+}

*P-MYOSIN
(VAS DEFERENS)
+ ACTIN

$\uparrow$ATPase

$\downarrow$ATPase

EGTA

FIGURE 3

259

It should be noted that both mechanisms by which Ca^{2+} may regulate smooth muscle actin-myosin interaction require the Ca^{2+} dependent phosphorylation of myosin. The major difference between them involves the mechanism for decreasing the actin-activated myosin ATPase activity. In one case (A), a lowering of the Ca^{2+} concentration results in myosin dephosphorylation and the consequent loss of actin-myosin interaction. In the second case (B), the lowered Ca^{2+} concentration does not result in myosin dephosphorylation but the actin-activated ATPase activity is inhibited. This myosin mediated suppression of ATPase activity is due to an inherent property of phosphorylated vas deferens smooth muscle myosin.

The two proposed mechanisms for Ca^{2+} regulation are by no means mutually exclusive and each may be operative under different conditions. For instance the dephosphorylation switch (A) may operate when full relaxation is required, whereas the direct inhibition of P-myosin by a lowering of Ca^{2+} may act as a modulator under conditions in which the phosphatase is inhibited. Again, it should be emphasized that the above model for smooth muscle regulation by Ca^{2+} and its relation to contraction-relaxation must be regarded as speculative until the proper experiments are carried out.

The role of Ca^{2+} in regulating actin-myosin interaction in non-muscle cells is less well understood since kinase activity in these cells is independent of Ca^{2+} and as demonstrated in Table 3, the purified myosin ATPase activity is not decreased by the removal of Ca^{2+}. This is not meant to imply that Ca^{2+} does not play a regulatory role in actin-myosin interaction in these cells, but that the mechanism for such a role, if one exists, must be different from that found for smooth muscle. It is possible that other proteins such as the recently described troponin-C-like protein that has been isolated from a number of non-muscle cells[19] may play an important role in regulating the effect of Ca^{2+}.

Other proteins may play a role in controlling actin-myosin interaction in non-muscle cells and smooth muscle. Tropomyosin has been shown to markedly increase the actin-activated ATPase activity of phosphorylated vas deferens[5] and gizzard[21] myosin. Undoubtedly additional proteins and ligands, still to be identified, will be shown to alter actin-myosin interaction either directly or through an effect on the kinase or phosphatase.

Phosphorylation-dephosphorylation of myosin has been shown to act as a general regulatory mechanism controlling the actin-myosin interaction in both non-muscle and smooth muscle cells. Important differences exist in the two systems, particularly in the role Ca^{2+} plays in regulating the actin-myosin interaction. Further work in elucidating the various mechanisms for controlling actin-myosin interaction should increase our understanding of how contractile proteins function in both muscle and non-muscle cells.

REFERENCES

1. Greaser, M.L. and Gergely, J.   (1971)   J. Biol. Chem. $\underline{246}$, 4226-4233.
2. Szent-Gyorgyi, A.G., Szentkiralyi, E.M., and Kendrick-Jones, J.   (1973)   J. Mol. Biol. $\underline{74}$, 179-203.
3. Kendrick-Jones, J., Szentkiralyi, E.M., and Szent-Gyorgyi, A.G.   (1976)   J. Mol. Biol. $\underline{104}$, 747-775.
4. Adelstein, R.S., and Conti, M.A. (1976) Cell Motility: Cold Spring Harbor Conferences on Cell Proliferation (Cold Spring Harbor Laboratory, Cold Spring Harbor, N.Y.) Vol. 3, 725-738.
5. Chacko, S., Conti, M.A., and Adelstein, R.S.   (1977)   Proc. Natl. Acad. Sci. USA $\underline{74}$, 129-133.
6. Adelstein, R.S., and Conti, M.A. (1975)   Nature $\underline{256}$, 597-598.
7. Scordilis, S.P. and Adelstein, R.S.   (1977)   Nature (in press).
8. Cohen, I. and Cohen, C.   (1972)   J. Mol. Biol. $\underline{68}$, 383-387.
9. Barylko, B., Conti, M.A., and Adelstein, R.S.   (1977)   Biophysical J. $\underline{17}$, 270a.
10. Adelstein, R.S., Chacko, S., Barylko, B., Scordilis, S.P. and Conti, M.A. (1976) Current Topics in Intracellular Regulation II, eds. Perry, S.V., Margreth, A. and Adelstein, R.S. (Elsevier, The Netherlands) 153-163.
11. Perrie, W.T. and Perry, S.V. (1970) Biochem. J. $\underline{119}$, 31-38.
12. Daniel, J.L. and Adelstein, R.S. (1976) Biochemistry 15, 2370-2377.
13. Fairbanks, G.T., Steck, T.L. and Wallach, D.F.H. (1971) Biochemistry $\underline{10}$, 2606-2617.
14. Anker, H.S. (1970) FEBS Lett. $\underline{7}$, 293.
15. Sobieszek, A.   (1977) The Biochemistry of Smooth Muscle, Stephens, N.L. ed. University Park Press, Baltimore, 413-443.
16. Gorecka, A., Aksoy, M.D., and Hartshorne, D.J. (1976) Biochem. Biophys. Res. Commun. $\underline{71}$, 325-331.
17. Sobieszek, A. (1977) Eur. J. Biochem. $\underline{73}$, 477-483.
18. Morgan, M.W., Perry, S.V., Ottaway, J. (1976) Biochem. J. $\underline{157}$, 687-697.
19. Watterson, D.M., Harrelson, W.G. Jr., Keller, P.M., Sharief, F. and Vanaman, T.C. (1976)   J. Biol. Chem. 251, 4501-4513.
20. Small, V. and Sobieszek, A. (1977) Eur. J. Biochem. (in press).
21. Sobieszek, A. and Small, V. (1977) J. Mol. Biol. $\underline{112}$ (in press).
22. Adelstein, R.S., Conti, M.A., Daniel, J.L., and Anderson, W. (1975) Bio-chemistry and Pharmacology of Platelets. Ciba Foundation Symposium $\underline{35}$ (new series) 101-119.
23. Lowey, S., Slayter, H.S., Weeds, A.G. and Baker, H. (1969) J. Mol. Biol. $\underline{42}$ 9-29.
24. Scordilis, S.P., Anderson, J.L., Pollack, R., and Adelstein, R.S. (1977). J. Cell Biol. (in press).

Acknowledgements

    The authors wish to acknowledge the excellent technical assistance of William Anderson, Jr. and James M. Miles. We are very grateful to Exa Murray for typing the manuscript.
    S.C. is on leave from the Department of Pathobiology, School of Vet. Med., Univ. of Pa., Philadelphia, Pa.; B.B. is on leave from the Nencki Inst. of Experimental Biol. Warsaw, Poland and S.P.S. is a Post-doctoral Fellow of the Muscular Dystrophy Association, Inc.

Conformational Dynamics of Fluorescence-Labeled Troponin C

Herbert C. Cheung

Biophysics Section, Department of Biomathematics
University of Alabama in Birmingham
Birmingham, Alabama   35294

SUMMARY

Troponin C (TN-C) has been studied with two fluorescence probes. Present data suggest that TN-C labeled with a sulfhydryl probe exists in two conformations in the absence of Ca^{++}. Fluorescence (tyrosine) stopped-flow traces of the reaction between Ca^{++} and TN-C at 20° give single first order rate constants which are dependent upon the final ratio of Ca^{++} to TN-C. This reaction is attributed to an isomerization induced by Ca^{++} binding. The removal of bound Ca^{++} by EDTA from this protein follows first order kinetics with a rate constant of 1.24 sec.$^{-1}$. Two first order reactions are observed with rate constants of 1.2 and 6 sec.$^{-1}$ for the reaction of EDTA with the Ca^{++} complex of SH-labeled TN-C, as monitored by the extrinsic fluorescence.

INTRODUCTION

Regulation in vertebrate skeletal muscle is mediated by a complex of troponin and tropomyosin located on the actin filament. In the resting state actin and myosin are unable to interact because of the absence of Ca^{++} and generation of force is prevented. The inhibition of this interaction apparently results from the blocking of specific sites on the actin filament by tropomyosin. Troponin consists of three components. It is generally believed that binding of Ca^{++} by the calcium-binding component, TN-C, results in a conformational change which causes tropomyosin to move, thus exposing the blocked sites on actin and allowing interaction with myosin to occur. Troponin C has four Ca^{++} sites[1], two with a low affinity and two with a high affinity for Ca^{++}. In this communication we summarize some of our recent findings on the solution dynamics of TN-C as deduced from fluorescence probe and stopped-flow studies.

MATERIALS AND METHODS

Troponin from rabbit skeletal muscle was prepared according to Ebashi, et al.[2] and troponin C isolated by the method of Perry and Cole[3]. The separated TN-C, about 100µM, was first dialyzed against 0.1M KCl, 20 mM Tris, 1 mM EDTA, pH 7.5. It was further dialyzed against the same KCl-Tris buffer but containing 10µM EDTA (standard buffer) and unless stated otherwise, this solution was used as Ca^{++} free TN-C. Labeling of TN-C with dansyl chloride (DNS) and the SH probe 1-(iodoace-tamidoethyl)aminoaphthalene-5-sulfonic acid (1,5I-AEDANS) was carried out at 4°

overnight and the unreacted probes were removed by dialysis against the standard buffer. Fluorescence polarization was measured in a ratio photometer and fluorescence lifetime was measured in a pulsed instrument[4]. Digitized stopped-flow traces were obtained from a Durrum D-110 spectrometer interfaced with a minicomputer.

RESULTS AND DISCUSSION

Unlike DNS, 1,5I-AEDANS was not rigidly attached to TN-C. Its mobility, however, was eliminated upon addition of Ca^{++}. TN-C appears to be asymmetrical since the harmonic means of the rotational relaxation times of both fluorescent conjugates were considerably larger than the value expected for an equivalent spherical molecule.

Listed in Table 1 are the lifetime data of 1,5I-AEDANS.

Table 1

Lifetime of 1,5I-AEDANS attached to TN-C, 20°

$[Ca^{++}]/[TN-C]$	τ_1 (nsec.)	C_1 (amplitude)	τ_2 (nsec.)	C_2 (amplitude)
0	16.59 ± 0.45	0.62	8.14 ± 0.83	0.38
1	15.24 ± 0.10		--	
2	15.21 ± 0.12		--	
9	15.22 ± 0.13		--	

Two distinct lifetimes were observed for the Ca^{++} free sample. Since TN-C has only one sulfhydryl group, these data indicate two conformations of the labeled TN-C in the absence of Ca^{++}. The data of the Ca^{++} complexes were better fitted by a single lifetime although we cannot rule out entirely a second lifetime with a very small amplitude. The emission maximum of the attached probe was 485 nm in 1 mM EDTA and was blue-shifted to 480 nm upon addition of Ca^{++}. The fluorescence data indicate that calcium binding is accompanied by a change in the fluorescence conformation of TN-C and that this change involves a region of the protein in which the single cysteine is located.

When TN-C was mixed with Ca^{++} in a stopped-flow apparatus, the observed kinetics monitored by the tyrosine fluorescence was first order. The rate constants for the combination with 1.43 and 5.2 moles of Ca^{++} at 20° were 340 and 160 sec.$^{-1}$, respectively. The difference reflects the fact that the two sets of sites have different affinity for the cation. Our observed rates are considerably faster than that reported by Ilo, et al. (13.7 sec.$^{-1}$)[5]. Since their instrument had a dead time of 20 msec., they could not have observed rates in excess of 40-50 sec.$^{-1}$. The binding of Ca^{++} by the SH-labeled protein was monitored by its extrinsic emission. Under conditions identical to those in which the tyrosine emission was measured, the reaction occurred within the mixing time (ca. 2 msec.). When bound Ca^{++} was removed from unlabeled TN-C by reacting with EDTA, a single first order rate constant of 1.24 sec.$^{-1}$ was obtained; the same rate was observed for the reaction with TN-C samples containing 1-4 moles of Ca^{++}. For the SH-labeled TN-C-Ca^{++} complex, the EDTA reaction was monitored by its extrinsic fluorescence and showed two first order rate constants of 1.2 and 6 sec.$^{-1}$. In both combination and removal reactions, the two fluorophores do not sense the same events.

We tentatively suggest that the fast first order rate constant observed upon mixing Ca^{++} with TN-C is associated with an isomerization of the calcium complex. Of the two rate constants observed in the reaction of SH-labeled Ca^{++} complex of TN-C with EDTA, the fast one (6 sec.$^{-1}$) is associated with calcium removal and the slow one (1.2 sec.$^{-1}$) with a conformational change of Ca^{++} free TN-C. The Ca^{++} removal step apparently was too rapid to be sensed by the tyrosine emission in un-labeled TN-C. The slow first order EDTA reaction (rate constant 1.24 sec.$^{-1}$) observed from the tyrosine fluorescence may be identical with the conformational change step observed with labeled TN-C. The present data are not sufficient to rule out other more complex mechanisms.

REFERENCES

1. Potter, J. D. and Gergely, J. (1975) J. Biochem. 250, 4628.
2. Ehashi, S. Wakabayshi, T., and Ebashi, F. (1971) J. Biochem. 69, 441.
3. Perry S. and Cole, H. A. (1974) Biochem. J. 141, 733.
4. Harvey, S. C. and Cheung, H. C. (1977) Arch. Biochem. Biophys. 179, 391.
5. Ilo, T., Mihashi, K., and Kondo, H. (1976) J. Biochem. 79, 689.

ACKNOWLEDGMENT

This work was supported in part by AM-14589 of the U. S. National Institutes of Health.

Regulation of Calcium-Binding by Magnesium

Jos A. Cox, Wlodzimierz Wnuk and Eric A. Stein

Department of Biochemistry, University of Geneva,

P.O.Box 78 Jonction, 1211 Geneva 8
Switzerland

INTRODUCTION

Cellular stimulation generally increases intracellular Ca^{2+} levels, resulting in binding of Ca^{2+} to specific regulatory proteins. In brain, Ca^{2+} binds for instance to a modulator protein which enhances the activity of both adenyl cyclase (1) and phosphodiesterase (2). In muscle, Ca^{2+} binds to phosphorylase kinase which stimulates glycogen breakdown (3), and simultaneously to troponin C, which triggers muscle contraction (4); moreover, Ca^{2+} also binds to low molecular weight sarcoplasmic calcium-binding proteins - SCP - (5) which are named parvalbumin in the instance of vertebrates (6).

Binding of calcium to regulatory proteins, studied in the absence of Mg^{2+}, does not describe "in vivo" behavior, and cannot explain the control of physiological processes by Ca^{2+}. Indeed, it has been shown that Mg^{2+} can modify the binding of regulatory ligands: in rabbit troponin C, Mg^{2+} lowers the affinity for Ca^{2+} of 2 out of 4 binding sites by straight competition (7). Moreover Mg^{2+} induces positive cooperativity in the binding of AMP to fructose diphosphatase (8).

The characterization of various SCP's reveals that the effect of physiological levels of Mg^{2+} on Ca-binding is not limited to the simple competition observed in parvalbumins, which act as biological Ca-Mg exchangers; in invertebrate SCP's, Mg^{2+} often elicits a more refined response that brings about a genuine regulation of Ca-binding, namely cooperativity.

METHODS

Binding isotherms were obtained by equilibrium dialysis against 25 mM tricine, pH 7.4, 80 mM KCl, the free metal ion concentration being buffered by 0.1 mM EDTA or EGTA. Ca^{2+} and Mg^{2+} levels were determined by atomic absorption. The intrinsic binding constants were evaluated according to Cornish-Bowden and Koshland (9) followed by least-square curve fitting with computer.

RESULTS

Three effects of Mg^{2+} on Ca-binding have been observed:

1° Mg^{2+} competes for the calcium binding sites in parvalbumin. This protein contains 2 high affinity Ca-binding sites that can also accommodate Mg^{2+} (10). Mg^{2+} reduces the affinity for Ca^{2+} according to the rule of straight competition (Table 1). The binding of Ca^{2+} is noncooperative, irrespective of the presence or absence of Mg^{2+}.

266

TABLE 1

APPARENT Ca-BINDING CONSTANT (K_{app}) AS A FUNCTION OF Mg^{2+}

Mg^{2+} (mM)	2	37.5	75
K_{app} (M^{-1})	2.5×10^6	9×10^4	5×10^4

The data obey the competition equation, $K_{app} = K_{Ca}/(1 + K_{Mg}[Mg^{2+}])$ with $K_{Ca} = 10^9 \ M^{-1}$ and $K_{Mg} = 2.3 \times 10^5 \ M^{-1}$.

2^0 $\underline{Mg^{2+} \text{ enhances positive cooperativity in crustacean SCP}}$. The dimeric crayfish SCP (6) contains 6 Ca-binding sites. In the absence of Ca^{2+} the protein binds 4 Mg^{2+} (Fig. 1A). Binding of Ca^{2+} to the first 2 sites does not influence the amount of Mg bound, but uptake of Ca^{2+} by the next 4 sites causes a simultaneous release of equivalent amounts of Mg^{2+} (Fig. 1B).

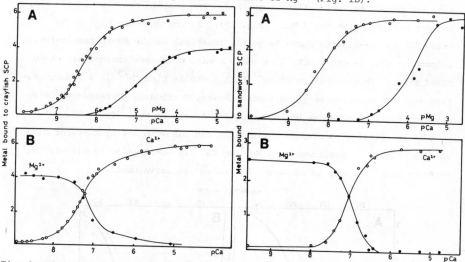

Fig. 1. METAL-BINDING TO CRAYFISH SCP.   Fig. 2. METAL-BINDING TO SANDWORM SCP.

A.   Ca-binding in the absence of Mg^{2+} (O); Mg-binding in 1 mM EGTA (●);

B.   Ca-binding and Mg-release in 1 mM Mg^{2+}.

Taking into account both specificity and affinity for Ca^{2+}, one can distinguish 3 pairs of sites in crayfish SCP:

2 Ca-specific sites $\Big\}$	with higher affinity for Ca^{2+}
2 Ca-Mg sites	
2 Ca-Mg sites	with lower affinity for Ca^{2+}

Table 2 shows that Mg^{2+} enhances positive cooperativity in all 4 higher affinity Ca-binding sites, but does not affect the negative cooperativity in
the remaining 2 sites.

TABLE 2

INTRINSIC Ca-BINDING CONSTANTS OF CRAYFISH AND SANDWORM SCP

SCP from	Mg^{2+}	K'_i (M^{-1})	1	2	3	4	5	6
Crayfish	None	$\times 10^{-8}$	1.2	1.5	2.4	5.1	0.50	0.08
	1 mM	$\times 10^{-6}$	3.7	9.1	17	29	2.7	0.42
Sandworm	None	$\times 10^{-8}$	1.7	1.7	1.7			
	1 mM	$\times 10^{-6}$	2.4	16	38			

3° $\underline{Mg^{2+}}$ $\underline{induces\ positive\ cooperativity\ in\ annelid\ SCP}$. Sandworm SCP (11) is an
acidic (pI = 4.3) monomeric protein of 20,000 mol.wt; its amino acid composi-
tion resembles that of dogfish troponin C. The protein can bind 3 Ca^{2+} and,
in EGTA buffered systems, 3 Mg^{2+} (Fig. 2A). In the presence of 1 mM Mg^{2+}
(Fig. 2B) the Ca-binding curve is nearly identical to the Mg-release curve,
suggesting that sandworm SCP, like parvalbumin, possesses exclusively Ca-Mg
mixed sites. In the absence of Mg^{2+} both proteins bind Ca^{2+} in a noncooperative way. In contrast to parvalbumin however, Ca-binding to sandworm SCP becomes strongly cooperative at physiological levels of Mg^{2+} (Table 2).

It is interesting to compare parvalbumin and sandworm SCP by circular dichroism (Fig. 3): replacement of Mg^{2+} by Ca^{2+} is accompanied by allosteric
conformational changes in sandworm SCP but not in parvalbumin.

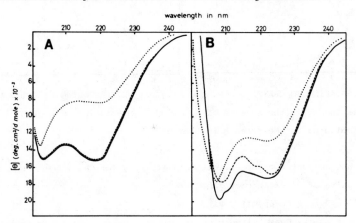

Fig. 3. CIRCULAR DICHROIC SPECTRA OF PARVALBUMIN (A) AND SANDWORM SCP (B).
Ca-form (———); Mg-form (‐‐‐‐); metal-free form (········).

DISCUSSION

Our results show that in certain invertebrates, Mg^{2+} induces or enhances positive cooperativity in Ca-binding to SCP's. This allows the saturation of the protein by much smaller fluctuations of free Ca^{2+} than is the case for the simple competition between Ca^{2+} and Mg^{2+} observed in parvalbumin. Cooperativity may be of particular significance for regulatory proteins, since minute Ca^{2+} currents become sufficient to trigger physiological processes.

If one assumes that 1° invertebrate SCP's are the counterpart of parvalbumins in vertebrates, and 2° the metal-binding properties of all SCP's are instrumental to their physiological function, then parvalbumins might have lost some of their regulatory potential in vertebrate muscle as compared to SCP's in muscle from invertebrates. This raises the question whether the exact role of these proteins might not be easier to elucidate in an ancient invertebrate SCP rather than in parvalbumins.

ACKNOWLEDGEMENT

This work was supported by the Swiss NSF grant Nr. 3.725.72.

REFERENCES

1. Brostrom, C.O., Huang, Y., McL. Beckenridge, B. and Wolff, D.J. (1975) Proc.Natl.Acad.Sci.USA 72, 64.

2. Cheung, W.Y. (1971) J.Biol.Chem. 246, 2859.

3. Heilmeyer, L.M.G., Meyer, F., Haschke, R.H. and Fischer, E.H. (1970) J.Biol. Chem. 245, 6649.

4. Ebashi, S. and Endo, M. (1968) Prog.Biophys.Mol.Biol. 18, 123.

5. Cox, J.A., Wnuk, W. and Stein, E.A. (1976) Biochemistry 15, 2613.

6. Pechère, J.F., Capony, J.P. and Demaille, J. (1973) Syst.Zool. 22, 533.

7. Potter, J.D. and Gergely, J. (1975) J.Biol.Chem. 250, 4628.

8. Nimmo, H.G. and Tipton, K.F. (1975) Eur.J.Biochem. 58, 575.

9. Cornish-Bowden, A. and Koshland, D.E. (1975) J.Mol.Biol. 95, 201.

10. Lehky, P., Comte, M., Fischer, E.H. and Stein, E.A. (1977) Anal.Bioch. (in press).

11. Cox, J.A., Winge, D.R., Wnuk, W. and Stein, E.A. (1976) Fed.Proc. 35, 1363.

Similarity of Ca^{2+}-Induced Conformational Changes in Troponin-C, Protein Activator of Cyclic Nucleotide Phosphodiesterase and Their Tryptic Fragments

W. Drabikowski, J. Kuźnicki, Z. Grabarek

Department of Biochemistry of Nervous System and Muscle,
Nencki Institute of Experimental Biology,
3 Pasteur St., Warsaw, Poland

INTRODUCTION

Several authors[1,2] have pointed recently out to a similarity between troponin-C (TN-C) and Ca^{2+}-dependent protein activator (PA) of 3',5'cyclic nucleotide phosphodiesterase (PDE). Both proteins bind 4 moles of Ca^{2+} per mole and exhibit several common features induced by Ca^{2+} binding, i.e. an enhancement of tyrosine fluorescence, increase of mobility in urea gel, interaction with troponin-I and resistance to splitting by trypsin[3].

MATERIALS AND METHODS

TN-C and TN-I were prepared from rabbit skeletal muscles according to Drabikowski et al.[4], and PA and PDE from bovine brain according to Watterson et al.[2].

RESULTS

During digestion of both TN-C and PA with trypsin in the presence of Ca^{2+} essentially no change of fluorescence intensity takes place. Under these conditions the digestion of both proteins leads to the accumulation of two large peptides resistant to further splitting[3,5]. In the present work these peptides were isolated and their properties were compared with those of the corresponding parent protein. One of the peptide from TN-C (peptide I) contains two calcium binding sites (1 and 2), and the second one (peptide II) sites 3 and 4. The peptides obtained from PA are not identical in size and charge but share similar properties with the corresponding TN-C peptides and therefore are also refered to as the peptide I and II.

Peptide II from both proteins retains most of the properties of the native molecule. Its tyrosine fluorescence intensity depends on $[Ca^{2+}]$ on exactly the same manner as that of the corresponding native molecule, so that apparent binding constant of Ca^{2+}, calculated from the transition midpoint of the fluorescence changes, is for each peptide II exactly the same as that for the original

270

molecule (about 5×10^7 M^{-1} for TN-C and its peptide II[6] and about 1×10^7 M^{-1} for PA and its peptide II[5]).

Similarly to TN-C and PA, peptides II show in alkaline urea gel an increase of mobility in the presence of Ca^{2+} and an interaction with troponin-I, although the affinity of the peptide II from PA to troponin-I seems to be somewhat weaker. Peptides I from both proteins do not interact with troponin-I[5].

All these results furnish new evidence for a pronounced structural similarity between troponin-C and protein activator.

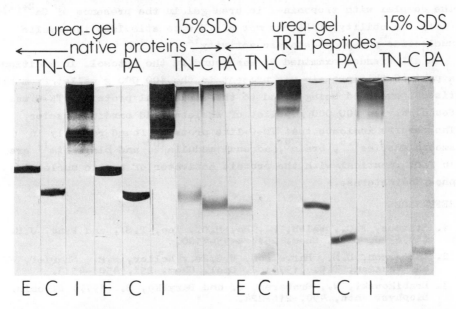

Fig. 1. Electrophoretic properties of TN-C, PA and their tryptic peptides. E - 1 mM EGTA, C - 0.1 mM $CaCl_2$, I - 0.1 mM $CaCl_2$ + troponin-I

It has been reported recently[7,8,9,10,11] that in addition to skeletal and cardiac muscles several other tissues contain TN-C or TN-C-like proteins. In view of the close similarity between TN-C and PA one could expect that TN-C-like proteins are in fact PA. Therefore in the present work the following tissues were analyzed for the presence of PA and/or TN-C: skeletal muscles (rabbit), cardiac muscles (bovine), smooth muscles (rabbit uterus and chicken gizzard), adrenal medulla (bovine), brain (bovine) and platelets (bovine). The tissue homogenates in 0.1 M KCl, imidazole, pH 7.5, were fractionated by centrifugation at 100 000 g. Both, supernatants containing cytosol, and pellets consisting of structural proteins, were fractionated on Sephadex DEAE A-50. The fraction of acidic proteins eluted between 0.2-0.8 M KCl was collected and subjected to a preparative urea gel electrophoresis. The band revealing a common property of both TN-C and PA, i.e. the change of mobility in urea gel depending on Ca^{2+}, was isolated and analyzed. The following properties were used to distinguish between TN-C and PA: (i) the difference in mobility on 15% SDS polyacrylamide gel due to different molecular weight (16 000 for PA as compared to 18 000 for TN-C)[5], (ii) the difference in mobility of the complex with troponin-I in urea gel in the presence of Ca^{2+}(5), (iii) the ability of PA but not of TN-C to stimulate the cyclic nucleotide phosphodiesterase activity[12].

In all tissues examined PA was found in the cytosol. In addition, a part of this protein was present in the 100 000 g pellet from all tissues examined being bound to the structural proteins. TN-C was found only in 100 000 g pellet of skeletal and cardiac muscles. The results indicate that TN-C-like proteins found recently in smooth muscles[7,8], brain[9], adrenal medulla[10], and platelets[11] are in fact identical with the protein activator of cyclic nucleotide phosphodiesterase.

REFERENCES

1. Stevens, F.C., Walsh, M., Ho, H.C., Teo, T.S., and Wang, J.H. (1976) J. Biol. Chem. 251, 4495-4500.

2. Watterson, D.M., Harrelson, W.G.Jr., Keller, P.M., Sharief, F. and Vanaman, T.C. (1976) J. Biol. Chem. 251, 4501-4513.

3. Drabikowski, W., Grabarek, Z. and Baryłko, B. (1977) Biochim. Biophys. Acta, 490, 216-224.

4. Drabikowski, W., Dąbrowska, R. and Baryłko, B. (1973) Acta Biochim. Polon. 20, 181-199.

5. Drabikowski, W., Kuźnicki, J. and Grabarek, Z. Biochim. Biophys. Acta, in press.

6. Leavis, P., Drabikowski, W., Rosenfeld, S., Grabarek, Z. and Gergely, J. (1977) Feder. Proc. 36, 831.

7. Head, J.F., Weeks, R.A., Perry, S.V. (1977) Biochem. J. 161, 465-471.

8. Head, J.T. and Mader, S. (1975) J. Cell Biol. 67, 162a.

9. Fine, R., Lehman, W., Head, J. and Blitz, A. (1975) Nature 258, 261-262.

10. Kuo, I.C.Y. and Coffee, C.J. (1976) J. Biol. Chem. 251, 6315-6319.

11. McGowan, E.B., Speiser, S. and Stracher, A. (1976) Biophys. J. 16, 162a.

12. Wang, J.H., Teo, T.S., Ho, H.C. and Stevens, F.C. (1975) Adv. Cyclic Nucleotide Res. 5, 179-194.

Dantrolene Sodium Effects on Caffeine Contractures
in Frog Sartorius Muscle

K.O. Ellis and F.L. Wessels
Scientific Affairs Department
Norwich Pharmacal Company
Division of Morton-Norwich Products, Inc.
Norwich, New York 13815
U.S.A.

Dantrolene sodium (DS) is a skeletal muscle relaxant which has been shown to have a site of action within the muscle itself[1]. This action is on some aspect of the excitation-contraction (EC) coupling process. There is no effect on excitation of the muscle membrane, but the contractile response is markedly depressed. It has been hypothesized that DS dissociates EC coupling by inhibiting the release of Ca^{++} from the sarcoplasmic reticulum (SR)[1,2,3,4]. In these studies caffeine, which is known to release Ca^{++} from the SR, was used to more accurately define the action of DS on skeletal muscle.

Isolated paired frog sartorius muscles (one experimental, one control) were used to study each experimental condition. Contractures produced by caffeine (8 mM) were measured in grams tension/gram tissue wet weight. Drug effects were measured on the initial fast phase (T_I) of the contracture and the maximal tension (T_M) developed.

DS had the greatest effect on T_I (60% decrease versus a 20% decrease on T_M, see Table I). Depolarization of the muscle (80 mM KCl for 15 min) resulted in a 50% decrease in T_I, but had no effect on T_M. Procaine HCl (P) decreased T_I 55% and T_M 75%. Tetrodotoxin (TTX) and zero Ca^{++} Ringer's had no effect on T_I or T_M. In depolarized muscles DS and P totally abolished T_I, but had no greater effect on T_M. In zero Ca^{++} Ringer DS had the same effect on T_I while P abolished T_I and neither had a greater effect on T_M.

The results indicate that DS has an effect on the release of a caffeine sensitive Ca^{++} supply in skeletal muscle. The primary effect of DS is on T_I with a lesser effect on T_M. P effects both T_I and T_M. Prior depolarization of the muscle produces an effect on T_I similar to DS and P. T_I appears to be a caffeine sensitive Ca^{++} supply which is not dependent on membrane excitation or external Ca^{++} for release by caffeine, as it is not affected by TTX or zero Ca^{++} Ringer's.

It is concluded that DS's effect in skeletal muscle is to decrease the release of a Ca^{++} supply which is membrane potential dependent. We propose that under normal conditions this membrane potential dependent Ca^{++} supply serves as the trigger for the release of SR Ca^{++} and that dantrolene sodium's effect in skeletal muscle is on this trigger Ca^{++}.

273

TABLE I

DANTROLENE SODIUM EFFECTS ON CAFFEINE CONTRACTURES IN

FROG SARTORIUS MUSCLE

Experimental Condition[1]	Caffeine (8 mM) Contracture Response	
	$T_I{}^2$	$T_M{}^2$
Control	27.8 ± 4.3	53.7 ± 8.1
Dantrolene Sodium (35 mM)	10.8 ± 3.0	44.2 ± 10.7
Control	28.9 ± 1.4	95.4 ± 15.5
Depolarized (80 mM KCl)	14.4 ± 1.1	91.9 ± 5.4
Control	24.8 ± 3.6	54.8 ± 7.6
Procaine HCl (5 mM)	11.2 ± 1.2	14.3 ± 5.0
Control	26.5 ± 3.3	81.1 ± 13.6
TTX (3 µm)	26.3 ± 1.6	82.8 ± 13.5
Control	38.5 ± 4	110 ± 7
Zero Ca^{++} Ringer	34.5 ± 3.6	112.1 ± 3.2
Depolarized (80 mM KCl)	8.9 ± 1.5	50.2 ± 8.9
Dantrolene Sodium (35 mM)	0	39.9 ± 10.5
Depolarized (80 mM KCl)	24.9 ± 5.1	79.4 ± 28.3
Procaine HCl (5 mM)	0	17.1 ± 9.7

[1] n = 4 pair of muscles for each experimental condition.

[2] Mean values expressed in grams tension/gram tissue wet weight ± SEM

REFERENCES

1. Ellis, K.O. and J.F. Carpenter (1974) Arch. Phys. Med. Rehabil., 55:362-369.
2. Brocklehurst, L. (1975) Nature, 254:364.
3. Van Winkle, W.B. (1976) Science, 193:1130-1131.
4. Morgan K.G. and S.H. Bryant (1977) J. Pharmacol. Exp. Ther., 201:138-147.

Troponin-C-Like Modulator Protein from Vertebrate Smooth Muscle

James F. Head, Susan Mader and Benjamin Kaminer
Department of Physiology
Boston University School of Medicine

INTRODUCTION

Phosphodiesterase activator proteins from bovine heart and brain and rat testis
have been shown to be chemically similar to the calcium binding troponin-C subunit of vertebrate striated muscle[1-4]. The brain and testicular proteins have
also been shown to be able to substitute functionally for troponin-C in forming
troponin-like complexes with other troponin subunits and subsequently being able
to restore calcium sensitivity to actomyosin ATPase[4,5]. The potentially multi-
functional role of the brain protein, including its ability to activate adenylate
cyclase[6], has led to it being designated a troponin-C-like "modulator protein"[3].

It has been shown by Head et al[7,8] that a troponin-C-like protein is also
present in vertebrate smooth muscle. In the present study we describe the purification and properties of this protein and show it to closely resemble the
brain modulator protein. We also describe the purification from chicken gizzard
muscle of a phosphodiesterase which consists principally of an 80,000 dalton subunit. Both the troponin-C-like modulator protein and the phosphodiesterase are
prepared from smooth muscle using affinity columns of rabbit skeletal muscle
troponin components coupled to Sepharose-4B.

RESULTS

We have isolated the modulator protein from chicken gizzard and rabbit uterus
using essentially the affinity chromatographic method of Head et al[8]. This procedure involves dispersing the tissue under study in 8M urea and passing the extract over an affinity matrix of rabbit skeletal muscle troponin-I coupled to
Sepharose-4B. The column is washed with 8M urea, 20mM Tris-HCl, 1mM $CaCl_2$ pH 7.8
to remove unbound material, and finally eluted with the same buffer but including
5mM EGTA in place of $CaCl_2$. As shown in Fig. 1a when gizzard extracts are applied to the column a protein showing a single band of approximately 18,000 daltons
on SDS gels, is recovered from the EGTA eluate. This protein possesses many properties common to vertebrate striated muscle troponin-C. It is a highly acidic
protein (Asx + Glx/Lys + Arg = 3.2), low in tyrosine (2 mol/mol), and histidine
(1 mol/mol) having no tryptophan, but 8 phenylalanines/mol. The high Phe: Tyr
+ Tryp ratio gives a low U.V. absorption spectrum with phenylanine fine structure
which is common among troponin-C and troponin-C-like proteins. Equilibrium
dialysis of gizzard modulator protein shows 4 Ca^{2+} ions bound per mole, 2 at each
of 2 classes of sites having stability constants of 1×10^5 M^{-1} and 6×10^5 M^{-1}. Alkaline urea gel electrophoresis in the presence and absence of Ca^{2+} shows the protein to undergo a calcium dependent mobility change similar to that shown by

275

striated muscle troponin-C, which is indicative of a calcium induced conformational change[9]. In accordance with the results of Amphlett et al[5] using brain modulator protein we find the smooth muscle modulator protein will form calcium dependent complexes with both striated muscle troponin-I and troponin-T on alkaline non-urea gel electrophoresis.

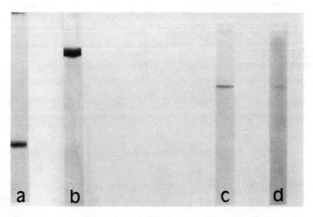

Fig. 1 Gel electrophoreses of gizzard modulator protein and phosphodiesterase:
a) SDS gel 10 μg gizzard troponin-C-like modulator protein b) SDS gel 20 μg gizzard phosphodiesterase c) Alkaline gel 10 μg gizzard phosphodiesterase d) As c) but stained for phosphodiesterase[11].
SDS gels according to Weber and Osborn[13]. Alkaline gels on 8% acrylamide, 40% glycerol, 25 mM tris-80 mM Glycine pH 8.6.

The smooth muscle troponin-C-like modulator is distinguished from striated muscle troponin-C by its ability, common to all such modulator proteins, to activate phosphodiesterase. Also in common with the rat testicular protein[4] we find that on 15% SDS gels the smooth muscle protein migrates with a slightly lower apparent molecular weight than striated muscle troponin-C, about 17,500 daltons. Comparison of the tryptic fingerprints of chicken brain and gizzard modulator proteins and chicken skeletal muscle troponin-C shows the skeletal protein to be largely distinct from the modulators while the modulator proteins themselves show considerable identity.

To determine if sufficient modulator protein is present in smooth muscle for it possibly to play a direct role in the calcium regulation of contraction at the myofilament level, we determined the amount of this protein in gizzard and its ratio to actin and tropomyosin. The results of this study (Table 1) show that, in proportion to actin and tropomyosin, there is about half as much modulator protein in gizzard as there is troponin-C in striated muscle. Although this makes a direct regulatory role for the modulator protein unlikely in smooth muscle contraction, it shows that the amount present is greater than one would reasonably expect for a protein activator of any single enzyme system other than myosin.

TABLE I

QUANTITATION OF ACTIN, TROPOMYOSIN AND TN-C-LIKE PROTEIN IN WHOLE GIZZARD

	Rabbit Skeletal Muscle			Chicken Gizzard Muscle		
	Actin	TM	TN-C	Actin	TM	TN-C-Like Protein
%Age of Total Protein	12	2.7	0.67	11.6	2.8	0.4
Weight Ratio	18.5	4.2	1	29	7.0	1
Molar Ratio	7	1	1	7	1.1	0.6

Gizzard actin and tropomyosin determined by densitometry of SDS gel electrophoreses. Troponin-C-like protein determined from alkaline gel electrophoreses.

As part of our studies on cross-functional interactions between the calcium regulatory systems for contraction and control of cyclic nucleotide levels, we have found that gizzard phosphodiesterase will bind in a calcium dependent manner to rabbit skeletal muscle troponin-C coupled to Sepharose-4B. Crude activator-free phosphodiesterase was prepared from gizzard essentially following the method described by Watterson et al[3] for the brain enzyme. This crude phosphodiesterase preparation was applied in 20 mM Tris-HCl, 1 mM $CaCl_2$ pH 7.8 to a troponin-C affinity column, prepared by the method of Syska et al[10], and washed through in the same buffer. To avoid any non-specific binding, we eluted the column with salt steps up to 0.3 M NaCl prior to elution with 5 mM EGTA. The EGTA eluate was found to have phosphodiesterase activity and to contain principally a protein of 80,000 daltons as shown by SDS gel electrophoresis (Fig. 1b). Alkaline non-urea gel electrophoresis also showed a single band when stained for protein (Fig. 1c) and an equivalent non-urea gel stained for phosphodiesterase[11] with C-AMP or C-GMP also gave one band at the same position (Fig. 1d) (High background stain results from non-specific lead sulphide precipitate)

The smooth muscle phosphodiesterase we have isolated is evidently different from that isolated from bovine brain by Watterson and Vanaman[13] which contain three components 60,000, 40,000 and 18,000 daltons. The smooth muscle and brain enzymes may therefore represent different polymorphs.

REFERENCES
1. Wang,J.,Teo,T.,Ho.,H.,Stevens,F.(1975) Adv. Cyclic Nuc. Res. 5, 179-194.
2. Stevens,F., Walsh,M.,Ho,H.,Teo,T,Wang,J. (1976) J.Biol.Chem.251,4495-4500.
3. Watterson, D.M., Harrelson, W.G., Keller, P.M.,Sharief, F., Vanaman, T.C.(1976) J. Biol. Chem. 251, 4501-4513.
4. Dedman, J.R., Potter, J.D., Means, A.R. (1977) J. Biol. Chem. 252, 2437-2440.
5. Amphlett, G., Vanaman, T., Perry, S.V. (1976) FEBS Lett. 72, 163-168.
6. Brostrom, C.O., Huang, Y.C., Breckenridge, B., Wolff, D.J. (1975) PNAS 72,64-68.
7. Head, J.F. and Mader, S. (1975) J. Cell Biol. 67, 162a.
8. Head, J.F., Week, R.A. and Perry, S.V. (1977) Biochem. J. 161, 465-471.
9. Head, J.F. and Perry, S.V. (1974) Biochem. J. 137, 145-154.
10. Syska, H., Perry, S.V. and Trayer, I.P. (1974) FEBS Lett. 40, 254, 257.
11. Goren, E.N., Hirsch, A.H., Rosen, O. (1971) Anal Biochem. 43, 156.
12. Watterson, D.M., Vanaman, T.C. (1976) Biochem. Biophys. Res. Comm. 73, 40-46.
13. Weber, K. and Osborn, M. (1969) J. Biol. Chem. 244, 4406-4412.

Subcellular Localization of Calcium in Tracheal Smooth Muscle in Response to Therapeutic Concentrations of Theophylline

Ralph C. Kolbeck, William A. Speir, Jr., Edwin D. Bransome, Jr.

Department of Medicine, Medical College of Georgia

Augusta, Georgia 30901

INTRODUCTION

A number of investigators have suggested that the mechanism of action of methylxanthine bronchodilators is competitive inhibition of cyclic nucleotide phosphodiesterase (PDE), thereby increasing intracellular accumulation of cyclic adenosine 3', 5'-monophosphate (cAMP) in respiratory smooth muscle. This hypothesis is based upon data from _in vitro_ studies utilizing high concentrations of theophylline (10^{-3}M or greater) which are toxic _in vivo_ (1). There is, however, evidence that the action of theophylline on a number of tissues may not involve PDE and cAMP (2-4). To answer this question in regard to bronchodilation, we have studied the effects of therapeutic concentrations of theophylline (5-20 µg/ml serum; $3 - 5 \times 10^{-5}$M) on tissue levels of cAMP, cyclic guanosine 3', 5'-monophosphate (cGMP) and calcium, as well as on the intracellular distribution of calcium, in tracheal smooth muscle.

MATERIALS AND METHODS

Incubated tracheal smooth muscle:

Tracheal tissue was obtained from male mongrel dogs (average body weight was 19.3 ± 0.4 kg). Prior to surgical procedures the dogs were anesthetized with 12 mg/kg sodium pentobarbital. The anterior aspects of the neck and thorax were opened and the trachea, extending from larynx to bronchial bifurcation, was excised and immersed in oxygenated Krebs-Ringer solution. The smooth muscle strip was dissected free and incubated for 6 minutes in 37°C physiological solutions containing tracer amounts of ^{45}Ca and either 0,5,10 or 20 µg/ml theophylline. The incubation was terminated by immersing the strips in liquid nitrogen.

Total tissue analysis:

Cyclic-AMP and -GMP determinations were accomplished by procedures for radioimmunoassay using an iodinated cyclic nucleotide derivative as a tracer (5). Isotopic calcium was measured by liquid scintillation techniques and protein analysis was according to the method of Lowry (6).

Selected frozen strips were pulverized and vacuum dessicated at -60°C. Subcellular components were isolated without ion diffusion by ultracentrifugation in non-polar density gradients. Isotopic and total calcium analysis of the purified subcellular fractions was carried out using liquid scintillation counting and atomic absorption spectrometry.

RESULTS

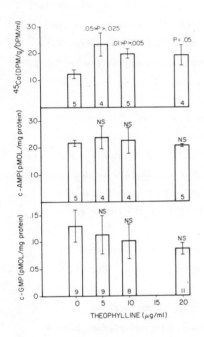

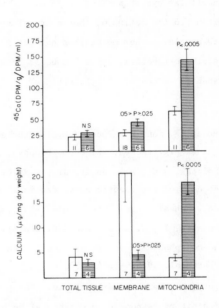

Fig. 1. Dose reponse relationship of theophylline and tissue levels of c-AMP, c-GMP and calcium in incubated dog tracheal strips.

Fig. 2. The effects of 10 μg/ml theophylline on the amount of ^{45}Ca and total calcium associated with several subcellular fractions.

Total tissue analysis (Fig. 1):

Therapeutic concentrations of theophylline had no effect on control levels of either c-AMP or c-GMP. Isotopic calcium uptake, however, was very sensitive to the stimulatory influence of theophylline. Concentrations as small as 5 μg/ml were sufficient to stimulate ^{45}Ca uptake to plateau levels.

279

<u>Subcellular calcium analysis (Fig. 2):</u>

Ten μg/ml of theophylline caused a 78.5% decrease in membrane-bound calcium and a 381.1% increase in mitochondrial-bound calcium without significant change in total tissue calcium. Both membrane and mitochondrial fractions exhibited increased ^{45}Ca turnover rates (53.1% and 126.3%, respectively).

DISCUSSION

Therapeutic concentrations of theophylline caused a redistribution of intracellular calcium without a loss of total tissue calcium or any change in intracellular cyclic nucleotide levels. Cyclic nucleotides may thus not be important to the action of therapeutic levels of methylxanthines in respiratory smooth muscle. Other mechanisms not yet identified, which may involve effects on calcium metabolism, appear to be of greater importance in at least one therapeutic effect of methylxanthines.

REFERENCES

1. Murad, F., and H. Kimura. Cyclic nucleotide levels in incubations of guinea pig trachea. Biochem. Biophys. Acta 343:275-286, 1974.

2. Kolbeck, R.C., W.A. Speir, Jr., W.R. Grabenkort, and E.D. Bransome, Jr. Effects of therapeutic concentrations of theophylline on calcium exchange in tracheal rings. Chest 70:431-432, 1976.

3. Polacek, I., J. Bolan, and E.E. Daniel. Accumulation of adenosine 3', 5'-monophosphate and relaxation in the rat uterus <u>in vitro</u>. Can. J. Pharmacol. 49:999-1004, 1971.

4. Levy, B., and B.E. Wilkenfield. The potentiation of rat uterine inhibitory responses to noradrenaline by theophylline and nitroglycerine. Br. J. Pharmacol. 34:604-612, 1968.

5. Steiner, A. L., D.M. Kipnis, R. Vtiger, and C. Parker. Radioimmunoassay for the measurement of adenosine 3', 5'-cyclic phosphate. Proc. Nat. Acad. Sci. 64:367-373, 1969.

6. Lowry, O.H., N.J. Rosebrough, A.L. Farr, and R.J. Randall. Protein measurement with the Folin phenol reagent. J. Biol. Chem. 193:265-275, 1951.

(Aided in part by U.S.P.H.S. Grant # 5 K07 HL 70057, Pulmonary Academic Award and a grant from the Georgia Lung Association).

Localization of High and Low Affinity Ca^{2+}-Binding Sites on Fragments of Troponin C Produced by Trypsin or Thrombin Degradation

P.C. Leavis[*], W. Drabikowski[**], S. Rosenfeld[*], Z. Grabarek[**] and J. Gergely[*]
*Dept. of Muscle Research, Boston Biomed. Res. Inst.,
20 Staniford St., Boston, MA 02114

**Dept. of Biochem. Nervous System & Muscle,
Nencki Institute of Exptl. Biology
Warsaw, Poland

Each of the four Ca^{2+} binding regions of troponin C (TnC) consists of a Ca-binding loop (Fig. 1, black areas) flanked by two α-helical segments (Fig. 1, stippled areas)[1,2]. Although these regions have tentatively been located in the primary structure, two questions remain to be answered: (1) which regions correspond to the two Ca^{2+}-Mg^{2+} binding sites and which to the Ca^{2+}-specific sites[3] (the Ca^{2+} affinity of the former is higher by about a factor of 100); (2) what is the role of the two classes of sites in maintaining the structure of TnC and in regulating the actin–myosin interaction? We have attempted to address these questions by studying TnC fragments containing intact Ca^{2+} binding sites, either singly or in various combinations, and the associated α-helices (Fig. 1). Ca^{2+}-binding to the fragments has been monitored either directly using a Ca^{2+}-specific electrode (Orion Research), or indirectly by following changes in their fluorescence and circular dichroism (CD).

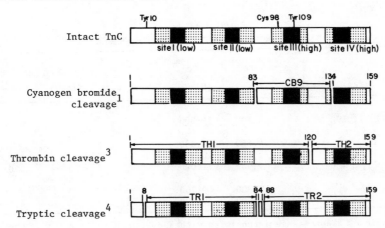

Fig. 1. Fragmentation of TnC. For details see Text.

Tryptic fragments: TR1, containing sites I and II, shows no change in fluorescence or CD upon Ca^{2+} addition. However, using the Ca^{2+}-specific electrode we could show that TR1 binds two Ca^{2+}. TR2, on the other hand, shows a two-fold enhancement of Tyr fluorescence and CD upon Ca^{2+} addition with a transition midpoint in both cases at 17 nm Ca^{2+}. Both changes are complete upon binding of two Ca^{2+}. In the presence of 2 mM Mg^{2+}, the transition midpoint of the Ca^{2+} titrations is shifted to 0.1 μM Ca^{2+}, indicating Mg^{2+} competition for these sites. It is to be noted that the slope of the titration curve is steeper than expected for binding to a single site, indicating interaction between sites III and IV. All the changes in TR2, containing only sites III and IV, are similar to those occurring upon saturation of the Ca^{2+}-Mg^{2+} sites of intact TnC.

Thrombin fragments: TH1 contains sites I-III. Addition of Ca^{2+} produces a 20% fluorescence enhancement and a 20% increase in $[\theta]_{222}$ with a transition midpoint at 5 μM Ca^{2+}. Stoichiometric titrations indicate that one Ca^{2+} is required for the spectral changes, although two additional Ca^{2+} can be bound as determined with the Ca^{2+}-specific electrode. TH2, containing only site IV, binds one Ca^{2+}, as indicated by both an increase in Phe fluorescence (TH2 contains 2 Phe, residues 148 and 151, as its only fluorophors) and in $[\theta]_{222}$. The transition midpoint for these changes occurs at 40 μM Ca^{2+}.

CB9 fragment: CB9 containing site III alone binds Ca^{2+} as indicated by a Tyr fluorescence enhancement and $[\theta]_{222}$ increase with a midpoint at 4.5 μM, in agreement with Nagy et al.[5].

Table I summarizes the results for all of the fragments studied:

TABLE I

Ca^{2+} BINDING PROPERTIES OF TnC AND ITS FRAGMENTS

Peptide	Sites	F/F_o*	$-Ca^{2+}$ $[\theta]_{222}$	$+Ca^{2+}$ $[\theta]_{222}$	Mg^{2+} Competition	No. of Sites	$[Ca^{2+}]_{\frac{1}{2}}$ (μM)**
TnC	I,II,III,IV	1.5	9,600	14,800	+	2	0.05
						2	3.3
TR1	I,II	1.0	14,800	14,800	−	2	>10
TR2	III,IV	2.0	5,900	12,900	+	2	0.02
TH1	I,II,III	1.2	9,000	11,000	−	1	5.0
						2	>10
TH2***	IV	2.0	5,800	9,700	−	1	40
CB9	III	1.5	5,000	9,800	−	1	4.5

*F/F_o, ratio of fluorescence intensities with and without Ca^{2+}; λ_{ex} = 280 nm; λ_{em} = 300 nm, except for TH2, λ_{ex} = 260 nm and λ_{em} = 295 nm.

**$[Ca^{2+}]$ at spectral transition midpoints.

***$[\theta]_{222}$ values are taken from ref. 5.

Interaction of fragments with other troponin subunits (TnI and TnT): Ca^{2+} dependent formation of complexes of TnC fragments labeled with a dansyl fluorophor either at Cys 98 or at the amino terminus with the other troponin subunits has been detected by increases in dansyl fluorescence, either excited directly or by energy transfer from Trp residues on TnI or TnT. TnI binds to TR2, TH1 and CB9, while TnT binds to TR2 and TH2, all with a one-to-one stoichiometry. The TnI binding to TR2 and TH1 has also been verified by urea gel electrophoresis[6].

Conclusions: The high Ca^{2+}-affinity Ca^{2+}-Mg^{2+} sites of TnC can be identified with sites III and IV, as shown by the binding properties of fragment TR2, which contains both of these sites. It appears that the Ca^{2+}-Mg^{2+} binding character- istics of these sites found in intact TnC are not present in fragments that contain sites III and IV separated from each other, viz. TH1, CB9, and TH2. Sites I and II behave as low Ca^{2+}-affinity sites and appear to be unaffected by the presence or absence of adjoining regions of the molecule. The behavior of sites III and IV indicates that binding properties of a site very strongly depend on interaction among peptide segments in different regions (tertiary structure) and cannot be exclusively explained in terms of the primary structure of the region under consideration.

By comparing the peptides that interact with TnI and TnT we conclude that the binding site of TnC for TnI is located between residues 88-119 and that for TnT between residues 120-159. The localization of these binding sites in this III-IV domain may be the basis of the previously reported enhancement of the Ca^{2+}-affinity in complexes of TnC[7] and the presumably reciprocal stabilization by Ca^{2+} of subunit interactions.

REFERENCES

1. Collins, J.H. (1974) Biochem. Biophys. Res. Comm. 58, 301.

2. Kretsinger, R.H. and Nockolds, C.E. (1973) J. Biol. Chem. 248, 3313.

3. Leavis, P.C., Rosenfeld, S., Drabikowski, W., Grabarek, Z. and Gergely, J. (1977) Fed. Proc. 36, 831.

4. Drabikowski, W., Grabarek, Z. and Barylko, B. (1977) Biochem. Biophys. Acta 490, 216.

5. Nagy, B., Potter, J.D. and Gergely, J. (1976) Biophys. J. 16, 71a.

6. Drabikowski, W., Kuznicki, J. and Grabarek, Z. (1977) Biochim. Biophys. Acta (in press).

7. Potter, J.D. and Gergely, J. (1975) J. Biol. Chem. 250, 4628.

Tropomyosin and Troponin-Like Complex from Bovine Brain

C. Mahendran and S. Berl
Department of Neurology, Mount Sinai School of Medicine,
Fifth Avenue and 100th Street, New York, New York 10029, U.S.A.

INTRODUCTION: The Mg^{2+}-ATPase activity (EC 3.6.1.3) of skeletal muscle is sensitive to micromolar concentration of Ca^{2+} due to the presence of the tropomyosin-troponin complex. The Mg^{2+}-ATPase activity of brain actomyosin too, is sensitive to micromolar free Ca^{2+} (1). A crude regulatory protein complex was isolated from bovine brain cortex which conferred Ca^{2+} sensitivity on Mg^{2+}-ATPase activity of brain and skeletal muscle actomyosins from which the calcium regulatory complex had been removed (2). These findings suggested that the crude regulatory complex may contain a tropomyosin-troponin-like complex similar to that of striated muscle actomyosin: hence their isolation was attempted from this protein preparation.

PREPARATION OF PROTEINS: Muscle myosin, actin and tropomyosin were prepared by the methods described by Richards et al. (3), Spudich and Watt (4) and Bailey (5) respectively. Brain tropomyosin was prepared from crude regulatory proteins by applying the same principle for the preparation of muscle tropomyosin (5,6). Troponin-like complex from brain was isolated by a modification of the method of Eisenberg and Keilly (6,7).

RESULTS AND DISCUSSION: Electrophoretic analysis of the crude regulatory protein preparation on polyacrylamide gel (Fig. 1) shows bands four of which could account for tropomyosin and troponin-like proteins.

BRAIN TROPOMYOSIN AND TROPONIN-LIKE FRACTIONS: The tropomyosin preparation showed essentially a single band on gel electrophoresis (Fig. 2). Similar to chick brain (8) the molecular weight of bovine brain tropomyosin is lower than that of muscle tropomyosin and formed paracrystals when dialyzed against 50mm $MgCl_2$ and 50mM Tris-HCl.

The troponin-like complex showed three major components on polyacrylamide gel (Fig. 3). A minor band of molecular weight 40,000 also copurified, as has been observed with the muscle troponin preparation (9).

284

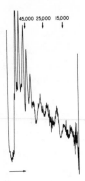

Fig. 1. Densitometric tracing of the electrophoretic pattern of the brain crude regulatory proteins on sodium dodecyl sulphate polyacrylamide gel (12%). The positions of molecular weight markers are indicated at the top.

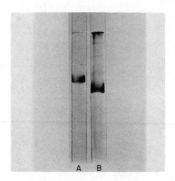

Fig. 2. Electrophoretic pattern of tropomyosins on sodium dodecyl sulphate polyacrylamide gel (12%) (A) from skeletal muscle; (B) from brain cortex. Migration is from top to bottom.

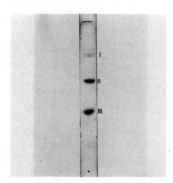

Fig. 3. Electrophoretic pattern of tro-ponin-like fraction on sodium dodecyl sulphate/polyacrylamide gel (12%). The estimated molec-ular weights are: I = 30,000; II = 17,500; III = 14,500.

CONTROL OF MUSCLE ACTIN-MYOSIN INTERACTION BY TROPOMYOSIN AND TROPONIN-LIKE FRAC-TIONS FROM BRAIN: The Mg^{2+} stimulated ATPase activity of deregulated muscle actin plus myosin was inhibited by either brain tropomyosin or the troponin-like fraction (Table 1). The effect of latter was greater than the former. Similar observations have been made with preparations from skeletal muscle (4,7). This inhibitory effect may be due to troponin-I in the troponin-like complex. Either tropomyosin or the troponin-like complex did not confer Ca^{2+} sensitivity. However, the presence of both did generate significant Ca^{2+} sensitivity (Table I) as well as increase the Mg^{2+} ATPase activity. Ca^{2+} sensitivity could not be due to hydrolysis of ATP by contaminating enzymes in brain tropomyosin or troponin-like fraction which require Ca^{2+}, as brain tropomyosin and troponin-like fraction alone did not have any ATPase

285

TABLE 1

Effect of Brain Tropomyosin and Troponin-like Fraction on
Deregulated Muscle Actin-Myosin Interaction

Protein Components	% Mg^{2+}-ATPase Activity		Inhibition (%)
	10^{-5}M Ca^{2+}	1mM EGTA	
MM + MA + BTM + F(Tn)	100	40	60
MM + MA	88	94	-6
MM + MA + BTM	74	76	-2
MM + MA + F(Tn)	55	43	12

Assay system: 1 mM $MgCl_2$; 1 mM ATP, 0.06M KCl, 25 mM Imidizole – HCl pH 7.0, 37° for 30 min, MM: MA = 1.3; MA: BTM F (Tn) = 3:1:2. 100% ATPase activity ranged from 0.41 - 0.63 umole Pi/min/mg protein. MM muscle myosin; MA, deregulated muscle actin; BTM, brain tropomyosin; F(Tn), troponin-like fraction, EGTA, ethylene glycol-bis-(β aminoethyl ether)-N,N'-tetracetate. The results are the average of two preparations assayed in duplicate.

activity or generate activity with either myosin or actin. These observations

unequivocally indicate that brain tropomyosin in combination with a troponin-

like fraction from brain can replace the tropomyosin-troponin complex of muscle,

thus conferring Ca^{2+} sensitivity on the interaction between actin and myosin.

REFERENCES:

1. Mahendran, C., Nicklas, W.J. and Berl, S. (1974) J. Neurochem:23:497-501.

2. Mahendran, C. and Berl, S. (1976) J. Neurochem:26:1293-1295.

3. Richards, E.G., Chung, L.S., Menzel, B.B. and Olcott, H.S. (1967) Biochem: 6:528-540.

4. Spudich, J.A. and Watt, S. (1971) J. Biol. Chem:246:4866-4871.

5. Bailey, K. (1948) Biochem. J.:43:271-275.

6. Mahendran, C. and Berl, S. (1977) Proc. Natl. Acad. Sci. (In press).

7. Eisenberg, E. and Keilley, W.W. (1974) J. Biol. Chem:249:4742-4748.

8. Fine, R.E., Blitz, A.L., Hitchcock, S.E. and Kaminer, B. (1973) Nature New Biol:245:182-185.

9. Lee, L.W. and Watanabe, S. (1975) J. Biol. Chem:245:3004-3007.

The L_2 Light Chain of Skeletal Myosin Enhances Calcium-Sensitivity

Suzanne Pemrick
Departments of Medicine and Biochemistry
Mount Sinai School of Medicine of the
City University of New York, New York, New York 10029

INTRODUCTION

There are several instances when Ca++-free troponin is unable to inhibit formation of actomyosin complexes. This occurs at: a) low MgATP concentrations when myosin is nucleotide-free[1]; b) millimolar MgATP concentrations when myosin has been treated with N-ethylmaleimide at low ionic strength[2,3]. Although these conditions are varied, their effect is the same: presumably to increase the Ca++-affinity of troponin[1,2]. This may be viewed as an indirect regulatory function for vertebrate-striated myosin. Another example of the ability of myosin to alter thin filament cooperativity is presented in this report: the L_2 light chain stabilizes a conformation of myosin which enhances Ca++-sensitivity and actin-interaction[4].

METHODS

Actin, myosin and reconstituted thin filaments were prepared from rabbit psoas muscle[3]. L_2-deficient myosin was prepared by treatment with 5,5'-dithiobis(2-nitrobenzoic acid)(DTNB) and EDTA[5]. The MgATPase activity was assayed in a total volume of 2.0 ml of the following media: 20 mM imidazole-Cl (pH 7, 25°), 1.0 mM $MgCl_2$, 30 mM KCl, 1-2 mM EGTA/CaEGTA buffer, and 0.6 mM MgATP to start the reaction. When the ATPase activity was measured as a function of ATP concentration, creatine phosphokinase (2 mg/ml) and creatine phosphate (5 mM) were present in the medium. The liberation of creatine[3] or inorganic phosphate was measured. Actomyosin was preformed at 0.6 M KCl by first mixing regulated actin followed by addition of myosin to a total volume of 3.0 ml. Ca++ was regulated by varying the ratio of EGTA/CaEGTA assuming a Kd of 0.19 μM at pH 7 for Ca++ and EGTA. Ca++-sensitivity was defined as the Ca++ concentration required for half-maximal activation of the actin-activated ATPase. The per cent relaxation was defined previously[3,4].

RESULTS AND DISCUSSION

DTNB-treatment of myosin released approximately 50% of the L_2 light chain without decreasing the relative amount of either the L_1 or the L_3 light chain. Following prolonged dialysis with dithiothreitol, the K-(18), Ca-(1.6) and Mg-(0.086 sec⁻¹site⁻¹) ATPase activities were restored which is indicative that the hydrolytic site had not been irreversibly modified by exposure to DTNB.

At a 1:1 ratio of actin (A) to myosin "heads" (M = 0.6 μM) removal of 50% of L_2 increased the Ca++ required for half-maximal activation from 0.15 to 0.85 μM and maximal activation decreased from 1.5 to 1.1 sec⁻¹site⁻¹(Fig. 1). Data from 3 separate experiments on 2 batches of myosin and 3 preparations of L_2-deficient

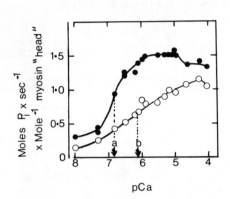

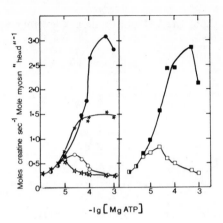

Fig. 1. Actin-interaction verses Ca++ for (●) native, and (○) L_2-deficient myosin. Arrows point to Ca++ at half-maximal activation for a, native and b, L_2-deficient myosin. 1:1 ratio of A:M = 0.6 uM myosin "heads" and 0.75 uM regulated actin.(redrawn from Pemrick, 1977[4]).

Fig. 2. Comparison of actin-interaction verses ATP concentration for (●,○) native, (★,✰) L_2-deficient, and (■, □) reassociated myosin (L_2+DTNB-myosin), in the presence (closed symbols) and absence (open symbols) of Ca++. (redrawn from Pemrick, 1977[4]).

myosin and thin filaments indicated a mean and standard error for Ca++-sensitivity (A:M = 1:1) of 0.18 ± 0.04 µM Ca++ for acto-native myosin and 0.82 ± 0.05 µM Ca++ for acto-L_2-deficient myosin (p≤ 0.005). In addition, the per cent relaxation varied by less than 10% in 4 out of 6 experiments.

The effect of actin and myosin concentration on Ca++-sensitivity was compared for native and L_2-deficient myosin. At a 12:1, and 5:1 ratio of A:M, actin (0.75 µM) was in considerable excess over myosin and therefore, cooperative interactions between actomyosin complexes (MA or A-M.ADP.P) and regulated actin were minimal. Under these conditions Ca++-sensitivity was increased by increasing the concentration of myosin and was insensitive to partial removal of L_2. At a 12:1 ratio of A:M, Ca++-sensitivity was 0.78 ± 0.12 µM Ca++ and 0.87 ± 0.03 µM Ca++ for acto-native and acto-L_2-deficient myosin respectively. Increasing the concentration of myosin "heads" 2.5-fold (A:M = 5:1) increased Ca++-sensitivity 1.5-2-fold: to 0.50 ± 0.06 µM Ca++ for acto-native myosin and 0.53 ± 0.05 µM Ca++ for acto-L_2-deficient myosin.

The inhibitory effect of 50% removal of L_2 on Ca++-sensitivity required concentrations and ratios of actin and myosin in which more than 20% of the actin monomers were capable of interacting with myosin "heads". Loss of L_2 decreased Ca++-sensitivity (p ≤ 0.005, one-tailed paired difference test) and maximal activation at a 0.6:1, 1:1, and 2.5:1 ratio of A:M. Ca++-sensitivity was optimal at a 2.5:1 ratio of A:M: 0.086 ± 0.001 µM Ca++ for acto-native myosin and 0.23 ± 0.01 µM Ca++ for acto-L_2-deficient myosin. Ca++-sensitivity decreased slightly at a 1:1 and 0.6:1 ratio of A:M but was similar at these two ratios. Half-maximal activation for a 0.6:1 ratio of A:M occurred at 0.16 ± 0.04 μM Ca++ for native myosin and at 0.83 ± 0.04 μM Ca++ for L_2-deficient myosin (compare with A:M = 1:1 in Fig. 1). Therefore, decreasing the ratio of A:M from 1:1 to 0.6:1 did not increase either Ca++-sensitivity or the inhibitory effect of 50% removal of L_2 in the homologous system. In contrast, increasing the concentration of total myosin 25% by addition of native myosin to the 1:1 ratio of acto-L_2-deficient myosin increased Ca++-sensitivity 2.5-fold. This suggests a qualitative response of the thin filament to the type of myosin: native or L_2-deficient.

Reassociation of L_2 to L_2-deficient myosin (L_2+DTNB-myosin) was less than complete (final value = 1.8 moles L_2/mole myosin), but was sufficient to demonstrate at a 1:1 ratio of A:M, restoration of Ca++-sensitivity (from 0.85 to 0.15 μM Ca++) and actin-interaction (from 1.5 to 2.9 $sec^{-1}site^{-1}$ in Fig. 2) to native levels.

The simplest explanation for these results was that partial removal of L_2 decreased the steady state concentration of actomyosin complexes, thereby, decreasing the Ca++-affinity of troponin. In order to decide which of two general actomyosin complexes (MA or A-M.ADP.P) was regulated by L_2, actin-interaction was followed as a function of increasing ATP concentrations (Fig. 2). At low MgATP concentrations, sufficient myosin was nucleotide-free and capable of forming "rigorlike" complexes (MA) to destroy relaxation (Fig. 2 and [1]). The ATP concentration at which some relaxation was apparent is a qualitative index of the ability of myosin to respond to ATP by dissociating from actin. Since relaxation was observed at 50 μM MgATP for the 3 species of myosin, there was little indication of a change in the population distribution among "native" conformations of MA complexes. Thus, a small percentage of MA complexes could have persisted at millimolar ATP, but were not of sufficient concentration to alter relaxation (compare the 3 species of myosin in Fig. 2).

Beyond 50 μM MgATP, it can be assumed that myosin was saturated with ATP, since the MgATPase activity in the absence of actin had plateaued to a similar value for the 3 species of myosin. Partial removal of L_2 altered actin-interaction only in the presence of Ca++, above 50 μM MgATP where myosin existed as an M.ADP.P intermediate. Therefore, it is concluded that removal of 50% of L_2 has decreased the steady state concentration of A-M.ADP.P complexes.

This ability of L_2 to enhance actin-interaction as active complexes, at low free Ca++ (from 0.05 to 0.5 μM Ca++) suggests that this light chain may play a role in regulating the resting tension of the system.

REFERENCES

1. Bremel, R.D. and Weber, A. (1972) Nature New Biol. 238, 97-101.
2. Weber, A. and Bremel, R.D. (1971) Contractility of Muscle Cells and Related Processes, Prentice-Hall, Englewood Cliffs, N.J., p. 37.
3. Pemrick, S. and Weber, A. (1976) Biochemistry 15, 5193-5198.
4. Pemrick, S. M. (1977) Biochemistry, in press.
5. Weeds, A., and Lowey, S. (1971) J. Mol. Biol. 61, 701-725.

VITAMIN D- AND VITAMIN K-DEPENDENT CALCIUM-BINDING PROTEINS

Vitamin D-Dependent Calcium-Binding Proteins

R. H. Wasserman and J. J. Feher
Department of Physical Biology
Cornell University
Ithaca, New York

INTRODUCTION

Vitamin D has been recognized for many years as essential for the proper metabolism of calcium and phosphorus in most vertebrate species.[1] Its effects on the intestinal absorption of calcium are well-documented, and recognition of its impact on phosphate absorption is of more recent origin. In addition to the intestine, other target organs of vitamin D are the kidney, the shell gland of the laying hen, the bone, and possibly the brain.[2]

Recent investigations by several groups provided evidence that vitamin D is a hormone rather than a bona fide fat soluble vitamin.[3-5] The basis of this proposition is, as follows. The vitamin, acquired nutritionally or via a u.v.-dependent reaction in the skin, undergoes a series of hydroxylations, yielding the most biologically active derivative, 1,25-dihydroxycholecalciferol [1,25(OH)$_2$D$_3$]. The formation of 1,25(OH)$_2$D$_3$ by a kidney-based hydroxylase system is feed-back regulated and, as an example of this, the rate of production of the hormone is accelerated under conditions of calcium or phosphorus stress, whether the stress is of nutritional or physiological origin.

The health-related impact of the renal 1α-hydroxylase reaction is apparent in patients with kidney failure. Often the degree of formation of 1,25(OH)$_2$D$_3$ is suppressed in these individuals, leading to abnormal calcium and bone metabolism. The administration of 1,25(OH)D$_3$ or 1α(OH)D$_3$ to such patients has proven to be of therapeutic value in certain instances.[7]

Physiologically, vitamin D increases both the active and passive transport of calcium across the intestine and, at the cellular level in the intestine, vitamin D appears to affect the movement of Ca^{+2} across the brush border membrane into the cell, as well as an energy-dependent exit reaction.[1,2] Part of the translocation of calcium might occur via the paracellular route, i.e., the so-called "shunt-path".[8]

Molecular Effects of Vitamin D: Several intestinal macromolecules are influenced by vitamin D or its active metabolites. One such macromolecule is the vitamin D-dependent calcium-binding protein (CaBP).[1] Others include the vitamin D-stimulated calcium-dependent adenosine triphosphatase (CaATPase)[9,10] and alkaline phosphatase.[11,12] In addition, 1,25(OH)$_2$D$_3$ stimulates an intestinal adenylate cyclase directly or indirectly,[13,14] and there is an increase in the amount or turnover of a 80,000-90,000 dalton brush border protein[15,16], as discussed at a recent meeting on vitamin D.

Vitamin D-Dependent Calcium-Binding Proteins: Vitamin D-dependent CaBPs were isolated from several species and these proteins fall into two molecular weight classes, one class with a molecular weight of 28,000 daltons (A) and the other, about 10,000 daltons (B)[2] (Table 1). Class A proteins are present in avian in-

TABLE 1

Molecular Weight Classes of Vitamin D-Dependent CaBP*

(A) 28,000 Daltons

Avian

Intestine
Kidney
Shell Gland
Brain

Mammalian

Bovine Brain
Rat Kidney
Rat Brain

(B) 10,000 Daltons

Mammalian Intestine (in general)
Bovine Kidney
Guinea Pig Kidney

* For original references, see ref. 2.

testine, kidney, shell gland and brain; bovine brain; rat kidney and human kidney. Class B proteins occur in mammalian intestine (in general); bovine kidney; and guinea pig kidney.

It is of interest that antisera produced against chick CaBP cross-reacts immunologically with mammalian kidney (rat, human) and mammalian brain (rat, bovine)[2] (Table 2). However, antisera produced against bovine, pig and guinea pig intestinal CaBPs are species specific and no cross-reactivity by the Ouchterlony double diffusion procedure has been observed between the mammalian intestinal proteins of different species.

Some properties of chick and bovine CaBP are summarized in Table 3. The binding characteristics of the chick protein were determined by equilibrium dialysis and Scatchard plot analysis[17]. Four high affinity sites with a k_a of about $2 \cdot 10^6 M^{-1}$ and several low affinity sites are present. The smaller bovine protein apparently has two high affinity sites with a k_a of about 10^6-$10^7 M^{-1}$ (see following report for details). The isoelectric points of chick and bovine proteins are 4.2-4.3 and 4.7, respectively, and each type has about 30% α-helicity.

Calcium Absorption and CaBP: Evidence for the involvement of intestinal CaBP

TABLE 2*

Immunological Cross-Reactivity

1. Antisera Against Chick CaBP

Chick Intestine
Chick Kidney
Chick Shell Gland
Chick Brain

Rat Kidney
Human Kidney

Rat Brain
Bovine Brain

2. Antiserum Against Bovine Intestine CaBP

Bovine Intestine
Bovine Kidney

* For original references, see ref. 2.

in calcium transport has come from correlative data in which the rate of absorption of Ca was shown to be directly related to the concentration of the protein under a wide variety of conditions. A correlation of significance is the temporal relationship between CaBP formation and Ca transport after $1,25(OH)_2D_3$ is given

TABLE 3

Properties of Vitamin D-Dependent CaBP

	Chick	Bovine
Molecular wt.	27,000	9,700
Ca-binding		
K_a	$2 \cdot 10^6 M^{-1}$	$10^6 - 10^7 M^{-1}$
Sites	4	2
pI	4.2-4.3	4.7
α-Helicity	20-40%	(30%)*

* Pig CaBP (Dorrington et al, 1974)

to rachitic chicks.[18] As shown in Fig. 1, CaBP in minute quantities can be detected in intestinal tissue before a significant enhancement of Ca absorption occurs. Thereafter, there is a parallel increase in both parameters. Thus, CaBP was present coincidental with a change in the transport reaction, contrary to a recent report by Spencer et al.[19] An even earlier event is the transient stimulation of cyclic AMP, and replicates in vivo the pattern observed by Corradino,[14] using embryonic chick intestine in organ culture. Alkaline phosphatase, also implicated in vitamin D action, did not change in activity during the course of this short term experiment, but did increase at later time periods (not shown).

This prolonged lag of response of the alkaline phosphatase to $1,25(OH)_2D_3$ was also shown by Norman.[5]

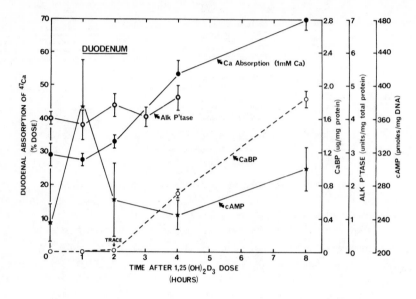

Fig. 1.   Temporal Response of Rachitic Intestine to $1,25(OH)_2D_3$. $1,25(OH)_2D_3$ (1 μg) i.v. at zero time.   Values represent mean $\pm$ S.E.M. of 6 chicks.   From Wasserman et al.[18]

A more direct experimental approach comes from other studies on the embryonic chick intestine in organ culture.[20]   The embryonic intestine does not contain CaBP, although this protein can be induced in vitro by the addition of $1,25(OH)_2D_3$. Accompanying the appearance of CaBP is an increase in Ca^{+2} uptake by the tissue. When purified CaBP is added to the culture medium without $1,25(OH)_2D_3$, there is also a stimulation of Ca uptake.   This observation tends to substantiate a significant role of CaBP in Ca translocation and most likely has a bearing on its mechanism of action.

In this laboratory, two recent experiments yielded information indicating that the concentration of the vitamin D-dependent CaBP in intestinal tissue cannot account for the total intestinal response to vitamin D.   Previously, hydrocortisone in pharmacological doses was shown to inhibit the vitamin D-mediated absorption of Ca in the rat.[21,22]   CaBP, as measured by the Chelex ion-exchange assay, was either unchanged or increased by glucocorticoid treatment.[21,22]   Our study also showed the inhibitory effect of the hormone on Ca absorption in rachitic chicks given either vitamin D_3 or $1,25(OH)_2D_3$.   However, in this experiment, the concentration of CaBP was also suppressed and to about the same degree as Ca absorption. Thus, the response of the chick to glucocorticoid treatment differed from that reported for the rat. In the context of the present dis-

TABLE 4

Effect of Hydrocortisone, Vitamin D_3 and $1,25(OH)_2D_3$ on Ca Absorption, and Intestinal CaBP and Alkaline Phosphatase[*],[+]

Grp.	Vit.D Activity	Hydrocortisone	^{47}Ca Absorption	CaBP[**]
			(%)	(µg/mg)
1	D	−	77 ± 10	32 ± 3
2	D	+	50 ± 5	23 ± 3
3	$1,25(OH)_2D_3$	−	69 ± 4	9 ± 2
4	$1,25(OH)_2D_3$	+	36 ± 3	5 ± 1

[*] From J. J. Feher and R. H. Wasserman (in preparation).

[+] Values = mean ± SEM. Six chicks per group. Vitamin D_3 (500 I.U.) was given 72 hr., and $1,25(OH)_2D_3$ (1 µg) 24 hr., before experiment. Hydrocortisone acetate (2 mg) given i.m. 48 hr. and 24 hr. before experiment.

[**] Soluble CaBP.

cussion, non-correlative evidence comes from a comparison of the relationship of CaBP concentration to Ca absorption in those groups given vitamin D_3 or $1,25-(OH)_2D_3$ (Table 4, Grps. 1 and 3). These data suggest that considerably less CaBP is required to support Ca absorption after $1,25(OH)_2D_3$ than after vitamin D_3. These differences in response cannot be attributed with certainty to the different forms of vitamin D, the reason being that the vitamin D_3 data were generated at 72 hr. after dose, and that for the $1,25(OH)_2D_3$ group at 24 hr. Therefore, the differential response might be due more to time-dependent phenomenon rather than a differential response to the particular metabolite of vitamin D_3. Whichever, it is clear that more units of Ca are transported per unit of CaBP in one situation than in the other.

A plot of data points from several experiments in the hydrocortisone series (Fig. 2) reveals that, within the gross treatment groups [$1,25(OH)_2D_3$ vs. vitamin D_3], there is a direct correspondence between Ca absorption and the intestinal concentration of CaBP. However, the slopes are obviously different which supports the idea that another factor besides the total concentration of intestinal CaBP is involved in the intestinal transport of calcium.

The effect of an inhibitor of protein synthesis, cycloheximide, on the action of $1,25(OH)_2D_3$ was recently examined. Cycloheximide depressed the synthesis of intestinal CaBP and alkaline phosphatase in response to $1,25(OH)_2D_3$, but no significant reduction of Ca absorption occurred. These results are similar to

297

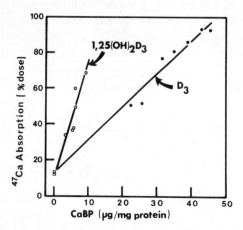

Fig. 2. Relation of Ca absorption and intestinal CaBP. Either $1,25(OH)_2D_3$, vitamin D_3 or vehicle given to rachitic chicks before experiment. Note different slopes derived from the two forms of vitamin D activity. From Feher and Wasserman (in preparation).

those previously reported by Zolock et al[23] and suggest that $1,25(OH)_2D_3$ is exerting an effect that is not totally dependent upon the concentration of CaBP or alkaline phosphatase. This would also imply that the function of $1,25(OH)_2D_3$ does not entirely depend upon the synthesis of other proteins in the intact chick. One might infer a direct action of $1,25(OH)_2D_3$ on the appropriate intestinal membranes but, at the same time, it should be pointed out that studies in organ culture clearly showed that inhibitors of protein synthesis (actinomycin D, α-amanitin) inhibit both the $1,25(OH)_2D_3$-stimulated synthesis of CaBP and $1,25(OH)_2D_3$-dependent Ca uptake.[24] A primary or essential role of CaBP in the chick cannot be eliminated by the cycloheximide experiment since some CaBP was present. Again, the cycloheximide studies are consistent with the view that another factor or condition in addition to CaBP is involved in Ca absorption.

The Bound Form of CaBP: The majority of the chick CaBP is readily soluble in aqueous buffer upon homogenization and appears in the supernate after centrifugation. However, a variable fraction remains bound to some component of the debris. This was disclosed by washing the debris repeatedly by sequential centrifugation steps until the supernate was free of immunologically-detectable CaBP. Solubilization of the washed debris by Triton X-100 releases the bound CaBP. When released, the protein is identical to the soluble form in molecular size, electrophoretic mobility and immunological reactivity.[25]

Bound CaBP could represent the entrapment of soluble CaBP by vesicles formed during homogenization and this possibility was essentially eliminated because sonication and hypoosmotic treatment of washed debris released only a small portion of the bound protein.

Further evidence of two forms of CaBP comes from equilibrium centrifugation

298

TABLE 5

Effect of Cycloheximide and 1,25(OH) D on Ca Absorption,
and Intestinal CaBP and Alkaline Phosphatase*,+

Grp.	$1,25(OH)_2D_3$	Cyclo	^{47}Ca Absorption (%)	CaBP+ ($\mu g/mg$)	Alk. P'tase+ (μ/mg)
1	–	–	9.0 + 0.8	–	0.29 + 0.02
2	+	–	27.6 + 4.6	9.2 + 1.7	0.42 + 0.04
3	–	+	11.9 + 1.2	–	0.15 + 0.02
4	+	+	34.8 + 6.2	3.2 + 0.5	0.18 + 0.03

* From R. H. Wasserman (in preparation).

+ Values = mean + SEM. Six chicks per group. $1,25(OH)_2D_3$ (0.150 µg) was given orally 23 hr. before experiment. Cycloheximide (20 µg) was injected i.p. at 24, 20, 16 and 12 hr. before experiment.

**+ Soluble forms.

studies of intestinal homogenate on sorbitol gradients (Fig. 3). Clearly, two primary peaks of CaBP are seen, one immunoreactive without solubilization and one immunoreactive only after Triton X-100 solubilization. In this same experiment, various marker enzymes were measured and it is evident (Fig. 4 and 5) that bound CaBP is not associated with Na,K-ATPase, CaATPase, cytochrome C oxidase, or acid phosphatase. There does appear to be some coincidence with a small proportions of the alkaline phosphatase and glucose-6-phosphatase content of the intestine. Whether this represents a meaningful association remains to be delineated. In addition, intestinal brush borders purified by the method of Forstner et al[26] did not contain bound and soluble CaBP.

Attention was given to the possibility that bound CaBP is the nascent peptide associated with membrane-bound polysomes in the microsomal fraction. However, the bulk of the RNA-containing fraction on the sorbital gradient does not coincide with the fractions containing bound CaBP, and bound CaBP is not released by RNAase.

Thus, bound CaBP does not appear not to be associated with the intestinal brush border, basal-lateral membranes, mitochondria or polysome-containing microsomes, but a definitive localization of bound CaBP at the subcellular level cannot yet be made.

Speculation on the mode of action of CaBP is limited by lack of definitive information on its cellular localization. Histological studies, using fluorescein-labeled antibody, suggested that CaBP is present in goblet cells and the brush border region.[27,28] Another study suggested that CaBP is localized within the cytoplasm of the intestinal cell and not on the brush border or in the goblet cell.[29] These differences must be resolved before a meaningful

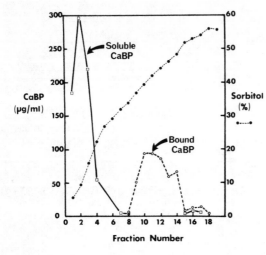

Fig. 3. Separation of soluble and bound intestine CaBP on a sorbitol gradient. (equilibrium centrifugation) From Feher and Wasserman (in preparation).

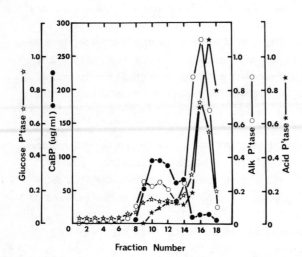

Fig. 4. Marker enzymes in sorbitol gradient fractions. Same experiment as in Fig. 3. From Feher and Wasserman (in preparation).

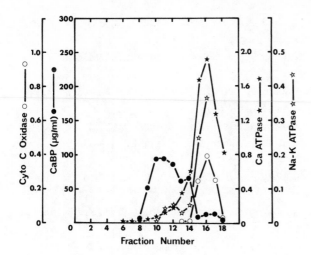

Fig. 5. Marker enzymes in sorbitol gradient fractions. Same experiment as in Fig. 3. From Feher and Wasserman (in preparation).

hypothesis of action of CaBP can be offered. (Supported by NIH Grant AM-04652 and NIDR Training Grant DE-90-14).

REFERENCES

1. Wasserman, R. H. and Taylor, A. N. (1976). In: Parathyroid Gland, Endocrinology Series, Handbook of Physiology (Sect. 7, Vol. VII), ed. G. D.
   Aurbach (American Physiological Society, Washington, D.C., pp. 137-155.

2. Wasserman, R. H., Fullmer, C. S. and Taylor, A. N. (1977) In: Vitamin D,
   ed. D.E.M. Lawson (Academic Press, London), in press.

3. DeLuca, H. F. (1976) Am. J. Clin. Nutrition, 29, 1258-1270.

4. Kodicek, E. (1974) Lancet, 1, 325.

5. Norman, A. W. (1974) In: Vitamins and Hormones, Vol. 32 (Academic Press,
   New York), pp. 325-384.

6. Haussler, M. (1974) Nutrition Reviews, 32, 257.

7. Coburn, J. W., Brickman, A. S., and Hartenbower, D. L. (1975) In: Vitamin
   D and Problems Related to Uremic Bone Disease, ed. A. W. Norman et al.
   (Walter de Gruyter, Berlin), pp. 219-240.

8. Wasserman, R. H., Taylor, A. N. and Lippiello, L. (1975) In: Calcium
   Metabolism, Bone and Metabolic Bone Diseases, ed. F. Kuhlencordt and
   H.-P. Kruse (Springer-Verlag, Berlin), pp. 87-90.

9. Holdsworth, E. S. (1970) J. Membrane Biol., 3, 43-53.

10. Norman, A. W., Mircheff, A. K., Adams, T. H., and Spielvogel, A. (1970).
    Biochim. Biophys. Acta, 215, 348-359.

11. Melancon, M. J., Jr., and DeLuca, H. F. (1970) Biochemistry, 9, 1658-1664.

12. Haussler, M. R., Nagode, L. A., and Rasmussen, H. (1970) Nature 228, 1199–1201.

13. Neville, E., and Holdsworth, E. S. (1969). FEBS Letters, 2, 313–316.

14. Corradino, R. A. (1974) Endocrinology, 94, 1607–1614.

15. Lawson, D. E. M. (1977) Third Annual Workshop on Vitamin D, Asilomar Conference, Jan. 9–13.

16. Rasmussen, H., Max, E., and Goodman, D. B. P. (1977) Third Annual Workshop on Vitamin D, Asilomar Conference, Jan. 9–13.

17. Bredderman, P. J., and Wasserman, R. H. (1974) Biochemistry 13, 1687–1694.

18. Wasserman, R. H. (1977) Third Annual Workshop on Vitamin D, Asilomar Conference, Jan. 9–13, 1977.

19. Spencer, R., Charman, M., Wilson, P., and Lawson, D. E. M. (1976) Nature 263, 161–163.

20. Corradino, R. A., Fullmer, C. S., and Wasserman, R. H. (1976), Arch. Biochem. Biophys., 174, 738–743.

21. Kimberg, D. V., Baerg, R. D., Gershon, E., and Graudirsius, R. T. (1971), J. Clin. Invest., 50, 1309–1321.

22. Krawitt, E. L., and Stubbert, P. R. (1972) Biochim. Biophys. Acta, 274, 179–188.

23. Zolock, D. T., Morrissey, R. L., and Bikle, D. D. (1977) Third Annual Workshop on Vitamin D, Asilomar Conference, Jan. 9–13.

24. Corradino, R. A. (1973) Nature, 243, 41–43.

25. Feher, J. J., and Wasserman, R. H. (1976) Fed. Proc., 35, 339.

26. Forstner, G. G., Sabesin, S. M., and Isselbacher, K. J. (1968) Biochem. J. 106, 381–390.

27. Taylor, A. N., and Wasserman, R. H. (1970) J. Histochem. Cytochem. 18, 107–115.

28. Lippiello, L., and Wasserman, R. H. (1975) J. Histochem. Cytochem. 23, 111–116.

29. Morrissey, R. L., Empson, R. N. Bucci, T. J., and Bikle, D. D. (1976) Fed. Proc. 35, 339.

Bovine Intestinal Calcium-Binding Protein:
Cation-Binding Properties, Chemistry and Trypsin Resistance

C. S. Fullmer and R. H. Wasserman
Cornell University, Ithaca, N. Y.

INTRODUCTION

During the isolation and storage of native bovine intestinal calcium-binding protein (BCaBP), the molecule undergoes alterations that result in the appearance of multiple forms, as visualized by analytical gel electrophoresis. These alterations are apparently irreversible and the products so formed are not susceptible to further change under similar conditions. These have been termed the major (native), minor A and minor B components, and are designated in order of increasing anodal electrophoretic migration rates. These forms differ only slightly in composition, and are similar in molecular size, calcium-binding activity and immunologic reactivity[1].

Along another avenue of investigation, preliminary studies designed to establish the proteinaceous nature of BCaBP showed the calcium-binding activity of partially purified preparations to be entirely susceptible to proteolytic digestion by pronase but resistant to the action of trypsin[1]. Trypsin resistance was later found to be a function of the calcium content of the protein. Protein saturated with calcium resisted hydrolysis by trypsin whereas identical preparations, pretreated with EDTA to remove bound calcium, were completely hydrolyzed to the constituent peptides with concomitant loss of calcium-binding activity[2].

Derivatives of CaBP on Storage and Trypsinization: Additional experiments[2] were conducted to establish the nature of this calcium-conferred "protection" of BCaBP from tryptic digestion. BCaBP (major component), essentially free of bound calcium, was incubated with and without trypsin and 1 mM added $CaCl_2$ at $37^\circ C$ in N-ethylmorpholine acetate buffer (pH 8.1). At various times, samples were removed and analyzed for calcium-binding activity, immunological reactivity and electrophoretic properties. The results (Figs. 1 and 2) confirmed earlier data. BCaBP incubated with trypsin but no added calcium was rapidly hydrolyzed as indicated by loss of activity and disappearance of protein bands on acrylamide gels after electrophoresis. BCaBP incubated with trypsin plus 1 mM $CaCl_2$ did not lose calcium-binding activity or immunological reactivity but did show an alteration in electrophoretic mobility. Over the time course of the experiment, the protein band corresponding to the native molecule disappeared but was replaced by two new bands corresponding in position to the minor A and minor B components noted earlier. The preparation containing no trypsin or added calcium showed a similar alteration, but of a lesser magnitude. The speculation here was that small quan-

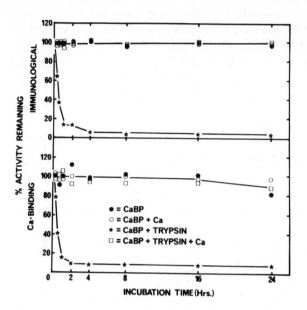

Fig. 1. The effect of 1 mM added Ca^{++} on the tryptic degradation of BCaBP. Immunologic and Ca^{++}-binding activities expressed as % activity remaining relative to zero time control values. [Reprinted from Biochimica et Biophysica Acta, 412 (1975) 256-261].

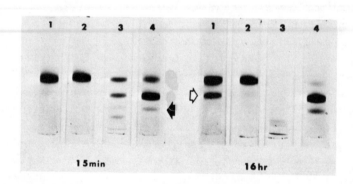

Fig. 2. Results of analytical disc gel electrophoresis performed in the presence and absence of 1 mM added Ca^{++} after 15 min and 16 hr incubation with trypsin. Open and closed arrows indicate positions of bands tentatively identified as minor A and minor B components, respectively. (1) CaBP only; (2) CaBP + 1 mM Ca^{++}, (3) CaBP + trypsin; (4) CaBP + trypsin + 1 mM Ca^{++}.

tities of trypsin or tryptic-like enzymes were carried over in the purification process that slowly modified the protein upon storage.

Molecular Changes of Storage Derivations: More recently[3], comparative investigations were undertaken of the structures of the major, minor A and minor B components of BCaBP. End group determinations by the dansyl chloride procedure[4] established that the NH_2-terminus of the major component was blocked and, therefore, not amenable to direct sequencing techniques. Unexpectedly, the minor A and minor B components showed the presence of dansyl-lysine and dansyl-serine residues, respectively.

In addition, the three components were subjected to direct automated Edman degradation in a Beckman model 890 C sequencer using the DMAA (N,N-Dimethyl-N-Allylamine)-peptide program. Residues from each cycle were determined as the PTH (phenylthiohydantoin) amino acids by gas-liquid chromatography and thin layer chromatography on polyamide sheets. All residues were also hydrolyzed in 6N HCl at $120°C$ for 24 hours and identified by amino acid analyses. No sequence was obtained for the major BCaBP component, confirming the presence of an NH_2-terminal blocking group. On the other hand, unambiguous sequences of the NH_2-terminal regions of the minor A and minor B peptides were obtained. As shown in Fig. 3, these sequences differed only by the absence of the NH -terminal lysine in the minor B component.

```
                              NBS
                               ↓
Minor A:
             5         10          15          20          25          30          35
       K-S-P-E-E-L-K-G-I-F-E-K-Y-A-A-K-E-G-D-P-N-Q-L-S-K-E-E-L-K-L-L-Q-T-E-
Minor B:
         -S-P-E-E-L-K-G-I-F-E-K-Y-A-A-K-E-G-D-P-N-Q-L-S-K-E-E-L-K-L-L-Q-T-E-
Major:
(BG,A,K,K,S,P,E,E,L,K,G,I,F,E,K)-Y-A-A-K-E-G-D-P-N-Q-L-S-K-E-E-L-K-L-L-Q-T-E-
```

Fig. 3. Results of automated sequencing performed on the minor A, minor B and major components of BCaBP. Results combine direct sequencing from the N-terminus (minor A and minor B) and sequencing following NBS oxidation from the minor A, minor B and major components. BG is blocking group. (For details, see ref. 3).

The existence of a single tyrosine and the absence of tryptophan and histidine permitted the use of the highly specific N-bromosuccinimide (NBS) oxidation as means of peptide cleavage at the tyrosine residue[5]. The reaction was employed to provide a single point of cleavage allowing additional comparative sequence determinations, as well as to circumvent the blocking group of the major component.

The major, minor A and minor B components were incubated in 50% acetic acid with a ten-fold molar excess of NBS at room temperature for 16 hrs. Cleavage products from each component were separated on a Sephadex G-50 (medium) column (1.5 cm x 85 cm), equilibrated with 5% acetic acid. Fractions were analyzed for ultraviolet absorbance (279 nm) and fluorescence after reacting the fractions with Fluorescamine reagent (Hoffmann-LaRoche, Inc.). Two peptides, one large and one small, were easily separated by this procedure[3].

Amino acid compositions of the smaller NBS fragments from either the minor A or the minor B components (Table I) showed these to be NH_2-terminal and corresponded exactly to the compositions derived from the sequencing studies (residues 1-12 for minor A and 1-11 for minor B components). Also, the smaller NBS peptide from the minor B component, when subjected to 9 cycles of Edman degradation, yielded a sequence identical to that previously determined for the intact minor B protein. Amino acid composition of the smaller NBS peptide generated from the

TABLE I

AMINO ACID COMPOSITIONS OF THE N-TERMINAL NBS PEPTIDES FROM BOVINE CaBPs

	Major	Minor A	Minor B
LYS	3.6	2.8	2.1
SER	1.0	0.9	0.8
GLU	3.0	3.0	2.8
PRO	0.9	1.0	0.8
GLY[a]	1.0	1.0	1.0
ALA	0.5	0.1	0
ILE	0.9	0.9	0.8
LEU	1.0	1.0	0.9
TYR[b]	-	-	-
PHE	0.8	0.9	0.9

a) Number of residues computed on the basis of 1 glycine.

b) C-terminal tyrosine assumed for all peptides but destroyed during NBS reaction.

major component indicated the probable presence of an additional alanine and lysine residue beyond that in the minor A component.

The larger (C-terminal) NBS peptides from the three components showed the presence of N-terminal alanine following dansylation. Automated Edman degradation established the sequence identity of the large (C-terminal) NBS fragments from the three components for at least an additional 25 residues, following the point of cleavage and permitted the extended sequence determination of the major component to about 43 residues.

Protection Against Trypsinization of BCaBP by Various Cations: The displacement of radiocalcium from BCaBP under equilibrium dialysis conditions was examined as a measure of the relative affinity of BCaBP for a number of cations. BCaBP was dialyzed exhaustively against N-ethylmorpholine acetate buffer (pH 8.1) and determined to contain 1.2 moles of Ca^{++} per mole protein. ^{45}Ca was added to the

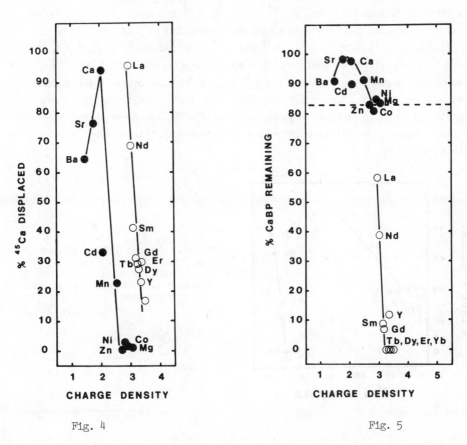

Fig. 4   Fig. 5

Fig. 4. Displacement of bound ^{45}Ca from BCaBP by various cations.

Fig. 5. Protection of BCaBP from complete tryptic digestion by various cations at 1 mM concentration. Dashed line indicates level of protection afforded by 1.2 moles Ca^{++}/mole CaBP with no additional cation added.

307

protein and dialysis allowed to continue for 24 hr against 1 mM final concentrations of the test cations (as chlorides). ^{45}Ca was analyzed in both compartments
and the percentage ^{45}Ca displaced expressed as a function of the cation charge
density (charge/ionic radius in Å). In a companion experiment, the same BCaBP
preparation was incubated with trypsin (trypsin:CaBP, 1:10, by wt.) for 16 hr at
$38°C$ in N-ethylmorpholine acetate buffer (pH 8.1) in the presence of the cations
at 1 mM concentration. Percentage BCaBP remaining at the end of the incubation
period was determined by immunoassay. The results (Figs. 4 and 5) indicate that
the divalent cations Ca^{++}, Sr^{++}, Ba^{++}, Cd^{++} and Mn^{++} are effective in displacing
bound ^{45}Ca and also in conferring additional protection to that afforded by the
existing bound Ca^{++}. Zn^{++}, Ni^{++}, Co^{++} and Mg^{++} were not effective in displacing
bound ^{45}Ca and offered no additional protection from tryptic digestion. The lanthanide series elements displaced bound ^{45}Ca as a function of their ionic radii
and, as a consequence, apparently interfered with the protection by Ca^{++} from
trypsin. In Fig. 6, percent ^{45}Ca displaced plotted vs. the percent cation protection suggests a qualitative difference in interaction between BCaBP and the
di- and trivalent cations.

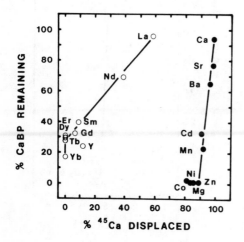

Fig. 6. Percent ^{45}Ca displaced
from BCaBP by various cations
vs. the ability of those cations
to confer protection from complete
tryptic digestion.

The ability of these same cations to protect CaBP from trypsin was also
examined in preparations containing no bound Ca^{++} (Fig. 7). In this case, Ca^{++},
Cd^{++}, Sr^{++} and Ba^{++} were extremely effective whereas Mn^{++} and Co^{++} were less so.
La^{+++} was the only trivalent cation which afforded any protection.

The relative relationship of Ca^{++} concentration and the degree of protection
from tryptic digestion was investigated. BCaBP ($1 \times 10^{-5}M$) was incubated with
trypsin, as described previously, in the presence of varying concentrations of

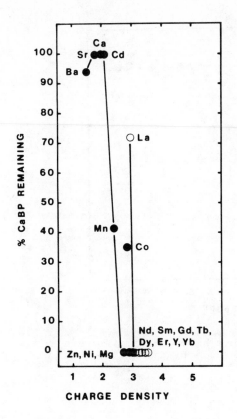

Fig. 7. Protection of Ca^{++}-free BCaBP from complete tryptic digestion by various cations.

CaCl$_2$. After 16 hr, the incubation mixtures were analyzed for immunological re-activity and Ca^{++}-binding activity. The results (Fig. 8) show that there is no appreciable loss of either activity until the ratio of total Ca^{++} to CaBP concen-trations in the incubation mixture approaches a value of 2, indicating that 2 moles of Ca^{++}/mole of CaBP are required to confer complete protection from tryptic digestion.

Calcium-Binding Properties of BCaBP: Experiments designed to determine the number of Ca^{++}-binding sites on BCaBP as well as their affinities for Ca^{++} are underway. Initially, to determine optimum conditions for binding, the effect of pH and salt concentration were studied by equilibrium dialysis techniques. Vary-ing concentrations of NaCl and KCl in 1 mM Pipes buffer (pH 6.8) were found to exert minimal effects on Ca^{++}-binding over the range, 0.01 to 0.2 M (Fig. 9). The effect of pH on Ca^{++}-binding was examined in the same manner. One mM acetate, Mes, Hepes, Pipes, Tris, and Tris-glycine buffers were prepared in 0.15 M KCl.

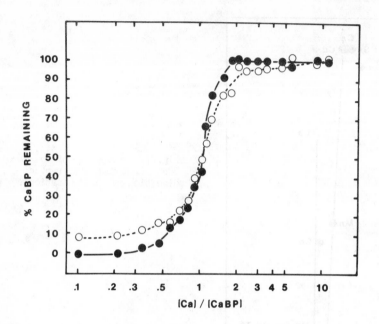

Fig. 8. Effects of varying Ca^{++} concentrations in the incubation media on the tryptic digestion of BCaBP. (●—●) Immunological reactivity, (○--○) Calcium-binding activity.

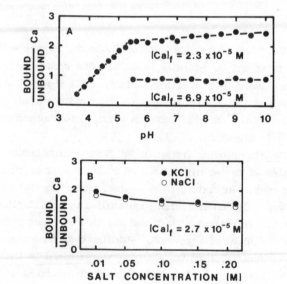

Fig. 9. Effect of pH (A) and Salt concentration (B) on Ca^{++} binding activity of BCaBP.

The results (Fig. 9) show that Ca^{++}-binding activity does not change significantly in the range pH 5.5 to 10, but declines at lower pH values.

Preliminary determination of the number of Ca^{++}-binding sites associated with BCaBP by equilibrium dialysis and Scatchard plot methodology[6] is shown in Fig. 10. BCaBP concentration was 3.1×10^{-5}M in 1 mM Pipes, 0.15 M KCl buffer (pH 6.8). The results suggest that each molecule of BCaBP binds 2 atoms of Ca^{++} with high affinity ($\sim 10^6$ M^{-1}) and an additional 2-3 atoms with lesser affinity ($\sim 10^4$ M^{-1}). However, additional points are required at low free Ca^{++} levels to make an accurate estimate of the high affinity constant.

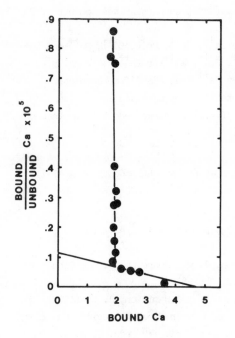

Fig. 10. Ca^{++}-binding data for BCaBP by Scatchard plot methodology.

SUMMARY

The results reported herein and elsewhere[3], appear to confirm the contention that the minor A and minor B components of BCaBP arise from the native molecule by limited and relatively specific cleavage by trypsin or tryptic-like enzymes. Protection against total tryptic digestion of the protein, with consequent loss of Ca^{++}-binding activity and immunological reactivity, is conferred upon the molecule by Ca^{++}, presumably via alteration of the three-dimensional structure in which internal lysine residues are masked. The only substantial points available for cleavage, in the presence of bound Ca^{++}, are those peptide bonds immediately adjacent to the lysine residue at position 1 of the minor A component. It may be observed that the partial sequence reported herein differs markedly from the same peptide region reported earlier[7]. On the basis of more recent studies[3], the prior sequence data is now known to be in error.

Preliminary data suggest that 2 moles of Ca^{++} are bound per mole of CaBP with high affinity and 2-3 with lesser affinity. The results obtained from tryptic digestion in the presence of varying Ca^{++} levels show that 2 moles of Ca^{++}/mole of BCaBP are required to confer trypsin resistance, supporting the concept of 2 high affinity Ca^{++}-binding sites.

^{45}Calcium bound to BCaBP is displaced by Ca^{++}, Sr^{++}, Ba^{++}, Cd^{++}, Mn^{++} and the lanthanide series elements, but not by Mg^{++}, Co^{++}, Ni^{++} or Zn^{++}. $Strontium^{++}$, Ba^{++}, Cd^{++}, Mn^{++} and La^{+++} partially protect Ca^{++}-free BCaBP from complete hydrolysis by trypsin and the degree of protection and displacement of bound Ca^{++}, in general, decrease with divergence of cation charge density from that of Ca^{++}.

ACKNOWLEDGEMENTS

This research was supported by NIH grant AM-04652 and ERDA Contract E(11-1)-2792. Manuscript is in preparation.

REFERENCES

1. Fullmer, C.S. and Wasserman, R. H. (1973) Biochim. Biophys. Acta, 317, 172-186.

2. Fullmer, C.S., Wasserman, R. H., Hamilton, J. W., Huang, W. Y. and Cohn, D. V. (1975) Biochim. Biophys. Acta, 412, 256-261.

3. Fullmer, C. S., Wasserman, R. H., Cohn, D. V. and Hamilton, J. W. (1977) In: Proceedings Third Annual Workshop on Vitamin D, In Press (Elsevier, New York).

4. Gray, W. R. (1972) In: Methods in Enzymology, Vol. XXV, eds. C. H. W. Hirs and S. N. Timasheff (Academic Press, New York) p. 121.

5. Ramachandran, L. K. and Witkop, B. (1967) In: Methods in Enzymology, Vol. XI, C. H. W. Hirs, ed. (Academic Press, New York) p. 283.

6. Scatchard, D. (1949) Ann. N.Y. Acad. Sci., 51, 660-672.

7. Huang, W. Y., Cohn, D. V., Hamilton, J. W., Fullmer, C. S. and Wasserman, R. H. (1975) J. Biol. Chem., 250, 7647-7655.

Prothrombin: A Vitamin K-Dependent Calcium-Binding Protein*

J. W. Suttie, T. L. Carlisle, and L. Canfield
Department of Biochemistry
College of Agricultural and Life Sciences
University of Wisconsin-Madison
Madison, Wisconsin 53706 U.S.A.

INTRODUCTION

Prothrombin, the plasma zymogen of the enzyme thrombin, is a calcium-binding protein which is dependent on vitamin K for its hepatic synthesis. In contrast to the vitamin D-dependent calcium-binding proteins which have been discussed in this symposium, vitamin K does not control the de novo rate of synthesis of prothrombin, but rather the vitamin is required for a posttranslational conversion of non-calcium-binding liver microsomal precursors of prothrombin to the calcium-binding plasma forms. This conversion involves the carboxylation of specific glutamyl residues in the amino-terminal region of the precursor proteins to form γ-carboxy-glutamyl (Gla) residues in the completed protein. As the structure, activation, and biosynthesis of prothrombin have recently been reviewed in detail (1), only limited references will be provided here.

INACTIVE FORMS OF PLASMA PROTHROMBIN

The realization that the vitamin K-dependent reaction in prothrombin biosynthesis had anything to do with the calcium-binding properties of the molecule came much later than the discovery that there was a calcium-dependent association of prothrombin and the other vitamin K-dependent clotting factors (VII, IX, and X) to a phospholipid surface during their action. The identification of a circulating form of plasma prothrombin which lacked this ability was essential to an understanding of the action of this vitamin. The possibility that there were two forms of prothrombin present in the plasma of patients receiving anticoagulant therapy, one of which might be a circulating form of a precursor protein involved in the formation of prothrombin, was first clearly stated by Hemker et al. (2). They were studying a coagulation anomaly in patients receiving anticoagulant therapy which they attributed to the presence of an inactive form of plasma prothrombin. This hypothesis was strengthened by subsequent observations (see ref. 1) that the plasma of man or cattle treated with coumarin anticoagulants contained a protein which was antigenically similar to prothrombin, but which lacked biological activity.

Stenflo (3) demonstrated that plasma from dicoumarol-treated cows also contained two proteins with the antigenic properties of prothrombin which could be

---

*Supported in part by the College of Agricultural and Life Sciences, University of Wisconsin-Madison, in part by Research Grant AM-14881 and Training Grant DE-07031 from the National Institutes of Health.

separated in the presence, but not in the absence, of calcium ions. This observation provided the first direct demonstration that the calcium-binding properties of prothrombin were altered as a result of the action of this vitamin K antagonist. This "abnormal" form of bovine prothrombin was subsequently isolated and studied in both our laboratory and Stenflo's. Both laboratories reported (4,5) that normal bovine prothrombin bound Ca^{2+} in a cooperative fashion, indicating that specific protein structure changes were occurring during the binding process. The "abnormal" prothrombin bound relatively little Ca^{2+} and did not show any evidence of a cooperative binding process. These studies unambiguously related the action of vitamin K in the liver to some difference in the molecular structure of the protein which was reflected in an alteration of Ca^{2+} binding affinity of the plasma protein. Investigations in both laboratories (see ref. 1) established that

TABLE 1

PROPERTIES OF ABNORMAL BOVINE PROTHROMBIN*

Property	Comparison to Prothrombin
M.W. and CHO	Indistinguishable
Amino acid composition and sequence	Apparently identical
Immunochemical determinants	Similar or identical
Electrophoretic mobility	Similar without Ca^{2+}
Hydrodynamic properties	Indistinguishable
Circular dichroism spectra	Indistinguishable
Adsorption to Ba salts	Very low
Ca^{2+} binding	Very low
Ca^{2+}-dependent PL binding	Lacking
$[X_a, V, PL, Ca^{2+}]$ activation	Lacking or very low
Trypsin activation	Apparently identical
E. carinatus venom activation	Apparently identical

*For details see (1).

the prothrombins isolated from normal cattle, and those treated with vitamin K antagonists, were indistinguishable in molecular weight, amino acid composition after acid hydrolysis, and carbohydrate composition. The "abnormal" prothrombin was not converted to thrombin under usual activation conditions (phospholipid, Ca^{2+}, X_2, V); but, if snake venom enzymes which could activate prothrombin were used in place of the plasma derived activators, both the normal and "abnormal" prothrombin could be converted to thrombin. The thrombin formed from either form of prothrombin was indistinguishable in biological activity. The properties of these two proteins are summarized in Table 1. Stenflo (6) demonstrated that the difference in electrophoretic mobility which existed between the two prothrombins in the presence of Ca^{2+} was a property related to differences in the amino-terminal region of prothrombin, and Nelsestuen and Suttie (7) succeeded in iso-

314

lating a calcium-binding peptide from a tryptic digest of normal prothrombin which could not be isolated from the "abnormal" prothrombin. The specific structural difference between the normal and abnormal plasma prothrombin was shown by Stenflo (8) and by Nelsestuen et al. (9) to be the presence of γ-carboxyglutamic acid residues in normal prothrombin. Magnusson et al. (10) subsequently determined the location of ten of these previously unidentified amino acids in bovine pro-

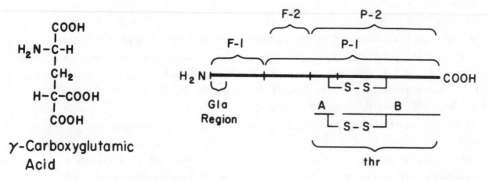

Fig. 1. Structure of γ-carboxyglutamic acid and a diagramatic representation of the structure of prothrombin. All of the Gla residues are located in the first 33 residues from the amino-terminal end of the molecule. Six of the ten Gla residues in bovine prothrombin are present as Gla·Gla sequences.

thrombin (Fig. 1). The calcium-dependent phospholipid interactions were not investigated during the early studies of the abnormal prothrombin; but, more recently, it has been demonstrated (11) that the rate of activation of the abnormal prothrombin by Factor X_a and Ca^{2+} is not stimulated by phospholipid addition, as is normal prothrombin activation (Fig. 2).

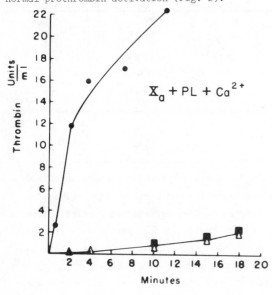

Fig. 2. Effect of phospholipid on prothrombin activation. Normal prothrombin and the abnormal bovine prothrombin were activated with Factor X_a and Ca^{2+}. ■——■ Normal and abnormal prothrombin, no phospholipid. △——△ Abnormal prothrombin with phospholipid. ●——● Normal prothrombin with phospholipid. For details see (11).

315

ISOLATION AND PROPERTIES OF PROTHROMBIN PRECURSORS

Studies of the abnormal bovine plasma prothrombin revealed that it was possible to rapidly generate a thrombin-like activity from it by treatment with E. carinatus venom. This suggested that the hypothesized liver precursor might also be detected by thrombin generation. Utilizing this as an assay, it has been possible to isolate two proteins from the livers of Warfarin-treated rats which have the properties predicted for a prothrombin precursor. The first (12) is a glycoprotein which is both immunochemically similar to prothrombin and has a molecular weight indistinguishable from rat prothrombin. Both electrophoretic and isoelectric focusing analyses indicate that this protein (precursor I) is less negatively charged (pI = 5.8) than rat plasma prothrombin (pI = 5.0). Specific proteolysis of precursor I yielded fragments indistinguishable from those formed by similar proteolysis of prothrombin. This protein does not adsorb to $BaSO_4$, and its rate of activation to thrombin by Factor X_a and Ca^{2+} was not stimulated by the addition of phospholipid. This protein appears to be identical to prothrombin with the exception that it does not contain sialic acid residues and does not contain γ-carboxyglutamic acid. More recently (13) a second protein with very similar properties but with an isoelectric point of 7.2 has been isolated from the same microsomal preparations. The increased basic nature of this protein is a property of the amino-terminal region of the molecule, but the chemical alterations responsible for the shift in pI have not been determined. The properties of these proteins and the abnormal bovine plasma prothrombin are summarized in Table 2.

TABLE 2

PROPERTIES OF RAT PLASMA PROTHROMBIN AND LIVER PRECURSORS

Property	Prothrombin	Precursor I	Precursor II
M.W. (SDS gel)	85,000	85,000	85,000
pI	5.0	5.8	7.2
Neutral sugars	yes	yes	yes
Sialic acid	yes	no	no
Ca^{2+}-dependent PL binding	yes	no	no
Adsorbs to $BaSO_4$	yes	no	no
II_a formed [X_a,V,PL,Ca^{2+}]	yes	no	no
II_a formed (snake venoms)	yes	yes	yes

The physiological roles of the 5.8 and 7.2 forms of precursor are unknown. Either might be the immediate protein substrate for the carboxylating enzyme, an intermediate in the synthesis of that substrate, or a degradation product. Alternatively, both of these proteins might be substrates for carboxylation. It is possible that the chemical modification which is manifest as a difference in isoelectric point of these two proteins might not be related to carboxylation, but might be involved in a more general process such as transport within the intracel-

lular membranes. To approach these questions, an immunochemical method for the
detection of prothrombin-related proteins in extracts of subcellular fractions has
been developed. Liver microsomes prepared from a homogenate made 1 mM in phenylmethylsulfonylfluoride (PMSF) were solubilized in 0.1 M Tris pH 9.5 containing 2%
Triton X-100 and 1 mM PMSF. After the addition of 1 mM benzamidine·HCl, the remaining pellet was removed by centrifugation for 45 min at 165,000 x g, and the
supernatant was dialyzed for 3 h against 50 mM Imidazole, 50 mM NH_4Cl, 1 mM benz-
amidine·HCl at pH 7.8. The prothrombin-related proteins were adsorbed to a
Heparin-agarose A-5m column, washed with three or more column volumes of the
dialysis buffer, step eluted with 0.5 M NH_4Cl buffer, and concentrated about ten-
fold in an Amicon A-25 miniconcentrator. Laurell crossed immunoelectrophoresis
was performed against rabbit anti-rat prothrombin IgG. Parallel electrophoreses
in the first dimension in buffer containing either 2 mM calcium lactate or 2 mM
NaEDTA allowed qualitative measurement of calcium binding.

Two electrophoretic forms (A and B) of antiprothrombin cross-reacting material
result from the application of this method to Warfarin-treated rat liver microsomes (Fig. 3a). Peak B, with higher anodic mobility, predominates. Both peaks

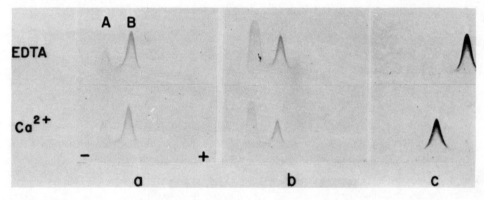

Fig. 3. Crossed immunoelectrophoretic analysis of rat liver precursor proteins
and rat plasma prothrombin. a) Extract of Warfarin-treated (5 mg/kg, 18 h) rat
liver microsomes, 1.0 E. carinatus venom (ECV) unit; b) As in a), 1.0 ECV U pI
7.2 precursor added; c) Rat plasma prothrombin, 1.0 ECV U.

A and B accumulate during Warfarin treatment or vitamin K deficiency. Their mobility does not change markedly when EDTA replaces calcium in the electrophoresis
buffer (Fig. 3a) reflecting a lack of the characteristic calcium binding of plasma
prothrombin (Fig. 3c). The anodic mobility of each is lower than that of plasma
prothrombin, consistent with a more basic isoelectric point. If the microsomal
extract and purified pI 7.2 precursor are mixed prior to concentration on a
heparin agarose column, the area of peak A markedly increases (Fig. 3b). Peak A
thus corresponds to the pI 7.2 precursor. Peak B has not yet been identified.
The properties of peaks A and B are consistent with those of the prothrombin pre-

cursors: they represent noncalcium-binding proteins apparently more basic than plasma prothrombin which accumulate in the cell during vitamin K deficiency.

Vitamin K administration to Warfarin-treated rats results in a rapid disappearance of both peaks A and B (Fig. 4). A calcium-binding peak (peak C), also

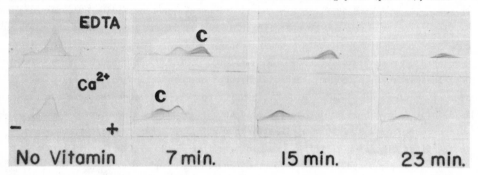

Fig. 4. Time course of microsomal prothrombin synthesis. Rats treated with Warfarin for 18 h were given 1 mg vitamin K_1 intracardially and killed at the times indicated. Microsomal extracts were prepared as described in the text.

with a lower mobility than plasma prothrombin, concurrently appears. Peak C is maximal between 7 and 15 min and almost disappears by 30 min, probably reflecting the rapid carboxylation of precursor and the subsequent secretion of prothrombin. The vitamin K-dependent in vivo disappearance of both peaks A and B suggests that both represent physiological intermediates which can be converted to prothrombin. If peak C represents prothrombin resulting from the carboxylation of precursor in vivo, it should be possible to demonstrate an in vitro vitamin K-dependent incorporation of radioactive HCO_3^- (see vitamin K-dependent carboxylase section below) into this peak. Incubation of microsomes from vitamin K-deficient rats results in the vitamin K-dependent diminution of both peaks A and B. A calcium-binding peak similar to peak C is formed, and radioactivity from the added $H^{14}CO_3^-$ can be detected in this peak. No radioactivity can be detected in the decreased amounts of peaks A and B that remain. This supports the identification of peak C as the major carboxylated intracellular form of prothrombin.

VITAMIN K-DEPENDENT CARBOXYLASE

Prior to the identification of Gla residues in prothrombin, Shah and Suttie (14) had developed an in vitro system that produced biologically active prothrombin when vitamin K was added to a postmitochondrial supernatant prepared from vitamin K-deficient rats. After the demonstration that prothrombin contained Gla residues, it was apparent that this system should also carry out a vitamin K-dependent incorporation of radioactive HCO_3^- into glutamyl residues of the microsomal precursor to form radioactive Gla residues in prothrombin (Fig. 5). This activity was demonstrated by Esmon et al. (15) who were able to show the formation of radioactive prothrombin after incubation with $H^{14}CO_3^-$ and who directly demon-

318

strated that the radioactivity was present in Gla residues. This $\underline{in\ vitro}$ system has subsequently been studied in a number of laboratories (see ref. 1). The activity can be solubilized in various detergents, and the properties of the vitamin K-dependent carboxylase as they are now understood are summarized in Table 3. It

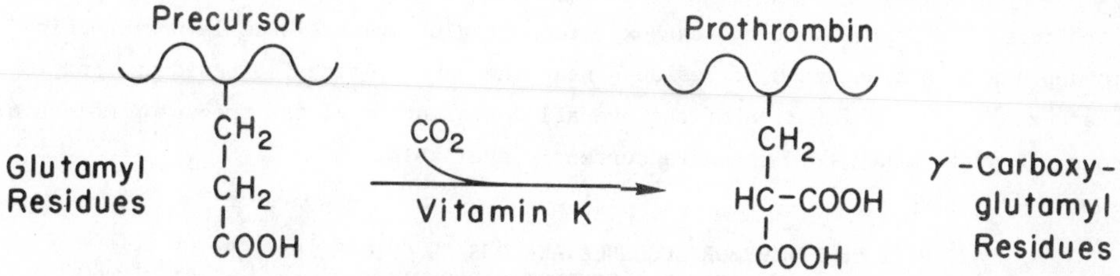

Fig. 5. The reaction catalyzed by the microsomal vitamin K-dependent carboxylase.

is readily apparent that this carboxylation reaction possesses properties considerably different from most carboxylase systems in that it requires O_2 and the reduced form of vitamin K, but appears to have no requirement for ATP. Presumably, the energy needed to drive the carboxylation comes from the reoxidation of the reduced vitamin, but this has not been directly demonstrated. The available data suggests that the vitamin must be functioning either in a manner analogous to biotin to activate (or transfer) $CO_2(HCO_3^-)$ in this lipid rich environment, or that it must somehow function to labilize the hydrogen on the γ-position of the glutamyl residues, so that CO_2 may attack this position. The available data are not sufficient to distinguish these two possibilities.

TABLE 3

PROPERTIES OF THE VITAMIN K-DEPENDENT CARBOXYLASE

Absolute requirements	Known inhibitors	Noninhibitory conditions
Vitamin K and NAD(P)H	Chloro-K	ATP analog (AMPP(NH)P**
$\underline{or}$ vitamin KH_2	Warfarin*	Avidin
O_2	Sulfhydryl poisons	Cyt P_{450} inhibitor
HCO_3^- (CO_2)	Spin-trapping agents	EDTA
Presence of precursor	Anaerobic conditions	

There is not complete agreement in the published literature on all points; the properties assigned represent the author's evaluation of the consensus of the published literature.
*Only when intact microsomes are present.
**Some inhibition when intact microsomes are present.

Studies of the mechanism of this reaction have been hampered by the lack of a suitable substrate for the carboxylase system and by the crudeness of the microsomal system. Some progress has been made in solving both of these problems. Suttie $\underline{et\ al.}$ (16) have demonstrated that the pentapeptide Phe·Leu·Glu·Glu·Val, which would be homologous to residues 5-9 of the bovine prothrombin precursor,

will serve as a substrate for the vitamin K-dependent carboxylase, and that the
Glu residues in this peptide are carboxylated to Gla residues upon incubation with
solubilized microsomes and vitamin K. Preliminary studies of the substrate specificity for various peptide sequences have now been carried out, and it can be seen
(Table 4) that the carboxylase can distinguish between peptides of rather similar
structure. To what extent the carboxylation of glutamyl residues in prothrombin
is dependent on a sequence of residues near the carboxylated Glu residues, and to
what extent it is a function of the overall conformation of the precursor molecule,
cannot be determined from the data currently available.

TABLE 4

ACTIVITY OF PRECURSOR SEQUENCE ANALOGS AS SUBSTRATES FOR THE
VITAMIN K-DEPENDENT CARBOXYLASE

Peptide added	Carboxylation (dpm/ml)
Phe·Leu·Glu·Glu·Ile	36,800
Phe·Leu·Glu·Glu·Leu	31,700
Phe·Glu·Leu·Glu·Leu	0
Phe·Gly·Glu·Glu·Leu	1,080
Phe·Glu·Ala·Leu·Glu·Ser·Leu	780

All peptides were added at 1 mM. Incubation conditions have been described
(16). Previously unpublished peptides were synthesized by S. R. Lehrman and D. H.
Rich. A number of commercially available Glu-containing peptides, Glu·Ala, Ala·
Glu, Leu·Glu·Glu, Glu·Ala·Ala, and Gly·Gly·Glu, were found to be inactive as substrates in this system.

Attempts to purify this microsomal carboxylase have now begun. The vitamin K-
dependent carboxylase seems to be an intrinsic membrane protein. In contrast to
extrinsic or peripheral membrane proteins which are loosely attached to the membrane and easily dissociated, intrinsic membrane proteins are tightly associated
with the membrane, require membrane disruption for their solubilization, and have
been difficult to purify to homogeneity. Many intrinsic membrane proteins are inactive in the absence of specific lipids and some also require the addition of
specific proteins for their activity. The vitamin K-dependent carboxylase requires
greater than 0.5% detergent for solubilization from microsomes in an active form
(15), and concentrations of Triton X-100 in excess of 1% are routinely employed in
our laboratory. These high detergent concentrations complicate purification of
the enzyme by many standard enzyme purification techniques. The carboxylase activity can be assayed using the synthetic peptides as substrates, and some preliminary
attempts to purify the enzyme have been made. Gel filtration of the solubilized
microsomal preparation results in a broadened peak of carboxylase activity roughly
paralleling that of protein concentration. Essentially no purification results
from the application of standard ion exchange chromatographic techniques, as very
little of the detergent-solubilized protein is retained on these columns. When
the detergent is removed from solubilized preparations prior to ion exchange

chromatography, protein is retained on the columns, but most of the enzyme activity is lost. These observations suggest that the vitamin K carboxylase may be part of an enzyme complex, and that other components, probably specific lipids and proteins, will likely be required to retain activity.

It is possible, however, to remove a considerable amount of protein from the solubilized system and yet retain a significant amount of activity. We have developed such a system which provides a five-fold purification over the solubilized microsomal preparation. This system is outlined in Table 5. Microsomes

TABLE 5

PARTIAL PURIFICATION OF VITAMIN K-DEPENDENT CARBOXYLASE

Fraction	Carboxylase (dpm/ml)	Protein (mg/ml)	Specific activity (dpm/mg)
Microsomes	120,000	25	4,500
Detergent extracted microsomes	100,000	10	10,000
$(NH_4)_2SO_4$ (0-40%)	100,000	5	20,000
QAE A-50	72,000	3	24,000

Male rats were injected I.P. with 5 mg/kg Warfarin and fasted overnight. After 18 h the rats were killed and microsomes prepared from liver homogenates as previously described (17) except that KCl concentrations were increased to 0.5 M and 1 mM DTT was included in all buffers. Microsomes were suspended in 0.25 M sucrose, 0.025 M Imidazole, pH 7.2, containing 1 mM DTT and 0.5 M KCl and 0.1% Triton X-100 was added. Partially solubilized microsomes were sedimented and resuspended in the 1.0% Triton in the same buffer. The solution was made 40% in $(NH_4)_2SO_4$ and centrifuged at 11,000 x g. The material floating on the surface was suspended in 10 mM KCl, 1 mM DTT, 0.025 M Imidazole, 10% glycerol, 1% Triton X-100, dialyzed against the same buffer, and applied to QAE A-50 columns previously equilibrated with the same buffer. The column was eluted with increasing KCl concentrations, and the fractions containing the carboxylase activity were collected and pooled.

are isolated from liver homogenates in buffers containing a high salt concentration, resuspended in the same buffers to which small amounts of detergent have been added, and sedimented by ultracentrifugation. The microsomal preparations are then solubilized in the same buffers containing high detergent concentrations and precipitated by ammonium sulfate. This fraction, which contains no more than 20% of the original protein present in the microsomes, retains about 80% of the carboxylase activity originally present in the microsomes (Table 5). These manipulations have removed many of the peripheral membrane proteins, including most of the microsomal-bound vitamin K reductase activity, as this fraction has little activity when vitamin K and NADH are substituted for the reduced vitamin in the assay. A further separation of protein may be achieved by passage of this preparation through ion exchange columns, increasing the overall purification to about five-fold. Further purification of the enzyme will apparently require a more detailed knowledge of its lipid and protein requirements and the utilization of some type of affinity chromatographic procedure. Studies directed toward this end are in progress.

REFERENCES

1. Suttie, J. and Jackson C. (1977) Physiol. Rev. 57, 1-70.
2. Hemker, H., Veltkamp, J., Hensen, A. and Loeliger, E. (1963) Nature 200, 589-590.
3. Stenflo, J. (1970) Acta Chem. Scand. 24, 3762-3763.
4. Nelsestuen, G. and Suttie, J. (1972) Biochemistry 11, 4961-4964.
5. Stenflo, J. and Ganrot, P. (1973) Biochem. Biophys. Res. Commun. 50, 98-104.
6. Stenflo, J. (1973) J. Biol. Chem. 248, 6325-6332.
7. Nelsestuen, G. and Suttie, J. (1973) Proc. Nat. Acad. Sci. USA 70, 3366-3370.
8. Stenflo, J., Fernlund, P., Egan, W. and Roepstorff, P. (1974) Proc. Nat. Acad. Sci. USA 71, 2730-2733.
9. Nelsestuen, G., Zytkovicz, T. and Howard, J. (1974) J. Biol. Chem. 249, 6347-6350.
10. Magnusson, S., Sottrup-Jensen, L., Petersen, T., Morris, H. and Dell, A. (1974) FEBS Letters 44, 189-193.
11. Esmon, C., Suttie, J. and Jackson, C. (1975) J. Biol. Chem. 250, 4095-4099.
12. Esmon, C., Grant, G. and Suttie, J. (1975) Biochemistry 14, 1595-1600.
13. Grant, G. and Suttie, J. (1976) Biochemistry 15, 5387-5393.
14. Shah, D. and Suttie, J. (1974) Biochem. Biophys. Res. Commun. 60, 1397-1402.
15. Esmon, C., Sadowski, J. and Suttie, J. (1975) J. Biol. Chem. 250, 4744-4748.
16. Suttie, J., Hageman, J., Lehrman, S. and Rich, D. (1976) J. Biol. Chem. 251, 5827-5830.
17. Esmon, C. and Suttie, J. (1976) J. Biol. Chem. 251, 6238-6243.

Calcium Function in Vitamin K-Dependent Proteins

Gary L. Nelsestuen
Department of Biochemistry
College of Biological Sciences
University of Minnesota
St. Paul, MN 55108

Summary: The requirements for interaction of prothrombin with membranes have been investigated with respect to calcium requirements, protein structural changes, and lipid structure requirement. It was shown that calcium functions in two distinct roles which include an essential protein transition and the actual formation of the protein-membrane complex. Three or four calcium ions are necessary for effecting the protein transition and a similar number are involved in protein-membrane complex formation. The membrane must contain 8 acidic phospholipid molecules per prothrombin-binding site. Factor X is qualitatively similar to prothrombin in its membrane-binding properties but is quantitatively different in nearly all respects.

The vitamin K-dependent structure in some proteins of higher animals, γ-carboxyglutamic acid,[1-3] arises from the post-ribosomal carboxylation[4] of a number of glutamic acid residues in the amino terminal region of the 4 vitamin-dependent blood-clotting proteins, prothrombin* and factors VII, IX and X. These proteins are quite homologous and are presumed to be descendent from a common ancestral gene.[5] The carboxylated glutamate residues are essential for the calcium and membrane-binding properties of these proteins. In prothrombin there are 10 of these residues in the first 31 amino acids of the amino terminus[1-3,6-9] and there are about 12 in the amino terminal portion of factor X.[10,11] In addition, at least 6 of these amino acids are found in pairs in the prothrombin sequence.[8,9] This high concentration of carboxyl groups has prompted the proposal that γ-carboxyglutamic acid itself or as an adjacent pair represents the calcium-binding sites on these proteins. Experimental evidence, however, has clearly established that γ-carboxyglutamic acid residues do not provide particularly tight calcium binding sites and that intact secondary and/or tertiary protein structure is essential to provide both the calcium and membrane-binding functions of the vitamin K-dependent proteins.[12] This report summarizes the present knowledge of these functions and how γ-carboxyglutamic acid may participate.

MATERIALS AND METHODS

The preparation procedure for bovine prothrombin,[13,14] prothrombin fragment 1[15] and bovine factor X[10] are as described elsewhere. Phospholipids were purchased from the Sigma Chemical Co. and are all >95% pure based on the suppliers

---

*Protein nomenclature is as recommended by the Task Force on the Nomenclature of Blood-clotting Zymogens and Zymogen Intermediates. Prothrombin fragment 1 is the 23,000 dalton amino terminal region containing the γ-carboxyglutamic acid residues and membrane-binding site. Prethrombin 2 is the 40,000 dalton carboxyl terminal region and is the direct thrombin precursor. Prothrombin fragment 2 is the 12,000 dalton peptide which, together with prethrombin 2, comprises prethrombin 1.

estimates. Phospholipid vesicles were formed by mixing the desired ratio of phospholipids in organic solvent, drying and sonicating in buffer solution for 3 x 1.5 min intervals. Single bilayer vesicles were isolated by chromatography on Sepharose 4B according to the method of Huang.[16] The average radius of the vesicles used was 150-180 Å.[17]

Several types of measurements were made in the collection of the data reported here including intrinsic protein fluorescence,[18] direct protein-membrane binding by column chromatography techniques,[18] fluorescence energy transfer from protein amino acids to a chromophore covalently attached to the phospholipids of the membrane[19] and relative 90° light scattering using a Perkin Elmer Hitachi model MPF-2A fluorescence spectrophotometer as the light scattering photometer.[20] Typical results for protein fluorescence, fluorescence energy transfer and relative light scattering measurements are shown in figures 1A, 1B and 1C respectively. The relative molecular weight for the phospholipid-protein complex in figure 1C was calculated from the relationship:

$$\frac{I_{s_2}}{I_{s_1}} = (\frac{\partial n_2/\partial c_2}{\partial n_1/\partial c_1})^2 \, (\frac{M_2}{M_1})^2$$

(1)

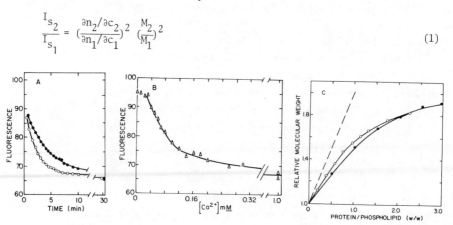

Figure 1. Techniques used for protein transition and protein-membrane association measurements. Figure 1A gives a plot of prothrombin fragment 1 versus time after addition of 1.2 mM calcium at temperatures of 24° (-•-) and 32° (-o-). This monitors a protein transition essential for prothrombin-membrane interaction[18] discussed further below.

Figure 1B shows the method of measurement of prothrombin-membrane interaction by fluorescence energy transfer. Manganous ion (0.2 mM) was added to catalyze the prothrombin transition and calcium was added to form the prothrombin-membrane complex. The membrane contains N-2,4-dinitrophenyl-ethanolamine which acts as the acceptor of fluorescence energy from the protein.

Figure 1C gives the measurement of relative molecular weight of the protein-membrane complex for prothrombin (-•-) and factor X (-o-) as a function of protein/phospholipid ratio in 1 mM calcium ions. This was determined by relative light scattering measurements and equation 1.[17] The dotted line gives the theoretical curve if 100% of the protein bound to the phospholipid.

where I_{s_2} and I_{s_1} are the light scattering intensities from the protein-phospho-
lipid complex and the phospholipid itself, ∂n/∂c is the refractive index increment
for these components and M_2/M_1 is the molecular weight ratio. From the data presented in figure 1C, calculation of free and bound protein concentrations can be
made and equilibrium constants calculated from double reciprocal plots of this
data.[20]

RESULTS AND DISCUSSION

Calcium binding to prothrombin. A number of workers have reported the calcium
binding properties of prothrombin and its isolated domains which are designated
prothrombin fragment 1 and prethrombin 1. The results of some of these reports
summarized in Table I are somewhat heterogeneous. It would appear advantageous to
correlate calcium-binding to other types of studies rather than to attempt to promote one or the other of these studies on the basis of the reliability of the data
gathered since these studies showed experimental errors less than the variations
between reports.

TABLE I

SUMMARY OF CALCIUM-BINDING BY PROTHROMBIN*

| | Number of sites | | | |
	ref. 21	ref. 22	ref. 23,24	Terbium binding (ref. 25)
Prothrombin	10(0.53)	N.D.	11(0.32,2.0)	11
Fragment 1	15(0.68)	10(0.63)	6(0.63)	7
Prethrombin 1	N.D.	N.D.	4(3.2)	4**
Thrombin	N.D.	N.D.	0	0**
Cooperativity	-	+	+	

*Numbers in parentheses give the reported millimolar calcium concentrations
where the sites are half-filled. Where two numbers are given, two classes of
binding sites were given.

**Martin, R. B., personal communication.

With respect to cooperativity of calcium binding to prothrombin, the observed
cooperativity of a calcium-dependent protein transition described below (Hill
coefficient of about 2.6) requires that calcium binding be cooperative. This
cooperativity may be restricted to 3 calcium ions.[18]

The greatest discrepancy in the number of calcium-binding sites is with respect
to fragment 1, the γ-carboxyglutamic acid-containing region of prothrombin. The
results obtained with terbium, a metal ion which will often substitute for calcium
due to similar size and coordination properties (last column, Table I) indicate 7
binding sites on fragment 1 and 4 sites on prethrombin 1 (which are specifically
located in the fragment 2 region of prethrombin 1 since thrombin contains no
sites (Table I). These results corroborate the binding site distribution reported
by Bajaj et al.[23] Direct comparison of terbium-binding with calcium binding must
naturally be done cautiously due to the possibility of terbium occupying multiple

calcium sites.[28] There appear to be only 2 high affinity sites for binding Gd(III) to prothrombin fragment 1.[29] The presence of calcium-binding sites on prethrombin 1 is inconsistent with the reports that abnormal prothrombin (des-γ-carboxyglutamate prothrombin) contains one or fewer calcium binding sites.[26,27] The abnormal prothrombin differs only in that vitamin K action has been blocked during synthesis so that the amino terminal glutamic acid residues are not carboxylated. It should be identical to normal prothrombin with respect to the prethrombin 1 region. The calcium-binding sites on the fragment 2 region of prothrombin are of very low affinity (Table I) however, and may not have been detected in the studies on abnormal prothrombin. Bajaj et al.[23] have presented data which they interpret to indicate that the calcium sites on the fragment 2 region of prothrombin actually function in the interaction of prothrombin with blood-clotting factor V in the prothrombinase reaction.

In another attempt to identify the calcium-binding sites in prothrombin fragment 1, Argos[30] has analyzed the primary sequence of fragment 1 attempting to locate calcium-binding sites on the basis of the structural requirements of Kretsinger.[31] He located 7 potential calcium-binding sites without consideration of the γ-carboxyglutamic acid-containing region. This finding must be interpreted in light of the failure of abnormal prothrombin to bind calcium ions at all. The abnormal prothrombin contains the identical primary amino acid sequence in these regions so if these regions do constitute calcium-binding sites, the γ-carboxyglutamic acid residues must function to potentiate them. However remote, the present evidence does not eliminate the possibility that γ-carboxyglutamic acid functions only indirectly in calcium binding to prothrombin.

Due to the discrepancy in the reported number of calcium binding sites on prothrombin fragment 1, it would seem advantageous to survey possible causes for the differences. First of all, we have found that fragment 1 can lose its functional property of membrane binding without detected change of molecular weight.[18] Perhaps these modified protein molecules also display modified calcium-binding properties. Another explanation could be dimerization of fragment 1 reported by Prendergast and Mann.[24] This is effected by calcium ions but is unrelated to the protein transition reported below since it has considerably different calcium concentration requirements. It is also protein concentration dependent while the transition is not. This dimerization could affect the calcium binding depending on the preparation and the protein concentrations used in the binding measurements.

In summary, despite extensive independent evaluation of calcium binding to prothrombin there remain some ambiguities. The basis for the observed differences will require investigations where potential differences in protein populations, their degree of aggregation, native conformation, etc. are carefully taken into account. From the combined data we can, however, conclude that i) the vitamin K-dependent blood-clotting proteins contain a number of calcium binding sites (20 sites have been reported for factor X,[22]), ii) some of these sites are filled cooperatively, iii) the affinity of these sites is very low and would be ineffective at the calcium concentrations found in most tissues outside of blood plasma, and iv) most or all of these sites are dependent in some way on γ-carboxy-glutamic acid.

Calcium-dependent prothrombin transition. When exposed to calcium ions, pro-thrombin undergoes an unusual protein transition which can be monitored by a de-crease in intrinsic protein fluorescence.[18] This transition is limited to the prothrombin fragment 1 region for which the intrinsic fluorescence change is about -35%. The fluorescence change is only about -6% for intact prothrombin (see fig. 1B at zero calcium). The transition is unusual with respect to its time course and activation energy. The slow transition is first order and requires several minutes for completion (see fig. 1A) at room temperature. An activation energy of about 20 kcal gives a reaction half-life of about 100 min. at 0°.[18] Direct measurement of protein-membrane binding by column chromatography techniques con-ducted at 0° demonstrated that membrane-binding shows the same kinetics as the transition and it was concluded that the protein transition must precede protein-membrane binding.[18] Other methods of measuring protein-membrane binding, includ-ing fluorescence energy transfer from the protein to a chromophore in the phospho-lipid membrane[19] and relative light scattering,[20] have corroborated this fact. The time-course for the reaction seen in figure 1A is biphasic which appears to be the result of two different populations of prothrombin molecules. Some molecules (about 25%) undergo a rapid transition accounting for the initial drop in protein fluorescence when calcium is added and the remaining molecules undergo a slow change. The explanation of two different populations of molecules rather than two sequential protein transitions (which could also account for the observed biphasic change) is suggested by fluorescence energy transfer experiments where protein fluorescence quenching by transfer of excitation energy to a chromophore in the phospholipid is used as a measure of protein-membrane binding. These experiments showed biphasic protein-membrane binding with, again, about 25% of the prothrombin molecules binding rapidly and the remainder binding with the slow reaction kinetics of the transition seen in figure 1A (Nelsestuen, unpublished data). Protein-mem-brane binding measured by the technique of relative light scattering also show biphasic membrane-binding kinetics (Nelsestuen, unpublished data). One view which we have expressed is that the population of molecules undergoing the slow transi-tion are normal and those undergoing the rapid transition are modified, perhaps by virtue of contamination by heavy metals.[19] Indeed, heavy metals are capable of changing the rate of the transition for all of the prothrombin molecules.[19] Studies attempting to separate these two populations of prothrombin molecules are in progress.

The cation specificity of the prothrombin transition is very low and all di- and tri-valent cations tested were effective with the exception of beryllium.[19] The group IIA cations (Mg, Ca, Sr, Ba) all gave similar kinetics for the transition and, by virtue of their divergent ionic radii, serve to illustrate the lack of steric demands of the binding sites required in the transition. The cation concentrations required for catalysis of the transition did vary considerably.[19] Other cations including lanthanides, zinc, iron and manganese also function in causing the prothrombin transition but at much lower concentrations and with modified kinetics such that the slow transition was partially or entirely eliminated. Based on studies with manganese, the metal-binding sites responsible for modifying the rate of the transition appeared to be sites not receptive to calcium ions.[19]

Equilibrium titration of the prothrombin transition (figure 1B) showed cooperativity with respect to calcium ions and the estimated Hill coefficient is 2.6. Based on this value and the total calcium sites filled at various stages of the transition, it was concluded that 3 or 4 calcium ions are required for catalysis of the protein transition.[18] Other workers have found that prothrombin fragment 1 has two high affinity sites for manganese.[32] It appears possible that two manganous ions will catalyze the prothrombin transition although correlation of these sites with the transition has not been made. This discrepancy in the number of required calcium versus manganous ions is not necessarily inconsistent since, in another instance, one terbium (III) ion has been shown to replace two calcium ions.[28] As expected from the observed cooperativity of the transition, more calcium is bound to prothrombin which has undergone the transition than that which has not.[18] The number of calcium ions bound per prothrombin molecule was observed to increase by about one after allowing the protein transition to occur. This is a minimum value due to the observation that about 25% of the molecules undergo the rapid transition and would be unaffected by the time-course of the calcium-binding measurements and some additional "normal" prothrombin molecules will have undergone the protein transition in the 30 min. (at 0°) required to make the calcium-binding measurement. The rate for binding of this additional calcium has not been correlated with the rate of the protein transition.

Prothrombin-membrane interaction. The important and well-documented function of calcium in blood-coagulation is in binding prothrombin, factor X and presumably the other vitamin K-dependent blood-clotting proteins to a membrane surface. Calcium binding to the protein and the protein transition described above are prerequisites for this protein-membrane interaction. The following equilibria serve as an adequate model for the observed properties of prothrombin-membrane interaction:[20]

$$P + iCa \overset{1}{\rightleftarrows} P_{iCa} \overset{2}{\rightleftarrows} P'_{iCa}$$
$$\overset{3}{+} \quad +mCa \overset{4}{\rightleftarrows} P' \cdot PL_{i+j+mCa} \qquad (2)$$
$$PL + jCa \overset{3}{\rightleftarrows} PL_{jCa}$$

328

Reactions 1 and 2 are calcium binding and the prothrombin transition described above. Reaction 3 is calcium binding to phospholipid which is also a prerequisite to prothrombin-membrane binding. Reaction 4 contains a calcium term (m) which is variable and decreases to zero at high calcium (>5 mM) concentrations. Consequently, where i and j are saturating, m is zero. The expression in equation 2 is used to clearly demonstrate the fact that the protein-membrane interaction (reaction 4) is calcium dependent at the calcium concentrations of greatest interest. By direct measurement, m was estimated to be 3 ± 1.5 at 0.5 mM calcium and 1 ± 1.5 at 1.2 mM calcium. This trend agrees with other experiments showing the variability of m.[20] The total number of calcium ions required for the protein-membrane complex was estimated to be 9 ± 1 at 0.5 mM calcium. This number could include non-functional calcium ions. At 1.2 mM calcium, there were 13 ± 1 calcium ions bound per molecule of membrane-bound prothrombin and it is felt that this number probably does include non-functional calcium ions.[20]

It is well known that the membrane must include acidic phospholipids for prothrombin-membrane association.[33] The results of studies using relative light scattering to determine the maximum protein/membrane ratio of the complex at saturating protein concentrations are given in figure 2 for prothrombin and factor X. It can be seen that where phosphatidyl serine is limiting (<15% PS), the capacity of the membrane is approximately proportional to phosphatidyl serine content. Based on a random distribution of phosphatidyl serine residues on the inner and outer surface of the vesicle, it was estimated from the data below 15% phosphatidyl serine (figure 2A) that 9 ± 1 phosphatidyl serine residues must be exposed per prothrombin-binding site. Since phosphatidyl serine does not distribute randomly between the inner and outer surface of the vesicle but prefers the inner surface[34] the value of 9 ± 1 is a maximum value, and, based on the inner and outer surface distribution observed in one instance, would be about 15% larger than the actual ratio.

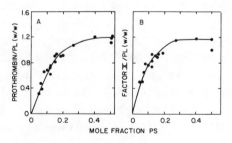

Figure 2. Maximum protein-binding capacity of membrane vesicles as a function of phosphatidyl serine composition.

Based on the calcium and phosphatidyl serine requirements for prothrombin-phospholipid interaction, two possible models for this interaction are shown in figure

329

3. The model in 3A shows protein binding by coordination bond(s) to membrane-bound calcium ions.   There are 4 additional calcium ions bound to the protein which are required for catalysis of the prothrombin transition. Model 3B shows ionic binding between protein and phospholipid. In this model, the calcium ions which catalyzed the protein transition must also participate in forming the protein-membrane complex. These models conform to the observations of Lim et al.[17] who showed that prothrombin binds to the membrane surface at one tip without detected penetration into the lipid region of the membrane. The remainder of the prothrombin molecule extends radially into solution.

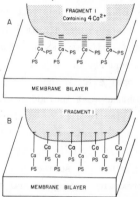

Figure 3. Models for prothrombin-membrane interaction. PS is phosphatidyl serine in the membrane bilayer. Ionic bonding is indicated by solid lines and coordinate bonding (no estimate of the number is implied) by dashed lines.

Factor X-membrane interaction. The properties of factor X-membrane interaction are qualitatively similar but quantitatively different from prothrombin. For example, factor X has the same equilibria given in equation 2 but differs with respect to i) the rate of the transition is too rapid to measure by conventional fluorescence techniques,[19] ii) the cation requirement is different in that higher concentrations of calcium are necessary and factor X is somewhat more selective for calcium among the group IIA cations,[19] iii) Reaction 3 also appears to require more membrane-bound calcium ions but iv) there appear to be only 4-6 phosphatidyl serine residues required per membrane-bound factor X molecule.[35] Like prothrombin, factor X has two high affinity binding sites for Gd(III) and appears to have a larger number of binding sites for Tb(III).[36] Due to the considerable similarities between the vitamin K-dependent blood-clotting proteins,[5] it will be interesting to determine the variability of these proteins in their membrane-binding characteristics.

In summary, we find that calcium functions in at least two very different capacities in the vitamin K-dependent, γ-carboxyglutamic acid-containing proteins. The first role is in calcium binding to the protein and catalysis of a protein transition. We know by comparison to the abnormal prothrombin that γ-carboxyglutamic acid must function in this role but direct binding of calcium to γ-carboxy-glutamic acid has not yet been demonstrated. The second role for calcium is in formation of the protein-membrane bond. The evidence suggests protein-membrane binding through calcium ions as shown in figure 3 but the groups on the protein which interact directly with these calcium ions are not known. The available evidence does not require that γ-carboxyglutamic acid residues be involved in this second calcium role at all.

ACKNOWLEDGMENTS

The author is indebted to Dr. Tong Lim and Dr. Victor Bloomfield for assistance in the theoretical analysis of relative light scattering measurements and for determining vesicle size by quasi-elastic light scattering and to Ms. Margaret Broderius for her excellent technical assistance. This work was supported in part by NIH grant HL 15728.

REFERENCES

1. Stenflo, J., Fernlund, P., Egan, W., and Roepstorff, P. (1974) Proc. Nat. Acad. Sci. USA 71, 2730-2733.

2. Nelsestuen, G. L., Zytkovicz, T. H., and Howard, J. B. (1974) J. Biol. Chem. 249, 6347-6350.

3. Magnusson, S., Sottrop-Jensen, L., Petersen, T. E., Morris, H. R., and Dell, A. (1974) FEBS Lett. 44, 189-193.

4. Esmon, C. T., Sadowski, J. A., and Suttie, J. W. (1975) J. Biol. Chem. 250, 4744-4748.

5. Fujikawa, K., Coan, M., Enfield, D., Titani, K., Ericsson, L., and Davie, E.W. (1974) Proc. Nat. Acad. Sci. USA 71, 427-430.

6. Zytkovicz, T. H., and Nelsestuen, G. L. (1975) J. Biol. Chem. 250, 2968-2972.

7. Howard, J. B., and Nelsestuen, G. L. (1974) Biochem. Biophys. Res. Commun. 59, 757-763.

8. Fernlund, P., Stenflo, J., Roepstorff, P. and Thomsen, J. (1975) J. Biol. Chem. 250, 6125-6133.

9. Morris, H. R., Dell, A., Petersen, T. E., Sottrup-Jensen, L., and Magnusson, S. (1976) Biochem. J. 153, 663-679.

10. Howard, J. B., and Nelsestuen, G. L. (1975) Proc. Nat. Acad. Sci. USA 72, 1281-1285.

11. Enfield, D. L., Ericsson, L. H., Walsh, K. A., Neurath, H., and Titani, K. (1975) Proc. Nat. Acad. Sci. USA 72, 16-19.

12. Nelsestuen, G. L., Broderius, M., Zytkovicz, T. H., and Howard, J. B. (1975) Biochem. Biophys. Res. Commun. 65, 233-240.

13. Ingwall, J. S., and Scheraga, H. A. (1969) Biochemistry 8, 1860-1869.

14. Nelsestuen, G. L. and Suttie, J. W. (1973) Proc. Nat. Acad. Sci. USA 70, 3366-3370.

15. Heldebrant, C. M., and Mann, K. G. (1973) J. Biol. Chem. 248, 3643-3652.

16. Huang, C. (1969) Biochemistry 8, 344-352.

17. Lim, T. K., Bloomfield, V., and Nelsestuen, G. L. (1977) submitted for publication.

18. Nelsestuen, G. L. (1976) J. Biol. Chem. $\underline{251}$, 5648-5656.

19. Nelsestuen, G. L., Broderius, M., and Martin, G. (1976) J. Biol. Chem. $\underline{251}$, 6886-6893.

20. Nelsestuen, G. L., and Lim, T. K. (1977) submitted for publication.

21. Bensen, B. J., and Hanahan, D. J. (1975) Biochemistry $\underline{14}$, 3265-3277.

22. Henricksen, R. A., and Jackson, C. M. (1975) Arch. Biochem. Biophys. $\underline{170}$, 149-159.

23. Bajaj, S. P., Butkowski, R. J., and Mann, K. G. (1975) J. Biol. Chem. $\underline{250}$, 2150-2156.

24. Prendergast, F. G., and Mann, K. G. (1977) J. Biol. Chem. $\underline{252}$, 840-850.

25. Brittain, H. G., Richardson, F. S. and Martin, R. B. (1976) J. Am. Chem. Soc. $\underline{98}$, 8255-8260.

26. Nelsestuen, G. L., and Suttie, J. W. (1972) Biochemistry $\underline{11}$, 4961-4964.

27. Stenflo, J., and Ganrot, P.-O. (1973) Biochem. Biophys. Res. Commun. $\underline{50}$, 98-104.

28. Matthews, B. W., and Weaver, L. H. (1974) Biochemistry $\underline{13}$, 1719-1725.

29. Furie, B. C., Mann, K. G., and Furie, B. (1976) J. Biol. Chem. $\underline{251}$, 3235-3241.

30. Argos, P. (1977) Biochemistry $\underline{16}$, 665-672.

31. Kretsinger, R. H. (1974) in Perspectives in Membrane Biology. S. Estrada and C. Gitler, Ed. (Academic Press).

32. Bajaj, S. P., Nowak, T., and Castellino, F. J. (1976) J. Biol. Chem. $\underline{251}$, 6294-6299.

33. Papahadjopoulos, D., and Hanahan, D. J. (1964) Biochim. Biophys. Acta $\underline{90}$, 436-439.

34. Berden, J. A., Barker, R. W., and Radda, G. A. (1975) Biochim. Biophys. Acta $\underline{375}$, 186-208.

35. Nelsestuen, G. L. and Broderius, M. (1977) submitted for publication.

36. Furie, B. C., and Furie, B. (1975) J. Biol. Chem. $\underline{250}$, 601-608.

Comparison of γ-Carboxyglutamic Acid-Containing Proteins from Bovine and Swordfish Bone: Primary Structure and Ca^{++} Binding

Paul A. Price, Allen S. Otsuka and James W. Poser
Department of Biology
University of California, San Diego
La Jolla, California 92093

SUMMARY

A partial amino-acid sequence of the γ-carboxyglutamic acid-containing protein of swordfish bone is presented. Comparison of this sequence with the calf bone structure shows that 21 of the 45 residues common to each are identical. The three γ-carboxyglutamic acid residues and the single disulfide bond are in identical relative positions in the two proteins. An important difference between the proteins is the absence of 4-hydroxyproline in the swordfish protein. Calcium binds weakly to the calf protein, with a minimum of three binding sites with a dissociation constant of 3 mM. Phosphate does not bind to the protein. The role of γ-carboxyglutamic acid in the binding interaction between the protein and hydroxyapatite crystals is discussed.

INTRODUCTION

We have described earlier the isolation from calf bone of a 5700 molecular weight protein that contains three γ-carboxyglutamic acid (Gla) residues[1]. This protein (Gla protein) is removed from calf cortical bone upon demineralization in EDTA and accounts for 1-2% of the total protein in calf bone. A similar Gla protein has been found in swordfish vertebrae and human tibia[1], and a Gla-containing protein has been reported in chicken tibia[2]. The abundance of this protein and its presence in all vertebrate bones which have been examined indicate that it has an important function in calcified tissues. We have described two important properties of the Gla protein[1]: its strong binding to hydroxyapatite crystals but not to amorphous calcium phosphate, and its inhibition of hydroxyapatite crystallization from supersaturated solution. These properties of the Gla protein may be conferred upon it by γ-carboxyglutamic acid, in analogy with the function of this vitamin K dependent amino acid in the Ca^{++} and BaSO$_4$ binding properties of the blood coagulation factors[3,4].

We have also reported the complete covalent structure of the Gla protein from calf bone[5]. Interesting features of this structure are the presence of Gla residues at positions 17, 21 and 24, a disulfide bond joining residues 23 and 29, and a residue of 4-hydroxyproline. To help determine which structural characteristics of the Gla protein are important to its function, we have undertaken the determination of the amino acid sequence of the Gla protein from swordfish vertebrae. We have also investigated Ca^{++} and phosphate binding to the calf bone Gla protein by gel filtration in order to understand better the mode of Gla protein binding to hydroxyapatite.

MATERIALS AND METHODS

Purified swordfish bone Gla protein was isolated from swordfish vertebrae by the procedure used for calf bone[5], and was reduced and S-aminoethylated prior to use. The amino acid sequence was determined by automatic sequence analysis with the Beckman Sequencer (Model 890B), and by manual Edman degradation of cyanogen bromide, chymotryptic and pronase peptides as described previously[5]. More detailed methods will be described when the completed sequence is published. Ca^{++} and phosphate binding to the purified, intact Gla protein from swordfish vertebrae was investigated by the gel filtration method[6,7] using a 0.4 by 50 cm Sephadex G25 column. $^{45}Ca^{++}$ and $H^{32}PO_4^{=}$ specific radioactivities were 10^{12} cpm per mole. Buffers contained 50 mM PIPES at pH 7.4 and were adjusted to an ionic strength of 0.15 M with KCl to minimize non-specific Ca^{++} association due to the high negative charge of the protein[1,5]. The average Gla protein concentration in fractions used for estimation of bound Ca^{++} atoms was 5×10^{-4} M, and usually 5 to 10 fractions were used from each column run to compute the mean and the standard deviation values in Table 1.

RESULTS

The partial amino acid sequence of the Gla protein from swordfish bone is compared with the complete covalent structure of the calf bone protein in Fig. 1. Identification of amino acid residues at positions 14, 24 and 37 in the swordfish structure is tentative; they are accordingly enclosed in parentheses. There is extensive homology between the two structures in the region of the Gla residues and the disulfide bond. In all, 21 of the residues in the partial swordfish structure are identical to the residues in the calf bone structure. Important differences are the absence of 4-hydroxyproline in the swordfish protein and the presence of four additional

N-terminal residues in the calf structure and 2 additional C-terminal residues in the swordfish protein.

C 1	Tyr	Leu	Asp	His	Trp	Leu	Gly	Ala	Hyp	Ala	Pro	Tyr	Pro	Asp	Pro
S					Ala	Thr	Arg	Ala	Gly	Asp	Leu	Thr	Pro	Leu	Gln

C 16	Leu	Gla	Pro	Lys	Arg	Gla	Val	Cys	Gla	Leu	Asn	Pro	Asp	Cys	Asp
S	Leu	Gla	(Ser)	Leu	Arg	Gla	Val	Cys	Gla	Leu	Asn	Val	(Ser)	Cys	Asp

C 31	Glu	Leu	Ala	Asp	His	Ile	Gly	Phe	Gln	Glu	Ala	Tyr	Arg	Arg	Phe
S	Glu	Met	Ala	Asp	Thr	Ala	Gly	Ile	Val	Ala	(Ala)	Tyr	Ile	Ala	Tyr

C 46	Tyr	Gly	Pro	Val		
S	Tyr	Gly	Pro	Ile	Gln	Phe

Fig. 1. Primary structure of the γ-carboxyglutamic acid-containing proteins from calf and swordfish bone. Calf sequence, C; swordfish sequence, S; numbers refer to calf protein sequence.

As can be seen in Table 1, Ca^{++} binds weakly to the calf bone Gla protein. Scatchard analysis[8] of these is data indicates that the calf bone Gla protein has a minimum of three Ca^{++} binding sites with an average dissociation constant of 3 mM and a maximum of 7 sites with an average dissociation constant of 7 mM. The calf bone Gla protein bound less than .05 phosphate molecules per Gla protein at an unbound phosphate concentration of 1 mM, which demonstrates the absence of phosphate binding to the Gla protein.

TABLE 1

CONCENTRATION DEPENDENCE OF Ca^{++} BINDING TO CALF BONE GLA PROTEIN

Concentration of Unbound Ca^{++} (mM)	Ca^{++} Bound per Gla Protein
0.2	0.18 ± .04
0.4	0.31 ± .03
0.5	0.46 ± .07
1.0	0.94 ± .10
2.0	1.41 ± .46
5.0	1.80 ± .57

Each value of Ca^{++} bound to the Gla protein is the mean of 5 to 10 samples from a single filtration experiment; error is standard deviation of the mean. Column, 0.4 x 50 cm of Sephadex G25; 25°; 5 x 10^{-4}M Gla protein; 50 mM PIPES pH 7.4 adjusted to 0.15 M ionic strength with KCl.

335

DISCUSSION

The unique feature of the bone Gla protein is the presence of three residues of γ-carboxyglutamic acid, a vitamin K dependent Ca^{++} binding amino acid first discovered in the blood coagulation factors [9,10,11]. Although the biological function of the bone Gla protein is unknown, it is possible that the function of the Gla residues in the protein is to coordinate the protein to the hydroxyapatite mineral phase of bone. The Gla residues in prothrombin are required for binding to $BaSO_4$, an analogous binding interaction. The interaction between the Gla protein and bone hydroxyapatite could be highly specific, requiring a set of Gla residues oriented precisely to form a binding site that is complementary to the spacing of Ca^{++} atoms in hydroxyapatite. Alternatively, the binding could be a non-specific association between Gla residues and a random set of Ca^{++} atoms in a local region of the crystal surface. The latter hypothesis would lead to the prediction that Gla proteins from the bones of widely different species could differ in the number and spacing of Gla residues in the structure and still retain the requisite affinity for the mineral phase of bone. It would not, however, account for the selective binding of the Gla protein to hydroxyapatite rather than to amorphous calcium phosphate[1], since both solid phases have Ca^{++} atoms on the surface to coordinate with randomly spaced Gla residues in a protein.

Comparison of the calf and swordfish bone Gla protein sequences strongly indicates that Gla residues precisely oriented in space are essential to the biological function of the protein. The region of the sequence containing the 3 Gla residues and the disulfide bond is the most highly conserved in the two proteins, with 14 of 19 residues identical between residues 16 and 34 in the calf sequence. Such structurally important features as the relative sequence positions of the Gla residues and the disulfide bond are identical in the two proteins. The number of negative charges in this region at pH 7 is also the same. The high degrees of homology between these proteins in the Gla region leads us to suggest that this portion of the structure may be a major interaction point with another biologically important substance. It is tempting to speculate that this substance is hydroxyapatite, and that the spacial homology of Gla residues in this region is important to the specificity of this binding interaction.

336

The nature of the binding of the purified Gla protein from bone to hydroxyapatite could be achieved by coordination of the protein with either Ca^{++} or phosphate groups on the surface of the crystal lattice. The results presented here clearly show that the protein can bind Ca^{++} but not phosphate. The affinity for Ca^{++} is weak, with at most a dissociation constant of 3 mM for 3 binding sites. Such weak interaction with Ca^{++} still permits rather strong interaction between the protein and the crystal, since this interaction represents the sum of the binding energies for each of 3 Gla residues. A stronger Ca^{++} binding to the same residues involved in coordinating the protein to hydroxyapatite would only competitively inhibit protein binding to the crystal surface, since the free Ca^{++} concentration in plasma is 1-2 mM.

ACKNOWLEDGEMENTS

We thank Joanne Kristaponis for assistance in the Ca^{++} binding study. This work was supported in part by U.S. Public Health Service Grant GM 17702-07.

REFERENCES

1. Price, P.A., Otsuka, A.S., Poser, J.W., Kristaponis, J. and Raman, N. (1976). Proc. Natl. Acad. Sci. USA 73,1447-1451.

2. Hauschka, P.V., Lian, J.B. and Gallop, P.M. (1975). Proc. Natl. Acad. Sci. USA 72,3923-3929.

3. Magnusson, S. (1971). In The Enzymes, ed. Boyer, P.D. (Academic Press, New York), 3rd ed., Vol. 3, pp. 277-321.

4. Nelsestuen, G.L., Broderius, M.B., Zytkovicz, T.H. and Howard, J.B. (1975). Biochem. Biophys. Res. Commun. 65,233-240.

5. Price, P.A., Poser, J.W. and Raman, N. (1976). Proc. Natl. Acad. Sci. USA 73,3374-3375.

6. Hummel, J.P. and Dreyer, W.J. (1962). Biochim. Biophys. Acta 63,530.

7. Price, P.A. (1972). J. Biol. Chem. 247,2895-2899.

8. Scatchard, G. (1949). Ann. N.Y. Acad. Sci. 51,660.

9. Stenflo, J., Fernlund, P., Egan, W. and Roepstorff, P. (1974). Proc. Natl. Acad. Sci. USA 71,2730-2733.

10. Nelsestuen, G.L. Zytkovicz, T.H. and Howard, J.B. (1974). J. Biol. Chem. 249,6347-6350.

11. Magnusson, S., Sottrup-Jensen, L., Peterson, T.E., Morris, H.R. and Dell, A. (1974). FEBS Lett. 44,189-193.

Purification and Calcium-Binding Properties of Osteocalcin, the γ-Carboxyglutamate-Containing Protein of Bone

Peter V. Hauschka and Paul M. Gallop

Department of Orthopaedic Research, Children's Hospital
Medical Center and Harvard Medical School, and Depart-
ment of Oral Biology, Harvard School of Dental Medicine,
Boston, Massachusetts 02115

Vitamin K-dependent posttranslational carboxylation of proteins (1) generates specific Ca^{2+} binding sites which are associated with the newly formed γ-carboxyglutamic acid (Gla) residues (2). This process has been elegantly studied in relation to prothrombin, where 10 Gla residues in the NH_2-terminal region (3-5) facilitate the binding of 6 to 10 Ca^{2+} with dissociation constants in the range $K_d = 6 \times 10^{-4} M$ (6-11). If the appropriate glutamic acid residues are not carboxylated, as occurs during vitamin K deficiency or antagonism with warfarin or dicoumarol-type anticoagulant drugs, then the resultant "acarboxyprothrombin" exhibits only weak, relatively nonspecific interaction with Ca^{2+} ions (2,7). Gla residues are essential not only for the binding of Ca^{2+} to the protein, but for association of the protein with phospholipid micelles or plasma membrane fragments through Ca^{2+}-mediated salt bridges (10-12). This latter process is mechanistically important for proteolytic activation of prothrombin and other vitamin K-dependent proteins during hemostasis (2, 10-13). At least seven different proteins containing Gla residues have now been isolated from blood plasma: prothrombin, factors VII, IX, and X, protein C (14), protein S (15), and protein Z (16).

Gla-proteins occur significantly in tissues other than blood. Bone, in particular, contains abundant quantities of a small Gla-protein, osteocalcin, which we and others have isolated (17-20). In addition, the kidney is a site of vitamin K-dependent Gla-protein synthesis (21). Numerous sites of pathological ectopic calcification (e.g. renal stones, atherosclerotic plaque) also contain Gla-proteins (22,23), as does urine (24).

Osteocalcin, the bone Gla-protein, is unrelated to the blood coagulation proteins by amino acid composition (17,19,20) and sequence (19,20). Furthermore, vitamin K dependence (25) and de novo biosynthesis of osteocalcin in isolated, cultured bone has been demonstrated by $NaH^{14}CO_3$ incorporation into Gla (26).

It is likely that osteocalcin plays an important role in the regulation of mineral deposition and/or remodelling processes in mineralized tissues (17). We

---

Definitions: Gla, γ-carboxyglutamic acid
Gla-protein, any protein with Gla residues in the polypeptide chain
Osteocalcin, the 6500 mw Gla-protein found in bone

report here the further characterization and Ca^{2+}-binding properties of this protein.

MATERIALS AND METHODS

 EDTA extraction of chicken metatarsal bones (9 week Penobscot; kindly provided by J. Fabish, Empire Kosher Poultry, Mifflintown, PA) was as described (17), with the addition of the protease inhibitors p-toluenesulfonylfluoride (3×10^{-5}M) and p-chloromercuribenzoate (10^{-4}M) to the 0.5M EDTA, pH 8.2 extraction solution.

 Amino Acid analysis was performed on both acid and alkaline-hydrolyzed samples by established methods (17,18,21) using the ninhydrin color factor measured for synthetic Gla (18,21).

 Agarose slab gel electrophoresis followed previous techniques (17), except that fixation, staining (0.5% amido black), and destaining were in 12.5% TCA at 0°.

 Isoelectric focussing was done at 10° in an LKB Uniphor column using pH 2.5 - 4.0 Ampholine stabilized by a sucrose density gradient. The current was monitored at 500V and the run terminated when it had reached a constant low value (~80 hr).

 Equilibrium dialysis was used to measure calcium binding by osteocalcin. Spectropor 3 tubing (3500 mw cutoff; Spectrum Industries, Los Angeles, CA) completely retained the protein (>90% recovery) during the dialysis period. Each knotted bag contained 0.5 ml osteocalcin solution at $1 - 2 \times 10^{-4}$M (0.65 - 1.3 mg/ml). Dialysis proceeded at 4° on a rotary shaker for 48 hr which was shown to be more than sufficient for attainment of equilibrium. Each dialysis flask contained 125 ml buffer, including $^{45}CaCl_2$ (New England Nuclear) at a final specific activity of 80 - 160 cpm/nmole, 5mM Tris-HCl pH 7.60, and other additives as specified for each experiment. After dialysis the protein concentration was determined both by absorbance at 276 nm ($\varepsilon = 3800$ $M^{-1}cm^{-1}$) and by acid hydrolysis and amino acid analysis of a 50 µl aliquot. ^{45}Ca was counted in the ^{14}C window of an Intertechnique liquid scintillation spectrometer; 400 µl aliquots of protein and equilibrated solvent were dispersed in 10 ml Insta-Gel (Packard) for counting.

 Terbium fluorescence was measured at 27° in 0.5 ml cuvettes on a Perkin-Elmer MPF-3 fluorometer (xenon lamp) with an excitation band width of 10 nm and an emission band width of 6 nm.

RESULTS AND DISCUSSION

 Neutral EDTA demineralization of bone powder extracts about 10% of the total protein and 95% of the total Gla. Most of the extracted Gla is nondialyzable. Remaining behind is the largely insoluble bulk of bone collagen which comprises some 90% of the bone protein and contains only traces of Gla (~0.05 res/1000 amino acid res). Gel filtration on Sephadex G-100 of the EDTA-solubilized materials resolves numerous components as shown in Fig. 1. Based on amino acid analysis of pooled fractions, virtually all (86%) of the Gla elutes with a prominent protein peak at

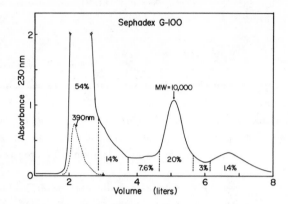

Fig. 1. Gel filtration on Sephadex G100 of an EDTA extract of chicken bone. 553 mg of protein was applied to a 10 x 100 cm column equilibrated with 0.1M NH₄HCO₃, pH 7.7, at 4° and flowing at 3.3 ml/min. Osteocalcin elutes as a prominent peak at an apparent molecular weight of 10,000, based on calibration with standard globular proteins. This peak contains 20% of the total applied protein and 86% of the total Gla. Small amounts of Gla-containing proteins are present in all other fractions, with about 7% eluting at the void volume coincident with the turbid 390 nm peak.

an apparent molecular weight of 10,000. This protein, whose further purification and characterization is described below, has been named <u>osteocalcin</u> because of its abundance in bone and its propensity for the binding of Ca^{2+} ions.

Osteocalcin from Sephadex G-100 columns contains hydroxyproline (8 res/1000) and has high 260 nm absorbance ($A_{280}/A_{260} \sim 1.1$). Chromatography on DEAE-Sephadex (Fig. 2) resolves osteocalcin from both the Hyp-rich collagenous peptides which contain no Gla (peak A) and the strongly absorbing nucleic acid contaminants (peaks E,F) (see also Fig. 4). It was essential to include 20mM $CaCl_2$ in the buffer to achieve this resolution, since in its absence osteocalcin is more nega-tively charged and elutes with the nucleic acid peak.

DEAE-Sephadex yields three distinct peaks of osteocalcin which, upon rechroma-tography, elute at reproducibly different NaCl concentrations (Fig. 2): peak B (0.19M), peak C (0.24M), and peak D (0.33M). Peak C is the major species (>60% of the total Gla-protein), and all studies referring to "osteocalcin" have been done with DEAE-peak C unless otherwise indicated. Table 1 shows the relative purifica-tion afforded by each step described above. After the DEAE-Sephadex step, the 99% pure protein may be considered homogeneous in one sense, in that it contains only Gla-rich polypeptides. The modest differences in amino acid composition and charge of various resolved species may result from several factors: proteolysis during preparation; incomplete carboxylation; or the existence of truly distinct and independently synthesized Gla-proteins in bone.

Table 2 compares the composition of chicken osteocalcin with a similar pro-tein isolated from bovine bone (19,20). There are notable differences in that the

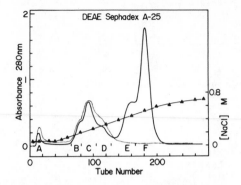

Fig. 2. DEAE-Sephadex A-25 chromatography of osteocalcin. 200 mg osteocalcin from Sephadex G-100 (Fig. 1) was applied to the column (1.2 x 60 cm) in 20 mM imidazole, 70 mM NaCl, 2mM EDTA, 20mM $CaCl_2$, pH 7.0 (INEC), flowing at 25 ml/hr; 3.3 ml fractions were collected. A gradient of 250 ml INEC and 250 ml INEC + 0.7M NaCl eluted osteocalcin (peaks B,C,D) free of nucleic acid (E,F) and collagenous peptides (A). Lowry protein assay ($\cdots\cdots$).

Fig. 3. Isoelectric focussing of osteocalcin. 8 mg of DEAE peak C protein (Fig. 2) was focussed at 10° in a 220 ml gradient formed with pH 2.5 - 4.0 Ampholine (LKB) and stabilized by sucrose. The pI values for the various bands are: A (3.40), B (3.50), C (3.60), D (3.68), and E(3.94). Absorbance at 278 nm (——); pH ($\cdots\cdots$).

chicken bone protein is devoid of Hyp, Cys ½, His, Lys, and Trp and contains Ser and twice as much Glu and Gly as the bovine protein. Table 3 shows the composition of the osteocalcin peaks from DEAE-Sephadex (Fig. 2). Peaks B and C are identical in amino acid composition, leaving unexplained the more anionic character of peak C. However, peak D, which is the most anionic species of osteocalcin, has a clearly reduced content of Arg, Phe, Tyr, Val, Gly, and Pro (Table 3). The absorbance properties of peaks C and D are in accord with the compositional differences. Both show an absorption maximum at 276 nm with A_{276}/A_{260} = 1.8 for C and 1.1 for D. The molar extinction coefficient for C is ε = 3800 $M^{-1}cm^{-1}$ (based on 57 res/molecule); for D, ε = 1750 (assuming 47 res/molecule).

Osteocalcin (DEAE-peak C) has a molecular weight of 6500 in the analytical ultracentrifuge. The Yphantis meniscus depletion method at 67,700 rpm showed no

Table 1: Purification of Chicken Bone Osteocalcin

stage	residues Gla 1000 amino acid res	fraction of protein which is osteocalcin
Whole Bone Powder	0.70	0.009
EDTA Extract	20	0.20
Sephadex G-100	60	0.80
DEAE Sephadex A-25	75	0.99

341

Table 2: Species Variation in Osteocalcin Composition

Amino Acid	Chicken	Bovine*
Gla	4	3
Hyp	-	1
Asp	6	6
Thr	-	-
Ser	3	-
Glu	7	3
Pro	7	6
Gly	6	3
Ala	6	4
Cys ½	-	2
Val	3	2
Met	-	-
Ile	1	1
Leu	4	5
Tyr	3	4
Phe	3	2
His	-	2
Lys	-	1
Trp	-	1
Arg	4	3
	57	49

*Price et al. (19,20)

Table 3: Heterogeneity of Chicken Osteocalcin on DEAE-Sephadex (residues/mole)

Amino Acid	peak C	peak D	Loss
Gla	4	4	-
Asp	6	6	-
Ser	3	3	-
Glu	7	7	-
Pro	7	6	1
Gly	6	4	2
Ala	6	6	-
Val	3	2	1
Ile	1	1	-
Leu	4	4	-
Tyr	3	1	2
Phe	3	1	2
Arg	4	2	2
	57	47	10

Peaks were pooled as shown in Fig. 2 and then rechromatographed. Peak B is identical in amino acid composition to peak C. The reduced content of certain residues in peak D is probably due to proteolytic cleavage of the parent osteocalcin molecule during extraction and purification.

evidence of polydispersity. The protein was run at 23° near its isoelectric point in 0.1M NaCl, 1mM acetic acid, pH 3.86, at c_o = 0.20 mg/ml.

Quantitative Edman degradation showed one NH_2-terminal residue for each 51-54 residues in peak C, giving an independent molecular weight estimate of 5800-6100. Tyr is the principal NH_2-terminal residue, but Ala, Asp, and Gly are also present as expected from the heterogeneity of the isofocussing pattern (Fig. 3). Edman degradation of DEAE-peak D protein gave an average chain length of 35-40 residues, with Ala as the principal NH_2-terminal residue.

Isoelectric focussing of osteocalcin reveals some heterogeneity (Fig. 3). The main species (peak B, pI = 3.50) is indistinguishable by amino acid composition from peaks A and C, and is identical to the values in Table 2. Peaks D and E differ, however, in that they contain a) 3 Gla and 8 Glu residues/molecule rather than 4 Gla and 7 Glu as found in the main species, and b) one residue each of His and Lys.

Calcium binding by osteocalcin was initially demonstrated by electrophoretic mobility (Fig. 5 and Ref. 17). The presence of 10mM Ca^{2+} causes a 1.9-fold decrease in anodal mobility of osteocalcin, compared to decreases of 2.9-fold for the F-1 fragment of human prothrombin but only 1.4-fold for serum albumin. Equilibrium dialysis studies confirmed that Ca^{2+} binds to osteocalcin (Figs. 6-9, Tables 4 and 5). Binding was examined at both low ionic strength and physiological

342

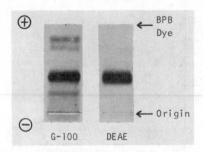

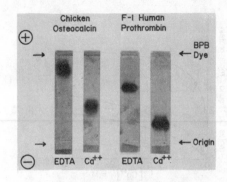

Fig. 4. Agarose slab gel electrophoresis pattern of osteocalcin after purification by Sephadex G-100 (left) and DEAE-Sephadex (right). Gel buffer was 50mM barbital 5mM calcium lactate, pH 8.6. 100 μg of protein was loaded in each slot.

Fig. 5. Calcium-dependent mobility of osteocalcin and the Gla-rich F-1 fragment of human prothrombin. Agarose slab gel buffer was 50mM barbital, pH 8.6, with 5mM EDTA or 10mM calcium lactate. 30 μg of osteocalcin or 10 μg F-1 was loaded in each slot.

ionic strength (150mM NaCl) in 5mM Tris-HCl, pH 7.60 (Figs. 6 and 7, Table 4). There are two distinct classes of Ca^{2+} binding sites in each environment. One class, consisting of 2 Ca^{2+} sites is present at low (K_d = 0.12mM) and high (K_d = 0.83mM) ionic strengths. The K_d for these sites is similar to that observed for other Gla-proteins (8-11). The Hill coefficient for these two sites (1.03) indicates non-cooperative binding. Another class of 2-3 Ca^{2+} is tightly bound to osteocalcin at low ionic strength (K_d = 24μM) but readily displaced by 150mM NaCl (Table 4, Figs. 6 and 9). The very weakly bound (K_d = 3mM) class of 3 Ca^{2+} which appears at high ionic strength may represent a similar collection of relatively nonspecific binding sites. The Ca^{2+} binding activity of osteocalcin is heat-stable (Table 5) and relatively insensitive to temperature (at 1mM $CaCl_2$, 22% less Ca^{2+} is bound at 27° than at 4°).

The pH dependence of Ca^{2+} binding is shown in Fig. 8. A smooth sigmoidal curve with an inflection point at pH 5.7 shows that titration of carboxyl groups in osteocalcin abolishes the affinity for Ca^{2+} at low pH. The high apparent pK_a for these carboxyl groups is due in part to polyelectrolyte effects for this highly anionic small protein. At lower ionic strength the pK_a should shift to even higher values. Thus with the steep portion of its Ca^{2+}-binding titration curve within the physiological range of pH and ionic strength, one function of osteocalcin may be to serve as a "calcium buffer" in bone.

Competition studies between Ca^{2+} and various metal ions are presented in Table 5. Ba^{2+} and Sr^{2+} do not compete significantly for the two Ca^{2+} sites on osteocalcin, while Mg^{2+} shows weak affinity. A variety of other cations show dramatic inhibition of Ca^{2+} binding, although it is not yet known whether this is due to

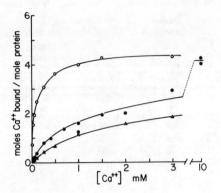

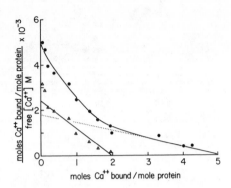

Fig. 6. Calcium binding by two osteocalcin fractions as a function of calcium concentration. Osteocalcin was dialyzed against 5mM Tris-HCl, pH 7.60, containing $^{45}CaCl_2$ in the range 5μM to 10mM: peak C from DEAE Sephadex (O); peak C with 150mM NaCl included in the buffer (●); peak D with 150mM NaCl included in the buffer (△).

Fig. 7. Scatchard plot of the calcium binding data for osteocalcin (peak C in 150 mM NaCl; see Fig. 6). After subtraction of the weak class of bound Ca^{2+} (n = 3, K_d = 3mM), the difference plot (△) shows 2.0 ± 0.3 sites for Ca^{2+} with K_d = 0.83mM.

Table 4:   Different Classes of Calcium Binding Sites in Osteocalcin*

Condition	n	K_d	n	K_d
5mM Tris-HCl, pH 7.60	2.4 ± 0.5	2.4 ± 1.0 × 10^{-5}M	1.9 ± 0.5	1.2 × 10^{-4}M
5mM Tris-HCl, pH 7.60 + 150mM NaCl	2.0 ± 0.3	8.0 ± 1.5 × 10^{-4}M	3.0 ± 0.5	3 × 10^{-3}M

*Protein peak C from DEAE-Sephadex (Fig. 2)

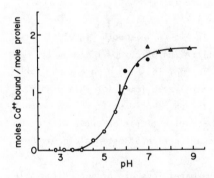

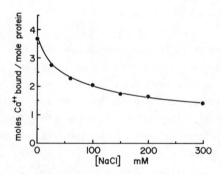

Fig. 8. Calcium binding by osteocalcin as a function of pH. Osteocalcin was dialyzed against 1mM $^{45}CaCl_2$, 150mM NaCl, and either 5mM acetate (O), 5mM imidazole (●), or 5mM Tris-HCl (△).

Fig. 9. Effect of ionic strength on calcium binding by osteocalcin. Osteocalcin was dialyzed against 1mM $^{45}CaCl_2$, 5mM Tris-HCl, pH 7.60, containing 0 to 300mM NaCl.

Table 5: Calcium Binding by Osteocalcin under Various Conditions

Condition	Ca^{2+} Bound (% of control)	Condition	Ca^{2+} Bound (% of control)
Control	100	Cu^{2+} 1000 μM	12
Boiled (10 min)	99	Cd^{2+} "	12
Mg^{2+} 1000 μM	70	$Fe^{2+,3+}$ "	2
Sr^{2+} "	90	La^{3+} "	0
Ba^{2+} "	101	La^{3+} 100 μM	0
Co^{2+} "	30	La^{3+} 10 μM	1
Mn^{2+} "	26	La^{3+} 1 μM	92
Ni^{2+} "	18	Tb^{3+} 200 μM	0
Zn^{2+} "	14	Tb^{3+} 20 μM	0

Osteocalcin at 10^{-4}M (0.65 mg/ml) was dialyzed 48 hr at 4° against 1mM $^{45}CaCl_2$, 150mM NaCl, 5mM Tris-HCl, pH 7.60. Metal ions were added as the chloride salts, except for magnesium acetate and terbium sulfate. Control samples bound 1.78 ± 0.15 moles Ca^{2+}/6500 g osteocalcin.

site competition or nonspecific conformational effects (as might be expected with Cu^{2+} for example). Trivalent lanthanide ions are known to be potent competitors for Ca^{2+} sites on many proteins (27). Both La^{3+} and Tb^{3+} completely inhibit the binding of Ca^{2+} to osteocalcin when present at only 1% of the concentration of Ca^{2+} (Table 5). Fluorescence studies with the Tb^{3+}-osteocalcin complex have demonstrated the characteristic green emission (556nm) of this bound lanthanide ion (27) when the protein is irradiated at the 280nm Tyr absorption maximum. Fluorescence titration of the apoprotein reveals an "induction region" where no emission occurs during addition of the first 2 equivalents of Tb^{3+} (Fig. 10). An abrupt 200-fold increase in fluorescence accompanies the addition of 2-3 more equivalents. Beyond 5 moles Tb^{3+}/mole osteocalcin, fluorescence intensity is unchanged. The "induction region" presumably represents binding of Tb^{3+} to sites remote from Tyr residues (27), in contrast to the other Tb^{3+} binding sites which allow strong fluorescence upon Tyr excitation. Which of these Tb^{3+} sites are identical to the two specific Ca^{2+} sites on the protein is not yet clear.

CONCLUSIONS

Osteocalcin is a highly anionic protein with a net charge of about -18 / 57 residues, based on composition and electrophoretic mobility in the presence and absence of Ca^{2+}. There are 2 Ca^{2+} binding sites per molecule for which the measured K_d falls within the range reported for Ca^{2+} binding to other Gla-proteins. It is assumed that the 4 Gla residues in osteocalcin participate directly in the binding of these 2 Ca^{2+} ions. Whether this is the case, and whether these residues are clustered in pairs or dispersed throughout the sequence remains to be discovered. The 3 Gla residues in bovine osteocalcin are not in pairs (20), while 6 of the 10 Gla residues in bovine prothrombin are paired (5).

Osteocalcin represents about 1% of the total protein in chicken bone. If the

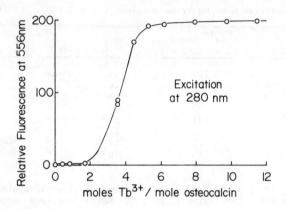

Fig. 10. Terbium (Tb^{3+}) fluorescence emission titration of osteocalcin. Protein at $2 \times 10^{-4}M$ was exhaustively dialyzed against 150mM NaCl, 5mM piperazine-HCl, pH 6.5. Steady-state fluorescence intensity at peak emission (uncorrected) was recorded after each volume addition of 5mM $Tb_2(SO_4)_3$. The excitation maximum for this fluorescence was 280nm.

function of osteocalcin involves direct interaction with Ca^{2+}, then it may behave as a "calcium buffer" as discussed above. Alternately, osteocalcin may regulate the steady state distribution of insoluble phases of calcium phosphate in bone. Preliminary studies with Dr. D.I. Hay have demonstrated that osteocalcin is a potent inhibitor of the transformation (28) of dicalcium phosphate dihydrate to basic calcium phosphate salts in vitro, and is thus similar to the salivary protein statherin (29).

Finally, osteocalcin may serve as an activation peptide which is proteolytically removed from an important bone proenzyme or bone procollagen in a Ca^{2+} and phospholipid-dependent event (analogous to the prothrombin $\longrightarrow$ thrombin conversion). It is intriguing that we find one molecule of osteocalcin for every 1-2 tropocollagen molecules in bone, and that hydroxyproline is a constituent of osteocalcin in some species (20). The search for higher molecular weight Gla-proteins in bone is in progress.

Acknowledgments. The authors benefited from discussion with Drs. J.B. Lian, M.J. Glimcher, and B. Holmquist and express gratitude to Dr. H.P. Traverso, M. Reid, C. Thibault, S. Klemm, D. Adams, M.T. Gows, and J. Komar for assistance with various aspects of this work. Access to the analytical ultracentrifuge was kindly provided by Drs. L. Bethune and W. Dafeldecker; Dr. D.I. Hay generously helped with the isofocussing experiments. Supported by NIH Grants AM 16754, AM 15671, and the John A. Hartford Foundation, Inc.; P.V.H. is the recipient of a Research Career Development Award from the National Institute for Dental Research (K04-DE00049).

REFERENCES

1. Esmon, C.T., Sadowski, J.A. and Suttie, J.W. (1975) J. Biol. Chem. 250, 4744-4748.
2. Fernlund, P., Stenflo, J., Roepstorff, P. and Thomsen, J. (1975) J. Biol. Chem. 250, 6125-6133.
3. Stenflo, J., Fernlund, P., Egan, W. and Roepstorff, P. (1974) Proc. Nat. Acad. Sci. USA 71, 2730-2733.
4. Nelsestuen, G.L., Zytokovicz, T.H. and Howard, J.B. (1974) J. Biol. Chem. 249, 6347-6350.
5. Magnusson, S., Sottrup-Jensen, L., Petersen, T.E., Morris, H.R. and Dell, A. (1974) FEBS Lett. 44, 189-193.
6. Nelsestuen, G.L. and Suttie, J.W. (1972) Biochemistry 11, 4961-4964.
7. Stenflo, J. (1973) J. Biol. Chem. 248, 6325-6332.
8. Benson, B.J., and Hanahan, D.J. (1975) Biochemistry 14, 3265-3277.
9. Bajaj, S.P., Butkowski, R.J. and Mann, K.G. (1975) J. Biol. Chem. 250, 2150-2156.
10. Henriksen, R.A. and Jackson, C.M. (1975) Arch. Biochem. Biophys. 170, 149-159.
11. Nelsestuen, G.L., Broderius, M., Zytkovicz, T.H., and Howard, J.B. (1975) Biochem. Biophys. Res. Commun. 65, 233-240.
12. Gitel, S.N., Owen, W.G., Esmon, C.T. and Jackson, C.M. (1973) Proc. Nat. Acad. Sci. USA 70, 1344-1348.
13. Esmon, C.T., Suttie, J.W., and Jackson, C.M. (1975) J. Biol. Chem. 250, 4095-4099.
14. Stenflo, J. (1976) J. Biol. Chem. 251, 355-363.
15. DiScipio, R.G., Hermodson, M.A., Yates, S.G. and Davie, E.W. (1977) Biochemistry 16, 698-706.
16. Prowse, C.V. and Esnouf, M.P. (1977) Biochem. Soc. Trans. 5, 255-256.
17. Hauschka, P.V., Lian, J.B. and Gallop, P.M. (1975) Proc. Nat. Acad. Sci. USA 72, 3925-3929.
18. Hauschka, P.V. (1977) Anal. Biochem. 80, (In Press).
19. Price, P.A., Otsuka, A.S., Poser, J.W., Kristaponis, J. and Raman, N. (1976) Proc. Nat. Acad. Sci. USA 73, 1447-1451.
20. Price, P.A., Poser, J.W., and Raman, N. (1976) Proc. Nat. Acad. Sci. USA 73, 3374-3375.
21. Hauschka, P.V., Friedman, P.A., Traverso, H.P. and Gallop, P.M. (1976) Biochem. Biophys. Res. Commun. 71, 1207-1213.
22. Lian, J.B., Skinner, M.S., Glimcher, M.J., and Gallop, P. (1976) Biochem. Biophys. Res. Commun. 73, 349-355.
23. Lian, J.B. and Prien, E.F., Glimcher, M.J., and Gallop, P. (1977) J. Clin. Invest. (in press June 1977).
24. Fernlund, P. (1976) Clin. Chim. Acta 72, 147-155.
25. Hauschka, P.V., Reid, M.L., Lian, J.B., Friedman, P.A., and Gallop, P.M. (1976) Fed. Proc. 35, 1354.
26. Lian, J.B., Hauschka, P.V., Glimcher, M.J., and Gallop, P.M. (1975) ORS abstract (22nd annual meeting).
27. Brittain, H.G., Richardson, F.S., and Martin, R.B. (1976) J. Am. Chem. Soc. 98, 8255-8260.
28. Grøn, P. (1973) Arch. Oral Biol. 18, 1379-1383.
29. Schlesinger, D.H. and Hay, D.I. (1977) J. Biol. Chem. 252, 1689-1695.

Activity of Intestinal Calcium-Binding Protein during Pregnancy in the Rat

Lindsay H. Allen and B.L. Lorens
Department of Nutritional Sciences
University of Connecticut
Storrs, Connecticut 06268

INTRODUCTION

It is well established in both humans and rodents that the efficiency of calcium absorption is increased during pregnancy. It is also known that the active transport of calcium is greater in gut-sacs from pregnant than from non-pregnant rats. This experiment was designed to determine whether intestinal calcium-binding protein (CaBP) is increased during pregnancy.

MATERIALS AND METHODS

Eighty female Wistar rats, weighing approximately 90 g, were maintained before and during pregnancy on a semi-purified diet designed to meet the nutritional requirements of the pregnant rat. This diet contained 0.6% Ca, 0.6% P, and 0.05 mg activated 7-dehydrocholesterol per kg. When they weighed 180 g, the rats were bred in sets of three. They were then sacrificed at gestational days 0 (controls), 12, 17, 19, 20, and 21. Mucosal tissue from the sets of three rats was homogenized and centrifuged after the method of Freund and Bronner (1). An aliquot of the homogenate was analyzed for DNA. After lyophilization of the supernatant, CaBP was separated on Sephadex G-75, using ammonium acetate buffer containing 0.2 mg 40Ca and 0.03 μc 45Ca per liter. Results were calculated as natoms calcium bound per mg protein in the CaBP peak, and as patoms calcium bound per μg DNA in the intestinal homogenate.

RESULTS

Calcium intake increased from 90 mg/day (controls) to 124 mg/day at the end of gestation. However, calcium consumed per 100 g body weight was constant throughout pregnancy.

Calcium bound per mg CaBP in pregnant animals was not increased over that of controls; the average calcium binding capacity was (as natoms Ca per mg CaBP $\pm$ SEM) 21.9 $\pm$ 2.9 for controls, and 15.2 $\pm$ 2.1 at 20 days of gestation, (Figure 1). There was no increase in calcium bound per mg intestinal mucosa. However, there was a significant increase (P<0.001) in the amount of calcium bound per cell, so that 264 patoms Ca/μg DNA were bound by mucosal tissue from non-pregnant controls, and 796 patoms Ca/μg DNA were bound at 21 days of gestation (Figure 2). There

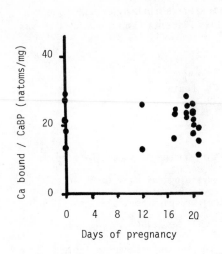

$$y=263.9-13.5x+1.8x^2$$
$$R^2=0.49, \; P<0.001$$

Fig. 1. Binding capacity of CaBP   Fig. 2. Calcium bound per cell

was also a significant increase (P<0.01) in the ratio of protein in the mucosal supernatant to DNA in the homogenate, suggesting that cellular hypertrophy occurred during pregnancy.

DISCUSSION

We conclude that since pregnant rats showed a significant increase in the amount of CaBP per mucosal cell, then there was an increase in the total amount of intestinal CaBP synthesized as a result of pregnancy. Since the amount of calcium bound per mg CaBP did not change, the increase in the total amount of CaBP may be non-specific response reflecting intestinal cell hypertrophy. We cannot conclude that an increased synthesis of CaBP per se is responsible for the increased efficiency of calcium absorption seen during pregnancy. However, the amount of CaBP per cell was greatest after the 17th day of gestation, which is at the same time as the greatest increase in maternal calcium absorption and calcium transfer to the fetus (2).

REFERENCES
1. Freund, T. and Bronner, F. (1975) Am. J. Physiol. 228, 861-869.
2. Chef, R. (1969) Compt. Rend. Soc. Biol. 163, 541-545.

Immunochemical Identity between
Rat Vitamin D-Dependent Calcium-Binding Protein in Duodenal Mucosa and a Calcium-Binding Protein in Placenta Tissue

M. Elizabeth H. Bruns, Aurora Fausto, and Louis V. Avioli
Washington University School of Medicine
The Jewish Hospital of St. Louis
216 S. Kingshighway
St. Louis, Missouri

INTRODUCTION

The steady-state concentration of the vitamin D-dependent calcium binding protein (CaBP) in small intestine parallels physiological variations in calcium absorption.[1] In birds, immunological identical proteins have been described in kidney, brain and shell gland extracts.[2-4] In mammalian studies using antiserum against pig intestinal CaBP, cross-reactivity was undetectable in brain and very low levels were present in kidney as compared to small intestine.[5] In the present study, we report the presence of a placental CaBP in rats with apparent identity to the vitamin D-dependent CaBP of the small intestine.

MATERIALS AND METHODS

Rabbit antibodies to a purified preparation of rat CaBP[6] were produced by injection of 0.38 mg CaBP in Freund's adjuvant followed by 0.1 mg at 4 and 7 weeks. The antibodies produced were monospecific as characterized by double immunodiffusion and immunoelectrophoresis using standard techniques.[7] Rat tissue extracts were screened for immunoreactive CaBP using 40,000 xg supernatant fractions (S-1) of total homogenates containing 15-20 mg protein/ml. CaBP was quantitated by a radial immunodiffusion[8] assay. The concentration of intestinal CaBP measured using this assay was approximately 10 fold higher in rats fed a low calcium diet containing vitamin D than in rats fed the same diet lacking the vitamin for 4 weeks. Sephadex G-100 chromatography, ^{45}Ca binding assays and disc gel electrophoresis were performed as described previously.[6]

RESULTS

Various rat tissue extracts (intestinal segments, liver, kidney, brain, fat, red blood cells, parathyroid, cartilage and placenta) were screened for immuno-reactive CaBP using a double immunodiffusion system. Using this assay system, immunoreactive material was detectable only in the mucosa of the proximal 12 cm of the small intestine and in placenta. The placental extract showed a line of complete identity with intestinal CaBP. Sephadex G-100 chromatography of 40,000 xg supernatant of placental homogenate revealed two peaks of ^{45}Ca binding, a high molecular weight peak (>60,000) and a low molecular weight peak (~10,000). Cross-reactivity was seen only in the low molecular weight peak, consistent with the known molecular weight of rat intestinal CaBP.[6] Polyacrylamide disc gel immuno-electrophoresis of the low molecular weight placental ^{45}Ca-binding peak showed a single arc precipitated by the antiserum. The corresponding protein band had a mobility identical to that of rat intestinal CaBP in the presence or absence of EDTA in the electrode buffers.

The levels of CaBP in placenta and maternal intestine increased progressively through gestation (Table I). The placenta level plateaued on day 19 when it was 8 times higher than on day 15. In the maternal intestine, CaBP concentration plateaued one day earlier than in placenta. These changes in CaBP closely parallel the rate of calcium accumulation reported for the rat fetus by Comar.[9]

TABLE I

CONCENTRATIONS OF IMMUNOREACTIVE CABP IN PLACENTA AND INTESTINE DURING GESTATION
The data are expressed as the Mean ($\pm$SE) and the number in the parentheses represents the number of experiments.

GESTATIONAL AGE	PLACENTA		MATERNAL INTESTINE	
(Days)	(µg CaBP/mg protein)		(µg CaBP/mg protein)	
0	-		3.3 $\pm$ 0.30	(18)
13	*	(4)	2.7 $\pm$ 0.06	(3)
15	0.10 $\pm$ 0.01	(4)	4.26 $\pm$ 0.48	(8)
18	0.22 $\pm$ 0.01	(3)	5.35 $\pm$ 0.65	(4)
19	0.80 $\pm$ 0.13	(4)	5.03 $\pm$ 0.64	(4)
20.5	0.80 $\pm$ 0.09	(7)	5.07 $\pm$ 0.42	(11)

*Trace amounts too low to be accurately measured.

SUMMARY

Placental tissue contains a protein similar or identical to the vitamin D-dependent CaBP of intestine based on immunological identity, apparent molecular weight on G-100 chromatography and mobilities in polyacrylamide disc gel electrophoresis in the presence or absence of EDTA.

The concentrations of placental CaBP and maternal intestinal CaBP paralleled the growth demands of the rat fetus and the rapid transfer of mineral to fetal skeleton during the last 4 days of gestation.[9] The data are consistent with the hypothesis that placental CaBP like duodenal CaBP reflects the rate of transcellular calcium transport.

This research was supported by National Institutes of Health Grant AM-11674 and Training Grant Am-07033.

REFERENCES

1. Wasserman, R.H. and Corradino, R.A. (1973) Vitamins and Hormones, 31-43.

2. Taylor, A.N. and Wasserman, R.H. (1972) Amer. J. Physiol. 223, 110-114.

3. Taylor, A.N. (1974) Arch. Biochem. Biophys. 161, 100-108.

4. Fullmer, C.S., Brindak, M.E., Bar, A., and Wasserman, R.H. (1976) Proc. Soc. Exp. Biol. Med. 152, 237-241.

5. Arnold, B.M., Kuttner, M., Willis, D.M., Hitchman, A.J.W., Harrison, J.E. and Murray, T.M. (1975) Can. J. Physiol. Pharmacol. 53:1135-1140.

6. Bruns, M.E., Fliesher, E.B. and Avioli, L.V. (1977) J. Biol. Chem. In press.

7. Clausen, J. (1972) Laboratory Techniques in Biochemistry and Molecular Biology (Editors, Work, T.S. and Work, E.) 1, 399-556, North-Holland, Amsterdam, Pergamon Press.

8. Mancini, G., Carbonara, A.O., Heremans, J.F. (1965) Immunochemistry 2, 235-254.

9. Comar, C.L. (1956) Ann. N.Y. Acad. Sci. 64, 281-298.

Vitamin D-Dependent Intestinal Calcium-Binding Protein:
A Regulatory Protein

T.S. FREUND and G. BORZEMSKY
Department of Biochemistry, School of Dentistry
Fairleigh Dickinson University
Hackensack, New Jersey 07601, USA

ABSTRACT

The rat intestinal vitamin D-dependent calcium binding proteins can be shown to increase the activity of the brush border alkaline phosphatase over a range of calcium from 0.1 μM to 0.1 mM. The differences observed between vitamin D-deficient mucosal alkaline phosphatase and the D-replete controls could be decreased by the addition of calcium binding proteins to the membranes, prior to assay.

INTRODUCTION

The vitamin D-dependent calcium binding proteins (CaBP), isolated from mammalian[1] and avian[2] intestinal tissues, have been implicated in the active movement of calcium although the mechanism of involvement is unknown.

The constant for the binding of calcium was calculated to be considerably higher[1] than the constant for transport across the intestine[3]. Thus the CaBP may not actually be a transporter of calcium but rather a modifier of the energy supplying systems.

CaBP has been reported to be localized primarily in the brush borders of duodenal mucosae[4], along with a calcium ATPase[5] and the intestinal alkaline phosphatase (AP)[6]. Both of these enzymes have been shown to be vitamin D-dependent and may function in the active transport of calcium.

We now report that CaBP may be a regulatory protein, which modulates the AP activity of rat intestinal brush borders.

MATERIALS and METHODS

Animals. Weanling male Sprague-Dawley derived rats, obtained from the Fairleigh Dickinson University Dental School Animal Care Facility, were divided into groups of six and fed one of several diets: 0.5% Ca and 0.5% P containing either no (II-D) or 2200 i.u. vitamin D_2/kg diet (II+D). Similar sets of animals were prepared with increased (1.5% Ca and 1.5% P; III±D) or decreased (0.06% Ca and 0.2% P; I±D) calcium.

Approximately 3-4 weeks later, when the plasma calcium of the vitamin D-deficient group were significantly lower, relative to their controls, the animals were sacrificed, their duodena removed and treated as previously described[1].

CaBP was isolated by Sephadex G-50 chromatography followed by rechromatography on the same column[1]. Samples were kept lyophilized until use. For purposes of calculation, a molecular weight of 11,000 daltons was assumed[1].

Brush borders were isolated from intestinal mocosae by the method of Forstner, et al.[7]. Similar results were obtained with 'crude' and 'purified' brush borders.

Alkaline Phosphatase (AP) activity was determined by measuring the hydrolysis of p-nitrophenyl phosphate as an increase in absorbance at 410nm at pH 8.0.

RESULTS

Table 1 illustrates the vitamin D-dependency of the brush border AP. The increase in AP activity in the presence of 0.275 μM CaBP is considerably less for the vitamin D-replete group than for the D-deficient samples.

The effect of calcium on the activity of the AP is shown in Fig.1. Here a suggestion at cooperativity for calcium activation is observed, which disappears in the presence of the CaBP. The CaBP activates the AP, but apparently only at the low calcium concentrations.

The method utilized to obtain the brush borders did have an effect on the results. Although 'crude' brush borders and 'purified' brush borders[7] gave similar data, extraction of the membranes with n-butanol failed to demonstrate a clear effect for the CaBP. Likewise, homogenized material did not demonstrate the observed CaBP activation.

Failure of an equivilent concentration of bovine serum albumin for the CaBP to activate the AP indicated a specificity for the CaBP in this particular system. CaBP, itself, did not possess any AP activity.

TABLE 1

ALKALINE PHOSPHATASE ACTIVITY IN RESPONSE
TO VITAMIN D AND CaBP*

RAT INTESTINAL SAMPLE	CaBP**	ALKALINE PHOSPHATASE*** µmoles/min./ mg protein
+D	+	0.48 (0.04)
	−	0.42 (0.04)
−D	+	0.44 (0.04)
	−	0.23 (0.03)
		(±S.E.M.)

*"purified" brush borders (50µg per assay)
**0.275 µM CaBP
***p-nitrophenyl phosphate hydrolysis

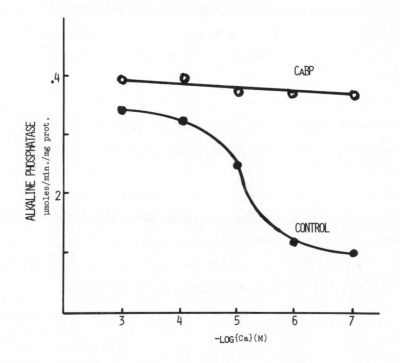

Fig.1. Effect of calcium on the alkaline phosphatase activity.

Determination of the K_M and the V_{Max} increased significantly (p<0.01) from 0.13 to 0.28 mM while the V_{Max} also increased from 0.153 to 0.375 µmole/min/mg protein.

DISCUSSION

The vitamin D-dependent calcium binding protein from rat intestinal mucosa has the ability to increase the activity of the mucosal alkaline phosphatase under the conditions examined. Apparently, this increased activation occurs primarily at calcium concentrations below 10 µM.

Since CaBP does markedly change the AP enzymatic activity, measured as K_M and V_{Max}, one could envision a considerable interaction between this protein and the brush border membranes. The particular interaction is, however, unknown, although it has been reported that in the presence of vitamin D, chick intestinal CaBP is bound to the AP enzyme system[8]. In addition it was reported a number of years ago[9], that chick CaBP also interacts with lysolecithin. It was suggested that this interaction might affect membrane permeability or membrane bound enzymes.

These results also suggest that the vitamin D-activation of alkaline phosphatase may not be an example of a direct dependency. Rather, the activation may be the result of increased calcium binding protein activity.

REFERENCES

1. Freund,T. and Bronner,F. (1975) Amer. J. Physiol. 228:861-869.

2. Wasserman, R.H. et al. (1968) J. Biol. Chem. 243:3987-3993.

3. Walling, M.W. and Rothman, S.S. (1970) J. Biol. Chem. 245:5007-5011.

4. Taylor, A.N. and Wasserman, R.H. (1970) J. Hist. Cyto. 18:107-115.

5. Melancon, M.J.Jr. and DeLuca, H.F. (1970) Biochemistry 9:1658-1664.

6. Norman, A.W., etal. (1970) Biochim. Biophys. Acta. 215:348-359.

7. Forstner, G.G., et al. (1968) Biochem. J. 106:381-390.

8. Moriuchi, S. and DeLuca, H.F. (1976) Arch. Biochem. Biophys. 174:367-372.

9. Wasserman, R.H. (1970) Biochim. Biophys. Acta. 203:176-179.

Vitamin D-Dependent Salivary Calcium-Binding Proteins

T,S,FREUND, J,VOLPE and A,WITKOWSKI
Department of Biochemistry, School of Dentistry
Fairleigh Dickinson University
Hackensack, New Jersey 07601, USA

ABSTRACT

Calcium binding proteins exhibiting a vitamin D-dependence have been found to be present in rat parotid and submandibular gland preparations. Samples tested exhibited the presence of a tight binding CaBP with K_d=0.5-2.5 μM and the maximum binding capacity ($\underline{n}$)=15.0-45.0 nmoles calcium bound/mg protein. Glands from vitamin D-deficient rats showed a significant decrease in their calcium binding capacity. Decreased dietary calcium showed only a slight increase in $\underline{n}$. Human salivary samples also exhibited the presence of similar proteins.

INTRODUCTION

A number of high affinity calcium binding proteins(CaBP), dependent upon the presence of vitamin D, have been isolated from mammalian and avian tissues. The presence of these proteins in gut[1,2], kidney[3,4], brain[5], egg shell gland[6] and parathyroid glands[7] suggests that they may be involved in the transmembranous movement of calcium. Tissues in which the movement of calcium appears significant may possess this class of protein.

Salivary stimulation with the resultant release of amylase and potassium is dependent upon both the extracellular calcium concentration and the rate of calcium uptake by isolated cells[8]. We thus examined the rat salivary glands for the existance of the vitamin D-dependent class of calcium binding proteins.

Animals.Weanling male Sprague-Dawley derived rats obtained from the Fairleigh Dickinson University Dental School Animal Care Facility were divided into groups of six and fed one of several diets: 0.5% Ca and 0.5% P containing either no (IID) or 2200 i.u. vitamin D_2/kg diet (II+D). Similar sets of animals were prepared with increased (1.5% Ca and 1.5% P; III±D) or decreased (0.06% Ca and 0.2%P; I±D)Calcium.

Approximately 3-4 weeks later, when the plasma calcium of the vitamin D-deficient group was significantly lower relative to their controls, the animals were sacrificed and their parotid and submandibular glands were removed.

Glandular samples.The glands were treated as previously described for intestinalCaBP[1]. The supernate (S-40) was decanted and then assayed for protein and calcium. The remainder of the supernate was treated for 15 minutes with an equal volume of Chelex-100 resin suspension[1]. Following centrifugation (12,000xg for 10 minutes) the supernate was analyzed for calcium binding and lyophilized material was then redissolved and further fractionated by chromatography on a Sephadex G-75 Column.

Chemical Assays. Calcium binding activity in terms of the binding constant (Kd) and maximum binding capacity ($\underline{n}$) was determined by varying the calcium concentration in the Quantitative Chelex Resin Procedure[1] or in each fraction eluted from the column by the micromodification of the competitive Chelex procedure[1].

RESULTS

The calcium binding constants and capacities, determined under several physiological conditions,are summarized in Table 1. For each of vitamin D-deficient and -replete glands, the deficiency always yielded a lower binding capacity. It should also be noted that the binding constant measured for the deficient glands markedly increased. Although not identified in the Table, when examined for statistical significance, differences at a level of $\rho < 0.01$ were obtained.

The protein nature of the binder was tested for, by digesting samples of supernate with either trypsin (0.02mg/ml) or pronase (0.01mg/ml). In all cases tested, the binding activity decreased. After one-half hour at $37^{\circ}C$ with pronase, greater than 50% of the calcium binding activity of a I+D S-40 sample has been destroyed. There was no significant difference in the Kd.

The absorbance and calcium binding curves shown in Fig.1 were obtained from parotid gland supernates that has been lyophilized, redissolved and chromatographed on Sephadex G-75. In the case of the material obtained from the vitamin D-replete controls (I+D) there were five peaks of binding activity, here numbered as 1 through 5. Peaks 4 and 5, on the basis of previous work[1,3], can be identified as representing ionic interference in the binding assay. Some of the binding in these

TABLE 1

VITAMIN D-DEPENDENT SALIVARY CALCIUM BINDING

Source	Type	Diet			Calcium Binding	
		%Ca	%P	Vit.D	Kd(μM)	n (nmoles bound / mg protein)
RAT	PAROTID	0.06	0.2	+	1.1(0.4)	39.5(4.5)
				−	15.6(3.2)	13.8(2.5)
		0.5	0.5	+	1.3(0.5)	33.5(3.1)
				−	14.2(4.1)	12.2(2.8)
		1.5	1.5	+	1.5(0.3)	32.0(3.0)
				−	12.2(3.6)	14.2(3.2)
	SUBMANDIBULAR	0.06	0.2	+	2.3(0.6)	24.8(3.0)
				−	18.8(2.5)	15.3(3.0)
		0.5	0.5	+	2.8(0.4)	22.4(3.2)
				−	13.3(1.8)	13.4(2.8)
		1.5	1.5	+	4.3(0.8)	20.5(2.5)
				−	16.4(3.2)	12.6(3.2)
HUMAN	MIXED	STIMULATED			1.4(0.3)	38.5(3.9)
	PAROTID	STIMULATED			1.0(0.2)	44.2(4.8)
	SUBMANDIBULAR	STIMULATED			4.2(0.6)	35.5(5.3)

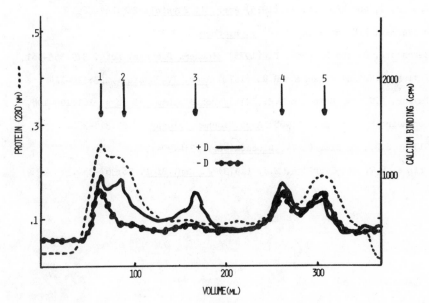

Fig.1. Chromatography of rat parotid gland homogenates on Sephadex G-75.

fractions may also be associated with the amylase activity.

DISCUSSION

It is apparent from the above results that rats possess a vitamin D-dependent CaBP in their salivary glands. Parotid glands contain more activity and show a greater vitamin D-dependence than their submandibular counterparts.

The fact that the deficient animals contained some CaBP is probably the result of the examination of the initial supernate rather than the pooled column fraction as was the case for the intestinal protein[1]. Clearly peak 1 is vitamin D-independent, as are peaks 4 and 5. The increase in the Kd observed during vitamin D-deficiency may be indicative of the predominance of the D-independent proteins under those conditions.

These proteins contain a very tight binding site and thus will be saturated under normal salivary conditions. Thus the presence of CaBP in the human salivary secretions may be indicative of salivary gland status.

The relationship between these proteins and the many other calcium binding proteins found in saliva is presently being examined.

REFERENCES

1. Freund, T. and Bronner, F. (1975) Amer. J. Physiol. 228:861-869.

2. Wasserman, R.H. et al., (1968) J. Biochem. 243:3987-3993.

3. Hermsdorf, C. and Bronner, F. (1975) Biochem. Biophys. Acta. 379:553-561.

4. Taylor, A.N. and Wasserman R.H. (1972) Amer. J. Physiol. 223:110-114.

5. Taylor, A.N. and Brindale M.E.(1974) Arch.Biochem. Biophys. 161:100-108.

6. Corradino R.A. et al., (1968) Arch.Biochem. Biophys. 125:378-380.

7. Oldham S.B., et al., (1974) Biochemistry 13:4790-4796.

8. Prince, W.T. and Berridge, M.J. (1973) J. Exp. Biol. 58:367-384.

Lanthanide Ions as Probes of the Metal-Binding Sites of the Vitamin K-Dependent Blood Coagulation Proteins

Barbara C. Furie, Michael Blumenstein, and Bruce Furie

Department of Medicine and Department of Biochemistry and Pharmacology
Tufts-New England Medical Center and Tufts University School of Medicine
Boston, MA   02111

Calcium ions play a central role in the blood coagulation cascade as cofactors of enzymatic proteolysis of the vitamin K-dependent blood coagulation proteins, prothrombin, factor X, factor IX, and factor VII.   In previous studies we have evaluated the metal binding properties of some of these proteins using lanthanide ions as substitutes for Ca(II).   Bovine factor X contains two high affinity metal binding sites which bind lanthanide ions with a K_d of 0.4 μM[1].   The activation of factor X by a protease in Russell's viper venom has an absolute requirement for Ca(II) for which lanthanide ions cannot substitute.   In contrast, lanthanide ions can replace Ca(II) in metal-dependent prothrombin activation by activated factor X[2].   Metal binding studies indicated that prothrombin, fragment 1, fragment 2, prethrombin 1, and prethrombin 2 have 2, 2, 0, 1 and 1 high affinity metal binding sites (K_d < 1 μM) respectively.

γ-Carboxyglutamic acid residues, present in the vitamin K-dependent proteins[3-5], may participate as metal ligands in the metal binding sites of these proteins[6,7]. In order to obtain a first approximation of the structural features of the metal binding sites of this class of Ca(II) binding proteins, we have studied the interaction of lanthanide ions with a γ-carboxyglutamic acid-rich peptide isolated from the N-terminal region of bovine prothrombin, fragment (12-44).

Fragment (12-44), containing eight γ-carboxyglutamic acid residues, was isolated from the tryptic digest of bovine prothrombin by barium citrate absorption and gel filtration.   Fragment (12-44) contains two high affinity metal binding sites (K_d 0.1 μM) as well as lower affinity metal binding sites, as measured by steady state rate dialysis using [153]Gd(III) at pH 6.8.   To evaluate the amino acid residues which participate in metal liganding, the interaction of paramagnetic

361

lanthanide ions with fragment (12-44) was studied by natural abundance ^{13}C NMR spectroscopy at 67.88 mHz at pH 5.39 at 22˚. The proximity between bound metal ions and carbon atoms in fragment (12-44) was estimated using Pr(III), Eu(III), and Gd(III) based upon the strategy that the magnitude of the change in the chemical shift or the transverse relaxation rate of resonances of the carbon nuclei induced by bound metals is related to the interatomic distances between bound metal and carbon nuclei.

The natural abundance ^{13}C NMR spectrum of fragment (12-44) obtained on a Bruker HX-270 NMR spectrometer is shown in Figure 1. For comparison the natural abundance ^{13}C NMR spectrum of DL-γ-carboxyglutamic acid at pH 5.4 contains resonances at 178.4 ppm, 178.0 ppm, 175.0 ppm, 55.7 ppm, 54.1 ppm and 31.3 ppm relative to TMS which were assigned to the γ_1-COOH, γ_2-COOH, α-COOH, C_γ, C_α, and C_β respectively. On this basis, we have assigned the envelope at 178 ppm ("a") to the γ-carboxyl carbons, some of the resonances in the multiplet at 55 ppm ("b") to the γ and α carbons, and the multiplet at 33 ppm ("c") to the β carbons of the eight γ-carboxyglutamic acid residues in the spectrum of fragment (12-44).

The titration of fragment (12-44) with Pr(III) led to a downfield shift of the carboxyl carbon resonances ("a"). The titration of fragment (12-44) with Eu(III) effected an upfield shift of the carboxyl carbon resonances. These shifts within the γ-carboxyl carbon envelope appeared selective, suggesting the non-equivalence of the relationship of the γ-carboxyl group of the eight γ-carboxyglutamic acid residues to bound metal ions. Titration of fragment (12-44) with Gd(III) resulted in the marked broadening of the γ-carboxyl carbon resonance envelope and selective broadening of many other resonances. As a control, fragment (12-44) was titrated with diamagnetic La(III); no appreciable changes in the ^{13}C NMR spectrum were observed.

The susceptibility of the γ-carboxyl carbons to perturbation by paramagnetic lanthanide ions suggests the close proximity of these carbon atoms to the bound metal ions. These data offer direct evidence that γ-carboxyglutamic acid residues participate in metal liganding in this γ-carboxyglutamic acid-rich fragment of bovine prothrombin. However, because we do not know the relationship between the

362

three-dimensional structure of fragment (12-44) compared to the region 12-44 in native prothrombin, caution must be taken in extrapolating these conclusions to the metal binding sites of prothrombin.

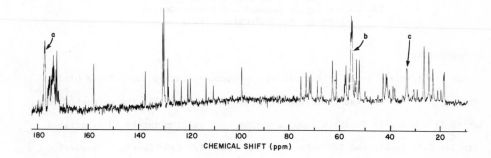

Figure 1. Natural abundance ^{13}C NMR spectrum at 67.88 mHz of bovine prothrombin fragment (12-44) at pH 5.39 and 22°. Chemical shifts are reported in ppm downfield of TMS.

Supported by grants from the National Institutes of Health (HL-18834) and the American Heart Association and its Massachusetts affiliate. NMR experiments were performed at the NMR facility of the Francis Bitter National Magnet Laboratory at the Massachusetts Institute of Technology, supported by NIH (RR 00995) and NSF (contract C 670). B.C.F. is the recipient of a Research Career Development Award (1 KO4 00235) from the NIH. B.F. is an Established Investigator of the American Heart Association and its Massachusetts affiliate.

REFERENCES

1. Furie, B.C. and Furie, B. (1975) J. Biol. Chem. 250, 601-608

2. Furie, B.C. et al., (1976) J. Biol. Chem. 251, 3235-3241

3. Stenflo, J. et al., (1974) Proc. Nat. Acad. Sci. U.S.A. 71, 2730-2733

4. Nelsestuen, G.L. et al., (1974) J. Biol. Chem. 249, 6347-6350

5. Magnusson, S. et al., (1974) FEBS Letters 44, 189-193

6. Nelsestuen, G.L. and Suttie, J. (1973) Proc. Nat. Acad. Sci. U.S.A. 70, 3366-3370

7. Stenflo, J. and Ganrot, P.-O. (1972) J. Biol. Chem. 247, 8160-8166

Calcium-Dependent Induction of Rat Calcium-Binding Protein *in vitro*

E. E. Golub, M. Reid, C. Bossak, L. Wolpert,
L. Gagliardi and F. Bronner
Department of Oral Biology
The University of Connecticut Health Center
Farmington, Connecticut 06032

INTRODUCTION

The steady-state level of the vitamin D-dependent intestinal calcium binding protein (CaBP) in the rat is regulated by calcium intake (1). The nature of regulation is a negative feedback loop such that high calcium intake is associated with low CaBP levels and low calcium intake with high CaBP levels. Moreover, shifting animals from a high to a low calcium intake leads to a marked rise in CaBP (1). Two possible complementary mechanisms may account for this adaptive phenomenon: a a systemic loop that involves a calcium-dependent modulation of the renal 25-hydroxyvitamin D-1-hydroxylase which leads to variations in the amounts of 1,25-dihydroxyvitamin D_3 [1,25-$(OH)_2$-D_3] that reach the target cells (2), and b a local regulatory mechanism, in which the cell calcium concentration modulates the CaBP level (1). To evaluate the relative contributions of these mechanisms to CaBP regulation, an isolated intestinal cell system was developed (3), where CaBP can be induced in vitro by the action of 1,25-$(OH)_2$-D_3. The studies reported here show that extracellular calcium inhibits the response of isolated intestinal cells to 1,25-$(OH)_2$-D_3.

EXPERIMENTAL

Duodenal tissue from vitamin D-deficient rats (plasma calcium: 4.30 ± 0.01 mg/dl) was incubated for 30 min (37C) in a medium (modified from 3) to which, where appropriate, 1,25-$(OH)_2$-D_3 (4.9×10^{-7}M), cycloheximide (0.4mM) or calcium (1mM) had been added. Cells were harvested and assayed for viability (trypan blue exclusion), cell number by DNA content (4), and for protein (5). CaBP analysis utilized Sephadex G-50 gel filtration (1) on columns that had been equilibrated with ^{45}Ca and ^{40}Ca (40,000 cpm/ml, 7 μM). The area of peak B ($v_e/v_o \simeq 1.5$) corresponds to the amount of calcium bound by the vitamin D-dependent CaBP (1, 6).

RESULTS

Addition of 1,25-(OH)$_2$-D$_3$ (4.9 x 10^{-7}M) for 30 min. to the isolation medium of the cells resulted in an increase in CaBP content equivalent to 47 n moles Ca bound per 10^6 cells in half the induction experiments tried. The effects of inhibitors were checked only in those experiments in which there was a marked effect of 1,25-(OH)$_2$-D$_3$ on CaBP induction. Table 1 compares the effect of 0.4mM cycloheximide added at the beginning of the experiment with induction experiments in which this inhibitor was not added. As can be seen, in these four paired experiments, the induction of CaBP by 1,25-(OH)$_2$-D$_3$ was repressed by cycloheximide.

TABLE 1

CYCLOHEXIMIDE INHIBITION OF CaBP INDUCTION IN ISOLATED DUODENAL CELLS

Expt. No.	CaBP induced patoms Ca$_{bound}$/10^6 cells (control)	CaBP induced patoms Ca$_{bound}$/10^6 cells (+ 0.4mM cycloheximide)	% Inhibition
1	42.4	32.7	23
2	24.6	8.2	67
3	50.1	16.0	68
4	20.4	18.8	8
MEAN	34.4	18.9*	42
SE	7.1	5.1	

8 animals per group per experiment.

*Significantly lower than control value
(t = 2.23, paired compairson, p < 0.05, one-tailed distribution)

Table 2 shows that the expression of the hormonal signal of 1,25-(OH)$_2$-D$_3$ was inhibited when the extracellular [Ca] was raised from 0.06 to 1mM. Cell isolation and cell number were unaffected by [Ca] of the isolation medium, as was CaBP content of duodenal cells from vitamin D-replete animals. Consequently, the effect of calcium in decreasing CaBP synthesis must have been direct.

TABLE 2

Ca INHIBITION OF CaBP INDUCTION IN ISOLATED DUODENAL CELLS

CaBP	0.06 mM			1.0 mM		
	Induced	Control	Δ	Induced	Control	Δ
patoms Ca bound per 10^6 cells	50.7	2.5	48.1	28.6	15.6	13.0*
SE	17.7	2.5	18.3	5.7	5.1	4.2
(n)	(6)	(6)	(6)	(7)	(7)	(7)

(n) is number of experiments, each utilizing cells from 8 animals.

*Significantly lower than 0.06 mM [Ca]
(t = 2.01, p < 0.05 for one-tailed distribution)

DISCUSSION

The mechanism by which calcium inhibits CaBP synthesis is not known. Calcium may act at the level of the hormone receptors, transcription, translation or post-translationally. Since these are intracellular events, a rise in extracellular calcium must either have altered the cell membrane and/or raised intracellular calcium.

Local regulation by calcium of CaBP synthesis minimizes the need for rapid and large changes in the plasma calcium and in the plasma levels of $1,25-(OH)_2-D_3$. Calcium thus provides a sensitive fine-tuning element for the regulation of CaBP synthesis.

REFERENCES

(1) Freund, T. and Bronner, F. Am. J. Physiol. 228:861–869, 1975.

(2) DeLuca, H. F. and Schnoes, H. K. Ann. Rev. Biochem. 45:631–666, 1976.

(3) Freund, T. and Bronner, F. Science 190:1300–1302, 1975.

(4) Burton, K. Biochem. J. 62:315–322, 1956.

(5) Lowry, O. H., Rosebrough, N. J., Farr, A. L., and Randall, R. J. J. Biol. Chem. 193:265–275, 1951.

(6) Hummel, J. P. and Dreyer, W. J. Biochim. Biophys. Acta 63:532, 1962.

[Supported by USPHS, NIH grant AM 14251, and The University of Connecticut Research Foundation].

The Distribution of Calcium
within the Goblet Cells of the Small Intestine

Bernard P. Halloran and James R. Coleman
Department of Radiation Biology and Biophysics
University of Rochester, Rochester, New York 14620

A vitamin D-dependent calcium-binding protein (CaBP) has been shown to be associated with sites of mucin accumulation within the goblet cells of the small intestine[1]. Presumably this protein is synthesized within the goblet cell and is eventually extruded into the intestinal lumen.

Calcium has also been found to be a constituent of goblet cell mucus. The question arises then as to the relationship between this calcium and CaBP.

In the present study a detailed map of the calcium distribution within the goblet cell was established using electron probe microanalysis.

Incubated and non-incubated duodenal tissue was fixed with a glutaraldehyde solution containing 1% K-oxalate. Thin sections of embedded tissue were cut, mounted on EM grids and analyzed using an ARL-EMX electron probe. Using conventional electron microscopy, a coordinated series of electron micrographs were taken before and after microprobe analysis to allow identification of specific cell morphology with calcium X-ray count rates.

Plate 1 shows a typical area before and after electron probe analysis. Note the series of small dark granules found throughout areas of mucin accumulation within the goblet cell. In addition, smaller granules (although barely visible) were seen either associated with what appear to be mitochondria or with the endoplasmic reticulum (RER).

Count data from the numbered sites in Plate 1 is presented in Table 1.

TABLE 1-CALCIUM X-RAY COUNTS IN GOBLET AND ABSORPTIVE CELLS
X-ray intensities are expressed as counts per 2 minutes.
Site numbers correspond to those in Plate 1.

Site No.	Description	Peak	Background	Pk - Bk
1	Absorptive cell cytoplasm	19	12	7
2	Absorptive cell cytoplasm	23	12	11
3	Goblet cell - RER	33	13	20
4	Goblet cell - RER	32	15	17
5	Goblet cell - mucin	29	12	17
6	Goblet cell - mucin	34	17	17
7	Goblet cell - granule	51	15	36

Significant amounts of calcium were found in the absorptive cells, the RER of the goblet cell and in areas of mucin accumulation within the goblet cell.

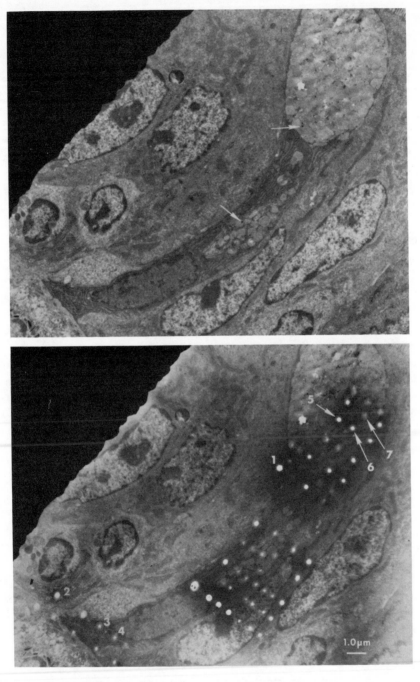

Plate 1. Goblet cell before and after microprobe analysis. Site numbers correspond to those in Table 1. Arrows indicate mucin granules.

In addition elevated concentrations of calcium were found in mucus areas containing granules.

A summary of the analysis shown on Plate 1 is presented in Table 2.

TABLE 2-AVERAGE CALCIUM X-RAY INTENSITIES

X-ray counts are expressed as counts per 2 minutes. (Peak-Background $\pm$ 1 S.D.)

Category	Counts/2 min.	No. of sites counted
Plastic	-1.2 $\pm$ 3.3	18
Absorptive Cell Cytoplasm	7.8 $\pm$ 4.3	20
Goblet Cell Cytoplasm	18.0 $\pm$ 7.0	6
RER of Goblet Cell	16.2 $\pm$ 3.6	5
Mucin	16.7 $\pm$ 6.1	15
Mucin Granules	35.5 $\pm$ 0.7	2

No detectable calcium signal was found over areas containing just embedding media (plastic), however, calcium was found in all tissue areas investigated. Higher concentrations of calcium were consistently found in goblet cells than in absorptive cells. The cytoplasmic fraction of the goblet cells as well as those sites definitively identified as goblet cell RER contained roughly the same amount of calcium as areas of mucin accumulation while mucin granules were substantially enriched in calcium above the surrounding mucus.

The results of this and similar analyses from both incubated and nonincubated tissues suggest that calcium is being sequestered during the process of protein synthesis, and is eventually extruded into the lumen. The similarity between the paths of calcium movement within the goblet cell and the probable path of CaBP is striking and it is possible that mucin associated calcium may represent a byproduct of CaBP secretion. On the other hand it has been shown that the concentration of calcium in goblet cells from normal and Vitamin D deficient rats is similar even though concentrations of CaBP are severly depressed in vitamin D deficient animals. This result supports the hypothesis that another calcium binding agent, perhaps a protein, is present in the goblet cell.

It is suggested that goblet cell calcium, rather than being a byproduct of CaBP secretion is rather a secretory product in itself and functions in some way to maintain calcium balance within the organism.

REFERENCES

1. Taylor, A.N. and Wasserman, R.H. (1970) J. Histo. and Cytochem. 18: 107.

Uptake of Calcium by Membrane Vesicles Isolated
from Pig Duodenal Brush Borders and Its Release by ATP

P.R. Hearn and R.G.G. Russell
Dept. Chemical Pathology, University Sheffield Medical School,
Beech Hill Road, Sheffield, S1O 2RX, U.K.

Nuffield Orthopaedic Centre, University of Oxford, Oxford, U.K.

In studies of mechanisms of transport of biological materials across the intestinal mucosa or across renal tubular epithelia it is difficult to distinguish between events occurring at the luminal (brush border) surface and those occurring at the basal-lateral cell surface. There has been some interest in developing methods for separating these membrane components[1,2] but there have been only a few studies of transport in them[3,4,5]. In the case of divalent cations such as Ca^{2+} there is virtually no information available. This report describes a method for preparing brush border membranes in a vesicular form, and their use to study the uptake and release of calcium ions.

METHODS

50 cm segments of fresh pig duodenum were opened and washed in ice-cold 0.9% NaCl. Mucosal scrapings were homogenised in a motor-driven Teflon-glass homogeniser (clearance 0.1-0.15 mm) in 30 mls of buffer B (250 mM sorbitol, 5 mM Tris-Cl, pH 7.4, 0.1 mM $CaCl_2$, 10 mM $MgCl_2$, 240 units/ml of pencillin and 125 units/ml of streptomycin) per gut. After centrifugation at 1,000 g for 15 min, the pellet was rehomogenised in 15 mls of this buffer B per gut segment, and centrifuged again at 1,000 g for 15 min. The combined supernatants were centrifuged at 14,000 g for 20 min to remove mitochondria. The supernatant was then spun at 45,000 g for 60 min and the resulting pellet was gently resuspended by hand in a Teflon-glass homogeniser and then finally centrifuged at 20,000 g for 30 min to yield the brush border vesicle (BBV) fraction. All these steps were performed on ice or in a refrigerated centrifuge. For transport studies, BBV were used within 5 hours of death of the animal.

To initiate studies of calcium transport, BBV resuspended in sorbitol buffer B were added in triplicate to equal volumes of the same buffer previously warmed to 37° containing ^{45}Ca in 1.9 mM $CaCl_2$ plus other additions as desired. The final concentration of Ca was 1 mM and of BBV around 0.1 mg protein/ml. Aliquots were sampled and BBV removed by Millipore (0.45 µ pore size) filtration. The filters were rinsed and dissolved and their ^{45}Ca content determined by liquid scintillation counting.

RESULTS AND DISCUSSION

Using alkaline phosphatase as a marker enzyme for brush border membranes the
BBV fraction contained a mean (± SEM) 10.4 ± 0.9% of the activity present in the
initial homogenate with a 10-fold enrichment in specific activity.   Contamination
of BBV by mitochondria (cytochrome C oxidase) or by endoplasmic reticulum (NADPH
cytochrome C reductase or NADH oxidase) was limited to less than 1% of initial
activity

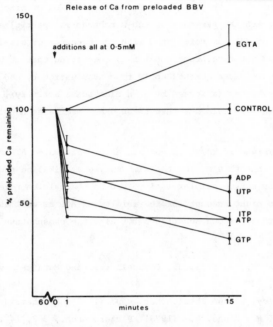

FIGURE 1

When BBV were incubated with 1 mM $CaCl_2$ there was a gradual and progressive
uptake of Ca which approached 100 nmoles/mg protein by 60 min at 37°.   This
uptake did not require added substrate.   Indeed addition of ATP caused release
rather than uptake of ^{45}Ca.   The uptake of ^{45}Ca was abolished by 5 mM EGTA, and
reduced to 35% by 0.5 mM EGTA.   This concentration (0.5 mM) of EGTA did not
remove calcium when added after uptake had occurred, suggesting that the ^{45}Ca
taken up was not surface bound.   When added at 60 min 0.5 mM ATP released 55%
of the ^{45}Ca within 1 min (Fig. 1).   GTP produced a similar effect whereas ADP,
ITP or UTP, each at 0.5 mM, or A23187 at 10 µg/ml, produced smaller effects on
Ca release.   The ATP-induced release of ^{45}Ca was reduced or abolished by
classical uncoupling agents (dicoumarol) or by inhibitors of electron transport
(azide or cyanide), suggesting that Ca release may be mediated by an ATP-
dependent electron transport system.   Further evidence to support this is that

NADH or NADPH both increased calcium uptake when added at 60 min, but only in the presence of dicoumarol or cyanide.

The properties of this system exhibit several unusual features. However, before attributing the uptake and release of calcium to the brush border membranes the question of interference by other cellular membrane systems needs to be considered. No intact mitochondria were seen by electron microscopy of the BBV preparation. Moreover ATP is known to promote uptake rather than release of calcium by mitochondria[6]. The low degree of contamination by mitochondrial marker enzymes may indicate the presence of sub-mitochondrial particles but the uptake and release of calcium correlated with the presence of alkaline phosphatase activity and not cytochrome C oxidase, suggesting that it was a property of the BBV. Similar arguments apply to contamination by endoplasmic reticulum and plasma membranes (assessed by 5'-nucleotidase activity), and in each case we are not aware of evidence that these other membranes handle calcium in the ways described.

The brush border membrane is known to contain a Ca^{2+}-activated ATPase[7,8] and there is some evidence that this may be the same enzyme as alkaline phosphatase[8]. It is possible that this enzyme complex is involved in the calcium release observed here, and this might account for the ability of ADP as well as ATP to cause release since both are known substrates for alkaline phosphatase[9].

REFERENCES

1. Louvard, D., Maroux, S., Baratti, J., Desnuelle, P. and Mutaftschev, S. (1973) B.B.A., 29, 747.

2. Douglas, A.P., Kerley, R. and Isselbacher, K.J. (1972) Biochem. J., 128, 1329

3. Hoffman, N., Thees, M., Kinne, R. (1976) Plugers Arch., 362, 147.

4. Murer, H., Hopfer, U. and Kinne, R. (1976) Biochem. J., 154, 597.

5. Eastham, E.J., Bell, J.I. and Douglas, A.P. (1977) Biochem. J., 164, 289.

6. Lehninger, A.L. (1970) Biochem. J., 119, 129.

7. Melancon, M.J. and DeLuca, H.F. (1970) Biochemistry, 9, 1658.

8. Russell, R.G.G., Monod, A., Bonjour, J.P. and Fleisch, H. (1972) Nature New Biology, 240, 126.

9. Fernley, H.N. and Walker, P.G. (1967) Biochem. J., 104, 1011.

The Amino Acid Sequence of a Calcium-Binding Protein
from Pig Intestinal Mucosa

T. Hofmann, M. Kawakami, H. Morris, A.J.W. Hitchman,
J.E. Harrison and K.J. Dorrington
Departments of Biochemistry and Medicine
University of Toronto, Toronto, Canada M5S 1A8

Porcine intestinal calcium-binding protein (CaBP) was isolated according to the procedure of Hitchman et al.[1]. The purity of the protein was confirmed by amino acid analysis and by assessing its characteristic spectral properties.[2] The amino acid sequence of this protein, shown in Figure 1, has now been determined from peptides obtained from tryptic, chymotryptic and thermolytic digests and from digests with the glutamic acid-specific protease from Staphylococcus aureus[3]. CaBP consists of 80 amino acids in a single chain with a blocked N-terminus which has been identified as acetyl-serine by mass spectrometry. Additional information was obtained by automatic sequencing of fragments produced by specific cleavages at tyrosine and arginine respectively. No evidence for γ-carboxyl glutamic acid, an amino acid that binds calcium in proteins of the blood-clotting system[4], was found with CaBP.

The partial sequence of the bovine intestinal CaBP, corresponding to the first 51 residues of the pig protein, was kindly provided by Dr. C.S. Fullmer and is also shown in Figure 1. The two sequences are identical except for 6 to 8 positions. Each of these differences can be attributed to single-base changes within each codon. Clearly the two proteins are homologous beyond a shadow of a doubt. These new data on the bovine protein confirm that the earlier sequence published by Huang et al.[5] is seriously in error.

Although a detailed analysis for possible homology of CaBP with other calcium binding proteins such as the parvalbumins of fish[6] and rabbit[7] is beyond the scope of this communication* it is noteworthy that porcine intestinal CaBP contains a sequence (residues 57-68) which is very similar to the E-F loop of carp muscle parvalbumin[8] (see Table 1). The E-F loop represents one of the strong Ca^{++} binding sites of parvalbumin, and by implication probably also of CaBP. This sequence is also very similar to four sequence segments found in both rabbit muscle troponin C and myosin alkaline light chains[9]. The aspect of the relation between calcium-binding sites of various calcium-binding proteins (E-F hand concept) is discussed elsewhere in this volume by Dr. R.H. Kretsinger.

---

* This is done, however, by Barker et al., pp. 73-75, this volume.

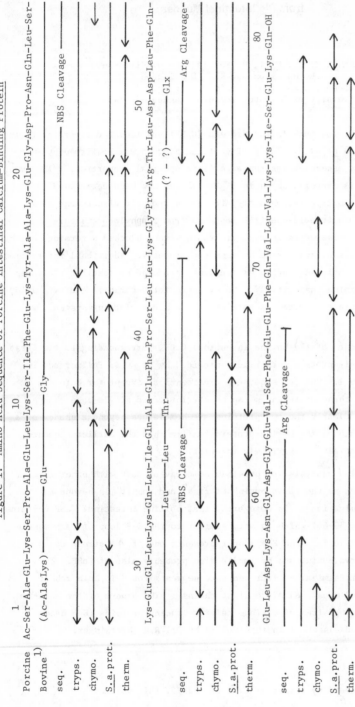

Figure 1. Amino Acid Sequence of Porcine Intestinal Calcium-Binding Protein

Porcine
Bovine[1]

Ac-Ser-Ala-Glu-Lys-Ser-Pro-Ala-Glu-Leu-Lys-Ser-Ile-Phe-Glu-Lys-Tyr-Ala-Ala-Lys-Glu-Gly-Asp-Pro-Asn-Gln-Leu-Ser-
(Ac-Ala,Lys)————————Glu——————————————————————————Gly

NBS Cleavage

seq.
tryps.
chymo.
S.a.prot.
therm.

Lys-Glu-Glu-Leu-Lys-Leu-Ile-Gln-Ala-Glu-Phe-Pro-Ser-Leu-Leu-Lys-Gly-Pro-Arg-Thr-Leu-Asp-Asp-Leu-Phe-Gln-
————Leu—Leu—Thr—————————————————————(? - ?)————Glx

NBS Cleavage   Arg Cleavage

seq.
tryps.
chymo.
S.a.prot.
therm.

Glu-Leu-Asp-Lys-Asn-Gly-Asp-Gly-Glu-Val-Ser-Phe-Glu-Glu-Phe-Gln-Val-Leu-Val-Lys-Lys-Ile-Ser-Glu-Lys-Gln-OH

Arg Cleavage

seq.
tryps.
chymo.
S.a.prot.
therm.

Footnotes: 1) The partial sequence of the bovine CaBP was kindly provided by Dr. C.S. Fullmer and is given in more detail elsewhere in this volume. In Figure 1 only those residues of the bovine protein that differ from the porcine one are written out. Abbreviations: seq. = sequencer, tryps. = trypsin, chymo. = chymotrypsin, S.a.prot. = Staph. aur. protease, therm. = thermolysin.

TABLE 1

Alignment of homologous sequences of porcine CaBP with sequence of one

calcium binding site of carp muscle parvalbumin (E-F loop).

The alignment was made independently by Dr. M.N.G. James and Dr. R.H. Kretsinger with the use of a computer programme developed by McLachlan[10]. Residues underlined are identical. x, y etc. are ligand sites as defined by Kretsinger[9].

	57										68
Porcine CaBP	Asp-Lys-Asn-Gly-Asp-Gly-Glu-Val-Ser-Phe-Glu-Glu
Carp Parvalbumin	Asp-Ser-Asp-Gly-Asp-Gly-Lys-Ile-Gly-Val-Asp-Glu
	90										101
	x	y	z	-y	-x		-z

The existence of one high-affinity calcium-binding site in porcine CaBP has been demonstrated in spectral studies but there is also evidence for a number of weaker sites. (Kells, D., Hofmann, T., & Dorrington, K.J., to be published).

Acknowledgements: We are grateful to Drs. M.N.G. James and R.H. Kretsinger for the alignment calculations shown in Table 1. This work was supported by grants from the Medical Research Council of Canada (MT-1982 and MT-4259) and a grant from the Connaught Fund, University of Toronto.

REFERENCES

1. Hitchman, A.J.W., Kerr, M.K. and Harrison, J.E. (1973). Arch. Biochem. Biophys. 155, 221-222.

2. Dorrington, K.J., Hui, A., Hofmann, T., Hitchman, A.J.W. and Harrison, J.E., (1974). J. Biol. Chem. 249, 199-204.

3. Houmard, J. and Drapeau, G.R. (1972). Proc. Nat. Acad. Sci. 69, 3506-3509.

4. Magnusson, S., Sottrup-Jensen, L., Peterson, T.E., Morris, H.R. and Dell, A. (1974). FEBS Letters 44, 189-193.

5. Huang, W.Y., Cohn, D.V., Fullmer, C., Wasserman, R.H. and Hamilton, J.W., (1975). J. Biol. Chem. 250, 7647-7655.

6. Nockolds, C.E., Kretsinger, R.H., Coffee, J. and Bradshaw, R.A. (1972). Proc Nat. Acad. Sci. U.S.A., 69, 581-584.

7. Enfield, D.L., Ericsson, L.H., Blum, H.E., Fischer, E.H. and Neurath, H. (1975) Proc. Nat. Acad. Sci. U.S.A. 72, 1309-1313.

8. Moews, P.C. and Kretsinger, R.H. (1975) J. Mol. Biol. 91, 201-228.

9. Kretsinger, R.H. (1976) Int. Rev. Cytol. 46, 323-393.

10. McLachlan, A.D. (1972) J. Mol. Biol. 64, 417-437.

Some Characteristics of Two Different Vitamin D-Dependent Calcium-Binding Proteins in Rat Intestinal Mucosa

Norimasa Hosoya, Toshikazu Yamanouchi and Sachiko Moriuchi
Department of Nutrition, School of Health Sciences
Faculty of Medicine, University of Tokyo
3-1, Hongo 7, Bunkyo, Tokyo, Japan

INTRODUCTION

Since the finding of a vitamin D dependent calcium binding protein (CaBP) in chick intestinal mucosa(1), similar proteins with calcium binding properties have been detected in the intestinal mucosa of several species (2-4). Concerning with the molecular size of rat intestinal CaBP, there have been some discrepancies among investigators (2-4). Recently, however, we demonstrated that there exist two different CaBPs differing in molecular size and tissue distribution (5). In order to gain further understanding on physio-logical significance of these CaBPs, some characteristics of these proteins were studied using polyacrylamide disc gel electrophoresis and Scatchard method. Furthermore, the effect of dietary Ca level and Cd on the CaBPs was observed.

MATERIALS AND METHODS

Animals: Female albino rats of the Wistar strain, weighing 40-50g, were fed vitamin D deficient diet (Ca: 0.47%, P: 0.3%) (6) ad libitum for 5-7 weeks. Vitamin D_3 (Tokyo Chem. Ind. Co.) was admin-istered orally to vitamin D deficient rats 48 hours before sacri-fice. When the effect of dietary Ca level and Cd on the production of CaBPs was observed, rats were maintained on the vitamin D defi-cient diet for 5 weeks in the absence or presence of 200 ppm of Cd. Animals were further divided by Ca level (0.43 or 0.002%) and vitamin D status. 100 IU of vitamin D_3 was dosed 5 times in 2 weeks. Animals were sacrificed 24 hours after the last dose of vitamin D_3.

Preparation of CaBp: Crude CaBPs were prepared from intestinal mucosa by the method of Wasserman et al (1). Two CaBPs were frac-tionated by ^{45}Ca equilibrated Sephadex G-100 column chromatography from jejunal mucosa supernatants and used for saturation analysis.

^{45}Ca binding activity assay: The ^{45}Ca binding activity was assayed by the method of ^{45}Ca equilibrated Sephadex G-100 column (2.5x61cm) (5) or Sephadex G-25 column (1.2x14cm) (2).

Electrophoresis: Polyacrylamide gel disc electrophoresis of supernatants was carried out according to the method of Davis (7) at pH 8.3.

^{45}Ca binding properties: ^{45}Ca binding properties of two CaBPs was analyzed by the method of Scatchard (8). Saturation analysis of ^{45}Ca binding activities of two CaBPs were carried out using ^{45}Ca equilibrated Sephadex G-25 column.

Protein determination: Protein concentration of sample was either monitored by absorbancy at 280 nm or determined by the methods of Lowry et al (9).

RESULTS AND DISCUSSIONS

Two different vitamin D dependent CaBPs are demonstrated in the rat intestinal mucosa, differing in molecular size and tissue distribution. The larger CaBP, found predominantly in the jejunum and ileum, had a molecular weight of 27,000. On the other hand the smaller CaBP, which was associated mainly with duodenum and jejunum, had a molecular weight of 12,500. Therefore, the electrophoretic mobility of two CaBPs was examined by the comparison with the electrophoretic pattern of duodenal, and ileal supernatants from vitamin D deficient and vitamin D_3 dosed rat. As a result, a new protein band was appeared at Rf 0.39 in duodenum by vitamin D_3 treatment and 0.81 in ileum. These results suggest that the protein of Rf 0.39 corresponds to smaller CaBP and that of Rf 0.81 to larger CaBP (Tab.1). From the density of protein band, larger CaBP was little in amounts compared with smaller CaBP.

^{45}Ca binding properties of two CaBPs were analyzed by Scatchard method using jejunal supernatants which contain both CaBPs. Two CaBPs were fractionated by Sephadex G-100 column chromatography. Three ^{45}Ca binding peaks were observed at void volume, Ve/Vo=1.84 and Ve/Vo=2.47. According to the elution orders, they were designated as PI, II and III. PII corresponds to larger CaBP and PIII to smaller CaBP. Fractions corresponding to PII and PIII were pooled and treated by Chelex 100 to remove ^{45}Ca and then the saturation analysis of ^{45}Ca binding was carried out. CaBP content in PII and PIII was estimated from the densitometric tracing of the electrophoretic gels of PII and PIII. As a result, smaller CaBP has one type of binding site, but larger CaBP has two types of binding sites differing in affinity (Tab. 1). The binding properties of larger CaBP was similar to chick duodenal CaBP (10) and that of smaller CaBP was similar to pig duodenal CaBP (4).

377

Harrison et al (11) reported that CaBP of molecular weight exceeding 25,000 is present in pig kidney and the CaBP is not so sensitive to the change in dietary Ca levels as intestinal CaBP. Deprivation of Ca from the diet decreased the production of smaller CaBP a little, but did not change the production of larger CaBP in response to vitamin D_3 compared with rats raised on normal Ca diet. Furthermore, larger CaBP was not infulenced by Cd feeding either, but ^{45}Ca binding activity of smaller CaBP was almost completely abolished by Cd feeding (Tab. 1). From these results, smaller CaBP is much more influenced by dietary factors and seems to be more important for the animals.

Table 1. Comparison of two CaBPs in rat intestinal mucosa

	larger CaBP	smaller CaBP
molecular weight	27,000	12,500
tissue distribution	jejunum, ileum	duodenum, jejunum
vitamin D dependency	+	++
electrophoretic mobility pH8.3	fast	slow
affinit constant (ki) M^{-1}	a 1.2×10^6	2.5×10^5
	b 2.2×10^4	
binding site (n)	a 3.6	1
	b 85	
response to dietary Ca		
normal	++	++
low	++	+
sensitivity to Cd	±	++

REFERENSES

1. Wasserman, R.H., Taylor, A.N. (1966) Science,152,791
2. Ooizumi, K., Moriuchi, S., Hosoya, N. (1970) J.Vitaminol,16,228
3. Drescher, D., DeLuca, H.F. (1971) Biochemistry,10,2302
4. Hitchman, A.J.W., Harrison, J.E. (1972) Can.J.Biochem.50,758
5. Moriuchi,S.,et al (1975)J.Nutr.Sci.Vitaminol.21,251
6. Suda,T.,DeLuca,H.F.,Tanaka,Y.(1970)J.Nutr.100,1049
7. Davis,B.J.(1964)Ann.N.Y.Acad.Sci.,121,Art.2,404
8. Scatchard,G.(1949)Ann.N.Y.Acad.Sci.,51,660
9. Lowry,O.H.,et al (1951)J.Biol.Chem.,193,265
10.Bredderman,P.J.,Wasserman,R.H.(1974)Biochemistry,13,1687
11.Harrison,J.E.,et al (1975)Vitamin D and Problems to Uremic Bone Disease,pp.75,Walter de Gruyter & Co. Berlin,New York

Identification of γ-Carboxyglutamic Acid in Urinary Proteins

J.B. Lian, M.J. Glimcher, and P.M. Gallop

Departments of Orthopaedic Surgery, Biological Chemistry and Oral Biology
Harvard Schools of Medicine and Dental Medicine
Children's Hospital Medical Center
300 Longwood Avenue, Boston, MA 02115, USA

INTRODUCTION

A protein of approximately 17,000 daltons isolated from calcium containing renal calculi[1] was shown to contain 45 residues of γ-carboxyglutamic acid (Gla) / 1000 res amino acid and showed no obvious relation to the clotting factors[2] or osteocalcin[3]. The protein in renal stones may have been produced de novo by the kidney where the capability for Gla synthesis has been demonstrated[4].

Concentration changes in calcium, oxalate[5] and inhibitors[6] present in the urine appear to contribute to calcium oxalate stone formation. Urine is normally supersaturated with respect to calcium oxalate, hence the continued growth of a stone can be explained although nucleation of a stone is not well understood. The calcium binding amino acid, Gla, in a matrix protein in calcium containing calculi may serve as a specific calcium buffering or nucleating substance. Alternately, it may be present as a passive urinary derived co-precipitant. The non-dialyzable proteins in the urine were studied to determine their relation to the Gla protein found in renal calculi. Comparisons are made with urinary proteins from normal individuals, patients on coumadin therapy and patients with soft tissue calcification disorders. In the latter group, Gla has been identified in proteins associated with ectopic subcutaneous plaques containing hydroxyapatite[7].

METHODS

Urines: Urines were collected from 5 males with a history of recurrent stone formation having microscopic crystals of calcium oxalate present in their urine. Urines were obtained from 2 female and 4 male patients on coumadin therapy whose prothrombin times ranged from 17-28 sec vs. 12-13 sec controls, from 2 patients (21 yr female and 11 yr male) with dermatomyositis and 1 patient with scleroderma (52 yr female). Normal urines from 4 males and 2 females were examined. 500-1000 ml portions of the urines were dialyzed exhaustively in Spectropor I membranes against distilled water at 5°C.

Gla analysis: Free Gla was measured in whole urine after deproteinization with 10% sulfosalicylic acid and dilution to a final volume of 1/2 the urine aliquot with 0.4 M citrate buffer, pH 2.0. Analysis was performed on the Beckman 121 M as previously described[3]. The program separates cysteic acid, urea, taurine, Gla and phosphoethanolamine respectively, and an as yet unidentified component eluting 1 min before Gla. Protein-bound Gla was determined after alkaline hydrolysis as previously described[3]. The Gla content in the urinary protein components separated by gel filtration is compared by the ratio res Gla/1000 res glutamic acid. The putative Gla peak derived directly from each urine sample or by alkaline hydrolysis of the urinary protein was confirmed as authentic Gla by acid and heat promoted decarboxylation to glutamic acid[8].

Chromatography: 200 mg of lyophilized non-dialyzable urinary material was fractionated on Sephadex G100 (25x90 mm) equilibrated in 0.1 M NH_4HCO_3, pH 8.0 and eluted with the same buffer at a flow rate of 12 ml/hr.

RESULTS

The 24 hr excretion of free and protein-bound Gla in the urines of the normal and selected patients is summarized in Table 1. Three of five stone formers had

TABLE I

TOTAL GLA EXCRETION (MICROMOLES) IN 24 HR URINES

	FREE GLA		PROTEIN-BOUND GLA	
	Range	Mean ± SD	Range	Mean ± SD
Normal Adults	21-38	31 ± 5	0.5-2.9	1.7 ± 0.7
Coumadin Patients	3-12	5.7 ± 3	0.04-0.6	0.21 ± 0.18
Stone Formers	34-89	56 ± 21	1.7-5.2	2.54 ± 1.5
Scleroderma	72		2.4	
Dermatomyositis	110,165	137± 25	3.8,6.1	4.9 ± 1.1

significantly higher free Gla levels than normal patients. In some stone formers and in patients with various ectopic calcifications, elevated protein excretion patterns contribute to the increase in urinary protein-bound Gla. In these cases, dodecyl sulfate gel electrophoresis shows the presence of low molecular weight bands which are not found in normal urines.

Fractionation on Sephadex G100 of the non-dialyzable urinary material from stone formers revealed 7 280nm UV absorbance peaks, three of which showed high Gla content. In normal urine only 5 major peaks appeared, and all had relatively low Gla content. The patients with scleroderma and dermatomyositis also excreted protein components with significant amounts of Gla. Table II compares the Gla content in various fractions from different patients.

DISCUSSION

The measurement of urinary Gla appears to be a sensitive parameter of the vitamin K dependent synthesis of Gla. In patients whose prothrombin times were elevated as a result of coumadin anticoagulant therapy, an 80-88% decrease in both free and protein-bound urinary Gla was observed. Patients with recurrent history of calcium oxalate stones excreted from 30-50% more free and protein-bound Gla in their urine as compared to normals. Four patients studied with

TABLE II

GLA IN URINE PROTEIN FRACTIONS SEPARATED BY SEPHADEX G100

(Gla res/1000 glutamic acid res)

G100 Peak	Stone Formers*		Dermatomyositis Patient	Normals**
	A	B		
I- Void	44	1	3	4
II	16	6	7	12
III	22	0	8	0
IV	102	188	6	3
V	20	290	25	0
VI	34	40	73	17
VII	0	0	24	0

* Two patients, A and B
** Average for three normal adults

certain types of soft tissue calcifications showed a 55-70% increase in free Gla excretion.

In all cases, the major amount of Gla is excreted as the free amino acid. The 3 patients studied with ectopic calcifications and 4 of the 5 stone formers have a higher excretion of protein bound Gla than normals. The marked increase in urinary Gla in these patients suggests a role for Gla proteins in mineral homeostasis. Further definition of the nature and origin of the urinary Gla proteins is essential and such studies are in progress. Since Gla containing proteins are now known to be made in liver, bone, and kidney, both free and protein bound urinary Gla excretion should serve as a sensitive index of vitamin K-dependent processes and could have considerable clinical significance.

REFERENCES

1. Lian, J., Prien, E., Gallop, P., Glimcher, M. (1977) J. Clin. Invest., June in press.
2. Suttie, J. and Jackson, C. (1977) Physiol. Rev. 57, 1-20.
3. Hauschka, P. (1977) Anal. Biochem. 80, in press.
4. Hauschka, P., Friedman, P., Traverso, H., Gallop, P. (1976) Biochem. Biophys. Res. Comm. 71, 1207-1213.
5. Robertson, W., Peacock, M., Marshall, R., Nordin, B. (1976) New Eng. J. of Med. 294, 249-252.
6. Dent, C. and Sutor, D. (1971) Lancet 2, 775-778.
7. Lian, J., Skinner, M., Gallop P., Glimcher, M. (1976) Biochem. Biophys. Res. Comm. 73, 349-355.
8. Hauschka, P., Lian, J., Gallop, P. (1975) Proc. Nat. Acad. Sci. USA 72, 3925-3929.

The authors wish to thank Drs. E.L. Prien, Jr., M. Skinner, L. Zemel and R. Levy for obtaining urines from their patients. This work supported in part by grants from the National Institutes of Health (AM 15671 and DE 04641) and from the New England Peabody Home for Crippled Children, Inc. J.B.L. is a Senior Investigator of the Arthritis Foundation.

381

Vitamin D-Stimulated Calcium-Binding Protein
from Rat Intestine: Development of a Radioimmunoassay

P. Marche[*], P. Pradelles[**], C. Gros[**] and M. Thomasset[*]
[*] INSERM, U.120 (Dir. H. Mathieu) 44 Chemin de Ronde 78110 Le Vésinet
France and [**] U.R.I.A. (Dir. F. Dray, Maître de Recherches INSERM)
Institut Pasteur 28 rue du Docteur Roux, 75724 Paris Cédex 15 France

Although intestinal vitamin D-stimulated calcium-binding proteins
(CaBPs) have been identified and characterized in various species
(1), it still remains to elucidate the exact role played by these
proteins in intestinal calcium transport. A first step towards this
goal was to develop specific assays which are sensitive enough to
measure CaBP content in tissue extracts and biological fluids. This
has been achieved by means of immunological tests for chicken, beef,
pig and human CaBPs (2-5). Recently a radioimmunoassay for the rat
duodenal CaBP has been developed (6). The purpose of this report is
to present the major characteristics of this assay and to discuss
the first results obtained with it.

Sensitivity of the assay is shown in figure 1 ; the standard
curve obtained with pure CaBP shows that the assay could measure as
little as 0.5 ng. The assay also enables the determination of CaBP
content directly in rat duodenal cytosol (a 100,000 g supernatant of
an homogenized duodenal mucosa) (figure 1).

The validity of the radioimmunoassay has also been checked. Since
the "biological activity" of the protein we are dealing with is to
bind calcium, we have first verified that values obtained with this
assay agree with those measured via the Chelex method (7). Results
are presented in Table 1. The agreement between the methods strongly
trends to confirm that rat duodenal CaBP possesses two high affinity
calcium-binding sites, as already described by others (7). In a
second set of experiments the dependence of the immunoreactive CaBP
on vitamin D has been assessed. Data obtained are shown in figure 2.
Even in vitamin D-deficient animals, a basal level of CaBP is still
present. Since intestinal calcium transport strictly depends on the
presence of vitamin D, our observation is consistent with recent
data (8) which raise the question of a direct or indirect
correlation between CaBP and intestinal calcium transport.

TABLE 1

COMPARISON BETWEEN CaBP CONCENTRATIONS DETERMINED BY THE CHELEX
ASSAY AND BY THE RADIOIMMUNOASSAY (RIA)

Sample (a)	CaBP (µg/ml) Chelex (b)	RIA	Sample (a)	CaBP (µg/ml) Chelex (b)	RIA
1	10.4	12.5	4	17.5	17
2	22.5	21.3	5	7.8	8.5
3	54	64	6	23.0	25.1

(a) Samples (peak B according to ref. 7) are from different groups
of rats. 6-20 animals per group.

(b) Values obtained assuming 2 Ca bound per CaBP molecule (MW=10000).

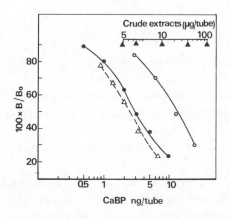
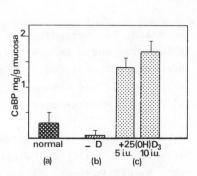

Fig. 1. Dose response curves obtained by dissolving 100 ng pure CaBP
in phosphate buffer (●—●) (standard curve) and in rat serum (△--△),
or by adding increasing amounts of proteins from duodenal cytosol of
rats (o—o) or pigs (▲—▲).

Fig. 2. Stimulation of CaBP by vitamin D
(a) 3-5 week old rats. Diet content : 0.85 % Ca, 0.80 % P, 4 i.u.
vitamin D_3 per gram diet.
(b) weanling rats were fed a vitamin D-deficient diet (0.50 % Ca,
0.36 % P) during 4 weeks. Then, hypocalcemic animals were kept for
an additional week on a vitamin D-deficient, low calcium (0.03 %)
diet before sacrifice.
(c) same diet as (b) but 5 and 10 i.u. $25(OH)D_3$, respectively, were
given i.v. 48 h before sacrifice.

The specificity of antibodies has been established. The assay does not detect the higher molecular weight calcium-binding protein peak of a Sephadex G75 column on which a rat duodenal cytosol had been loaded (6). In addition, we have also demonstrated that the rat renal CaBP described by Hermsdorf and Bronner (9) and the pig duodenal CaBP do not cross-react with the rat duodenal protein.

As already reported for porcine CaBP (4) an immunoreactive material has been found in rat serum (figure 1). Likewise, immunological reactivity has been detected in the placenta and the chorioallantoic membrane of a rat female on day 19 of pregnancy. Moreover, the immunoreactive material exhibited a molecular weight similar to that of intestinal CaBP and have been demonstrated to possess calcium-binding activity. Existence of these CaBPs, immunochemically identical to that of duodenum may be of importance for the understanding of the foeto-placental transport of calcium.

REFERENCES

1. Wasserman, R., Corradino, R., Fullmer, C. and Taylor, A. (1974) in Vitamins and Hormones (Harris, R., Munson, P., Diczfalusy, E. and Glover, J. eds)Vol. 32, pp. 299-324, Academic Press, New-York
2. Corradino, R. and Wasserman, R. (1971) Science 172, 731-733.
3. Fullmer, C. and Wasserman, R. (1973) Biochim. Biophys. Acta 317, 172-186.
4. Murray, T., Arnold, B., Kuttner, M., Kovacs, K., Hitchman, A. and Harrison, J. (1975) in Calcium-regulating hormones (Talmage, R., Owen, M. and Parsons, J. eds) Excerpta Medica American Elsevier, 371-375.
5. Piazolo, P., Hotz, J., Helmke, K., Franz, H. and Schleyer, M. (1975) Kidney International 8, 110-118.
6. Marche, P., Pradelles, P., Gros, C. and Thomasset, M. (1977) Biochem. Biophys. Res. Commun., in press.
7. Freund, T. and Bronner, F. (1975) Am. J. Physiol. 228, 861-869.
8. Spencer, R., Charman, M., Wilson, P. and Lawson, E. (1976) Nature 263, 161-163.
9. Hermsdorf, C. and Bronner, F. (1975) Biochim. Biophys. Acta 379, 553-561.

Supported by CNRS Grant ATP n° 2394.

Renal Calcium-Binding Protein: Possible Involvement in Tubular Calcium Reabsorption in Phosphorus-Deprived Rats

Monique Thomasset, Paulette Cuisinier-Gleizes and Henri Mathieu
I.N.S.E.R.M. Unité 120
44, Chemin de Ronde
78110 Le Vésinet
France

In contrast with the intestinal CaBP, the production of renal CaBP has not yet been related to either the vitamin D_3 intake or the renal calcium excretion[1].

We report here that the renal CaBP (rCaBP) is related to vitamin D_3 intake and could be involved in the tubular calcium reabsorption as well as the duodenal CaBP (dCaBP) in the intestinal calcium absorption.

MATERIALS AND METHODS

Animals and diets: Seven week-old male rats of the Sprague strain were kept for 2 weeks on a control diet (0.50 % Ca, 0.36 % P) or a phosphorus-deficient diet OP (0.50 % Ca, 0.03 % P). From the weanling the animals were either vitamin D-deficient or vitamin D_3-supplemented (1 or 10 i.u./g of dry diet). At the sacrifice time, serum and urinary calcium were determined, the mucosal or cortical tissue from 6 rats was isolated, pooled and fractionated.

CaBP content: A low molecular weight fraction was isolated from renal cortex by Sephadex G100 chromatography and from duodenal mucosa by Sephadex G75 chromatography. The CaBP content was obtained from the CaBP activity (maximum binding nmoles Ca^{++} bound/mg protein) quantitated by saturation analysis using a ^{45}Ca Chelex 100 assay[2]. The data were analyzed by the method of Scatchard.

RESULTS AND DISCUSSION

rCaBP (fig. 1, table 1): In spite of difficulties to quantitate rCaBP by using Chelex assay (apparent dissociation constant : $10^{-5}M$) our results show a direct relationship between the amount of rCaBP and the intake of vitamin D in OP rats. This is consistent with a stimulated $1,25-(OH)_2D_3$ biosynthesis in response to phosphorus deficiency[3].

As a marked hypercalciuria was noted in all OP rats and as the rCaBP activity was high in vitamin D-supplemented rats and

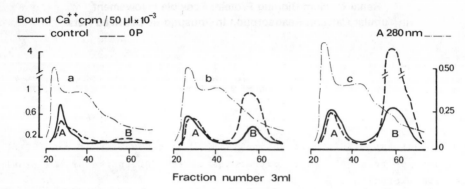

Fig. 1. Sephadex G100 elution profiles of renal cortex supernates (S100) from control and phosphorus-deficient rats (OP) fed either a vitamin D-deficient (a) or a vitamin D-supplemented diet : 1 i.u./g (b) and 10 i.u./g (c). Equal amounts of S100 protein (150 mg) were loaded onto the column (1.6 x 100 cm). Only the peak B (V_e/V_o = 2.1 ± 0.10) represents renal vitamin D-dependent CaBP.

TABLE 1. RENAL CALCIUM-BINDING PROTEIN (rCaBP)

Animals	rCaBP[2] nM Ca bound/ mg protein	Serum calcium[1] mg/100 ml	Urinary calcium[1] mg/24 h
0 vit. D			
Control	not measurable	5.9 ± 0.19	0.2 ± 0.09
OP	" "	9.9 ± 0.17	9.9 ± 1.01
		p<0.001	p<0.001
1 u. vit. D			
Control	2.5, 2.3	10.6 ± 0.17	0.9 ± 0.27
OP	5.3, 4.6	11.7 ± 0.13	17.5 ± 1.32
		p<0.001	p<0.001
10 u. vit. D			
Control	1.1, 2.6	10.2 ± 0.04	0.7 ± 0.10
OP	11.7, 7.7, 8.5	12.9 ± 0.15	12.2 ± 0.63
		p<0.001	p<0.001

(1) Means ± SEM of 18 rats in each group.
(2) Each determination was made on a pooled sample from 6 rats.

undetectable in vitamin D-deficient rats, the involvement of the rCaBP in the large hypercalciuria can be definitely ruled out.

When vitamin D-supplementation ranged from 1 to 10 i.u./g diet, while the serum calcium level was increasing, a decrease was noticed in the large hypercalciuria. Furthermore the fractional calcium excretion decreases (51±3.4, 37±5.4, 32±3.2) in parallel with the rise in vitamin D supplementation (0,15 and 30 i.u./day/OP rat, respectively). Such results suggest a direct action of vitamin D on the renal reabsorption of calcium and are consistent with other investigations[4]. This increase in calcium tubular reabsorption deserves to be related to the parallel increase in rCaBP.

Thus, the CaBP production probably resulting from the renal synthesis of $1,25-(OH)_2D_3$ stimulated by phosphorus-deprivation could represent the molecular basis of the calcium tubular reabsorption increased by vitamin D.

TABLE 2. DUODENAL CALCIUM-BINDING PROTEIN (dCaBP)

Animals	dCaBP (1) nM Ca bound/ mg protein	Calcium transport (2)	Initial supernate calcium (1) µg/mg protein
0 vit. D			
Control	5.2 ± 0.35	1.2 ± 0.19	not determined
OP	6.4 ± 0.75	1.3 ± 0.07	" "
	n.s.	n.s.	
1 u. vit. D			
Control	18.1 ± 2.54	2.0 ± 0.59	0.2 ± 0.04
OP	28.5 ± 1.65	3.5 ± 1.00	0.4 ± 0.11
	p<0.05	p<0.05	p<0.05
10 u. vit. D			
Control	22.5 ± 2.95	1.5 ± 0.73	0.2 ± 0.05
OP	42.7 ± 2.79	4.2 ± 0.78	0.5 ± 0.06
	p<0.05	p<0.001	p<0.05

(1) Means ± SEM of 4 pools of sample prepared from 6 rats each.
(2) Expressed as the ratio of serosal ^{45}Ca/mucosal ^{45}Ca ± SEM. There were 8 rats in each group.

dCaBP (table 2): A phosphorus depletion is associated with a parallel increase in duodenal CaBP, mucosal calcium concentration and intestinal calcium transport. These results have been previously discussed[5].

Acknowledgments.
This study was supported in part by INSERM Grant CRL 76.5.201.4. We are grateful to A. SUBRENAT for her expert technical assistance.

REFERENCES

1. HERMSDORF, C. and BRONNER, F. (1975) Biochem. Biophys. Acta 379, 553-561.
2. FREUND, T. and BRONNER, F. (1975) Am. J. Physiol. 228, 861-869.
3. HUGHES, M., BRUMBAUGH, P., HAUSSLER, M., WERGEDAL, I. and BAYLINK D. (1975) Science 190, 578-581.
4. STEELE, T., ENGLE, J., LORENC, R., TANAKA, Y., DUDGEON, K. and DELUCA, H. (1975) Am. J. Physiol. 229, 489-495.
5. THOMASSET, M., CUISINIER-GLEIZES, P. and MATHIEU, H. (1976) Biomedecine 25, 345-349.

EXTRACELLULAR
CALCIUM-BINDING PROTEINS

Structure and Biological Activities
of Salivary Acidic Proline Rich Phosphoproteins

Anders Bennick, Raymond Wong and Marilyn Cannon
Department of Biochemistry
University of Toronto
Toronto, Canada

INTRODUCTION

Saliva is the product of major as well as minor salivary glands and the largest volume of saliva is contributed by the parotid and submaxillary glands. Human saliva contains a complex mixture of proteins[1]. Amylase is a major component of parotid saliva[2] and a number of other biological activities have been described, but they only account for a small amount of the total protein[3]. It is therefore apparent that there is a major proportion of protein for which the biological activity has not been clarified, but it is possible that one of the functions of salivary proteins is related to maintaining the integrity of the teeth.

Amino acid analysis of unfractionated salivary proteins has demonstrated an unusual high amount of proline. Human parotid proteins contain 33 moles of proline/100 moles of amino acids and the corresponding figures for submaxillary and sublingual saliva are 16 and 17 moles/100 residues[4]. The proline content of proteins with known biological activities in saliva is much lower than that of the unfractionated proteins and it is therefore apparent that saliva contains substantial amounts of "proline rich proteins" with undefined biological activity. Studies in several laboratories have demonstrated the presence in human saliva of acidic proline rich proteins[3,5,6] as well as basic proline rich proteins[7,8,9] and glycosylated proline rich proteins[10,11,12,13]. Proline rich proteins have also been found in rat and monkey saliva[14,15,16,17]. In order to gain a further understanding of the structure and function of the acidic proline rich proteins, two major components named protein A and C were purified[3,18,19].

STRUCTURE OF PROTEIN A AND C

Amino acid analysis demonstrated a high degree of similarity between the two proteins[18,19]. Proline accounted for 24% of the residues in protein A and 27% of the residues in protein C and together proline, glycine and glutamic acid constituted 70% of the residues in protein A and 75% of the residues in protein C. Both proteins lacked threonine, tyrosine, tryptophan and sulfur containing amino acids and based on their minimum molecular weight they contained an identical amount of alanine, valine, leucine, isoleucine and phenylalanine. γ-Carboxyglutamic acid was absent from both proteins[19], but they each contained two mol of organic phosphorous/mol of protein, presumably in the form of phosphoserine, since the organic phosphorous was labile to alkaline hydrolysis and digestion with alkaline phosphatase[18,19]. A small amount of glucose was the only sugar present in the protein preparations, but this is likely a contaminant[19]. The molecular weights were determined by ultracentrifugation to be 9900 for protein A and 16300 for protein C, values which are in good agreement with the minimum molecular weight determined from the composition[18,19]. No secondary structure including polyproline could be detected by circular dichroism and nuclear magnetic resonance, and there were no conformational changes in buffers with low pH or high ionic strength[18,19]. The charge on acidic residues does therefore not prevent the formation of secondary structure. The relationship of the two proteins was further explored by means of specific antisera to protein A and C. While protein A and C reacted in an identical manner with antiserum to protein A, there was partial identity of the reaction of protein A and C with antiserum to protein C, due to the presence of an additional antigenic determinant in protein C. It could also be demonstrated that there were minor components with electrophoretic mobility similar to those of protein A and C which reacted with the antiserum, but no other immune reactive proteins could be detected in unfractionated saliva[19,20].

It was also found that both protein A and protein C had a blocked N-terminal, but whereas the C-terminal amino acid in protein A was arginine it was glutamine in protein C[18,19]. These results suggested that the difference between protein A and C could at least in part be due to the presence of an additional length of peptide at the C-terminal end of protein C. This has been confirmed by amino acid sequence analysis of protein A and C.

For amino acid sequence analysis the proteins were digested with enzymes and by chemical methods and the resulting peptides purified by gel filtration and paper electrophoresis. The peptides were sequenced automatically on a Beckman Model 890 C Sequenator[21] and by manual methods[22]. Carboxypeptidases were used for determination of C terminal sequences.

The methods of sequencing have been summarised in Fig. 1 and the partial sequence of protein A and C has been given in Fig. 2.

There are several interesting features of the sequence of protein A and C. Of the N-terminal 33 residues only 1 is proline, but there are 13 proline residues in the C-terminal 32 residues of protein A. The N-terminal 27 residues contain 7 of the 9 aliphatic or aromatic amino acid residues and both phosphoserine residues are located in the N-terminal 31 residues. The sequence pro-gln-gly-pro-pro-gln-gln-gly-gly-his-pro is duplicated in protein A and a part of this sequence i.e. pro-gln-gly-pro-pro-gln-gln-gly-gly-his is triplicated in protein C. While there are several tri and tetrapeptide sequences which are identical to

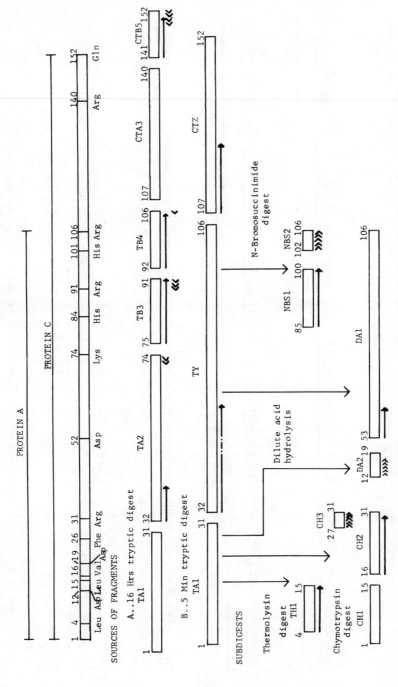

Fig. 1. Summary of the determination of the partial sequence of protein A and C. The intact protein A and C are illustrated at the top where specific cleavage points are shown. Automatic sequencing ———▶ . Manual sequencing ❯ and ❮ . Unidentified residue —·— °

1
(GLX$_{1-2}$ASX$_1$)LEU-ASN-GLU-ASP-VAL-SER-GLN-GLU-ASP-VAL-PRO-LEU-
                                                    12   15 16   19
VAL-ILE-SER-ASP-GLY-GLY-SER-ASP-GLU-GLN-PHE-ILE-ASP-
                                        26

31
GLU-GLU-ARG-GLY-PRO-PRO-LEU-GLY-GLY-GLN-SER-GLN-PRO-SER-ALA-GLY-ASP-GLY-ASN-GLN-ASN-ASP-GLY-PRO-GLN-GLX-GLY-
                                                                    52

74
PRO-PRO-GLN-(GLX$_{5-6}$PRO$_{2-3}$GLY$_{2-3}$)-GLY-LYS-PRO-GLN-GLY-PRO-PRO-GLN-GLN-GLY-GLY-HIS-PRO-PRO-PRO-PRO-GLN-GLY-ARG-PRO-
                                                84   91

101   106
GLN-GLY-PRO-PRO-GLN-GLN-GLY-GLY-HIS-PRO-ARG-PRO-PRO-ARG-PRO-PRO-GLN-GLY-ARG-PRO-GLN-GLY-PRO-PRO-GLN-GLY-GLY-HIS-GLN-GLN-GLY-

140
PRO-PRO-PRO-(GLY$_5$PRO$_{4-5}$GLN$_4$LYS$_1$ARG$_1$)-PRO-GLN-GLY-PRO-PRO-GLN-(GLN,GLY)-SER-PRO-GLY-GLN
                                                                                    152

THI   DA2   CH2   CH3
TY,TA2   DA1
TB3   NBS1
TB4   NBS2   CTZ
CTB5

Fig. 2. The partial amino acid sequence of protein A and C. Protein A terminates at residue 106 and protein C at residue 152. Specific cleavage points are indicated by the number of the residue. Automatic sequencing ⌐⌐. Manual sequencing ⌐⌐. Residue determined by means of carboxypeptidase digestion ⌐⌐. Tentative identification...... The abbreviations for the peptides are the same as used in Fig. 1.

sequences in collagen[23], the characteristic occurrence of glycine in every 3rd
position in collagen has not been found in protein A and C. The longest oligo-
proline sequence identified is-$(pro)_4$- and this is in agreement with the absence
of polyproline structure in protein A and C[24].

It is apparent that there is a very close genetic relationship between protein
A and C, and it is possible that protein A is formed from protein C by postriboso-
mal modification. The presence of other proteins immunologically related to
protein A and C indicates that there may be several genetically related proline
rich proteins present in saliva. Based on similarities in staining properties
Azen has proposed a genetic relationship of the proline rich proteins and has
suggested that the phenotypes are the expression of four autosomal codominant
alleles[25].

When the acidic and basic proline rich proteins were compared by peptide
mapping (D.L. Kauffman, unpublished results) a considerable degree of similarity
was observed. Moreover it was found by automatic sequencing of an undegraded
basic human proline rich protein that it contained a sequence of gly-lys-pro-gln-
gly-pro-pro-?-gln-gly-gly also found on protein A and C. It is possible that the
acidic and basic proline rich proteins belong to the same family. If this is
correct, intriguing questions arise regarding the biosynthetic relationship of
these proteins and the significance of such a multiplicity of related proteins in
the secretion.

BIOLOGICAL ACTIVITIES OF PROTEIN A AND C

After the initial purification protein A and C were tested for most of the
biological activities known to be present in saliva, but none of these could be
attributed to protein A and C[3].

Calcium binding studies by means of equilibrium dialysis demonstrated that the
apparent dissociation constant was approximately $2x10^{-4}$ M for both proteins[20,26].
Protein A bound 664 nmol of Ca/mg of protein, but protein C only bound 190 nmol
of Ca/mg of protein. This corresponds to 6.6 mol of Ca/mol of protein A and 3.0
mol of Ca/mol of protein C. Analysis of the binding data showed no cooperativity
of binding or heterogeneity of the binding sites. Heating of the proteins to 60°
or digestion with trypsin or collagenase had little if any effect on the Ca-binding, suggesting that the Ca binding sites were located within short stretches of
peptide chain and that the tertiary structure of the protein may not be important
for Ca binding. The amount of Ca bound to both proteins decreased with increasing
ionic strength of the buffer and removal of 80-90% of the organic phosphate from
the proteins caused a 60% reduction in the amount of protein bound Ca compared to
undigested protein[26]. The organic phosphate is therefore necessary for optimal
binding of Ca to the proteins. Differences were found in the pH dependence of

Ca binding. There was a sigmoid increase in Ca binding to protein A above pH 3.5 with a midpoint at pH 4.9 suggesting that aspartic or glutamic acid residues are necessary for Ca binding[20]. In protein C there was a gradual increase in Ca binding between pH 5.0 and 9.0[26]. Ultracentrifugation demonstrated that the value of $S°_{20,w}$ for protein A was 1.28S and this value was not affected by the presence of Ca. In the case of protein C, $S°_{20,w}$ was 1.29S in the absence of Ca but 1.46S in the presence of Ca[26]. The difference in the magnitude of the sedimentation coefficients for protein C suggests that there may be a conformational change or aggregation of protein C in the presence of Ca, although circular dichroism and nuclear magnetic resonance failed to detect any conformational changes in protein A and C in the presence of Ca.

It is clear that although protein A and C show many similarities in Ca binding a number of differences are also apparent. The reason for these differences are not clear at present. It may be that there are differences in the primary structure which have not yet been detected or it may be that the presence of the additional peptide chain in protein C has an effect on the Ca binding capacity of the protein.

Since the concentration of ultrafiltrable and presumably inorganic Ca varies from 0.64 mM in parotid saliva to 0.96 mM in submaxillary saliva[27] it can be seen that under physiological calcium concentrations protein A and C would be almost saturated with Ca. It would also be expected that changes in ionic strength, pH and Ca concentration in saliva could lead to a change in the amount of Ca bound to protein A and C. No effect of the type of buffering ions on Ca binding to the proteins has been observed[26]. Buffers with the same ionic strength, pH and ultra-filtrable Ca concentration as found in salivary secretions were therefore used to estimate the amount of protein bound Ca under physiological conditions. The amount of Ca bound to protein A varied from 35 nmol of Ca/mg of protein to 99 nmol of Ca/mg of protein.

The proteins therefore appear to have the capacity to participate in Ca exchange in the oral cavity. This would be particularly interesting if it occurred at the tooth surface. Hay[28] has demonstrated that there is a selective adsorption of acidic proline rich proteins to hydroxyapatite. To quantitate this adsorption we have used a quantitative immunoassay to determine in salivary secretions the affinity of protein A, C and immunologically related proteins for hydroxyapatite.

This was done by adding varying amounts of hydroxyapatite to a given volume of parotid or submaxillary saliva and determining the amounts of adsorbed proteins. In this manner it was possible to determine the maximum amount of adsorbed protein and the amount of hydroxyapatite necessary for half maximal adsorption, which is an expression of the affinity of the proteins for hydroxyapatite. While an average of 92% of the acidic proline rich proteins became bound to hydroxyapatite, an average of only 59% of the total protein was bound. Compared to unfractionated protein and amylase there was a higher affinity of the acidic proline rich proteins for hydroxyapatite in all the salivary secretions which were tested. The immunoreactive proteins accounted for approximately 42% of all the protein adsorbed to hydroxyapatite. The presence of these proteins in the protein layer adsorbed to teeth in vivo, the so called pellicle, could also be demonstrated. The proteins were present in pellicle formed in 2 hours as well as pellicle removed from extracted teeth.

If the proteins are digested with trypsin prior to adsorption to hydroxyapatite it can be demonstrated that the composition of the peptide adsorbed to hydroxyapatite corresponds closely to that of the N-terminal tryptic peptide (Table 1).

TABLE 1

LOCATION OF HYDROXYAPATITE BINDING SITE IN PROTEIN A AND C

A tryptic digest of protein A or C was adsorbed to hydroxyapatite and the amino acid composition of the adsorbed peptide was determined and compared with the composition of the N-terminal-tryptic peptide of protein A or C.

Amino Acid	Adsorbed Peptide Recovered from Hydroxyapatite		N-Terminal-Tryptic Peptide of Protein A or C
	Protein A Digest	Protein C Digest	
Lys	0.1	-	-
His	0.2	-	-
Arg	0.9	1.0	1.0
Asp	6.7	6.4	7.9
Thr	0.2	0.3	-
Ser	3.3	2.9	2.2
Glu	7.5	8.9	9.2
Pro	1.0	1.6	1.4
Gly	3.2	3.6	2.3
Ala	0.3	-	-
Val	2.7	2.0	2.9
Ile	1.6	1.7	1.8
Leu	2.0	2.3	2.2
Tyr	0.1	-	-
Phe	1.0	1.0	1.0

Compositions in Molar Ratios, Assuming Phe = 1.0

The binding site must therefore be located within this peptide. Removal of phosphate from the proteins by alkaline phosphatase before adsorption to hydroxyapatite causes a diminished binding of protein. Serine phosphate is therefore involved in the binding to hydroxyapatite.

While the proteins of newly formed pellicle has a high proline content it is low in the old pellicle[29] and this may be due to proteolytic modification of the proteins in the young pellicle. It is for example possible that the proline rich part of protein A and C is removed after adsorption by proteolytic enzymes originating from cellular elements in saliva such as desquamated epithelial cells, leukocytes and microorganisms. Since the proteins contain multiple binding sites for Ca it is possible that they retain Ca binding activity after adsorption to hydroxyapatite. If that is the case the Ca binding characteristics of protein A and C are such that they might e.g. influence the release of Ca from the enamel under decalcifying conditions.

Amsterdam et al[30] have demonstrated the association of proline rich proteins with the secretory granule membrane in rat salivary glands, but this apparently is a weak association [31]. Nevertheless it is possible that the function of these proteins is related to the glandular secretory mechanism. In this connection we have collaborated with Dr. A. Katz on the immunohistochemical localisation of the acidic proline rich proteins in human salivary glands. The results demonstrated that the proteins were located in the serous acini of the parotid gland and in the serous demilumen of the submaxillary gland. The proteins were not present in the mucinous acini of the submaxillary glands and they were not found in the labial glands which are pure mucinous glands. If the proteins are necessary for secretion there must be a difference in this process in serous and mucinous acini.

CONCLUSION

Our present knowledge suggests that the acidic proline rich proteins may be important in Ca metabolism in the oral cavity and it is also possible that they are necessary to maintain the integrity of the dental enamel. In this connection it is also likely that the ability of these proteins to inhibit precipitation of calcium phosphate from supersaturated solutions as demonstrated by Hay and Grön[32] is of physiological importance. The striking chemical characteristics of the proteins may be related to the fact that the proteins have to exert their function in an environment where there can be relatively large variations in pH and ionic strength. The high proline content of the proteins would ensure a conformational stability which would not so easily be subject to variations with changes in pH and ionic strength.

ACKNOWLEDGEMENT:

The experimental work described in this paper has been supported by Grant
No. MT 4920 to A.B. from the Canadian Medical Research Council. R. Wong is a
Canadian Medical Research Council Postdoctoral Fellow.

REFERENCES

1. Meyer, T.S. and Lamberts, B.L. (1965) Biochim. Biophys. Acta 107, 144–145.
2. Schneyer, L.H. (1956) J. Appl. Physiol. 9, 453–455.
3. Bennick, A. and Connell, G.E. (1971) Biochem. J. 123, 455–464.
4. Leach, S.A., Critchley, P., Kolendo, A.B. and Saxton, C.A. (1967) Caries Res.
   1, 104–111.
5. Oppenheim, F.G., Hay, D.I. and Franzblau, C. (1971) Biochemistry 10,
   4233–4238.
6. Friedman, R.D. and Merritt, A.D. (1975) Amer. J. Hum. Gen. 27, 304–314.
7. Armstrong, W.G. (1971) Caries Res. 5, 215–227.
8. Levine, M.J., Ellison, S.A. and Bahl, O.P. (1973) Arch Oral Biol. 18, 827–837.
9. Levine, M. and Keller, P.J. (1977) Arch. Oral Biol. 22, 37–41.
10. Levine, M.J., Weill, J.C. and Ellison, S.A. (1969) Biochim. Biophys. Acta 188,
    165–167.
11. Friedman, R.D., Merritt, D. and Bixler, D. (1971) Biochim. Biophys. Acta 230,
    599–602.
12. Arneberg, P. (1974) Arch. Oral Biol. 19, 921–928.
13. Degand, P., Boersma, A., Roussel, P., Richet, C. and Biserte, G. (1975) FEBS
    Lett. 54, 189–192.
14. Fernandez-Sorensen, A. and Carlson, D.M. (1974) Biochem. Biophys. Res. Commun.
    60, 249–256.
15. Wallach, D., Tessler, R. and Schramm, M. (1975) Biochim. Biophys. Acta 382,
    554–564.
16. Keller, P.J., Robinovitch, M.R., Iversen, J. and Kauffman, D.L. (1975)
    Biochim. Biophys. Acta 379, 562–570.
17. Jacobsen, N. and Arneberg, P. (1976) Comp. Biochem. Physiol. 54B, 423–425.
18. Bennick, A. (1975) Biochem. J. 145, 557–567.
19. Bennick, A. (1977) Biochem. 163, 229–240.
20. Bennick, A. (1976) Biochem. J. 155, 163–169.
21. Hermodson, M.A., Ericsson, L.H., Titani, K. and Neurath, H. (1972)Biochemistry
    11, 4493–4502.
22. Percy, M. and Buchwald, B.M. (1972) Anal. Biochem. 45, 60–67.
23. Dayhoff, M.O. (1972) Atlas of Protein Sequence and Structure. Vol. 5,
    National Biomedical Research Foundation, Washington.
24. Deber, C.M., Bovey, F.A., Carver, J.P. and Blout, E.R. (1970) J. Amer. Chem.
    Soc. 92, 6191–6198.

25. Azen, E.A. and Denniston, C.L. (1974) Biochem. Genetics 12, 109-120.
26. Bennick, A. (1977) Biochem. J. 163, 241-245.
27. Grön, P. (1973) Arch. Oral Biol. 18, 1365-1378.
28. Hay, D.I. (1973) Arch. Oral Biol. 18, 1517-1530.
29. Mayhall, C.W. (1970) Arch. Oral Biol. 15, 1327-1341.
30. Amsterdam, A., Schramm, M., Ohad, I., Salomon, Y. and Seliger, Z. (1971) J. Cell Biol. 50, 187-200.
31. Robinovitch, M.R., Keller, P.J. and Kauffman, D.L. (1975) Biochim. Biophys. Acta 382, 260-264b.
32. Hay, D.I. and Grön, P. (1977) in Microbial Aspects of Dental Caries (Stiles H.M., Loesche, W.I. and O'Brien, T.C., eds.) Information Retrieval Inc., Washington.

Human Salivary Statherin: A Peptide Inhibitor
of Calcium Phosphate Precipitation

Donald I. Hay and David H. Schlesinger,
Chemistry Department, Forsyth Dental Center, 140 Fenway, and
the Endocrine Unit, Massachusetts General Hospital,
Boston, Massachusetts

SUMMARY

The macromolecular fraction of human saliva inhibits precipitation from solutions supersaturated with respect to calcium phosphate salts. This unusual property enables saliva to exist in a stable state, yet be supersaturated with respect to the basic calcium phosphate salts which form the dental enamel. The resulting "stabilized supersaturation" provides important protection for the teeth. Two kinds of active macromolecule have been identified, the most active of which is a phosphopeptide, named statherin (from σταθεροιοω = I stabilize). At 10^{-6}M, statherin stabilizes a standard assay solution, which is supersaturated with respect to dicalcium phosphate dihydrate, for 24hr beyond the time at which precipitation normally occurs (2hr). It possesses a polar structure, 10 of the 12 charged groups present being located in the NH_2-terminal 13 residues. Inhibitory activity appears to be associated with this charged segment of the molecule.

INTRODUCTION

The calcium phosphate salts of most mineralized tissues exist in equilibrium with their constituents, in protected and precisely controlled environments. A contrary situation exists with respect to the mineral of the teeth, which is exposed to variable and sometimes damaging conditions. Despite this, the mineral of the surface enamel is stable [1], and, with the exception of effects attributable to abrasion or to attack by the acidic end-products of bacterial metabolism, will retain its form for decades. This stability depends in part on the degree of saturation of the saliva with respect to the calcium phosphate salts of the enamel. It is usually argued that the fluid which surrounds the teeth is supersaturated with respect to the enamel mineral [2], since saliva recalcifies both naturally [3] and artificially [4] demineralized enamel. Also, recent studies of the status

401

of calcium and phosphate ions in the salivary secretions [5], and the calculation of their degree of supersaturation, indicate that human salivas are usually super-saturated with respect to basic calcium phosphate salts, and in some instances are saturated with respect to dicalcium phosphate dihydrate [6]. In spite of this strong evidence indicating supersaturation, saliva lacks the following properties which normally characterize the supersaturated state. Firstly, spontaneous precip-itation of calcium phosphate salts from normal glandular salivas is unusual [7], secondly, mineral deposits do not form spontaneously on teeth, and thirdly, precipitation from saliva seeded with a solid calcium phosphate is considerably retarded, compared with precipitation from equivalent inorganic solutions [8]. Recently, we explained the absence of these properties by showing that the macro-molecular fraction of saliva strongly inhibits precipitation from solutions super-saturated with respect to calcium phosphate salts, and that this unusual property is absent from some other complex protein mixtures, such as sera from humans and other mammals [9]. We now describe some properties of one of the specific salivary macromolecules which are responsible for this inhibitory effect.

MATERIALS AND METHODS

Measurement of precipitation-inhibition. Stock phosphate; 0.014M KH_2PO_4, 0.056M KCl, 0.003M NaN_3, adjusted to pH 6.8 with KOH. Stock calcium; 0.028M $CaCl_2$ prepared from 2.803g $CaCO_3$ and 10ml 6N HCl and diluted to 1 liter with 0.003M NaN_3. Samples were dissolved in or dialyzed against the following buffer to ensure constancy of ionic composition and pH; 0.05M imidazole chloride, 0.15M NaCl, 0.003M NaN_3, pH 7.0. The assay system consisted of stock phosphate, 2.5ml, and sample, 0.5ml mixed in 12x75mm polystyrene tubes (type 2054, Falcon Products, Oxnard, Cal.), 0.5ml of stock calcium added, the tubes capped and mounted on an end-over-end shaker (Model 343 RotoRack, Fisher Scientific, Inc.) rotating at 7rpm for 24hr at 20°C. The solutions were filtered through 13mm diameter, 0.45µm pore size filters (Type MF, Millipore Corp. Bedford, Mass.) and analyzed for calcium by atomic absorption spectrophotometry [10] and phosphate by the method of Lowry and Lopez [11]. In the absence of inhibitors, the calcium concentration fell from 4mM to close to 0.5mM, and phosphate from 10mM to 6.5mM. The percentage of precipitable calcium or phosphate remaining in solution at 24hr, was taken as an arbitary, but convenient measure of the degree of inhibition.

Saliva collection and fractionation. Human parotid and submandibular salivas were collected using conventional methods [12,13], dialyzed and lyophilized. The salivary proteins were fractionated using DEAE Sephadex and a chloride gradient, as previously described [14], except that the tris chloride buffer was replaced by 0.05M pH 7.0 imidazole chloride to eliminate tris interference with the assay. All fractions were examined for inhibitory activity, analyzed for total protein by a biuret method [15], examined by disc electrophoresis [16] staining with amido-black, and their chloride content determined using an electrometric method (Model CM10 chloride titrator, Radiometer, Copenhagen, Denmark).

Purification of statherin. Statherin was isolated from human parotid saliva [14] and purified by salt precipitation, gel permeation chromatography on BioGel P6 (BioRad Laboratories, Richmond, Cal.) and anion exchange chromatography on BioRad DEAE agarose using a chloride gradient. The product had the same composition as material obtained by a more complex method of purification [17], and was shown to be pure by immunoelectrophoresis, disc electrophoresis and by amino-terminal analysis [18].

RESULTS AND DISCUSSION

Fig. 1 shows the results of assaying fractions obtained by chromatography of human submandibular saliva using DEAE Sephadex. A small amount of non-specific inhibitory activity is associated with many of the fractions, but reproducible and complete inhibition of precipitation was found only with fractions containing statherin, which eluted in fractions 36-40. Statherin has an unusual composition, containing a considerable proportion of aromatic (7 tyrosine, 3 phenylalanine/mole) and acidic residues (3 glutamate, 1 aspartate/mole), a high proportion of proline and glutamine (each 7 residues/mole), and a molecular weight, based on its amino-acid sequence [18], of 5380 daltons. The relationship between the concentration of purified statherin, and its inhibitory activity, is shown in Fig. 2. Over the range of 2-5µg/ml, its inhibitory activity increased from near zero to 100%. In order to compare relative inhibitory activities, the concentrations giving 50%

403

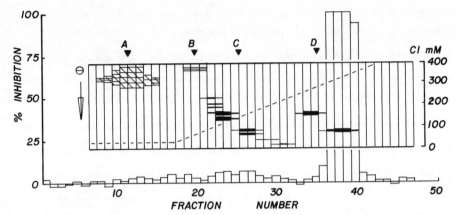

Fig. 1. Electrophoretic patterns and inhibitory activity of fractions obtained by chromatography of human submandibular saliva on DEAE Sephadex. 70-75% of the total protein eluted in the non-retarded fraction A, most of the constituents of which are not detected by the electrophoretic system used here. Secretory IgA (B), a complex group of anionic proline-rich proteins and some constituents characteristic of submandibular saliva (C) and a histidine-rich peptide (D) eluted before statherin (frs. 36-40) with which strong inhibitory activity was always associated. These fractions contained between 1 and 3% of the total biuret positive material.

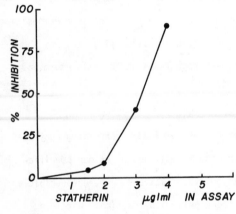

Fig. 2. Relationship between statherin concentration and % inhibition at 24hr. Complete inhibition for this time occurs at a concentration of 5μg/ml or 1μM.

inhibition were obtained by interpolation. A mean value of 3.25μg/ml±0.25μg/ml was obtained from four different preparations.

20mg/liter yields of statherin have been obtained from pooled, stimulated parotid saliva, with recoveries estimated at better than 90%. Previous studies [9] suggest that statherin concentrations in saliva vary over a threefold range, so indicating a range of 10-30μg/ml. These estimates of the salivary concentrations of statherin are substantially greater than the concentrations required to inhibit precipitation from the highly supersaturated assay system used for this study.

Since saliva has never been reported to reach this degree of supersaturation [2,6] it may be anticipated that typical sterile glandular salivas may remain stable for many hours or days, providing no effects operate to diminish inhibitory activity, or to markedly increase the degree of supersaturation.

The mechanism by which statherin, and numerous other compounds [19-23], inhibit both spontaneous and seeded precipitation, and phase transformation of calcium phosphate salts, is not completely understood [24]. Clearly, the low concentration at which inhibitors affect these processes, must preclude simple ion binding as a possible mechanism of action. Thus, with the molar ratios of statherin to calcium and phosphate being 1 to $4x10^3$ and 1 to 10^4 respectively, in the assay system used for this study, the activities of these ions would not be significantly affected by the binding of a few moles of ion/mole of inhibitor.

In order to place the activity of statherin into perspective, and perhaps reveal some aspects of its mechanism of action, its activity was compared with the activities of several other known inhibitors, and a number of model compounds. Table 1 shows that statherin has an inhibitory activity (active at 1μM for 24hr) similar to activities found for some other macromolecular inhibitors, such as phosvitin (3μM), and low molecular weight inhibitors, such as EHDP (5μM) and pyro-phosphate (7.5μM). It has been proposed that these inhibitors act by binding to the initial aggregates of ions, or nuclei, formed in the early stages of precipi-tation, and also to growth sites, such as kinks and dislocations on seed crystal surfaces, and prevent crystal growth. This mechanism is consistent with the known strong interactions which occur between these inhibitors and apatitic surfaces [30]. Statherin exhibits a similar strong interaction, in that it may be select-ively adsorbed from glandular salivas, in nearly pure form, by means of appropr-iate hydroxyapatite to saliva ratios [25]. It seems likely, therefore, that it acts by a similar mechanism.

The binding of proteins to apatitic surfaces occurs through both anionic and cationic groups [26], but inhibitory activity appears to be most usually associa-ted with anionic rather than with cationic molecules. However, the large difference between the activities of polyaspartates and polyglutamates was suprising, as was

405

TABLE I

RELATIVE ACTIVITIES OF SOME INHIBITORS OF CALCIUM PHOSPHATE PRECIPITATION

INHIBITOR		CONCENTRATION[a]	% INHIBITION[a] AT 24HR.
Statherin	5,380	1.0×10^{-6}M	100
Phosvitin[b]	34,000	3.0 "	100
poly-Aspartate[b]	7,700[c]	1.0 "	100
"	22,000[c]	0.1 "	100
poly-Glutamate[b]	8,100[c]	10.0 "	20
"	49,000[c]	10.0 "	30
"	74,000[c]	10.0 "	35
poly-Arginine[b]	43,000[c]	10.0 "	20
poly-Lysine[b]	31,000[c]	10.0 "	15
poly-Asparagine[b]	9,700[c]	10.0 "	0
EHDP[d,e]		5.0 "	100
Pyrophosphate[d]		7.5 "	100
Phytate[b,d]		150 "	60
Trimetaphosphate[d]		300 "	100
Phosphoserine[b]		500	50
Aspartate[b]		10.0×10^{-3}M	0
Glutamate[b]		10.0 "	0
Arginine[b]		10.0 "	0
Lysine[b]		10.0 "	0

[a]Either, minimum concentration in the assay system which gave 100% inhibition, or
the highest concentration tested, and its corresponding inhibitory effect.
[b]Supplied by Sigma Chemical Co. St. Louis, Mo.
[c]Molecular weight as stated by supplier
[d]Sodium salts.　　[e]Ethane-1-hydroxy-1,1-diphosphonate.

the modest difference between the activities of the polymers of the basic aminoacids, and the polyglutamates. Previous studies have indicated distinct differences
in the way aspartate and glutamate interact with apatitic surfaces [27], though
whether this difference is reflected in the behaviour of the polymers is not known.

The importance of charged residues, and the important contribution made by the
polymeric form of the inhibitors, was shown by the inactivity of polyasparagine,
and the monomeric acidic and basic amino acids, and the low activity of phosphoserine. Thus, aspartate was inactive at 10^{-2}M (1.33mg/ml) while polyaspartate (MW
22,000) was active at 10^{-7}M (0.002mg/ml). Related results have been reported for
the effect of glutamate and polyglutamate on the transformation of amorphous to
crystalline calcium phosphate [28], and the strong interactions which occur in

406

coprecipitates of anionic polymers and calcium phosphates have been demonstrated by electron spin resonance [29]. The relatively high concentrations of pyrophosphate required to inhibit precipitation reflects the high degree of supersaturation of the assay system used for these studies. Lower concentrations, (10^{-6}M or less) are more usually reported [19].

The consistent way in which inhibitory activity is associated with anionic compounds, particularly with those containing multiple or polymeric phosphate groups [19,23,28] suggests an important role for the highly charged amino-terminus of statherin [18]. The sequence of the amino-terminal third of this molecule is

$$\overset{1}{\text{ASP}}\text{-SER}(\text{PO}_3\text{H}_2)\text{-SER}(\text{PO}_3\text{H}_2)\text{-GLU-}\overset{5}{\text{GLU}}\text{-LYS-PHE-LEU-ARG-}\overset{10}{\text{ARG}}\text{-ILE-GLY-ARG-}$$

with this segment containing ten of the twelve charge centers present. It seems likely that this highly polar structure will be important in its mechanism of action, and that the two vicinal phosphoserine residues will play a particularly important role. These two residues could act in a manner similar to that proposed for the highly inhibitory multidentate diphosphonates [23] by providing a unique configuration for binding to nuclei and to growth sites on calcium phosphate crystal surfaces. In fact, recent experiments in which statherin was degraded using trypsin or clostridiopeptidase B, yielded active amino-terminal fragments (residues 1-6 and 1-9 respectively). Their specific activities, however, were reduced, compared to the parent molecule. If confirmed, these preliminary results indicate a role for the tyrosine and proline-rich carboxyl two-thirds of this unusual molecule.

ACKNOWLEDGEMENTS

Ms. Deborah Belensz is thanked for her expert technical assistance. These studies were supported by grant DE 3915 from the National Institute for Dental Research, and contract NO1 CB 53868 from the National Cancer Institute.

REFERENCES

1 Soggnaes,R.F.,Shaw,J.H. and Bogroch,R.(1965) Amer.J.Physiol. 180. 408-420.

2 McCann,H.G.(1968) Dental Caries Research. Academic Press. New York. 55-73.

3 Backer-Dirks,O. (1966) J.Dent.Res. 45. 503-511.

4 Koulourides,T.(1968) Ann. N.Y. Acad. Sci. 153. 84-101.

5 Grøn, P. (1973) Archs. Oral Biol. 18. 1365-1378.

6 Grøn, P. (1973) Archs. Oral Biol. 18. 1385-1392.

7 Mukherjee, S. (1968) J. Periodont. Res. 3. 236-247.

8 Rathje, W. (1956) J. Dent. Res. 35. 245-248.

9 Grøn, P. and Hay, D.I. (1976) Archs. Oral Biol. 21. 201-205.

10 Brudevold, F., McCann, H. and Grøn, P. (1968) Archs Oral Biol. 13. 877-885.

11 Lowry, O.H. and Lopez, J.A. (1946) J. Biol. Chem. 162. 421-428.

12 Shannon,I.L.,Prigmore,J.R. and Chauncey,H.H. (1962) J. Dent. Res. 41. 778-783.

13 Schneyer, L.H. (1955) J. Dent. Res. 41. 778-783.

14 Hay, D.I. (1975) Archs. Oral Biol. 20. 553-558.

15 Davis, B.J. (1964) Ann. N.Y. Acad. Sci. 121. 404-427.

16 Itzhaki, R.F. and Gill, D.M. (1964) Anal. Biochem. 9. 401-410.

17 Hay, D.I. (1973) Archs. Oral Biol. 18. 1531-1541.

18 Schlesinger, D.H. and Hay, D.I. (1977) J. Biol. Chem. 252. 1689-1695.

19 Fleisch, H. and Bisaz, S. (1962) Nature. 195. 911.

20 Fleisch, H. et al. (1968) Calc. Tiss. Res. 2. 49-59.

21 Francis, M.D. (1969) Calc. Tiss. Res. 3. 151-162.

22 Termine, J.D. and Posner, A.S. (1970) Archs. Biochem. Biophys. 140. 307-317.

23 Meyer, J.L. and Nancollas, G.H. (1973) Calc. Tiss. Res. 13. 295-303.

24 Robertson, W.G. (1973) Calc. Tiss. Res. 11. 311-322.

25 Hay, D.I. (1973) Archs. Oral Biol. 18. 1517-1529.

26 Bernardi,G.,Giro,M.G. and Gaillard,C. (1972) Biochem.Biophys. Acta.278.409-420.

27 Kresak,M.,Moreno,E.C.,Zahradnik,R.T. and Hay,D.I. (In Press) J. Coll. Int. Sci.

28 Termine,J.D.,Peckauskas,R.A. and Posner,A.S. (1970)Arch.Biochem.Biophys.140.318.

29 Peckauskas,R.A.,Termine,J.D. and Pullman,I. (1976) Biopolymers. 15. 569-581.

30 Jung,A.,Bisaz,S. and Fleisch,H. (1973) Calc. Tiss. Res. 11. 269-280.

Non-Collagenous Proteins of Bone and Dentin Extracellular Matrix and Their Role in Organized Mineral Deposition[a]

Arthur Veis, Maureen Sharkey[b] and Ian Dickson[c]
Northwestern University Medical School
Department of Biochemistry
Chicago, Illinois 60611

INTRODUCTION

The deposition of calcium hydroxyapatite within collagenous matrices to form bone and dentin involves specific interactions between calcium, phosphate and collagen, locating the earliest crystals or crystal nuclei within the collagen fibers (1). Long before this interaction was so clearly defined it had been postulated that the mineralization process might be regulated by one or more of the non-collagenous components of the matrix (2, 3). The involvement of proteoglycans or glycosaminoglycans (4-8), phospholipids (9-12), glycoproteins (13-19) and phosphoproteins (19-23) have all been proposed. Although no direct evidence linked any of these components to mineralization the glycoproteins and phosphoproteins present in bone and dentin in greatest amount appeared to deserve further study. They share many common properties, including having high concentrations of anionic residues.

The non-collagenous proteins (NCP) of bone and dentin are present in two pools: one set soluble and extractable at neutral pH during or following demineralization; the other set closely associated with the collagen and released only after degradation of the collagen. The extractable NCP of bone and dentin from animals of the same species differ in composition (13, 24, 25). Even the permanent and deciduous teeth of the same animal contain different extractable NCP (26). The collagen-associated NCP of bone and dentin released after collagenase digestion are similar in terms of contents of acidic residues (18) but are distinctly different in bone and dentin (27, 28).

In mature dentin highly phosphorylated proteins, called phosphophoryns, are the principal extractable NCP (20, 21). Phosphophoryns they are unique in containing a combined total of > 70 residue % seryl and aspartyl residues. A large fraction of the seryl residues are phosphorylated. Phosphophoryns have a large number of high affinity calcium binding sites (29). The collagen-associated NCP of mature bovine dentin are also phosphophoryns (27, 30). These data suggest a role for phosphophoryn in the in vivo binding, localization or transport of calcium ions in the mineralization process in dentin.

---

[a] Supported by grant DE01374, National Institute of Dental Research, NIH.
[b] Present address: National Institute of Dental Research, NIH, Bethesda, MD.
[c] Present address: Strangeways Laboratory, Worts Causeway, Cambridge, England.

We address ourselves here primarily to two questions. In view of the observations that permanent and deciduous teeth contain different NCP (26), are phosphophoryns present in immature deciduous teeth in the early stages of mineralization? Does bone exhibit the same type of relationship between some component of
the extractable and collagen-associated NCP as is the case with the phosphophoryns of dentin? A subsidiary concern is whether the collagen-associated NCP of
bone contain bound phosphorus. Answers to these questions should bring us closer
to depicting the nature and role of calcium binding NCP in mineralizing tissues.

MATERIALS AND METHODS

Bone and Dentin: Molars were obtained from fetal calves 5 to 7 months in utero.
Bovine cortical bone was from animals approximately 18 months old.

The teeth were washed initially overnight at 4°C with 10% ethylene diamine
tetraacetic acid (EDTA), pH 8.3 and then washed with deionized water. The cortical bone was treated by the method of Dickson (17). After cleaning the marrow
cavity and washing, spicules of cancellous or trabecular bone, and areas of vascu-
lar entry were dissected. The residual bone was pulverized in a mill cooled with
liquid nitrogen.

Extraction and Demineralization: The washed molars were demineralized in 0.5M
EDTA, pH 7.4, 4°C by the batchwise procedure of Veis et al. (20). The major portion of the NCP appeared in the first four extracts. EDTA was removed from the
extracts and residue by dialysis against distilled water.

The bone powder was extracted identically except that after each extraction
the bone matrix was collected by low speed centrifugation. The extracts were
combined and centrifuged at 30,000XG in an angle head rotor for 30 minutes. EDTA
was removed by dialysis and washing.

Fractionation of Soluble Proteins: A variety of gel exclusion and ion-exchange
procedures were used. The conditions for any particular run are given in the
appropriate figure legends. Sephadex G100, Biogel P10 and Biogel P60 were used
for molecular weight separations. Whatman DE52 diethylaminoethyl (DEAE) cellulose was used for anion exchange chromatography.

Cyanogen Bromide Degradation: EDTA insoluble bone protein was digested with
cyanogen bromide (CNBr) by the method of Volpin and Veis (31). After lyophilization of the digestion mixture the collagen peptides were dissolved in 0.1M acetic
acid. The acid insoluble residue was dissolved at neutral pH.

Analytical Procedures: Organic phosphorus was determined by a modification of
the method of Dryer et al. (32). Glassware was cleaned in boiling nitric acid
and rinsed with deionized water. Samples were wet ashed in a heating block at
150°C following the procedure of Ames and Nesbitt (33).

Samples for amino acid analysis were hydrolyzed in 6N HCl, in evacuated
vials purged with nitrogen, at 110°C for 22 hours. Analyses were carried out in a
JEOL 6AH analyzer by a single column technique. Tryptophan was determined after

hydrolysis of the protein at 115°C for 20 hours in 4N methanesulfonic acid containing 0.2% 3-(2 amino-ethyl) indole (Pierce Chemical Co.) (34).

Hexuronate contents were determined by the method of Bitter and Muir (35).

RESULTS

The Presence of Phosphophoryn in Fetal Bovine Dentin: The EDTA extract accounted for 12.7% of the organic matter of the fetal teeth. Small amounts of carbohydrate and sialic acid were present in the crude extracts. However, the major non-protein contribution to the weight was the result of the anionic character of the EDTA soluble proteins. After dialysis against water to remove the EDTA the proteins remained in sodium salt form and the ash content of the combined extracts was on the order of 20%.

The first EDTA extract was rich in phosphophoryn but, in contrast with studies of unerupted permanent molars (20), gel exclusion chromatography on Sephadex G100, Figure 1, yielded only about 20% of the total protein of extract 1 in the void volume. This fraction, however, contained the major portion of phosphophoryns, Table I. The second and succeeding extracts had lower contents of phosphophoryn relative to proteins containing large proportions of prolyl and glutamyl residues.

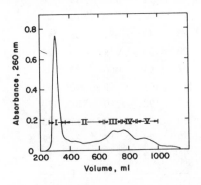

Figure 1. Sephadex G100 chromatography of the EDTA extract of fetal dentin. In 0.05M TRIS-HCl, pH 7.0. Fractions pooled as indicated.

Anion exchange chromatography of EDTA extract, not previously fractionated on Sephadex G100, yielded the phosphophoryn component only at an ionic strength >0.35, in accord with earlier observations (20, 28). The data, Figure 2 and Table I, are shown to emphasize the distinctly different compositions of the several other NCP of fetal teeth.

As shown in Table I, the two fractions identified as potential phosphophoryns on the basis of their amino acid content contained ∿1.5% P on a weight basis, corresponding to about 45 moles of phosphate per 10^5 g protein. Serine phosphate was the main phosphorus containing component in each case. Small amounts of phosphorus were present in every NCP fraction. Hexosamines, principally glucosamine and galactosamine were also present in every fraction.

Phosphophoryn collected from DEAE-chromatography fractionations, was pooled and purified further by Biogel P60 chromatography at pH 9. The purified phosphophoryn, component I of Figure 3, had the composition shown in Table II. Hydroxylysine is present in the purified phosphophoryn. No hydroxyproline could be

411

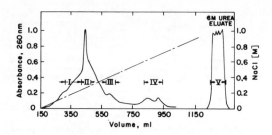

Figure 2. DEAE-cellulose fractionation of EDTA extract of fetal dentin. 0.05M TRIS-HCl, pH 8.3 in 0.001M EDTA. NaCl gradient indicated on right hand ordinate.

detected. Hydrolysis under conditions which preserve tryptophan showed the presence of this amino acid at the level of 2 residues per 10^3 residues. The purified phosphophoryn from the fetal teeth contains 2% organic phosphorus.

Extractable and Non-Extractable NCP of Cortical Bone: Although only a small quantity of protein

TABLE I

Amino Acid Compositions of Non-Collagenous
EDTA Soluble Proteins of Fetal Bovine Teeth Extracts and Fractions[a]

Amino Acid	EDTA Extract 1 Sephadex Fractions					EDTA Extract 2 DEAE-Cellulose Fractions				
	I	II	III	IV	V	I	II	III	IV	V
Hydroxyproline	--	--	--	--	--	--	--	--	--	--
Aspartic acid	259	132	107	138	87	33	48	321	66	47
Threonine	30	76	65	79	59	28	36	19	35	34
Serine[b]	309	78	60	75	113	57	64	366	93	88
Glutamic acid	43	80	78	96	108	157	109	40	90	153
Proline	35	74	59	51	38	258	192	40	157	170
Glycine	74	110	118	127	138	50	64	37	114	83
Alanine	45	110	115	110	127	30	38	17	23	29
Valine	28	78	68	53	44	39	50	5	34	40
Half cystine	6	T	T	T	T	--	--	3	T	--
Methionine[c]	5	22	36	26	32	34	43	5	33	34
Isoleucine	13	38	40	48	26	46	51	17	58	34
Leucine	32	66	75	65	68	100	94	22	65	84
Tyrosine	10	25	28	26	23	30	47	11	90	38
Phenylalanine	14	34	39	39	49	20	25	9	28	44
Hydroxylysine	T	--	--	--	--	--	--	--	--	--
Lysine	64	50	50	40	39	16	27	51	34	24
Histidine	12	6	18	6	16	82	76	17	49	46
Arginine	20	21	43	21	33	13	27	11	24	53
(% Phosphorus)[d]	(1.6)	(--)	(--)	(--)	(--)	(0.14)	(0.12)	(1.4)	(0.2)	(0.05)

[a] Residues of amino acid per 1000 total residues. [b] Sum of serine and phosphoserine. [c] Sum of methionine and methionine sulfoxide. [d] Percent dry weight. Note. T = Trace amount.

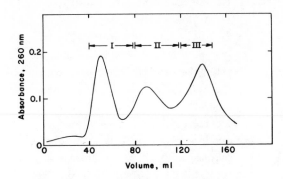

Figure 3. Biogel P-60 chromatography of fetal dentin DEAE-fraction III. 0.05M triethyl-ammonium bicarbonate, pH 9.0.

TABLE II

Amino Acid Composition of Biogel P-60
Purified Fetal Bovine Dentin Phosphophoryn

Amino Acid	Residues per Total 1000 Amino Acid Residues
Hydroxyproline	--
Aspartic acid	353
Threonine	14
(Serine + Phosphoserine)	404
Glutamic acid	31
Proline	28
Glycine	44
Alanine	12
Valine	9
Half cystine	trace
Methionine	7
Isoleucine	8
Leucine	15
Tyrosine	7
Phenylalanine	6
Hydroxylysine	1
Lysine	42
Histidine	10
Arginine	7
Tryptophan	2
(Phosphorus, w/w, %)	(2.0)

was extracted during EDTA de-mineralization of the cortical bone, the bone collagen is either inherently more soluble than dentin or is more readily degraded during extraction. The EDTA extracts contained readily measurable amounts of hydroxyproline. Degradation seems likely since some of the hydroxyproline is in low molecular weight peptides retained on Biogel P10 (exclusion limit 20,000). However, 80% of the bone extract was recovered in the void volume fraction. In spite of the presence of some collagen (that is, hydroxyproline) in this higher molecular weight fraction, the constituents were mainly a mixture of acidic proteins. These were partially resolved by ion exchange chromatography on DEAE cellulose as shown in Figure 4 and Table III. The undegraded collagen elutes in the column void volume. The more anionic components all have similar compositions, marked by equimolar concentrations of aspartic and glutamic acids, which account for ~25% of the total amino acid residues. The thoroughly dialyzed P10 void volume fraction distributed on the DEAE-cellulose contained 0.72% P (w/w) and it is noteworthy that phosphoserine was found to be present at the 20 to 30 mole/10^3 residue level in each of the acidic NCP fractions. The P10 void volume fraction also contained 2.04% hexuronate and each subcomponent also showed the presence of hexosamines.

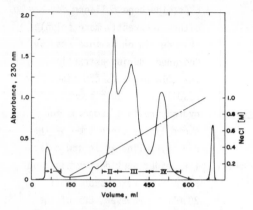

Figure 4. DEAE-cellulose chromatography of P10 void volume fraction of EDTA extract of cortical bone. 0.05M TRIS-HCl, pH 8.0. Ionic strength of NaCl gradient indicated on right hand ordinate.

Upon digestion of the EDTA insoluble bone matrix with CNBr, and after the resulting peptides were freed of excess CNBr by lyophilization, they were taken up in 0.1M acetic acid. As in the case of dentin matrix (31, 36), an insoluble residue remained, about 2.5% of the initial EDTA insoluble matrix. The insoluble peptides were typical of Type I bone collagen but the residue was less than 50% collagenous, with the composition shown in Table III. The collagen-associated residue is clearly closely related to the soluble EDTA extractable NCP and is similar to the fraction obtained upon periodate degradation of bovine cortical bone matrix (28), as shown in Table III. The residue contained 1.60% hexuronate and, most notably, 0.32% P. Again phosphoserine was detected upon amino acid analysis.

DISCUSSION

The straightforward extraction and isolation studies described above provide unequivocal evidence that fetal bovine dentin contains a phosphophoryn similar to that found in mature dentin. There is less phosphophoryn in the fetal teeth, ∿20% of the NCP as compared with ∿80% of the NCP in unerupted bovine molars (20) or in rat incisor dentin (21). The fetal phosphophoryn is also less highly phosphorylated than mature origin phosphophoryn. The phosphophoryn of mature bovine molars has a phosphorus content 2 to 3 times as high as in the fetal molars (29). In the mature tooth, enamel has a much higher mineral content than dentin, but in the fetal tooth at this stage in development it is the dentin which is more highly mineralized (37). The enamel proteins, rich in proline and glutamic acid, are present in large amount and are seen in Figure 2 and Table I as the components eluting at low ionic strength. These proteins are degraded and removed during the process of enamel maturation (38, 39). They do contain covalently bound phosphorus (40).

The finding that phosphophoryns are components of fetal dentin and thus are present at all stages of dentin development and mineralization is important in that it allows us to propose a coherent and internally consistent hypothesis for the mechanism of the collagen-epitactic deposition of calcium phosphates within

TABLE III

Amino Acid Compositions of Cortical Bone Fractions[a]

Amino Acid	Fractions from DEAE-Cellulose Chromatography of P10 Excluded Peak, Figure 4. Fraction Numbers				Acid Insoluble CNBr Digest		Periodate Bone[f]
	I	II	III	IV	Bone	Dentine[e]	
Hydroxyproline	96	11	--	7.5	30	34	T
Aspartic acid	54	114	118	116	99	192	100
Threonine	24	54	56	58	50	25	40
Serine[b]	44	91	81	91	69	176	76
Glutamic acid	81	129	117	119	113	70	133
Proline	105	78	73	72	83	60	83
Glycine	296	74	94	87	110	166	136
Alanine	122	73	67	75	76	68	88
Half cystine	--	9.3	3.7	13	11	7	N.D.
Valine	27	64	53	67	62	26	59
Methionine[c]	16	11	19	3.9	3.6[d]	2.4	7.3
Isoleucine	17	33	4.8	32	34	14	36
Leucine	26	74	84	77	73	33	86
Tyrosine	13	23	24	21	20	8.5	5.2
Phenylalanine	18	35	33	35	34	16	33
Hydroxylysine	4.3	1.5	--	--	T	2.5	T
Lysine	23	59	58	61	51	45	44
Histidine	3.6	27	29	24	24	13	11
Arginine	30	39	49	41	54	44	53
Homoserine	--	--	--	--	4.1	t	--

[a] Residues per 1000 amino acid residues. [b] Sum of serine and phosphoserine. Phosphoserine present in all samples. [c] Sum of methionine and methionine sulfoxides. [d] All uncleaved methionine was recovered as methionine sulfoxides. [e] Data from Dickson et al. (30) for comparison. Unerupted permanent molars. [f] Data from Shuttleworth and Veis (28).

dentin. The mechanism postulated is that the phosphophoryns interact directly and specifically with collagen at the mineralization front, the interaction being mediated by calcium ion binding to special sites on both collagen and phosphophoryn. At the same time, the large number of excess calcium binding sites on the phosphophoryns sequester additional calcium ions. The phosphophoryns in the complex undergo local conformational transitions which reproduce the calcium-phosphate lattice dimensions of one face of an hydroxyapatite crystal. These calcium-phosphophoryn-collagen complexes then nucleate collagen-structure-related crystal growth, localized within collagen fibrils.

This hypothesis is supported by data from several lines of investigation.

Recent studies of the post-synthesis processing of phosphophoryn indicate that the phosphophoryns may be secreted directly at the mineralization front or within the mineralized dentin (41-43), consistent with an earlier finding that phosphophoryn is present in mineralized dentin but not predentin (44). Phosphophoryns are calcium binding proteins, capable of complexing large numbers of calcium ions per molecule, and, at the same time forming insoluble complexes with calcium ion-phosphate group interactions comparable to that in hydroxyapatite (29). In in vitro systems of supersaturated calcium phosphate, the addition of phosphophoryn catalyzes the conversion of amorphous calcium phosphate to hydroxyapatite (41). Some phosphophoryn is bound firmly to dentin collagen (30, 31) and the dentin collagen matrix itself may have a few high affinity calcium ion binding sites (46) so that the binding of phosphophoryn may be mediated by bidentate calcium ions.

The organization of crystalline hydroxyapatite within bone collagen fibrils is similar to that within dentin collagen. Moreover, in each osteon of compact bone there is a zone of unmineralized osteoid. It therefore seems reasonable to postulate that the same series of events takes place, utilizing a specific bone non-collagenous calcium binding protein having the role of the dentin phospho-phoryn. Leaver et al. (13) and Shuttleworth and Veis (28) have pointed out the differences between bone and dentin NCP. There is no doubt, however, of the strong cation binding affinity of some of the bone NCP (47-49). By applying the same procedure for the examination of the residual collagenous matrix of bone after EDTA demineralization, exactly as carried out more extensively for the dentin collagen matrix, we have demonstrated above that a bone collagen-bone matrix NCP conjugate is present in which the NCP component is similar to one of the soluble NCP components. The collagen bound NCP contains organically bound phosphate but may also contain γ-carboxyglutamic acid (47) as the calcium binding component.

Although we are at the beginning of the direct and detailed study of the collagen associated NCP of bone we speculate that the mechanism of mineralization proposed for dentin will also apply to bone, except for tissue specific differences in the composition of the calcium binding protein. The essential requirements for a system in which there is epitactic deposition of mineral within collagen fibrils are a protein which binds to specific locales on the collagen fibril and, at the same time, has calcium binding sites whose dimensions are isomorphous with part of the hydroxyapatite structure and can thus initiate crystal growth.

REFERENCES

1. White, S.W., Hulmes, D.J.S., Miller, A. and Timmins, P. (1977) Nature 266, 421-425.

2. Sobel, A.E. (1954) Ann. N.Y. Acad. Sci. 60, 713-731.

3. Stack, M.E. (1954) Ann. N.Y. Acad. Sci. 60, 585-595.

4. Porter, K. (1971) J. Dent. Res. 50, 41-47.

5. DeBernard, B. (1968) Calc. Tiss. Res. Suppl. 2, 48-50.

6. Pugliarello, M.C., Vittur, F., de Bernard, B., Bonucci, E. and Ascenzi, A. (1973) Calc. Tiss. Res. 12, 209-216.

7. Hjertquist, S.O. and Vejlens, L. (1968) Calc. Tiss. Res. 2, 314-333.

8. Engfeldt, B. and Hjerpe, A. (1972) Calc. Tiss. Res. 10, 152-159.

9. Shapiro, I.M. (1970) Calc. Tiss. Res. 5, 13-20.

10. Eisenberg, E., Wuthier, R.E., Frank, R.B. and Irving, J.T. (1970) Calc. Tiss. Res. 6, 32-48.

11. Shapiro, I.M., Wuthier, R.E. and Irving, J.T. (1966) Archs. Oral Biol. 11, 501-512.

12. Wuthier, R.E. (1973) Clin. Orthop. 90, 191-200.

13. Leaver, A.G., Triffitt, J.T. and Holbrook, I.B. (1975) Clin. Orthop. 110, 269-292.

14. Andrews, A.T. de B., Herring, G.M. and Kent, P.W. (1967) Biochem. J. 104, 705-715.

15. Zamoscianyk, H. (1966) Dissertation, M.S. Degree, Northwestern University.

16. Shipp, D.W. and Bowness, J.M. (1974) Biochim. Biophys. Acta 379, 282-294.

17. Dickson, I. (1974) Calc. Tiss. Res. 16, 321-333.

18. Leaver, A.G., Holbrook, I.B., Jones, I.L., Thomas, M. and Sheil, L. (1975) Archs. Oral Biol. 20, 211-216.

19. Carmichael, D.J., Chovelon, A. and Pearson, C.H. (1975) Calc. Tiss. Res. 17, 263-271.

20. Veis, A., Spector, A.R. and Zamoscianyk, H. (1972) Biochim. Biophys. Acta 257, 404-413.

21. Butler, W.T., Finch, J.E. and Desteno, C.V. (1972) Biochim. Biophys. Acta 257, 167-171.

22. Spector, A.R. and Glimcher, M.J. (1972) Biochim. Biophys. Acta 263, 593-603.

23. Pieri, J., Mensteau, J. and Kerebel, B. (1975) Compt. Rend. Acad. Sci. Paris, Serie D, 281, 1905-1908.

24. Herring, G.M. (1968) Biochem. J. 107, 41-49.

25. Herring, G.M., Ashton, B.A. and Chipperfield, A.R. (1974) Prep. Biochem. 179-200.

26. Holbrook, I.B. and Leaver, A.G. (1976) Arch. Oral Biol. 21, 509-512.

27. Veis, A. and Perry, A. (1967) Biochemistry 6, 2409-2416.

28. Shuttleworth, A. and Veis, A. (1972) Biochim. Biophys. Acta 257, 414-420.

29. Lee, S.L., Veis, A. and Glonek, T. (1977) Biochemistry, in press.

30. Dickson, I.R., Dimuzio, M.T., Volpin, D., Ananthanarayanan, S. and Veis, A. (1975) Calc. Tiss. Res. 19, 51-61.

31. Volpin, D. and Veis, A. (1973) Biochemistry 12, 1452-1464.

32. Dryer, R.L., Thomas, A.R. and Routh, J.I. (1957) J. Biol. Chem. 225, 177-183.

33. Ames, A. and Nesbitt, F.B. (1958) J. Neurochem. 3, 116-126.

34. Liu, T.Y. and Chang, Y.H. (1971) J. Biol. Chem. 246, 2842-2848.

35. Bitter, T. and Muir, H. (1962) Anal. Biochem. 4, 330.

36. Scott, P.G. and Veis, A. (1976) Conn. Tiss. Res. 4, 117-129.

37. Termine, J. and Miyamoto, M. (1977) unpublished results, private communication.

38. Glimcher, M.J., Friberg, U.A. and Levine, P.T. (1964) Biochem. J. 93, 202-210.

39. Glimcher, M.J., Mechanic, G.L. and Friberg, U.A. (1964) U.A. Biochem. J. 93, 198-202.

40. Seyer, J.M. and Glimcher, M.J. (1971) Biochem. Biophys. Acta 236, 279-291.

41. Dimuzio, M.T. (1976) Doctoral Dissertation, Northwestern University Graduate School.

42. Dimuzio, M.T. and Veis, A. (1977) Calc. Tiss. Res., submitted.

43. Weinstock, M. and Leblond, C.P. (1973) J. Cell. Biol. 567, 838-845.

44. Carmichael, D.J. and Dodd, C.M. (1973) Biochim. Biophys. Acta 317, 187-192.

45. Nawrot, C.F., Campbell, D.J., Schroeder, J.K. and Valkenburg, M. (1976) Biochemistry 15, 3445-3449.

46. Li, Shu-Tung and Katz, E.P. (1977) Calc. Tiss. Res. 22, 275-284.

47. Price, P.A., Otsuka, A.S., Poser, J.W., Kristaponis, J. and Roman, N. (1976) Proc. Nat. Acad. Sci. U.S.A. 73, 1447-1451.

Calcium Binding and Molecular Pathology of Elastin

M. M. Long, D. W. Urry and W. D. Thompson

Laboratory of Molecular Biophysics
and the
Cardiovascular Research and Training Center
University of Alabama Medical Center
Birmingham, Alabama 35294

Connective tissue of skin, ligaments, lungs, and the arterial tree derives its resiliency, in large part, from the restorative force of the elastic fiber, which contains a single protein, elastin. The molecular architecture which ultimately endows elastin with its retractive properties also provides it with a scaffolding for its own pathology; the elastic fiber has been shown to be a site for both calcification and lipid deposition in the arterial wall.[1,2] In 1971,[3] the concept of neutral site binding and charge neutralization as an initiation mechanism for elastin calcification was developed, based on the parallel between cation selective polypeptide antibiotics composed of L residues, D residues and hydrophobic, sterically bulky side chains, and the amino acid composition of elastin. Elastin has few amino acid residues with functional side chains (2%)[4] but many glycines (33%) and hydrophobic residues (65%). Relative to this distribution of amino acids, the neutral site binding/charge neutralization theory proposed that the initiating binding site for calcium was comprised of carbonyl oxygens of the peptide backbone of the elastin matrix in specific spatial orientations. Once bound to formally neutral but polarizable acyl oxygens, the calcium ions positively charged the elastin meshwork and thereby set the stage for charge neutralization by polyvalent counterions, for example, phosphate. In an in vivo setting this could lead to calcification of the elastic fiber via hydroxyapatite growth. It is the purpose of this paper to furnish further evidence in support of both stages of this theory, i.e. direct evidence for calcium chelation by the peptide moiety's carbonyls of elastin repeat peptides and direct evidence for calcification by the repeat pentamer peptide.

Approximately one half of the amino acid sequence of tropoelastin, the precursor protein of elastin has been determined,[5,6] and found to contain three series of amino acid units; a tetramer, a pentamer, and a hexamer each occurring basically, four, six, and five times in single runs. In addition, by amino acid analysis and partial sequencing, Keller, et al.[7] have recently shown that a 80 residue peptide isolated from ligamentum nuchae may be made up of the repeating pentapeptide. Our laboratory has synthesized these peptides and, as predicted,[3] they form β-turns and bind calcium ions with high affinity and selectivity. In particular, the pentamer sequence Val-Pro-Gly-Val-Gly (VPGVG) and its high polymer preferentially bind calcium ions as shown in the following circular dichroism, proton magnetic resonance and carbon-13 magnetic resonance studies. Additional data collected with scanning electron microscopy and x-ray microprobe analysis of the cross-linked pentamer high polymer provide visual evidence for its calcifiability.

All of the repeat peptides of elastin, as outlined above, chelate calcium ions. The pentamer and its high polymer is a case in point.

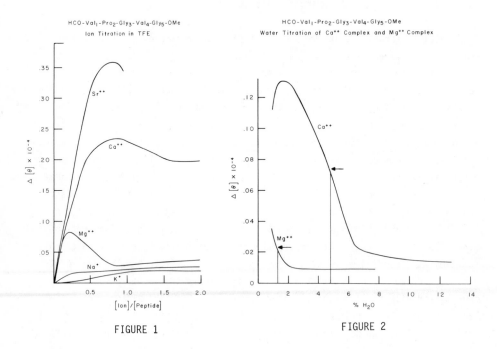

FIGURE 1   FIGURE 2

Figure 1 is an ion titration of HCO-V-P-G-V-G-OMe, where binding is monitored by following the sample's circular dichroism signal ($\Delta[\Theta]$) at 225 nm. $\Delta[\Theta]$ is the difference between the initial $[\Theta]$ value and the $[\Theta]$ value at any specific concentration of ion.[8] Experimentally, measured volumes of 0.1 M and 1.0 M stock aqueous salt solutions were mixed stepwise with a measured volume of peptide at 1 mg/ml in trifluoroethanol (TFE). Of the divalent ions, strontium bound the tightest, followed by calcium and magnesium. The initial slopes of the Ca^{++} and Mg^{++} curves were similar. Further in the titration, though, the magnesium plot became biphasic because as the aqueous ion aliquots were added to the TFE solution of peptide, the water concentration increased, and as it did, water oxygens competed successfully with the peptide for magnesium leading to decomplexation. Interaction with sodium and potassium was almost below the levels of detectability.

420

Although forward ion titrations with CD are rapid enough to allow many small incremental steps to be taken and require minimal amounts of salt and peptide, these titrations make difficult comparison of binding affinities for calcium and magnesium, for example, between two peptides. This is because different complexes can have different stoichiometries, different actual ion solubilization abilities and different sensitivities to the incrementing water concentrations as discussed above. The last problem can be taken advantage of, by using this water competition as a means to compare relative affinities. The more water a peptide-ion complex requires to back titrate to a specific point, the greater is its ability to compete with the water oxygens for the cation, i.e. the greater its affinity. Figure 2 is a plot of the CD data for the water titration of the pentamer-Ca^{++} complex and the pentamer-Mg^{++} complex. The complex is formed at an [ion]/[peptide] ratio of 0.25 so the peptide will be in excess. The arrow points to the midpoint in each complex's titration, where, it is interpreted, one-half of the ion is bound to the peptide and one-half to water. At this point the ratio $\frac{[H_2O]}{[C=O]}$ is calculated[9,10]

TABLE I

WATER TITRATION OF ION COMPLEXES OF REPEAT ELASTIN PEPTIDES

PEPTIDE	CALCIUM COMPLEX $\frac{[H_2O]}{[C=O]}$	MAGNESIUM COMPLEX $\frac{[H_2O]}{[C=O]}$
HCO-Val-Pro-Gly-Val-Gly-OMe	215	53
HCO-(Val-Pro-Gly-Val-Gly)$_n$-Val-OMe	350	97
HCO-(Val-Ala-Pro-Gly-Val-Gly)$_n$-Val-OMe	535	167

(Table I) as a way of normalizing for peptides with different numbers of residues and carbonyl groups. These data show the marked affinity of the pentamer for calcium and for calcium in relationship to magnesium. The CD data in Figures 1 and 2 provide evidence for calcium binding to this elastin repeat peptide; the following NMR data provide evidence that the calcium is bound to the carbonyls, i.e. direct evidence of neutral site binding.

Figures 3 and 4 depict the PMR ion binding studies of HCO-VPGVG-OMe in TFE with 3% water (to compensate for calcium's relative insolubility in TFE). The chemical shifts of each peptide NH and the formyl proton were followed as a function of [CaCl$_2$]/[Peptide] at constant [Peptide],[11] in Figure 3. The valine 1 and 4 and glycine 3 and 5 were not assigned because of their overlap at the initial point in the titration and the parameters necessary for the experiment. All the peptide NH's shifted downfield with calcium binding, indicating that the carbonyl of each respective peptide moiety interacted. These carbonyls are those

421

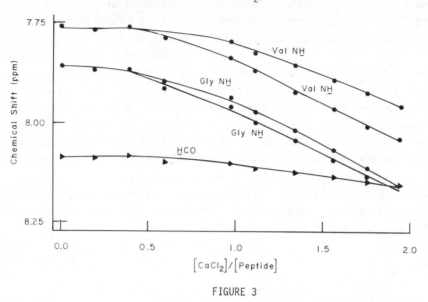

HCO-Val1-Pro2-Gly3-Val4-Gly5-OMe
CaCl2 titration
in TFE with 3% H2O
Chemical Shift (ppm)
7.75
8.00
8.25
Val NH
Val NH
Gly NH
Gly NH
HCO
0.0
0.5
1.0
1.5
2.0
[CaCl2]/[Peptide]

FIGURE 3

HCO-Val1-Pro2-Gly3-Val4-Gly5-OMe + CaCl2
in TFE
Chemical Shift (ppm)
7.75
8.00
8.25
Val NH
Val NH
Gly NH
Gly NH
HCO
0
20
40
60
Temperature °C

FIGURE 4

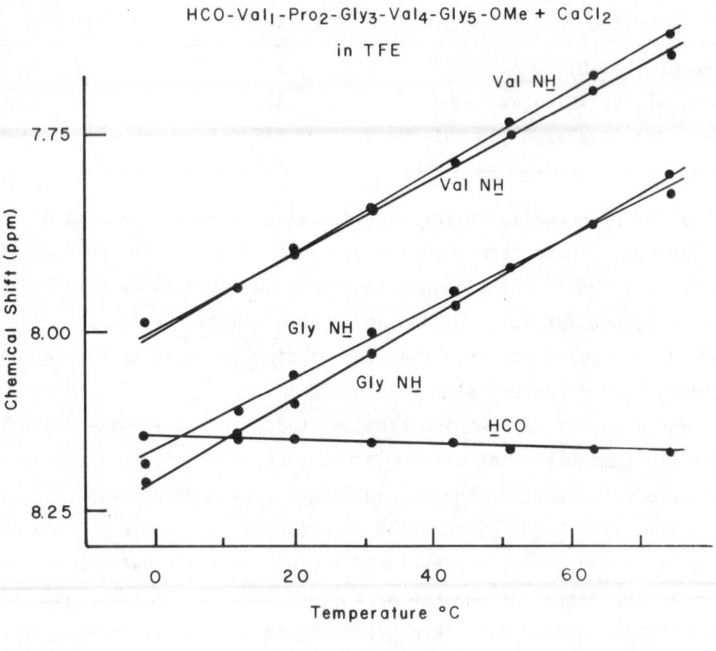

of the formyl group, Pro$_2$, Gly$_3$, and Val$_4$. There is no information about two carbonyls; the Val$_1$ carbonyl because of the imide nitrogen of Pro$_2$ and the Gly$_5$ carbonyl because it is part of the methyl ester. The temperature dependence of chemical shift of the pentamer-Ca^{++} complex is shown in Figure 4. The slopes are as follows in the order of their 0^0 intercepts: Gly N$\underline{H}$ -5.25 x 10^{-3} ppm/^{0}C, Gly N$\underline{H}$ -4.53 x 10^{-3} ppm/^{0}C, Val N$\underline{H}$ -5.15 x 10^{-3} ppm/^{0}C, and Val N$\underline{H}$ -4.77 x 10^{-3} ppm/^{0}C. Comparison with molecular standards of known solvent shielding and solvent exposure in TFE, gives the mole fraction of the shielded state of each peptide proton[12] as follows, again in order of their 0^0 intercepts: Gly N$\underline{H}$ 70%, Gly N$\underline{H}$ 90%, Val N$\underline{H}$ 70%, Val N$\underline{H}$ 80%. This high degree of solvent shielding could be due to hydrogen bonding or steric hindrance, limiting the NH's solvent exposure.

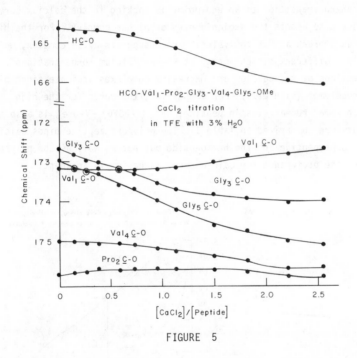

FIGURE 5

The stepwise titration of the pentamer's carbonyls is given in Figure 5[11] with assignments from reference 13. Four carbonyl carbon resonances shift downfield with calcium binding, a deshielding which could result from the electron withdrawing capacity of the calcium ions felt by both the carbonyl carbon and the amide proton of the peptide moiety.[11] These carbonyls are the formyl C=O, Gly$_3$ C=O, Val$_4$ C=O, and Gly$_5$ C=O. Calcium binding to the first three carbonyls was predicted by the PMR titration, no PMR data being available for the fourth. What is surprising is the Pro$_2$ $\underline{C}$-O data. Figure 3 indicated chelation by the Pro$_2$

carbonyl, while in Figure 5, there is seemingly, little indication of interaction, since there is little downfield shift. However the downfield shift could be obscured by changes in peptide conformation from the free to the totally bound state. In the absence of Ca^{++}, the Val$_1$ C-O is hydrogen bonded to the Val$_4$ N$\underline{H}$, placing the Pro$_2$ carbonyl in the end peptide moiety of a β-turn.[14] Because of upfield shifting due to magnetic anisotropy, a downfield shifting due to calcium binding could be balanced out. The actual upfield shift of the Val$_1$ C=O (for which there is no corresponding PMR data) probably does indicate, though, that this carbonyl is not greatly involved in calcium binding.

When polymerized, the pentamer interacts tightly with calcium (Figure 6). With this CD titration it is evident that strontium binds the best, followed by calcium and magnesium. Sodium and potassium show little interaction. As with the pentamer monomer, a sharp transition to an end point is lacking in the calcium curve, making it difficult to assess the stoichiometry of either complex. For the high polymer, there is a break at 0.5 equivalents which suggests a 2:1 complex, which then converts to a different stoichiometry at higher calcium concentrations. The water back titrations of the calcium and magnesium complexes indicate tight binding of the high polymer (Figure 7). For comparison, the curve for the high polymer of the hexamer repeat elastin peptide, HCO-(VAPGVG)$_n$-V-OMe, is also included. The data are summarized in Table I. The polypentapeptide binds calcium better than the pentapeptide and the hexapeptide but not as well as the polyhexapeptide. Again, the preference for calcium over magnesium is apparent.

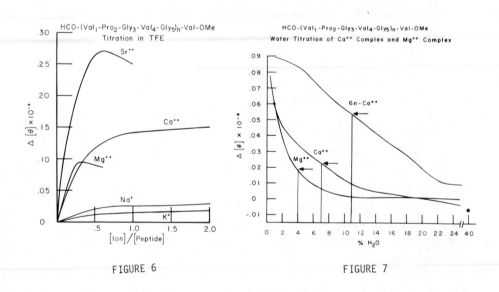

FIGURE 6   FIGURE 7

The NMR studies of the pentamer high polymer were restricted to the carbon-13 NMR (CMR) calcium titration of the peptide carbonyls (Figure 8). The PMR titration of the peptide N$\underline{H}$'s and subsequent temperature study of the complex are not included, because at 220 MHz the resonances were too broad to follow meaningfully. Again, as with the pentamer monomer, the CMR data provide definitive evidence for calcium binding to peptide carbonyls with the deshielding of Gly_3 $\underline{C}$=O, Val_4 $\underline{C}$=O, and Gly_5 $\underline{C}$=O upon calcium addition. Interpretation of the titration in terms of molecular structure is complicated by the biphasic aspect of the curve, suggesting two different complexes as a function of calcium ion activity. This was absent in the pentamer titration. There is a leveling near equivalences of 0.5 and 1.5.

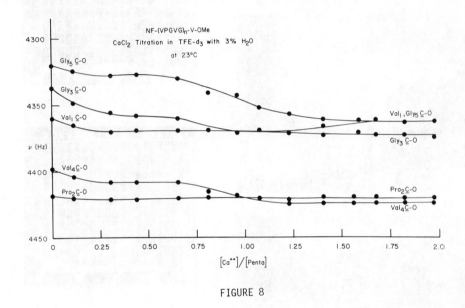

FIGURE 8

In summary, these two methods, CD and NMR afford substantive evidence that the repeat pentapeptide of elastin selectively and avidly binds calcium at neutral sites, i.e., at the acyl oxygens of the peptide moieties.

The second step in the neutral site/charge neutralization mechanism is the sequestering of polyvalent anions by the calcium chelated positively charged, elastin meshwork, and the subsequent growth of calcium phosphate deposits. The calcifiability of the pentamer high polymer, cross-linked to provide an insoluble matrix,[15] is a significant piece of data in support of the neutral site/charge neutralization theory. Figure 9a is an SEM photomicrograph of the cross-linked polypentapeptide which has been calcified in serum augmented with exogenous

425

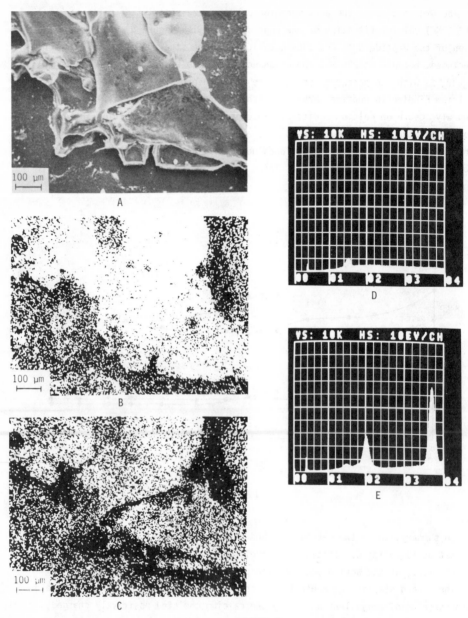

FIGURE 9

3.0 mM $CaCl_2$ and KH_2PO_4. Figure 9e is the x-ray elemental spectrum indicating
the presence of calcium at 3.7 keV, phosphorus at 2.0 keV, and aluminum (from the
coating) at 1.4 keV in the sample. This is in comparison to the spectrum, 9d,
which is an analysis of background support. Figures 9b and 9c are the calcium

and phosphorus maps respectively, demonstrating that calcium phosphate clearly deposited on the synthetic elastin framework. Calcification is not a surface phenomenon, it is a bulk property of the whole matrix as indicated by microprobe analysis of the thin, cross-sectioned calcified polypentapeptide, calcified from serum (with no additional calcium and phosphate) across dialysis tubing with 12,000 m.w. cut off.

The data presented in this paper support the neutral site/charge neutralization theory of the initiation of elastin calcification. They also provide a possible molecular framework to understanding in vivo elastin pathology.

ACKNOWLEDGMENTS: This work was supported, in part, by the National Institutes of Health, Grant No. HL-11310. The authors thank Dr. M. A. Khaled and Miss T. Trapane for assistance.

REFERENCES:

1. Wells, H. G., The Chemistry of Arteriosclerosis, in "Arteriosclerosis, A Survey of the Problem", 1933, ed. by E. V. Cowdry, McMillan, New York, pp 323-353.

2. Kramsch, D. M., Hollander, W. (1973) J. Clin. Invest., 52, 236-247.

3. Urry, D. W. (1971) Proc. Natl. Acad. Sci., 68, 810-814.

4. Starcher, B. C., Saccomani, G., Urry, D. W. (1973) Biochim. Biophys. Acta, 310, 481-486.

5. Foster, J. A., Bruenger, E., Gray, W. R. and Sandberg, L. B. (1973) J. Biol. Chem., 248, 2876-2879.

6. Gray, W. R., Sandberg, L. B. and Foster, J. A. (1973) Nature (London), 246, 461-466.

7. Keller, S., Mande, J., Birken, S., Canfield, R. (1976) Biochem. Biophys. Res. Commun., 70, 174-179.

8. Long, M. M., Ohnishi, T., Urry, D. W. (1975) Arch. Biochem. Biophys., 166, 187-192.

9. Urry, D. W., Starcher, B. C., Ohnishi, T., Long, M. M., Cox, B. A. (1974) Pathologie-Biologie, 22(8), 701-706.

10. Urry, D. W., Long, M. M., Ohnishi, T., Jacobs, M., Mitchell, L. W., in Calcium Transport in Contraction and Secretion, ed. by E. Carafoli, F. Clementi, W. Drabikowski and A. Margreth, North Holland Publishing Co., Amsterdam, pp 25-34, 1975.

11. Urry, D. W., Ohnishi, T. (1974) Bioinorg. Chem., 3, 305-313.

12. Urry, D. W., Long, M. M. (1976) CRC Crit. Rev. Biochem., 4, 1-45.

13. Urry, D. W., Mitchell, L. W., Ohnishi, T. (1974) Biochemistry, 13, 4083-4090.

14. Urry, D. W., Cunningham, W. D., Ohnishi, T. (1974) Biochemistry, 13, 609-615.

15. Okamoto, K., Urry, D. W. (1976) Biopolymers, 15, 2337-2351.

Role of Ca^{2+}-Binding Glycoprotein
in the Process of Calcification

B.de Bernard, N.Stagni, F.Vittur and M.Zanetti
Istituto di Chimica Biologica
Via Valerio 32, 34127 Trieste
Italy

The isolation of a glycoprotein with high affinity for Ca^{+2} ($K_d = 10^{-7}M$) from a calcifying cartilage (1) has raised the problem of establishing whether this molecule is somehow involved in the process of tissue mineralization. Our recent finding (2) that a similar glycoprotein, isolated from nasal septum, a non-calcifiable cartilage, is devoid of high affinity Ca^{+2}-binding sites, (see the figure) offers some grounds to our thesis that the former glycoprotein plays a specific role in calcification.

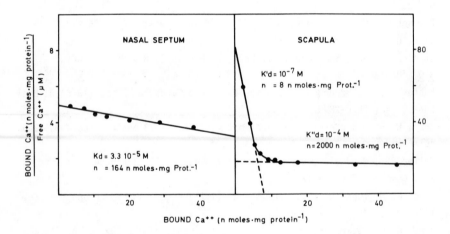

Fig.   Scatchard plots of Ca^{+2}-binding to glycoproteins from nasal septum and scapula cartilage. (For the experimental details, see (2)).

Even more promising in this sense is the observation that, beside binding Ca^{+2}, the glycoprotein exhibits an alkaline p-nitrophenyl phosphatase activity (E.C. 3.1.3.1.) (3,4). From the early report of Robison (5) to the most recent ones (6-14), the activity of an alkaline phosphatase has been suggested to be important for the process of cal-

Supported by CNR Italy

cification. In the attempt to better characterize the possible physiological role of a phosphatase activity associated to a Ca^{++}-binding
protein, various compounds were tested as substrates. These included
ATP and pyrophosphate, which are both highly implicated in the process of calcification (15, 16).

TABLE

SUBSTRATE SPECIFICITY OF CARTILAGE ALKALINE PHOSPHATASE

	pH 7.5			pH 9.0		
	-	Mg^{+2}	Ca^{+2}	-	Mg^{+2}	Ca^{+2}
p-nitrophenyl-phosphate	0.58	0.33	-	0.70	0.83	0.48
Glucose-6-phosphate	0.27	0.38	-	0.59	0.68	0.80
α-glycerophosphate	0.10	0.33	-	0.43	0.72	-
β-glycerophosphate	0.17	0.24	-	0.59	0.79	-
Pyridoxal-5-phosphate	0.05	0.13	-	0.27	0.42	-
Pyrophosphate	0.35	1.50	0.22	0.48	1.29	0.31
5'-AMP	0.40	0.40	0.95	0.96	0.92	0.50
ATP	0.53	1.10	0.90	0.50	1.41	1.55
Phosphatidic acid	0.00	0.00	-	0.07	0.15	-

Enzyme activity was measured as release of orthophosphate (17) from
various substrates at 5 mM concentration (with the exception of pyri-
doxal- 5 -phosphate = 3.75 mM; pyrophosphate = 2 mM; phosphatidic acid=
1 mg/ml). Each 50 μl assay contained 200 mM Tris·HCl (pH 7.5) or die-
thanolamine.HCl (pH 9.0) with and without 10 mM $MgCl_2$ or $CaCl_2$, and
40 μg glycoprotein, prepared according to Vittur et al. (18). The enzyme activity is expressed as nmoles of Pi/min/mg of protein.

The Table shows the ability of the glycoprotein to hydrolyze various substrates at both physiological and alkaline pH, in the absence and in the presence of Mg^{+2} and Ca^{+2}. The most important result is
that a high specific activity is observed with ATP and pyrophosphate
as substrates in the presence of 10 mM Mg^{+2}. When Ca^{+2} is used at the
same concentration, the ATPase activity is also stimulated (pH 9.0),
whereas the pyrophosphatase activity is slightly depressed.

An alkaline p-nitrophenyl phosphatase activity, covering ATP and
pyrophosphate hydrolysis, has been already reported in the literature, to the point that the three activities are considered by some investigators as catalytic expression of the same enzyme.

The interest of our finding is emphasyzed by the recent identification (11)of an alkaline phosphatase with the properties of ATPase

429

(E.C. 3.6.1.3.)and pyrophosphatase (E.C. 3.6.1.1.) in the Bonucci vesicles (19) of the cartilage, which are the site of early mineral deposition. Common features can be found in the catalytic properties of this enzyme (11) and those of our glycoprotein.

An important role of ATP and pyrophosphate in mineralization was suggested by Cartier and Picard (15). They found that a Mg^{+2}-requiring enzyme was present in sheep embryonic cartilage and that inactivation of the enzyme inhibited mineralization. These and other investigators (20) found that, among a number of possible phosphate donors, ATP was the most efficient. They postulated that calcium pyrophosphate was formed and that its hydrolysis by pyrophosphatase triggered mineralization. The presence of pyrophosphate in bone (21) and in mineral deposits in vitro (15) and the observation that pyrophosphate inhibits collagen-induced nucleation of crystals (16), provided the basis for the teory that pyrophosphate may inhibit calcification in vivo. The inhibiting effect of pyrophosphate was thereafter intensively studied by Fleisch and coworkers (22). Alcock and Shils (7) have shown that increased inorganic pyrophosphatase activity is associated with an increase in the uptake of ^{45}Ca by costal cartilage. The discovery of a glycoprotein which, from one side, hydrolyzes ATP and pyrophosphate and, from the other, binds Ca^{+2} with high affinity, strongly suggests that wherever early mineral deposits are formed the glycoprotein should be present.

REFERENCES

1. Vittur, F., Pugliarello, M.C. and de Bernard, B.(1972) Biochem.
   Biophys.Res.Commun., 48, 143-152.
2) Stagni, N., Vittur, F., Furlan, G., Zanetti, M., Picili, L., Co-
   lautti, I. and de Bernard, B. (1977) Biochemical Medicine, in press
3) Vittur, F. and de Bernard, B. (1973) FEBS Letters, 38, 87-90.
4) de Bernard, B. and Vittur, F. (1975) Colloques internationaux du
   C.N.R.S. N°230 Physico-chimie et cristallographie des Apatites
   D'Interet Biologique, pag.193-202.
5) Robison, R. (1923) Biochem.J., 17, 286-293.
6) Jibril, A.O. (1967) Biochim.Biophys.Acta, 141, 605-613.
7) Alcock, N.W. and Shils, M.E. (1969) Biochem.J., 112, 505-510.
8) Wöltgens, J.H.M., Bonting, S.L. and Bijvoet, O.L.M. (1970) Calc.
   Tiss.Res., 5, 333-343.
9) Arsenis, C., Rudolph, J. and Hackett, M.H. (1975) Biochim.Biophys.
   Acta, 391, 301-315.
10) Majeska, R.J. and Wuthier, R.E. (1975) Biochim.Biophys.Acta, 391,
    51-60.
11) Felix, R. and Fleisch, H. (1976) Calc.Tiss.Res., 22, 1-7.
12) Arsenis, C., Hackett, M.H. and Huans, S.M. (1976) Calc.Tiss.Res.,
    20, 159-171.
13) Jaffe, N.R. (1976) Calc.Tiss.Res., 21, 153-162.
14) Granström, G. and Linde, A. (1977) Calc.Tiss.Res., 22, 231-241.
15) Cartier, P. and Picard, J. (1955) Bull.Soc.Chim.Biol., 37, 1159-
    1168.
16) Fleisch, H. and Neumann, W.F. (1961) Amer.J.Physiol., 200, 1296-
    1300.
17) Baginski, E.S., Foà, P.P. and Zak, B. (1967) Clin.Chim.Acta, 15,
    155-158.
18) Vittur, F., Pugliarello, M.C. and de Bernard, B. (1972) Biochim.
    Biophys.Acta, 257, 389-397.
19) Bonucci, E. (1967) J.Ultrastruct.Res., 20, 33-50.
20) de Bernard, B. and Cartier, P. (1957) Bull.Soc.Ital.Biol.Sperim.,
    33, 1779-1782.
21) Perkins, H.R. and Walker, P.G. (1958) J.Bone Jt.Surg., 40 B, 333-
    339.
22) Fleisch, H., Schibler, D., Maerki, J. and Frossard, I. (1965)Na-
    ture, 207, 1300-1301.

Calcium Binding by Bone Cell Plasma Membranes

H. H. Messer
Division of Oral Biology
School of Dentistry
University of Minnesota
Minneapolis, Minnesota 55455

The skeleton serves as a major reservoir for calcium, with extensive movement of calcium between bone and extracellular fluid during bone formation, bone resorption and calcium homeostasis. The processes of calcium deposition in bone and calcium removal from bone are under cellular control, so that bone cells would be expected to possess efficient mechanisms for calcium transport. As part of a study of bone calcium transport, the calcium binding characteristics of bone cell plasma membranes are described here.

A plasma membrane fraction was obtained from bone cells derived from rat femoral and tibial metaphyses. The proximal tibial metaphysis and distal femoral metaphysis of 200 g rats were freed of adhering soft tissues, split longitudinally and rinsed with ice-cold 0.9% saline to remove marrow cells. The bones were then homogenized in a Waring blendor, filtered through cheesecloth to remove larger bone fragments, and re-homogenized in a glass homogenizer, in a cold (4°C) solution containing 0.25M sucrose, 0.02M NaCl, 5mM Na_2 EDTA, 1mM Mg^{2+} and 10mM imidazole. The plasma membrane fraction was isolated by a procedure similar to that described by Post and Sen[1] for renal plasma membranes (omitting the urea stage).

Calcium binding characteristics of the plasma membrane fraction were determined using a flow dialysis system[2] as described previously for calcium binding studies.[3] Protein concentration was approximately 1 mg/ml and the initial ^{45}Ca concentration approximately 0.6×10^{-5}M. Binding was measured at pH 8.0 and 22°C.

Two types of Ca^{2+}-binding sites were identified: high affinity sites with Ks = 2.6 ($\pm$ 0.1) $\times 10^{-5}$M and a capacity of 38 $\pm$ 4 nmoles Ca^{2+} per mg protein; and low affinity sites with Ks = 1.4 ($\pm$ 0.2) $\times 10^{-4}$M and a capacity of 90 $\pm$ 11 nmoles Ca^{2+} per mg protein.

Calcium binding was strongly pH-dependent. Very little binding occurred below pH 4, with a steep increase to a broad plateau at pH 6 - 10, and a further small increase to pH 12.

The affinity of Mg^{2+} and Sr^{2+} for the Ca^{2+}-binding sites was also examined by displacement of ^{45}Ca from the plasma membranes. The affinity of Mg^{2+} and Sr^{2+} were approximately 20-fold lower than that of Ca^{2+} for both the high and low affinity sites (Table 1).

TABLE 1

APPARENT Km VALUES FOR ^{45}Ca DISPLACEMENT BY Ca^{2+}, Mg^{2+} and Sr^{2+}

	Km_1 (mean ± S.E.)	Km_2 (mean ± S.E.)
Ca^{2+}	$1.1 \pm 0.1 \times 10^{-4}$	$3.5 \pm 0.4 \times 10^{-4}M$
Mg^{2+}	$2.3 \pm 0.3 \times 10^{-3}$	$6.8 \pm 0.8 \times 10^{-3}M$
Sr^{2+}	$4.0 \pm 0.2 \times 10^{-3}$	$5.6 \pm 0.4 \times 10^{-3}M$

The bone cell plasma membranes possess Ca^{2+}-binding sites with affinity, capacity and specificity comparable to other tissues or membranes in which Ca^{2+} transport is a major function (including sarcoplasmic reticulum[4], cardiac microsomes[5] and placental plasma membranes.[2]) Thus the binding sites described here could serve in the calcium transport mechanism of bone cells, assuming that the initial step in Ca^{2+} transport in a passive binding of Ca^{2+} to the membrane.

REFERENCES

1. Post, R.L. and Sen, A.K. (1967) Methods Enzymol. 10, 762-768.

2. Colowick, S.P. and Womack, F.C. (1969) J. Biol. Chem. 244, 774-777.

3. Shami, Y., Messer, H.H. and Copp, D.H. (1974) Biochim. Biophys. Acta 339, 323-333.

4. Chevallier, J. and Butow, R.A. (1971) Biochemistry 10, 2733-2737.

5. Repke, D.I. and Katz, A.M. (1972) J. Mol. Cell Cardiol. 4, 401-416.

The Distribution of Protein-Bound Calcium
in Normal Human Serum:
A Tightly Bound Subfraction Characterized

J. Wortsman, P. Franciskovich, R. Traycoff, L.E. Maroun
Southern Illinois University School of Medicine
P. O. Box 3926
Springfield, Illinois 62708

SUMMARY

The presence of nonexchangeable pools of calcium in human plasma was studied by successive ultrafiltrations. 7.5% of the total calcium could not be removed after extensive washes with 0.15 M NaCl. Treatment with EDTA or $(NH_4)_2SO_4$ gave similar results. This tightly bound calcium pool was characterized by ultracentrifugation on a linear sucrose gradient which showed most of the calcium co-sedimenting with albumin. A similar pattern was obtained when K-palmitate was added to either albumin or whole serum. The biological significance of this observation is unknown.

INTRODUCTION

Calcium in the body plays a central role in neuromuscular excitability. It has been shown that the ionized calcium pool is responsible for this action and constitutes about 40% of the total plasma calcium[1]. A significant proportion of the plasma calcium is found in a protein-bound form as albumin-calcium complexes which make up to 55% of the plasma calcium[2].

Little is known about the albumin-calcium complexes. Ultracentrifugation analyses have shown that the calcium-binding capacity of protein is approximately 0.10 mmole Ca per gram of protein and that the dissociation of the calcium-proteinate complex follows the mass law[2].

Studies performed by us in human subjects with chronic renal failure,[3] and by others in experimental animals,[4] have suggested the existence of nonexchangeable calcium compartments in plasma as demonstrated by radioactive calcium kinetics analyses. Since the only pool in the plasma calcium that could account for the disparity of the results observed was the protein-bound fraction, we decided to explore the possibility that this pool was non-homogenous. Using a combination of ultrafiltration and ultracentrifugation techniques, we have observed that a significant proportion of the calcium that is bound to albumin is found in a tightly bound fraction.

MATERIAL AND METHODS

Blood was drawn anaerobically for determination of ionized calcium (calcium flow-through electrode), total calcium (atomic absorption spectrophotometry), and ultrafiltration analysis. Ultrafiltration was performed in an Amicon apparatus using a Pellicon membrane with a cut-off point of 10,000 daltons. Two ml. of serum were filtered under a pressure of 80 psi. Sequential filtration cycles were performed after reconstituting the residual fraction up to its original volume. Ultracentrifugation was performed on a 200µL sample using a 10-40% sucrose gradient spun

434

at 50,000 rpm for 16 hours.

Albumin-calcium-palmitate complexes were prepared the following way: 7.6 milli-moles Palmitic acid were added to 15.3 millimoles of KOH. This mixture was incubated at $37^{\circ}C$ for 2 hours, mixed with 2.5 millimoles of $^{45}CaCl_2$ and then complexed with 0.65 millimoles of albumin.

RESULTS

Sequential washes of the original sample using 0.15 M NaCl showed that 7.5% of the calcium could not be removed. Similar observations were made using 0.1 M EDTA (Fig. 1), and ammonium sulfate (0.2 M - 1.4 M).

The tightly bound calcium fraction was shown to contain 3 subfractions sediment-ing at 2-3, 4-5 and 7-10S respectively (Fig. 2). Pure albumin sedimented at 4-5S, and $^{45}CaCl_2$ sedimented at 2-3S.

Potassium palmitate was added to whole serum to study the effect of free-fatty acids on the distribution of calcium. The addition of potassium palmitate produced a 70-80% decrease in ionized calcium. Precipitation of calcium as $Ca(OH)_2$ could only account for a 20-24% fall in ionized calcium (Table 1). Ultracentrifugation of this serum palmitate complex showed that most of the calcium co-sedimented in the 4-5S region along with the albumin marker (Fig. 3,4).

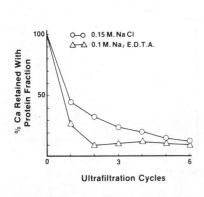

Fig. 1. Successive ultrafiltrations of protein-bound calcium. Volumes were re-constituted with 0.15 M NaCl; or 0.1 M Na_2 EDTA.

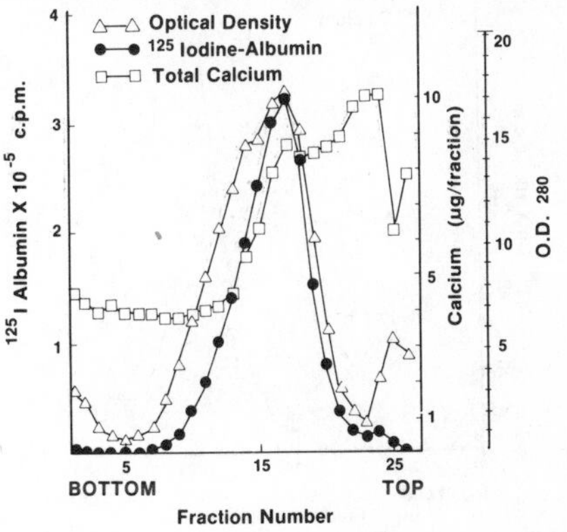

Fig. 2. Ultracentrifugation profile of tightly, protein-bound fraction.

DISCUSSION

Calcium in serum is present in three forms: ionized, protein-bound, and com-plexed. Forty percent of the calcium is ionized, and it is felt that this con-stitutes the physiologically active fraction. The roles of complexed calcium (5%) and protein-bound calcium (55%) are less well defined[1]. Evidence of a calcium-albumin complex in which calcium does not freely exchange with other compartments in plasma, could provide an explanation for the occasional finding of a heterogene-ous distribution of radioactive calcium in plasma[3,4].

435

The reaction produced by the addition of K-palmitate to physiologic concentrations of calcium and albumin resulted in a dramatic decrease in ionized calcium, suggesting that fatty acids may have a role in the strength of the calcium-albumin bond. This is further supported by the similarity in the ultracentrifugation profile between the tightly bound calcium fraction and the albumin-calcium-palmitate complex.

Subject No.	Basal Ca			Post KOH			Post K - Palmitate		
1	4.17	(±)	.031	3.20	(±)	.000	0.84	(±)	.006
2	4.36	(±)	.033	3.42	(±)	.009	0.86	(±)	.006
3	4.37	(±)	.033	3.44	(±)	.000	0.68	(±)	.007
4	4.56	(±)	.023	3.66	(±)	.009	1.10	(±)	.014

Table 1. Ionized calcium after K-palmitate (mg% ± S.D.)

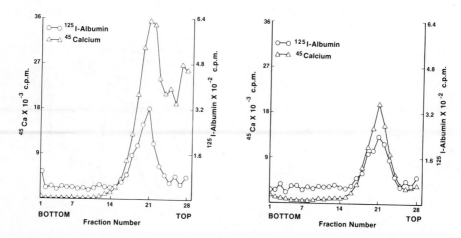

Fig. 3. Ultracentrifugation profile of human serum.

Fig. 4. Ultracentrifugation profile of human serum/K-palmitate.

REFERENCES

1. Potts, J.T., Deftos, L.J., (1974): In Duncan Diseases of Metabolism Vol. 2 W. B. Saunders, Philadelphia

2. Loken, H.F., Havel, R.J., et al. (1960): Ultracentrifugation Analysis of Protein-bound and Free Calcium in Human Serum. J. Biol. Chem. 235:3654

3. Wortsman, J., Benvenutto, X., et al. Calcium Dynamics in Chronic Renal Failure. To be published.

4. Giese, W., Comar, C.C., (1964): Existence of non-exchangeable Calcium Compartments in Plasma. Nature 202:31

INTRACELLULAR
CALCIUM-BINDING PROTEINS

Calcium-Binding Contractile Proteins in Protozoa

Lewis M. Routledge and W. Bradshaw Amos
Department of Zoology
Cambridge University
Cambridge, CB2 3EJ England

INTRODUCTION

In first level biology courses the Protozoa are often introduced by the characteristic organisms *Amoeba*, *Paramecium* and *Vorticella*. These typical protozoa exibit
interesting mechanisms of motility in which calcium is intimately involved (Figure
1a). Amoeboid movement has facinated scientists for centuries and the conflicting
theories concerning the site of contraction, for pseudopod formation, has produced
heated controversy. The biochemical and cytological studies[1] together with the
demonstration of actin and myosin filaments in *Amoeba* cytoplasm[2] indicate that
the same major elements responsible for muscle contraction are involved in amoeboid
movement. The low concentration of myosin relative to actin and the sol to gel
transition of actomyosin in cytoplasmic extracts suggest that thedetailed mechanism
of pseudopod formation is very different from that of muscle contraction. The sol
to gel transition of cytosol for *Acanthamoeba* requires MgATP and the contraction
of this gel requires calcium ion (Kd=10^{-7}M).[1]

Ciliary movements as illustrated by swimming in *Paramecium* (Figure 1b) are
also controlled by calcium. Naitoh and Kaneko[3] have shown that Triton-treated
Paramecium swim forward in MgATP solutions containing less than 10^{-7}M calcium ion
and backwards in solutions containing higher calcium ion concentration. This
reversal is the basis of the avoidance reaction in this organism in which stimulation of the front of the cell induces a membrane calcium current, leading to
local ciliary reversal. The site of calcium binding and the mechanism by which
calcium exerts its effect on the cilia are at present unknown.

A third mechanism of movement is the main topic of this paper; this is the
spasmoneme-based contraction (Figure 1c). The spasmoneme is a contractile organelle found in numerous ciliated protozoa, and is responsible for the very rapid
contraction of these organisms. The heterotrich ciliates, *Stentor*, *Condylostoma*
and *Spirostomum* undergo a change in body length with rates of up to 100 cell
lengths per second.[5] The peritrich ciliates, *Vorticella*, *Carchesium* and *Zoothamnium* have contractile stalks which coil helically or fold at rates of up to 170
lengths per second.[4] For comparison the very rapid mouse finger muscle only
contracts at 22 cell lengths per second.[6] The contraction of the spasmoneme takes
place in 2 to 10 msec. but the extension phase is slow, requiring several seconds
or even minutes before the extension is complete. The term 'spasmoneme' is used
due to the spasmodic contraction of the organelle and is preferred to the term
'myoneme' used in the older literature. As will be shown subsequently, the organelle does not resemble muscle either physiologically or biochemically and the
term 'myoneme' is thus misleading.

THE PHYSIOLOGY OF THE SPASMONEME CONTRACTION

The vorticellids are peritrich ciliates, the cell body bearing a ring of cilia which are used to create feeding currents. In *Vorticella* the cell body is anchored to the substrate by a stalk which contains the contractile organelle (Plate 1). The stalk is thrown into a tight helical coil when the animal contracts and this is presumed to be an escape reaction. Large vorticellids exist which are colonial, having several hundred cell bodies and contractile stalks attached to a common stalk. Two of these large colonial organisms *Carchesium polypinum* and *Zoothamnium geniculatum* have been used extensively for physiological and biochemical studies on the contractile process.

Birefringence: In striated muscle the *A*-band is highly birefringent (2.3×10^{-3}) due to the ordered arrangement of myosin filaments. This birefringence is present whether the sarcomere is contracted or relaxed. The spasmoneme is also birefringent with a value of 4×10^{-3} for the intact spasmoneme of *Carchesium*.[7] As first reported in 1940[8] the birefringence of the spasmoneme falls dramatically when the animal contracts, and in the fully contracted organelle the birefringence is abolished.[7,9]

The Energetics of Contraction: From high speed films it was found that *Vorticella* contracts in only 4 msec.[10] In *Zoothamnium* the organelle is large enough to measure accurately in high speed films; it contracts to 45% of its initial length in under 5 msec.[4] The tension per unit of spasmonemal cross-sectional area produced when a living specimen of *Carchesium* contracts has been measured[11] and is between 4×10^4 and 8×10^4 Nm^{-2}. This is similar to the indirect estimate made from the high speed film data.[7] When *Vorticella* contracts, the cell body, which may be approximated to a sphere of radius 20μm, is pulled 80μm at an average velocity of 23mm sec^{-1}. Assuming a viscosity of 10^{-3} Nsec m^{-2} (0.01 poise), the viscous drag on the body may be calculated from Stokes formula ($F = 6\pi a \eta \upsilon$) as 0.86×10^{-8}N. If the radius of the spasmoneme is taken as 0.5μm, the tension per unit cross-sectioned area becomes 1.1×10^4 Nm^{-2} (a minimum value since it takes no account of energy lost in overcoming the resistance of the stiffening fibers or the internal viscosity of the stalk).

These observed and calculated tensions in the spasmoneme are low in comparison to the maximum isometric tensions of muscles from a variety of animals, which are in the range 10^5 to 10^6 Nm^{-2},[12] but the spasmoneme tension is probably of the same order as the tensions in working muscles. It is the speed of shortening that marks out the spasmoneme as quite different from muscle. Because of this high rate of shortening the spasmoneme achieves a high power output (2.7 kW per kg of wet weight) but this is maintained for only a small fraction of the contraction/relaxation cycle. Repeated contractions can occur in *Vorticella* but they are separated by periods of extension at least 500 times longer than the contraction phase.

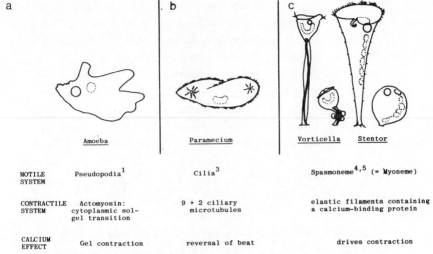

	Amoeba	Paramecium	Vorticella Stentor
MOTILE SYSTEM	Pseudopodia[1]	Cilia[3]	Spasmoneme[4,5] (= Myoneme)
CONTRACTILE SYSTEM	Actomyosin: cytoplasmic sol-gel transition	9 + 2 ciliary microtubules	elastic filaments containing a calcium-binding protein
CALCIUM EFFECT	Gel contraction	reversal of beat	drives contraction

Figure 1. Motile Systems in Protozoa

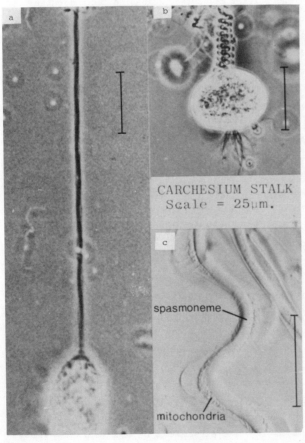

CARCHESIUM STALK
Scale = 25μm.

spasmoneme

mitochondria

PLATE 1. Contraction in glycerinated Vorticella. (a) Extended organism at 10^{-8} M Ca^{2+}. (b) Same organism contracted at 10^{-6} M Ca^{2+}, phase contrast optics, bar = 25 μm. (c) Partially extended stalk of *Carchesium*, mitochondria can be seen in the layer of cytoplasm around the spasmoneme. Nomarski differential interference optics.

<u>The calcium driven contraction</u>: Levine reported in 1956 that calcium and apparently also magnesium and manganese ions could induce contraction in glycerinated *Vorticella*.[13] He found that the contraction could be reversed by treatment with the calcium-chelating agent ethylene diamine tetra-acetic acid (EDTA) and, most significantly, the cycle of contraction and relaxation could be repeated many times by repeating the treatment, without the need to add ATP or any other apparent fuel. Hoffmann-Berling (1958) made similar experiments,[14] but added EDTA to the solution at a slightly lower concentration than the divalent ion under test in order to reduce the concentration of calcium ions which he suspected were present as a contaminant. With this precaution, he found that Mg^{2+} and Be^{2+} were not effective in inducing contraction. Less than 10^{-5} M Ca^{2+} was sufficient to cause contraction. Among the other alkaline earth ions, Sr^{2+} was effective, Ba^{2+} less so.

Hoffmann-Berling discovered that these effects of calcium were not prevented by cyanide or by the mercurial poison salyrgan (mersalic acid). He suggested that the molecular basis of the effect might be that calcium ions were able to neutralise negative charges on long polymer molecules, releasing the molecules from electrostatic repulsive forces and allowing them to fold thermokinetically. He observed more complex effects when ATP and magnesium ions were present in the solution but it was subsequently found[7] that these effects were abolished if the calcium level was adequately controlled by an EGTA buffer (EGTA = ethyleneglycol bis (β-aminoethyl ether) -N,N' -tetra-acetic acid). For the convenience of other workers, the apparent association constants of the calcium and magnesium complexes of EDTA and EGTA have been tabulated.[15]

In the presence of a Ca-EGTA buffer, the glycerinated *Vorticella* stalks coiled if the free calcium level was above 4×10^{-7} M and remained coiled indefinitely if the calcium level was kept high, implying that the spasmoneme maintained tension indefinitely under these circumstances, since the elastic sheath was present. If the calcium level was lowered to 10^{-8} M the stalks extended. In these glycerinated preparations the stalks took several seconds to contract and extend. It was confirmed that the stalks could be taken through many successive cycles of contraction and extension, as many as 35 being recorded over a period of 2 days at room temperature. Detergents (digitonin, saponin, tween 80), metabolic inhibitors (KCN, dinitrophenol, fluorodinitrobenzene) and mercurials (mersalic acid, parachloromercuribenzoate) were without effect on the number of cycles that could be obtained, even when present continuously throughout the experiment. Lanthanum and terbium induced contraction but the threshold concentrations were not measured.

Magnesium ions did not alter the threshold level for calcium, even when present in great excess, though a high concentration of magnesium (5×10^{-2} M) blocked contraction irreversibly. It seems unlikely that enough endogenous ATP was present to drive the contractile process in these experiments by some mechanism which is resistant to such a wide range of inhibitors. However, it may seem at first sight that the alternative hypothesis, that the difference in chemical potential of calcium ions which exists between the high and low calcium solutions serves as the source of energy, requires an unreasonably large amount of calcium to be bound by the contracting organelle. The following argument shows that this is not so.

The energy output of the spasmoneme in a single contraction can be calculated from the velocity and tension [7] as previously mentioned. It is 11 joule per kg of wet weight. The glycerinated spasmoneme progresses from almost full extension to contraction as the calcium level is increased over a 100 fold range (see Fig.5). It is reasonable to assume that the cell could vary its internal calcium ion level a 100-fold, for instance from 10^{-8} to 10^{-6} M, which would involve a chemical potential change of approximately 10^4 joule per mole of calcium ions. This is calculated from the equation

$$\Delta\mu_{Ca} = RT \log_e \frac{[Ca^{2+}] \text{ upper}}{[Ca^{2+}] \text{ lower}}$$

where $\Delta\mu_{Ca}$ is the change in chemical potential, R the gas constant and T the absolute temperature. To produce the observed output of energy the spasmoneme must bind at least $11 \div 10^4$ mole of calcium per kg of wet weight, or 0.04 g kg^{-1}: a modest amount. It can be seen that to produce such an effect, a cell need not generate a high concentration of calcium ions. It need only be able to vary the level over a large range, which the calcium pump in the sarcoplasmic reticulum of muscle is known to be capable of doing[16].

Calcium binding to the spasmoneme: When the concentration of free calcium ion was changed from 10^{-8} to 10^{-6} M, the glycerinated spasmoneme from *Zoothamnium* contracted to at least 60% of the extended length. This contraction was completely reversible. The organelle from *Z. geniculatum* is large (1 mm x 30-40 μm) when compared to that of a large *Vorticella* (1 mm x 1 μm). It can be handled with fine forceps, and its form was well preserved during preparation for microprobe analysis. The organelle could be seen clearly in both transmission and back-scattered electron images (Plate 2a). It was easy to position the organelle so that a rectangular median area could be selected for analysis (Plate 2d).

Line scans were also made across the organelle. In Plate 2b, the horizontal line is the line of scan and is also the baseline for the other traces, after subtraction of the background from the supporting film. In the extended organelle (Plate 2b), the upper trace M is proportional to the total mass present in the line of scan measured by an energy-dispersive silicon detector. Trace Ca is the signal obtained by a diffracting spectrometer set at peak Kα-radiation for calcium. Trace B is the signal from a diffracting spectrometer offset from the Ca Kα peak to record the background component. The amount of calcium present in the spasmoneme is given by the area between the traces Ca and B. In the contracted spasmoneme (Plate 2c), the background trace B was not significantly different from that obtained with the extended organelle and is omitted. Though the scales for the calcium and mass fraction counts are arbitrary, it can clearly be seen that the ratio of calcium to dry mass in the contracted organelle is much higher than in the extended organelle, even though both organelles were dried down in a calcium buffer containing the same amount of total calcium.

It is difficult to quantitate the calcium mass fraction from line scans because of the low level of X-ray counts from the small area involved. For quantitative measurements, small rectangular areas were selected along the organelle (Plate 2d) and scanned by an electron beam moving on a raster.

The amount of calcium in the contracted and extended spasmoneme at varying total calcium concentrations (including Ca-EGTA) is shown in Figure 2. Each point shows the mean and standard error of measurements for five or six spasmonemes, with four separate areas assayed within each spasmoneme. The calcium was presumed to arise from two sources: binding of calcium to sites within the organelle and accretion principally of Ca-EGTA during drying down. The effect of accretion should be directly proportional to the total calcium concentration in the buffer. The near linear decrease in calcium content of the extended spasmoneme with decreasing total calcium concentration in the buffer (Fig. 2) suggests that the calcium content of the extended organelle is indeed an accretion artifact. The difference in calcium content between contracted and extended organelles remains approximately constant and must represent a true calcium-binding. The amount of calcium bound (Table 1) over a 100-fold total variation in total calcium (20 µM-2 mM) was between 1.12 and 1.70 g calcium per kilogram dry mass of the spasmoneme. In order to relate these results to the hydrated state of the organelle, the dry mass concentration of a hydrated, glycerinated spasmoneme was measured in an interference microscope and was found to be 21 g/100 ml. The calcium binding therefore corresponds to an upper limit of 0.36 g/kg spasmoneme wet mass. This is more than eight times the amount needed to supply the work done against viscous forces on the chemical potential theory, which, as mentioned above, requires $0.04 \ g \ kg^{-1}$.

SPASMONEME PROTEINS

Our work on the protein composition of the spasmoneme has furnished the most direct evidence that this motile system is fundamentally different from others such as muscular contraction or flagellar beating.[18,19,20]

Zoothamnium geniculatum has proved the most convenient species for chemical work since the contractile material forms a high proportion of the total dry mass of the colony and a substantial part of it may be isolated by dissection after glycerination. The isolated preparation, consisting chiefly of the spasmoneme, but coated with a thin layer of cytoplasm, shows only a feeble periodic acid-Schiff reaction and probably consists almost entirely of protein. Unlike a myofibril, it is insoluble in KCl solutions but it is evidently not covalently

444

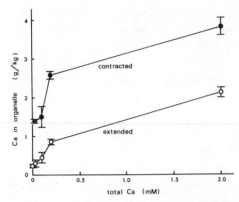

TABLE 1. *Microprobe measurements of calcium content of isolated, glycerinated spasmonemes of Zoothamnium geniculatum*

Total calcium ($M \times 10^{-6}$)	Calcium in organelle (g/kg dry mass)		Calcium bound during contraction (g/kg dry mass)
	contracted	extended	
0	–	0.21(0.04)	–
21	1.39(0.06)	0.27(0.07)	1.12(0.09)
104	1.58(0.27)	0.46(0.13)	1.12(0.30)
208	2.59(0.09)	0.89(0.06)	1.70(0.11)
2075	3.81(0.22)	2.12(0.12)	1.69(0.25)

Figures in parentheses are standard deviations.

Figure 2. Variation in calcium mass fraction of spasmonemes with total calcium concentration of the bathing solution. The free calcium ion concentration was adjusted to 10^{-6} M to produce contraction or to 10^{-8} M to produce extension. The standard error limits are indicated. (Reprinted, with permission, from Routledge *et al*. 1975.)

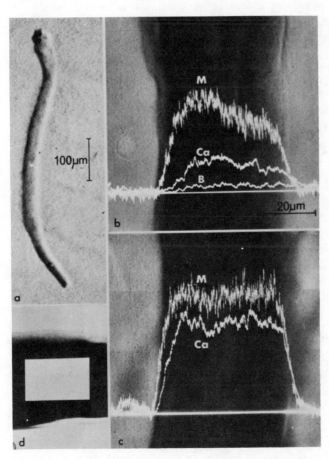

PLATE 2. (a) Secondary electron image of a spasmoneme formed in the JEOL JXA-50A micro-analyzer. The specimen was dried down from a solution containing 10^{-8}M free calcium ions and is extended. (b) Counting rates recorded during a linear scan across an extended spasmoneme and the supporting film. M is the total dry mass fraction; Ca, the calcium mass fraction; B, the background component (see text for details). The total concentration of calcium, including Ca-EGTA, was 0.2 mM. (c) As above, but with a contracted spasmoneme. It is evident that the calcium mass fraction is much greater in this case. (d) Transmission image of a spasmoneme with a superimposed image of the raster covering a rectangular median area of the specimen. The area is 40 x 26 μm^2. (Reprinted with permission from Routledge et al., 1975.)

cross-linked since it can be dissolved totally in 1% sodium dodecyl sulphate (SDS) and partially in 8 M urea or 3 M guanidine hydrochloride (GuCl). In order to determine the molecular weights of the proteins present, small numbers of isolated spasmonemes were dissolved in 2% SDS and subjected to electrophoresis in 15% polyacrylamide. Supplies of *Zoothamnium* are unfortunately seasonal and limited, but since the dry mass of the spasmonemal material from a single colony is as much as 0.5 μg, it has been possible to perform electrophoresis on small numbers of colonies. A microslab gel apparatus was devised for this purpose.[21] A discontinuous SDS gel system was used, the resulting pattern (Plate 3b) showed a prominent band corresponding to a molecular weight of 20,000 which contained 60% of the stainable material. The band was not dissociated into components of lower molecular weight by preheating or adding mercaptoethanol. Most of the remaining protein had molecular weights above 100,000. It is particularly significant that when the spasmoneme sample was run in parallel with actin and tubulin in an SDS slab gel, no bands co-migrating with these proteins were found in the spasmoneme protein, that is, the basic elements of myofibrillar and flagellar motility are lacking. Since, during contraction the spasmoneme equals striated muscle in power output per unit mass[7] it is extremely unlikely that traces of actin or tubulin below the limits of detection are responsible for the contraction. The high power output suggests that our attention should be directed to the major protein components and in particular to the 20,000 M.W. band which is characteristic of the spasmoneme. A recent study[20] has shown that a similar band predominates in the patterns from the spasmonemes of *Vorticella* and *Carchesium*.

Recent work has shown that the 20,000 M.W. band contains a distinct class of calcium binding proteins of a new type, which we have named *spasmin*.[18] All the major proteins of the spasmoneme were found to dissolve in a 3 M GuCl solution, leaving a remnant from which no protein could be extracted by SDS. Since the proteins remained in solution when freed of salt by dialysis, the spasmin could be subjected to isoelectric focusing. It proved to be quite acid, with an isoelectric point of 4.7-4.8. In this respect it resembles troponin C. When spasmin obtained by GuCl extraction was subjected to electrophoresis in 15% acrylamide gels without SDS it was resolved into a fast component (*A*) and a slightly slower component (*B*). *A* proved to have a slightly lower molecular weight than *B* when the two were run in an SDS gel. The relationship between *A* and *B* is at present not clear, but since the proportions of *A* and *B* remain approximately constant it seems more likely that two types of spasmin are present than that *A* is a degradation product of *B*.

Interaction with calcium on acrylamide gels: Initially, because only microgram quantities of material were available, an indirect method for calcium binding to the spasmoneme proteins was tested; the effect of calcium on the electrophoretic mobility of the proteins. A series of experiments was performed with polyacrylamide gels containing Ca-EGTA buffers. In a gel buffered at 10^{-8} M Ca^{2+}, the spasmin formed a complex leading band consisting of a prominent peak with a trailing shoulder (see Fig. 3). In another gel with the free calcium ion concentration increased to 10^{-6} M, the leading band was markedly retarded, while the mobility of spasmoneme proteins other than spasmin was unchanged. Also, the leading band became more distinctly divided into two principal components at the higher calcium level (Fig. 3). These effects were not observed with magnesium ions in the same concentration range, though a decrease in mobility occurred with 5×10^{-4} M Mg^{2+}. The decrease in mobility, which is induced by calcium with a high degree of specificity, must be due to an alteration of charge or conformation. The effect seems likely to have a bearing on the mechanism of contraction, since it occurs in the same range of concentrations of calcium as contraction in glycerinated preparations.

The microprobe measurements allow a comparison of the amount of calcium bound during contraction with the amount of protein present. If it is assumed that all the calcium is bound to the spasmin, which has a molecular weight of 20,000 and constitutes between 40 and 60% of the dry mass, the number of calcium atoms taken up by each spasmin molecule is between 1.4 and 2.1.

Calcium binding in urea: The high affinity calcium binding proteins from skeletal muscle, parvalbumin[22] and troponin-c[23] bind calcium, even in the presence of 6M urea, suggesting that the protein-calcium complex is highly stable. The electrophoretic mobility of the spasmoneme proteins from *Carchesium* and *Zoothamnium* was examined in 6M urea gels[23] in the presence and absence of calcium. On polyacrylamide electrophoresis in 6M urea, with 25 mM Tris-glycine buffer pH 8.5 and 2 mM EGTA, two prominent bands are observed in the gel. When EGTA is replaced by 2 mM $CaCl_2$ the first band of the pair increased in mobility by 25%. This effect was more clearly demonstrated in two dimensional polyacrylamide electrophoresis with calcium in one dimension and EGTA in the other.[20] Any protein which interacts with calcium and changes its mobility is displaced from the diagonal distribution of proteins in the second dimension. This technique clearly showed that one class of spasmoneme protein changes its mobility in the presence of calcium even in 6M urea and that the other class of protein is not affected by calcium. By the use of two dimensional gel electrophoresis,[20] with calcium-containing urea gels in the first dimension and SDS in the second dimension, it was shown that the fast peak was the 20,000 M.W. spasmin (*A*) and that the slower peak was a heterodimer of a 20,000 and 18,000 M.W. spasmin. This binding of calcium in 6M urea suggests a possible structural relationship between spasmin and the calcium binding proteins of striated muscle, which behave similarly.

Amino acid composition: An amino acid analysis of microgram samples of spasmin[18] has been obtained (by F. F. Yew), using sensitive fluorimetric methods. The spasmin was first eluted from SDS gels and hydrolysed in acid. In one method, the amino acids were separated on an ion exchange column and allowed to react with *o*-phthaldehyde to produce fluorescent compounds.

The results (see Table 2) appear to be similar for the spasmins *A* and *B*, but

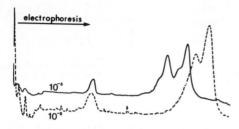

Figure 3. Reduction of the electrophoretic mobility of spasmin by free calcium ions: superimposed densitometer tracings of stained polyacrylamide gels which were run at 10^{-6} and 10^{-8} M free calcium ion concentrations. The concentrations were maintained by means of a Ca-EGTA buffer.

Table 2

Amino Acid Composition of the 20,000 MW Components from the Spasmoneme of *Zoothamnium*

Amino acid	Fast band (A)	S.E.	Slow band (B)	S.E.
Aspartic acid + asparagine	12.48	1.24	11.30	2.03
Threonine	7.95	0.17	6.20	0.94
Serine	15.05	4.15	12.58	2.25
Glutamic acid + glutamine	8.57	1.09	7.01	0.50
Proline	6.33	1.47	5.83	1.38
Glycine	9.32	2.07	8.90	2.08
Alanine	6.87	0.89	6.99	1.11
Valine	3.98	0.47	4.15	0.74
Cystine	0		0	
Methionine	0		0	
Isoleucine	2.82	0.26	4.08	0.53
Leucine	4.34	0.41	4.92	1.11
Tyrosine	3.89	0.80	3.30	0.76
Phenylalanine	1.95	0.37	2.20	0.74
Tryptophane	1.15	0.23	1.77	0.36
Lysine	6.95	1.10	7.85	1.35
Histidine	3.74	0.71	7.63	1.64
Arginine	4.59	0.71	6.33	2.42

Data from Amos, Routledge and Yew 1975. The results are expressed as residues per 100 residues. Proline was determined separately from the other amino acids. Each value is the average of six determinations. Standard errors are given.

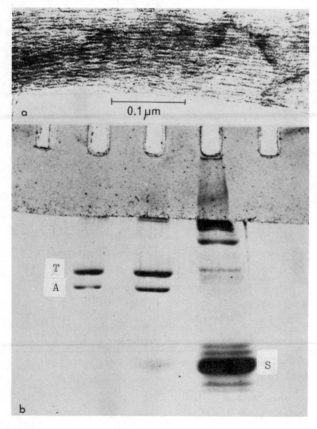

PLATE 3. (a) Fragment of a glycerinated spasmoneme of *Zoothamnium* negatively stained with uranyl acetate. The lighter areas probably represent filaments with a beaded structure.[19] (b) A 15% (w/v) polyacrylamide micro-slab gel, in which electrophoresis has been carried out on three samples. The right-hand one is a solution obtained by dissolving 10 *Zoothamnium* spasmonemes in SDS; on the left is a sample containing 0.2 μg each of actin (a) and tubulin (t); in the center is a mixture containing actin, tubulin and a small quantity of the spasmoneme sample. No band from the spasmoneme corresponds precisely to either tubulin or actin. The spasmin band (S; mol wt 20,000) is prominent. (Reprinted with permission from Amos, Routledge and Yew, 1975.)

it should be noted that these were not completely separated before analysis. Aspartic acid and glutamic acid, which are not distinguished by the method of analysis from asparagine and glutamine, were found to be abundant, as was serine. There were no significant unknown peaks in the chromatogram. We have found no correspondence between the amino acid composition of spasmin and of proteins from muscle. Unlike the calcium binding parvalbumin from carp muscle and troponin C, spasmin lacks cysteine.

Direct calcium binding measurements - Membrane binding assay: That calcium interacts with spasmin was shown indirectly by the change in mobility in calcium buffered acrylamide gels. This method was used in the initial studies because of the limited amount and heterogeneity of the spasmoneme proteins. Recently however, by using column chromatography coupled with ultrafiltration techniques it has been possible to both isolate and characterize the spasmoneme proteins.[24]

The spasmoneme proteins were extracted from between 300 and 1000 glycerinated colonies of Z. geniculatum, which had been freed of cell bodies. The extraction solutions (100 - 250μl) were 6M urea or 3M guanidine hydrochloride as in previous experiments and the samples either dialysed or applied directly to Biogel P60 columns (30 x 0.9 cm). The composition of the buffer and the elution profile is shown in Figure 4. Duplicate samples (50μl) were taken and analysed for calcium binding using UM 10 ultrafiltration membranes in an 8 channel pressure filtration cell (M.R.A. Boston, U.S.A.) using the procedure described by Paulus.[25] The difference in Ca^{45} bound at 10^{-8} and 10^{-6} M calcium ion in solutions containing identical total calcium concentration was taken as the amount of calcium bound to the protein. No calcium was bound to the peak of higher molecular weight proteins (>60,000) which appeared in the void volume, fraction 5 (Figure 5). The elution profile of the column was calibrated by standard proteins, including lysozyme, myoglobin, ovalbumin and bovine serum albumin. The calcium binding activity appeared as two peaks with apparent molecular weights of 52,000 and 28,000 with a trailing shoulder (Fig. 5). Samples from the column were concentrated and examined by gel electrophoresis, spasmins being present in both calcium binding peaks. When either the 52,000 or the 28,000 fractions were concentrated and rechromatographed on Biogel P150 columns they eluted with an apparent molecular weight of 28,000 and a Stokes radius of 2.3 nm. This strongly suggested that a monomer-dimer equilibrium exists between spasmin in solution and this was later confirmed by chemical cross-linking experiments.

Calcium binding to the isolated peaks was examined using ultrafiltration. A typical binding curve is shown in Figure 5 which is isolated spasmin monomer with a dissociation constant of 2×10^{-7} M Ca^{2+}. The number of moles of calcium bound per mole of spasmin (20,000) is between 2.42 and 3.70 with a mean value of 3.04 moles Ca^{2+}/mole spasmin. This preparation is still heterogenous, containing more than one spasmin isotype, which may bind differing amounts of calcium. As stated above, the electron microprobe measurements yielded a figure of 1.4 to 2.1 moles

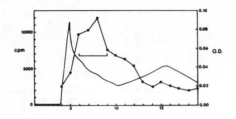

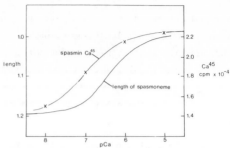

Figure 4. Elution profile of spasmoneme proteins from *Z. geniculatum* on a Biogel P60 column. Continuous line is the optical density at 280nm. Void volume fraction 5,total volume fraction 16: Buffer, 50mM KCl, 0.1 mM EGTA, 20 mM Imidazole pH 7.0. The calcium binding activity of the fractions as measured by ultrafiltration is indicated (o—o).

Figure 5. Calcium binding to isolated spasmin protein as a function of free calcium ion concentration (see text for details). The variation of unstressed length in the glycerinated spasmoneme of *Zoothamnium* is given along side for comparison.

Ca^{2+} bound/mole spasmin during contraction of the glycerinated spasmoneme, in fair agreement with the direct binding measurements. Figure 5 also contains a trace showing the length of the glycerinated spasmoneme at different calcium ion concentrations, the midpoint for the length change is 5×10^{-7} M Ca^{2+} which is also consistent with the calcium binding measurements.

CONCLUSION

Comparison with other contractile systems: We have shown both at the physiological and the biochemical level that the spasmoneme based contraction is entirely different from both actomyosin and tubulin based sliding filament contractility. Antibodies have also been prepared to spasmin and show no cross-reactivity with actin, myosin or tubulin.[26] The spasmoneme based contractile system represents a new contractile mechanism which has been characterized biochemically only in the vorticellid ciliate protozoa.[19] Ultrastructural and physiological studies indicate however that the spasmoneme-type contractile system is present in other ciliated protozoa including *Stentor*[5,9] and *Spirostomum*.[27]

The Mechanism of contraction: The exact mechanism of contraction of the spasmoneme is unknown but the available experimental evidence does point to a contraction mechanism. Cross-sections of the spasmoneme from Vorticellids show a highly refractile rod surrounded by a thin layer of cytoplasm and bounded by a plasma membrane (see also Plate 1c). The refractile rod is composed of 2 - 4 nm filaments with numerous 38 - 70 nm membrane bound tubules dispersed throughout the structure.[15] Calcium can be precipitated within the membranous tubules and these are believed to function in an analagous manner to sarcoplasmic reticulum.[28] Because of the rubbery texture of the spasmoneme it is difficult to obtain fragments thin enough for viewing after negative staining. Fragments can sometimes be obtained by soaking isolated glycerinated spasmonemes in distilled water before transferring them to saturated aqueous uranyl acetate as negative stain.

These appear to contain filaments of similar size to those seen in sections, they have an indistinct beaded appearance (Plate 3a) with a longitudinal periodicity of 3.5 nm.[15] On the grounds of abundance, it seems likely that spasmin is present in the filaments of which the spasmoneme is largely composed. Assuming a protein density of 810 daltons nm^{-3}, the diameter of a spherical spasmin molecule (20,000 daltons average M.W.) would be 3.6 nm, which compares well with the longitudinal spacing of 3.5 nm observed in negatively stained filaments. This suggests that the filaments may consist of linear aggregates of spasmin molecules. A linear arrangement has the attraction that it would result in the series summation of unit contractile events occuring independently in each molecule. This could explain the high shortening rate of the spasmoneme relative to muscle, where the units of series summation correspond to half sarcomers, and are therefore far fewer.

As yet no structural change in the filaments in the spasmoneme of vorticellids has been seen with the electron microscope. However, *Stentor* where the contractile fibers shorten by a factor of 7 rather than 3, contraction seems to involve a helical coiling of initially straight filaments[5,9]. A similar change in the filaments in vorticellids may occur as evidenced by a similar fall in birefringence on contraction[4,9].

In brief, the mechanism of contraction involves a stimulus, which causes release of stored calcium from sarcoplasmic reticulum-like structures. This calcium is bound to filaments of spasmin protein, which fold or coil rapidly. Extension, following contraction, involves a MgATP pump which removes calcium from the filaments causing them to extend actively and become aligned.

The origin of this mechanism of contraction is uncertain but the association of the spasmoneme with basal body structures[29] and the inhibition of stalk elongation by colchicine[30] suggest that spasmin may be a minor basal body protein who's function has become highly modified during the course of evolution.

In respect to the homology in amino-acid sequence, described for other high affinity calcium binding proteins during this symposium, it would be highly appropriate to examine the amino acid sequence of spasmin for calcium binding E - F hands.[31] This is a task which may, with some good fortune, be achieved. We have a suggestion however already, that this contractile system is not as highly conserved as that of actin and tubulin based contractility. Antibodies prepared to *Z. geniculatum* spasmin react only weakly with spasmin from *C. polypinum*[26] though ultrastructurally and in composition the spasmonemes appear to be similar.

ACKNOWLEDGEMENTS

We thank the Science Research Council for supporting this work. We would also like to thank Ms. V. Whittaker for excellent technical assistance on aspects of protein purification during this study.

451

REFERENCES

1. Pollard, T. L., Fujiwara, K., Niederman, R., and Maupin-Szamier, P. (1976). Evidence for the role of cytoplasmic actin and myosin in cellular structure and motility, in Cell Motility pp. 689 - 725. (eds. Goldman, R., Pollard, T. L. and Rosenbaum, T.) Cold Spring Harbor Laboratory.

2. Taylor, D. L., Condeelis, P. L., Moore, P. L. and Allen, R. D. (1973). The contractile basis of amoeboid movement. I. The chemical control of mobility in isolated cytoplasm. J. Cell Biol. 59:378 - 394.

3. Naitoh, Y. and Kaneko, H. (1972). ATP-Mg-reactivated Triton-extracted models of Paramecium: modification of ciliary movement by Ca^{2+}. Science 176:523-534.

4. Weis-Fogh, T. and Amos, W. B. (1972). Evidence for a new mechanism of cell motility. Nature 236: 301-304.

5. Huang, B. and Mazia, D. (1975). Microtubule and filaments in ciliate contractility, in Molecules and Cell Movement pp. 389-409 (eds. Inoue, S. and Stephens, R. E.) Raven Press.

6. Close, R. (1965). The relation between intrinsic speed of shortening and duration of the active state in muscle. J. Physiol. 180:542-559.

7. Amos, W. B. (1971). A reversible mechanochemical cycle in the contraction of Vorticella. Nature 229:127-128.

8. Schmidt, W. G. (1940). Die Doppelbrechung des stieles von Carchesium, inbesondere die optische-negative Schwankung seines Myonemes bei der Kontraktion. Protoplasma, 35:1-14.

9. Kristensen, B. T., Engdahl Nielsen, L. and Rostgaard, J. (1974). Variations in myoneme birefringence in relation to length changes in Stentor coeruleus. Exp. Cell Res. 85:127-135.

10. Jones, A. R., Jahn, T. L. and Fonseca, J. R. (1970). Contraction or protoplasm. IV. Cinematographic analysis of the contraction of some peritrichs. J. Cell Physiol. 75:9-20.

11. Rahat, M., Pri-Paz, Y. and Parnas, I. (1973). Properties of stalk "muscle" contractions of Carchesium sp. J. Exp. Biol. 58:463-471.

12. Prosser, C. L. and Brown, F. A. (1962). Comparative Animal Physiology, 2nd edn, p.433. W. B. Saunders Co.: Philadelphia & London.

13. Levine, L. (1956). Contractility of glycerinated vorticellae. Biol. Bull., III,319.

14. Hoffmann-Berling, H. (1958). Der Mechanismus eines neuen, von der Muskelkontraktion verschiedenen Kontraktionsyzklus. Biochim. Biophys. Acta, 27: 247-255.

15. Amos, W. B. and Routledge, L. M. (1976). The spasmoneme and calcium-dependent contraction in connection with specific calcium binding proteins. Symp. Soc. Exp. Biol. 30:273-301.

16. Hasselbach, W., Makinose, M. and Fiehn, W. (1970). Activation and inhibition

of the sarcoplasmic calcium transport. In Symposium on Calcium and Cellular Function, ed. A. W. Cuthbert, pp. 74-84. Macmillan: London.

17. Routledge, L. M., Amos, W. B., Gupta, B. L., Hall, T. A. and Weis-Fogh, T. (1975). Microprobe measurements of calcium-binding in the contractile spasmoneme of a vorticellid. J. Cell Sci., 19:195-201.

18. Amos, W. B., Routledge, L. M. and Yew, F. F. (1975). Calcium-binding proteins in a vorticellid contractile organelle. J. Cell Sci., 19 203-213.

19. Routledge, L. M., Amos, W. B., Yew, F. F. and Weis-Fogh, T. (1976). New calcium-binding contractile proteins, in Cell Motility pp. 93-113. (eds. Goldman, R., Pollard, T. L. and Rosenbaum, T.) Cold Spring Harbor Laboratory.

20. Routledge, L. M. (1977). Calcium binding proteins in the Vorticellid spasmoneme: extraction and characterisation by gel electrophoresis. J. Cell Biol. (in press).

21. Amos, W. B. (1976). An apparatus for microelectrophoresis in polyacrylamide slab gels. Anal. Biochem. 70:612.

22. Donato, H. and Martin, R. B. (1974). Conformations of carp muscle calcium-binding parvalbumin. Biochemistry. 13 4575-4579.

23. Head, J. R. and Perry, S. V. (1974). The interaction of the calcium-binding protein (troponin C) with bivalent cations and the inhibitory protein (troponin I). Biochem. J. 137:145-154.

24. Routledge, L. M. (1977). Calcium binding proteins in the Vorticellid spasmoneme: purification by column chromatrography and calcium binding properties. (in preparation).

25. Paulus, H. (1969). A rapid and sensitive method for measuring the binding of radioactive ligands to proteins. Anal. Biochem. 32 91-100.

26. Routledge, L. M. (1977). Antibodies to the contractile proteins of the Vorticellid spasmoneme. J. Protozool. abstracts British Section of Protozoologists (in press).

27. Hawkes, R. B and Holberton, D. V. (1975). Myonemal contraction of *Spirostomum*. II. Some mechanical properties of the contractile apparatus. J. cell. Physiol., 85:595-602.

28. Favard, P. & Carasso, N. (1965). Mise en evidence d'un reticulum endoplasmique dans le spasmoneme de cilies peritriches. J. Microscopie, 4 567-572.

29. Amos, W. B. (1972). Structure and coiling of the stalk in the peritrich ciliates *Vorticella* and *Carchesium*. J. Cell Sci., 10:95-122.

30. Rahat, M., Friedlaender, H. and Pimstein, R. (1975). Colchicine inhibition of stalk elongation in *Carchesium sp*: effect of Ca^{2+} and Mg^{2+}. J. Cell Sci. 19:183-193.

31. Kretsinger, R. H. (1977). The "EF-hand" concept and general implications of the calcium second messenger theory. pp. 63-72. In Calcium Binding Proteins and Calcium Function. (Wasserman et al., eds. New York: Elsevier North-Holland.

Mitochondrial Calcium-Binding Proteins

Ernesto Carafoli, Paolo Gazzotti, Klaus
Schwerzmann, and Verena Niggli, Laboratory of Biochemistry, Swiss Federal Institute of Technology (ETH), 8092 Zurich,
Switzerland

The translocation of Ca^{2+} across the inner mitochondrial membrane is
a complex process, which probably occurs through two separate path-
ways. One pathway is responsible for the influx of Ca^{2+} into the organelle, the other permits its efflux from it [1]. The influx pathway is probably energized by the negative electrochemical proton gradient maintained inside mitochondria by the operation of the respiratory chain [2], and is blocked by very low concentrations of the specific inhibitors La^{3+} [3] and ruthenium red [4,5]. The active influx
of Ca^{2+} into mitochondria is also abolished by uncouplers of oxidative phosphorylation (or by other agents that depress the energy
flow) and by the competitive uptake of other divalent cations like
Sr^{2+} [6]. The efflux pathway is stimulated specifically by Na^+ [7,1]
in heart and other mitochondrial types (but not, for instance, in liver or kidney mitochondria, where the mechanism of Ca^{2+} efflux is
still poorly understood). The Na^+-induced efflux pathway is not abolished by ruthenium red (Figure 1). The affinity of the influx route
for Ca^{2+} has been the subject of much debate, and appears now to be
conclusively fixed at values expressed by a K_m of about 10 μM [8].
The affinity of the efflux route(s) for Ca^{2+} has on the other hand
not yet been determined.

The specific inhibition of the influx pathway by low concentrations
of La^{3+} and ruthenium red, its competitive inhibition by Sr^{2+}, and
the high affinity of the system for Ca^{2+}, strongly indicate the existence of a specific molecular component involved in the transport of
Ca^{2+} into mitochondria. This component need not be a transmembrane,
mobile or immobile, Ca^{2+} carrier, since the kinetic parameters mentioned above could equally well be accounted for by the presence of

Abbreviations used:

EDTA: Ethylenediaminetetraacetic acid
SDS:　Na-dodecyl sulphate
RR:　 Ruthenium red

a specific, possibly superficial, Ca^{2+} receptor.

The specific effect of Na^+ on the efflux of Ca^{2+} from heart mitochondria, and the demonstration of the penetration of Na^+ into the organelle in exchange for the lost Ca^{2+} (1), on the other hand, convincingly prove the existence of a specific intramembrane Ca/Na carrier. One important characteristic of this carrier, which clearly sets it apart from that postulated for the influx pathway, is its insensitivity to ruthenium red (Figure 1).

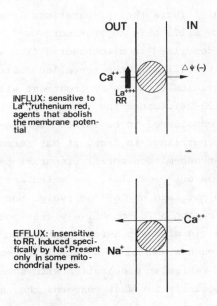

Figure 1. The transport of Ca^{2+} across the inner mitochondrial membrane

It is clear that the resolution of these systems for the transport of Ca^{2+} into their molecular components, and the reconstitution of the transport activity in artificial phospholipid membranes using solubilized components, would add greatly to the understanding of the mechanism of the transport of Ca^{2+}. This has been the case in other systems, one obvious example being the Ca^{2+}-dependent transport ATPase of sarcoplasmic reticulum (9-12). Reconstitution of various transport activities has now been obtained in a variety of membranes, including the inner mitochondrial one. Here it has been possible to reconstitute in artificial phospholipid bilayers the proton-pumping activity linked to the operation of the respiratory chain (13-15), and to the functioning of the "coupling" ATPase (16-18), as well as the transport of adenine nucleotides (19) and of inorganic phosphate (20). The proton-pumping activity has been reconstituted using impure submitochondrial fractions (the respiratory "complexes") and purified ATPase components. The transport of adenine nucleotides and phosphate has been achieved using partially purified fractions.

A description of the attempts to resolve the system(s) for the transport of Ca^{2+} across the inner mitochondrial membrane may begin with the findings by several Authors that preparations less complex than intact mitochondria are still able to transport Ca^{2+}. Active transport of Ca^{2+} has been described in mitochondria from which the outer membrane has been removed (21,22) (the so-called "mitoplasts"), and in submitochondrial vesicles obtained by treatment with detergents or by sonic oscillation (23-24). Since the ultrasonic submitochondrial vesicles are normally considered to be turned "inside-out", this last finding is rather puzzling. In fact, it has recently been found in this Laboratory that sub-mitochondrial preparations which contain practically only "inside-out" vesicles, and which are still capable of energy-coupling, do not take up Ca^{2+} actively. More recently, a submitochondrial preparation able to actively transport Ca^{2+} has been obtained with the aid of the non-ionic detergent lubrol (25); however, the preparation now exhibits an absolute requirement for inorganic phosphate. A vesicular preparation with somewhat similar characteristics, particularly for what concerns the phosphate requirement, has recently been obtained in this Laboratory using cholate as a solubilizing agent.

Interestingly, no success has been reported so far in attempts to resolve the Ca^{2+}-transport system below the level of these vesicular preparations. Attempts in this Laboratory to reconstitute the transport of Ca^{2+} by incorporating the so-called respiratory "complexes" into phospholipid bilayers, under conditions in which energy coupling, and other transport activities could be successfully reconstituted, have so far failed. The efforts have concentrated on complex III, which was reconstituted in vesicles formed by soybean phospholipids. The Ca^{2+} transport ability of these vesicles was tested using duroquinone as an electron donor, under a large variety of experimen-

tal conditions, including the presence or the absence of the mitochondrial Ca^{2+}-binding glycoprotein (see below), the inclusion of
Ca^{2+}-trapping anions inside the liposomes, the presence or the absence
of hydrophobic fractions obtained from the inner mitochondrial membrane (26). The failure of these attempts can be most conveniently
explained by postulating that the disassembly of the membrane which
takes place during the transition from submitochondrial vesicles to
the non-vesicular "complex" leads to the loss of one or more components that are evidently essential for the transport of Ca^{2+}.

The search for purified membrane components that could be parts of
the Ca^{2+}-transport system (possibly, the Ca^{2+} carrier) was initiated
in 1971, when Evtodienko et al. reported on the isolation from acetone powders of liver mitochondria of an ATPase which was specifically stimulated by Ca^{2+}, and could bind large amounts of it (27). Later,
other fractions were obtained from mitochondria with various treatments. They were purified and characterized to a variable extent, and
shown in all cases to possess the ability to bind Ca^{2+}. As shown in
Table 1, with one exception, they are protein in nature, and are wa-

Table 1
Mitochondrial Ca^{2+}-binding fractions

Author	Nature of Fraction	Affinity for Ca^{2+}
Evtodienko et al. (27)	Protein	high
Sottocasa et al. (28)	Glycoprotein	high
Gómez-Puyou et al. (29)	Glycoprotein	high
Kimura et al. (30)	Glycoprotein	low
Tashmukhamedov et al. (31)	Glycolipid	high

ter soluble. Very interestingly, they contain carbohydrates (in the
case of Evtodienko's Ca^{2+}-ATPase, no attempt to identify carbohydrates in the fraction has been reported). Of these various fractions, one has been purified to homogeneity in a cooperative effort
between this Laboratory and the Laboratory of G.L. Sottocasa (28),
and will therefore be described here in some detail.

The protein was solubilized by an osmotic shock procedure, applied
under conditions that minimized the leakage of proteins from the ma-

trix. The swollen mitochondria were removed by centrifugation, and the supernatant was suitably concentrated, and applied to a preparative polyacrylamide gel electrophoresis column. A fast-moving protein band was collected in 4 to 6 hours, after preliminary tests on analytical polyacrylamide gels had shown, by the use of specific stains, that this protein contained Ca^{2+}, and was glycoprotein in nature (28). It must be stressed at this point that at the time of these investigations no reports had yet appeared on the presence in mitochondria of Ca^{2+}-binding, carbohydrate-containing components. The decision to concentrate the attention on components of this type, was essentially due to the specific inhibitory effects of ruthenium red, a reagent supposedly specific for mucopolysaccharides and glycoproteins (32). Chemical analysis of the fraction showed that indeed it was a glycoprotein, containing about 10 % in weight sugars, including one sialic acid residue per mole. It also contained a variable amount of phospholipids. Molecular weight determinations yielded values varying from 30,000 to 33,000 depending on the animal species from which the glycoprotein was isolated. As expected from its fast migration velocity in gels, the protein was strongly acidic: more than one third of the aminoacid residues were contributed by glutamic and aspartic acids, with only negligible amounts of basic residues present. The protein contained rather large amounts of Ca^{2+} and Mg^{2+}, both of them firmly bound, since their level could be decreased only marginally by extended periods of dialysis. They could, however, be partly displaced by incubating the protein in the presence of excess La^{3+}. The glycoprotein was found to be contained in both the inner and outer membrane, and in the intermembrane space, but not in the matrix. Its presence in the intermembrane space, and its extraction with a mild osmotic shock, suggest its loose association wich the membrane environment. However, after the osmotic shock, additional glycoprotein could be extracted from the mitochondrial membranes by extensive sonication or by treatment with chaotropic agents. Evidently, part of the protein is either deeply embedded, and/or tightly bound to the membrane.

The protein bound Ca^{2+} at 2 classes of sites. The high affinity class contained 2 - 3 sites per mole of protein, and bound Ca^{2+} with an affinity corresponding to a K_d of between 0.1 and 1.0 μM. The low-affinity class contained about 20 - 30 sites per mole of protein, and had approximately 10 times lower affinity for Ca^{2+}. Interestingly, the reaction of the glycoprotein with Ca^{2+} was abolished almost completely by low concentrations of ruthenium red. Table 2 presents a summary of the properties of the glycoprotein.

Table 2

A summary of the properties of the mitochondrial calcium binding glycoprotein

Extraction	Osmotic shocks, additional extraction by sonication and by chaotropic agents.
Monomer MW	33,000
Carbohydrates	About 10 % (xylose, mannose, glucose, galactose, N-acetyl-glucosamine, N-acetyl-galactosamine, sialic acid).
Phospholipids	up to 33 %
Amino acids	About 1/3 glutamic and aspartic acid.
Tightly bound Ca^{2+} and Mg^{2+}	3 - 5 moles/mole
Ca^{2+} binding	2 - 3 sites/mole high affinity (K_d 0.1 μM). 20 - 30 sites/mole low affinity (K_d 10 μM).
Inhibition of Ca^{2+} binding	La^{3+}, ruthenium red.
Intramitochondrial location	Inner and outer membrane. Intermembrane space.

Also in line with a role of the glycoprotein in the transport of Ca^{2+} in vivo is the observation that no ability to bind Ca^{2+} with high affinity was found in a glycoprotein fraction isolated with the same technique from blowfly-flight muscle mitochondria, which, when isolated from adult flies, are essentially unable to transport Ca^{2+} actively (33). However, more direct, and perhaps conclusive, evidence in favour of the participation of the glycoprotein in mitochondrial Ca^{2+} translocation has very recently been obtained by Panfili et al. (34) who succeeded in preparing an antibody against the beef liver mitochondrial glycoprotein. Figure 2 shows that as little as 10 μg of the antibody, incubated for one hour with either intact mitochondria or mitoplasts, induced a very evident inhibition of the active transport of Ca^{2+}. Higher levels of antibody were required to obtain the same degree of inhibition in mitochondria with respect to mitoplasts, most

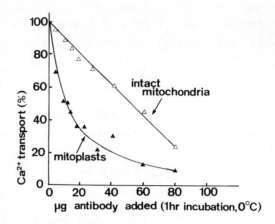

Figure 2. Inhibition of the transport of Ca^{2+} in liver mitochondria and mitoplasts by an antibody against the mitochondrial Ca^{2+}-binding glycoprotein. The preparation of the antibody, the preparation of mitochondria and mitoplasts, and the incubation conditions, are also described in (34). From Panfili et al.(34).

likely due to the outer membrane permeability barrier. Independent tests carried out by Panfili et al. (34) have shown that other energy-dependent (or independent) mitochondrial functions are not affected by levels of the antibody that produce a very evident inhibition of the transport of Ca^{2+}, thus confirming that the effect of the antibody is due to the interaction with a component specifically involved in the transport of Ca^{2+}.

It is thus rather logical to suggest that the glycoprotein has a role in the translocation of Ca^{2+} across the mitochondrial membrane. However, 2 possibilities could be envisaged: the protein could function as a carrier, which moves Ca^{2+} across the inner membrane to the matrix space, or it could act as a superficial Ca^{2+} receptor, or recognition site, located superficially on the inner membrane. This receptor could be associated only loosely, and reversibly, with the membrane domain, and could work in series with an (hypothetical) intramembrane carrier. (It is important to reiterate here that the sensitivity to inhibitors, and the other kinetic parameters of the Ca^{2+}-transport system, could be entirely accounted for by such a superficial receptor. As a consequence, then, the existence of a transmembrane Ca^{2+} carrier, even if still very probable, would not be an absolute necessity.) Experiments carried out on intact mitochondria by Sandri et al. (35), and on organic bulk phases by Carafoli (36) indeed indicate that the glycoprotein (or part of it) is loosely associated with the membrane, and suggest a mechanism for its reversible association with it. It has been found, 1) that the amount of glycoprotein which can be extracted from the membrane is increased greatly by enclosing EDTA in the extraction medium (Table 3) (35),

Table 3

Influence of Ca^{2+} on the degree of association of the glycoprotein with membranes

Conditions	Supernatant	Sediment	Total
	peak area (arbitrary units)		
Control (no EDTA)	75.4	85.3	160.7
+ EDTA	343.0	46.8	389.8

Experimental conditions: mitochondria from one rat liver (140 mg) were resuspended in 7 ml 10 mM triethanolamine-HCl buffer, pH 7.8 containing 10 ug rotenone. Aliquots of 1 ml were mixed with 0.1 mM EDTA, and left standing in ice for 10 minutes. Then, 1 ml 0.9 M sucrose, containing 2 mM $MgSO_4$ and 2 mM ATP, was added to each sample. After centrifugation for 30 minutes at 105,000 x g, the supernatant was subjected to polyacrylamide gel electrophoresis using 10 % gels. Staining was performed with toluidine blue as described by Sottocasa et al. (28). The sediments were suspended in 2 ml 0.2 M lithium di-iodo-salycilate in 50 mM Tris-HCl, pH 7.5 and extracted in this medium for 30 minutes, with gentle stirring at room temperature. After dialysis for 20 hours against the spacer gel buffer, the dialyzate was subjected to polyacrylamide gel electrophoresis and stained as described above. The densitometric traces were integrated by means of a Hewlett Packard Integrator Mod. 3770 B. (Modified from Sandri et al. (35)).

and, 2) that the glycoprotein, which is very hydrophilic, can be easily solubilized into apolar solvents by the addition of Ca^{2+} (Figure 3) (36). In agreement with the results of Gitler and Montal (37) on other hydrophilic proteins, it can be suggested that the protein becomes associated with the membrane via Ca^{2+} bridges (or penetrates into the membrane as a neutral species, due to the formation of ion pairs with Ca^{2+}). As Ca^{2+} is removed (to the outside by EDTA, or to the inside of the membrane by the pulling force of the negative-inside-membrane potential), the protein would be logically lost again

461

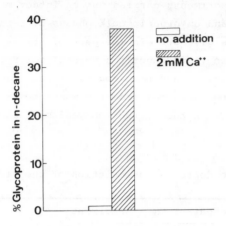

Figure 3. Ca^{2+}-induced solubilization of the Ca^{2+}-binding glycoprotein in n-decane. 2 mg of phosphatidyl-choline were dispersed in 4.0 ml distilled H$_2$O with the aid of a vortex-mixer (10 min.). 0.5 ml of an aqueous solution of the Ca^{2+}-binding glycoprotein (0.5 mg) were then added. The protein was extracted from osmotically-treated, soni-cated mitochondrial membranes, with 0.6 M lithium di-iodosalicylate (15 minutes, with gentle stirring, at room temperature). The inso-luble membranous material was removed by centrifugation, and the su-pernatant was dialyzed 24 hours at 0° C. The protein was then puri-fied by polyacrylamide gel electrophoresis (2). The tube was then stirred on the vortex-mixer for further 10 minutes. 1.5 ml n-decane were then layered on top of the aqueous medium; 0.5 ml of distilled H$_2$O plus or minus Ca^{2+} were added, and the mixing resumed for further 15 minutes. The separation of the phases was aided by centrifugation, and the protein remaining in the aqueous phase was then determined.

from the membrane into the intermembrane space. A general model for this mechanism is presented in Figure 4.

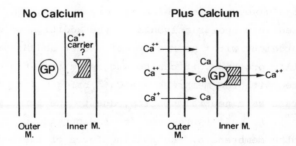

Figure 4. A model for the function of the Ca^{2+}-binding glycoprotein.

Experiments carried out on artificial phospholipid bilayers are in agreement with this interpretation. Using planar phospholipid bilayers, Prestipino et al. (38) have been able to show that the electrical conductance is increased by the addition of the glycoprotein, specifically in the presence of Ca^{2+}. The effect is abolished by ruthenium red (Table 4). However, when liposomes loaded with $^{45}Ca^{2+}$

Table 4

Effect of a mitochondrial Ca^{2+} binding glycoprotein on the electrical conductance of lipid bilayers

Addition	G_m $(\Omega^{-1}xcm^{-2}x10^{-8})$
Mitochondrial glycoprotein	10.1
Mitochondrial glycoprotein plus Ruthenium red	0.31
non-mitochondrial glycoproteins	0.09 - 0.28

The preparation of the bilayer and the details of the measurement are described by Prestipino et al. (1974). The G_m is the difference between the electrical conductance measured 22 to 33 minutes (depending on the different glycoprotein) after the addition of the glycoproteins, and that measured immediately before their addition. The glycoproteins were added at equal concentrations to both sides of the bilayer (1.48 to 5×10^{-7} M). The bathing solutions contained 4 mM $CaCl_2$ and 1 mM Tris-Cl, pH 7.8. Ruthenium red was 10^{-6} M.

were used, no efflux of Ca^{2+} (or, at best, a very limited efflux) was induced by the glycoprotein, added under a large variety of experimental conditions, designed to maximize its possible effect. The current interpretations of these results on artificial membranes is that the protein requires Ca^{2+} to become associated with the bilayer membrane, but does not significantly increase its permeability to it. The ruthenium red-sensitive increase in electrical conductance of the planar bilayer is evidently due to the penetration of the glycoprotein-Ca^{2+} complex into the bilayer. The electrical current, however, is not carried by Ca^{2+}, or carried by it only in a minor proportion.

Having thus concluded that the water-soluble glycoprotein probably operates as a superficial Ca^{2+}-recognition site, the problem becomes that of identifying, and possibly characterizing, the (hypothetical) Ca^{2+} carrier. The approach currently adopted in this Laboratory is based on the assumption that ruthenium red may bind not only to the water-soluble glycoprotein, but also to the other membrane components involved in the translocation of Ca^{2+}. A ruthenium red binding, water-soluble protein fraction was already isolated by Utsumi and Oda (39) from liver mitochondria treated with osmotic shocks. The approximate MW, as judged from molecular sieve experiments on Sephadex G-200, was higher than 68,000. On SDS polyacrylamide gels, the fraction dissociated into more than 6 bands. The fraction bound ruthenium red and contained sialic acid and hexosamines (56 nmoles and 200 nmoles per mg of protein, respectively). It thus very likely corresponds to an incomplete stage of purification of the Ca^{2+}-binding glycoprotein described above. In the approach followed in this Laboratory, mitochondria were first labelled with radioactive (^{106}Ru)-ruthenium red, and then the soluble proteins were removed by extensive sonication. The procedure, as expected, removed also a large part of the label, including the fraction of it bound to the Ca^{2+}-binding, water-soluble, glycoprotein. The membranous material left after sonication was further extracted with either lubrol or Triton X-100, a treatment that solubilized up to 90 % of the residual protein, but only about 20 % of the ^{106}Ru-label of the sonicated pellet. Thus, an insoluble protein fraction was obtained (Table 5) that was enriched

Table 5

Partial purification of a ruthenium red binding component from the membrane of rat liver mitochondria

	Total protein (mg)	Bound ^{106}Ru-RR (nmoles/mg prot.)	% of total added ^{106}Ru-RR
Mitochondria	342	0.1	50 %
Sonic particles	77	0.2	20 %
4 % Triton X-100 soluble fraction	69	0.06	10 %
4 % Triton X-100 insoluble fraction	6	1.4	10 %

Experimental conditions: rat liver mitochondria were resuspended (20 mg/ml) in a buffer containing equal amounts of 0.22 M mannitol, 0.07 M sucrose, 0.01 M Tris-Cl, pH 7.4, and 0.12 M KCl, 0.01 M Tris-Cl, pH 7.4. The suspension was sonicated for 3 minutes and then centrifuged to remove intact mitochondria. The supernatant was centrifuged at high speed (1000,000 x g, 30 minutes). The residual pellet was washed once and then resuspended in KCl-buffer (5 mg protein/ml). Triton X-100 was then added, to a final concentration of 4 % (v/v). The mixture was incubated for 15 minutes with gentle stirring, at 0° C, and then centrifuged (150,000 x g, 60 minutes). The pellet was resuspended in KCl-buffer, containing 1 mM dithiothreitol.

464

between 10 and 20 times in ruthenium red with respect to the original mitochondrial starting material, of which it represented, on a protein basis, only 1 - 2 %. When solubilized in SDS and submitted to polyacrylamide gel electrophoresis, this fraction could be dissociated into 6 to 8 major bands, ranging in molecular weight between 15,000 and 80,000 (Figures 5a and 5b). Of these 6 to 8 bands, only

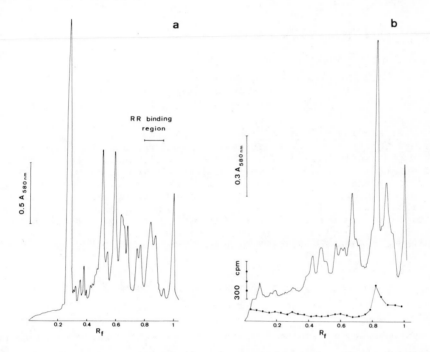

Figure 5a. SDS-polyacrylamide gel electrophoresis of inner membrane plus matrix vesicles (mitoplasts). Concentration of the stacking gel 4 % of the running gel, 8 %. Concentration of SDS, 0.1 %. The gel was stained with Amido Black. General conditions according to Laemmli (40).

Figure 5b. SDS-polyacrylamide gel electrophoresis of the Triton X-100 insoluble protein fraction. The lower trace shows the radioactivity pattern in a gel loaded with Triton X-100-insoluble protein fraction in the presence of ^{106}Ru-ruthenium red (1 nmole/mg protein). The gel was immediately sliced and the slices counted for radioactivity. Conditions as in Figure 5 a.

one, having a MW of about 20,000, and representing about 50 % of the total protein input, could be labelled with (^{106}Ru)-ruthenium red bound at 2 classes of sites: the high affinity class corresponded to about one nmole per mg of protein, and had a K_d of 0.3 µM (Figure 6). The Triton X-100 (or lubrol)-insoluble material also contained a va-

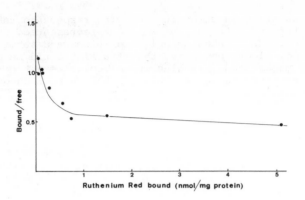

Figure 6. Binding of radioactive ruthenium red to the Triton X-100
insoluble protein fraction. The protein was resuspended (1 mg/ml) in
buffer containing 0.12 M KCl, 0.01 M Tris-Cl, pH 7.4, and variable
concentrations of [106]Ru-RR. The bound RR was measured by filtering
aliquots of the raction medium through millipore filters (pore size
0.22 μ, which arrests the insoluble protein aggregates) and assaying
the filters for radioactivity. The free RR was determined by centrifugation of aliquots of the reaction medium in a Beckman Airfuge
(100,000 x g, 15 minutes) and measuring the radioactivity in the supernatant. Incubation time, 5 minutes. Temperature 21° C.

riety of phospholipids, in approximately the same ratio to proteins
as in intact mitochondria, but no carbohydrates.

The hydrophobic, ruthenium red binding component clearly needs to be
purified further, and efforts in this direction are currently under
way in this Laboratory. Its role in the transport of Ca^{2+} across the
mitochondrial membrane is at the moment only a matter of conjecture,
but it is certainly interesting that a protein which is presumably
located in the apolar core of the membrane is able to bind the specific inhibitor of the transport of Ca^{2+} with very high affinity. The
hope that it may be the hypothetical Ca^{2+} carrier, or part of it, is
therefore rather reasonable.

One closing comment can be made on the exchange-diffusion, Na/Ca carrier which is responsible for the efflux of Ca^{2+} from heart mitochondria. As mentioned before, presently its identification rests mainly
on kinetic parameters. Efforts aimed at its isolation would require
a specific inhibitor, able to bind to the carrier with high affinity,
which at the moment is not available. A search for it is under way in
this Laboratory.

Acknowledgement

Parts of the original research described have been carried out with the financial assistance of the Swiss Nationalfonds (Grants No. 3.597.0-73 and 3.597.0-75).

References

1. Crompton, M., Capano, M., and Carafoli, E. (1976) Eur. J. Biochem. 69, 453-462.
2. Rottenberg, H., and Scarpa, A. (1974) Biochemistry 13, 4811-4817.
3. Mela, L. (1968) Arch. Biochem. Biophys. 123, 286-293.
4. Moore, C. (1971) Biochem. Biophys. Res. Commun. 42, 298-305.
5. Vasington, F.D., Gazzotti, P., Tiozzo, R., Carafoli, E. (1972) Biochim. Biophys. Acta 256, 43-54.
6. Carafoli, E. (1965) Biochim. Biophys. Acta 97, 99-106.
7. Carafoli, E., Tiozzo, R., Lugli, G., Crovetti, F., and Kratzing, C. (1974) J. Mol. Cell. Cardiol. 6, 361-371.
8. Crompton, M., Sigel, E., Salzmann, M., Carafoli, E. (1976) Eur. J. Biochem. 69, 429-434.
9. Racker, E. (1972) J. Biol. Chem. 247, 8198-8200.
10. Meissner, G., and Fleischer, S. (1974) J. Biol. Chem. 249, 302-309.
11. Warren, G.B., Toon, P.A., Birdsall, N.J.M., Lee, A.G., and Metcalfe, J.C. (1974) Proc. Natl. Acad. Sci. USA, 71, 622-626.
12. Racker, E., and Eytan, E. (1975) J. Biol. Chem. 250, 7533-7534.
13. Hinkle, P.C., Kim, J.J., and Racker, E. (1972) J. Biol. Chem. 247, 1338-1339.
14. Leung, K.H., and Hinkle, P.C. (1975) J. Biol. Chem. 250, 8467-8471.
15. Regan, C.I., and Hinkle, P.C. (1975) J. Biol. Chem. 250, 8472-8476.
16. Kagawa, Y., Kandrach, A., and Racker, E. (1973) J. Biol. Chem. 248, 676-684.
17. Shchipakin, V., Chuchlova, E., and Evtodienko, Y. (1976) Biochem. Biophys. Res. Commun. 69, 123-127.
18. Kagawa, Y. (1972) Biochim Biophys. Acta 265, 297-338.
19. Shertzer, H.G., and Racker, E. (1974) J. Biol. Chem. 249, 1320-1321.
20. Bauerjee, R.K., Shertzer, H.G., Kanner, B.I., and Racker, E. (1977) Biochem. Biophys. Res. Commun. 75, 772-778.

21. Chan, T.L., Greenawalt, J.W., and Pedersen, P.L. (1970) J. Cell. Biol. 45, 291-305.

22. Carafoli, E., Gazzotti, P. (1973) Experientia 29, 408-409.

23. Loyter, A., Christiansen, O.R., Steensland, H., Saltzgaber, J., and Racker, E. (1969) J. Biol. Chem. 244, 4422-4427.

24. Christiansen, O.R., Loyter, A., Steensland, H., Saltzgaber, J., and Racker, E. (1969) J. Biol. Chem. 244, 4428-4436.

25. Pedersen, P.L., and Coty, W.A. (1972) J. Biol. Chem. 247, 3107-33113.

26. Kagawa, Y., and Racker, E. (1971) J. Biol. Chem. 246, 5477-5487.

27. Evtodienko, J.V., Peskova, L.V., Shchipakin, V.N. (1971) Ukrainian J. Biochem. 43, 98-104.

28. Sottocasa, G.L., Sandri, G., Panfili, E., de Bernard, B., Gazzotti, P., Vasington, F.D., and Carafoli, E. (1972) Biochem. Biophys. Res. Commun. 47, 808-813.

29. Gómez-Puyou, A., Tuena de Gómez-Puyou, M., Baker, G., and Lehninger, A.L. (1972) Biochim. Biophys. Res. Commun. 47, 814-819.

30. Kimura, T., Chu, J.W., Mukai, R., Ishizuka, I., and Ymamkawa, T. (1972) Biochim. Biophys. Res. Commun. 49, 1678-1683.

31. Tashumkhamedov, B.A., Gagelgans, A.I., Mamatkulow, K., and Makhmudova, E.H. (1972) FEBS Lett. 28, 239-242.

32. Luft, J.H. (1971) Anatomical Record 171, 347-368.

33. Carafoli, E., Gazzotti, P., Saltini, C., Rossi, C.S., Sottocasa, G.L., Sandri, G., Panfili, E., and de Bernard, B. (1973) in Mechanism in Bioenergetics (G.F. Azzone, L. Ernster, S. Papa, E. Quagliariello, and N. Siliprandi, eds.) pp. 293-307, Academic Press, New York.

34. Panfili, E., Sandri, G., Sottocasa, G.L., Lunazzi, G., Liut, G., and Graziosi, G. (1976) Nature 264, 185-186.

35. Sandri, G., Panfili, E., and Sottocasa, G.L. (1976) Biochem. Biophys. Res. Commun. 68, 1272-1279

36. Carafoli, E. (1975) Molec. Cell. Biochem. 8, 133-140

37. Gitler, C., and Montal., M. (1972) FEBS Lett. 28, 329-332.

38. Prestipino, G.F., Ceccarelli, D., Conti, F., and Carafoli, E. (1976) FEBS Lett. 45, 99-103.

39. Utsumi, K., and Oda, T. (1974) Organization of Energy-Transducing Membranes (K. Nakao, and L. Packer, eds.) Univ. Park Press. Baltimore, p. 265.

40. Laemmli, U.K. (1970) Nature, 227, 680-685.

Properties of the Calcium-Sensitive
Bioluminescent Protein Aequorin

F. G. Prendergast, D. G. Allen,[*]
and J. R. Blinks
Department of Pharmacology
Mayo Foundation
Rochester, Minnesota

Introduction

Aequorin is the calcium-binding protein responsible for the luminescence of
the jellyfish Aequorea[1]. The bioluminescent systems of Aequorea and certain
other marine coelenterates are unusual in that they consist of single organic
components, termed photoproteins[2], which emit blue light in the presence of
calcium ions and do not require the presence of molecular oxygen or of any
other cofactor. Because of these unusual properties, the calcium-activated
photoproteins have been of great interest not only to biologists and biochemists
studying bioluminescent mechanisms but also to a rapidly increasing number of
investigators who need to detect calcium ions in concentrations, volumes,
locations, or timespans that preclude the use of conventional methods. Although
they have not been widely used for the purpose, the calcium-activated photoproteins also have properties that make them uniquely suited as models for the
study of various aspects of ion-protein interaction.

The Luminescent Reaction of Aequorin

The general nature of the luminescent reaction of aequorin is now fairly well
understood (for detailed review see refs. 3 & 4). The photoprotein consists
of a single polypeptide chain (the apoprotein, M.W. $\sim$ 20,000) to which a low
molecular weight chromophore is tightly bound. The chromophore is not covalently
bound, but is evidently trapped in a binding pocket from which it cannot escape
unless the protein undergoes a conformational change. (Although the chromophore
does not dissociate from the active photoprotein, it does so readily when the
protein is denatured.) The chromophore is an imidazolopyrazinone derivative
(Fig. 1a), very similar to the dissociable luciferins of certain other bioluminescent systems (notably those of Cypridina and Renilla)[5], and it is highly
probable that the basic chemical mechanisms involved in the generation of the
excited state are the same in all of these systems. The imidazolopyrazinone
nucleus is oxidized in the course of the reaction (Fig. 1), and in the case of
the Renilla and Cypridina systems molecular oxygen is required for luminescence.

---

[*] Present address: Department of Physiology, University College London,
London, England.

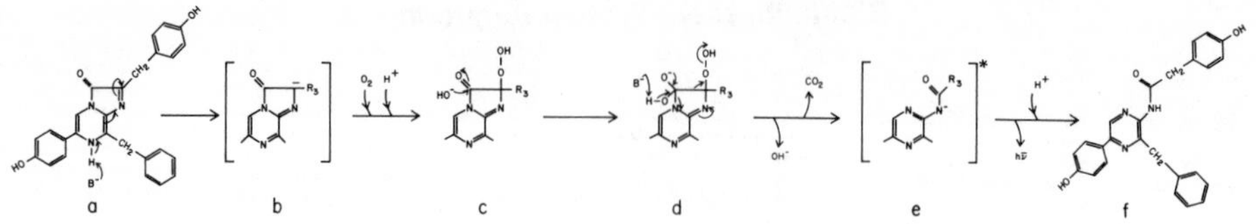

a   b   c   d   e   f

Fig. 1. A proposed reaction sequence for bioluminescence in aequorin. The
structures of the native chromophore (a) and of its oxidation product (f) are
known[6], and are shown in full. Intermediate steps have not been established,
but diagrams b,c,d, and e show a proposed[3] reaction sequence. The initial step
(a) is a proton abstraction effected by a hypothetical basic group (B⁻) on the
apoprotein. The oxygen required for the generation of the hydroperoxide (c) is
not exogenous O_2, but is presumably bound to the protein.

The luminescent reaction of aequorin does not utilize exogenous O_2, even though
the chromophore ends up as a product (Fig. 1f) identical to the oxyluciferin
of the <u>Renilla</u> system. The only plausible explanation for this is that
the required oxygen is bound somehow to the active photoprotein, and is kept
from oxidizing the chromophore (except at a very low rate) until calcium binds
to the protein. How the binding of calcium facilitates the reaction is not
clear, but it seems reasonable to suppose that it causes a conformational change
in the protein which apposes the bound oxygen to the chromophore, thus catalysing
the oxidation. Once the luminescent reaction has occurred, the oxidation
product (Fig. 1f) stays bound to the apoprotein as long as calcium is present;
this complex (in contrast to undischarged aequorin) is highly fluorescent, hence
spent aequorin is sometimes referred to as the blue fluorescent protein (BFP)[7].
When calcium is removed from the system, as by the addition of EDTA, the
oxidized chromophore dissociates from the apoprotein and fluorescence is lost.
Active aequorin can now be regenerated by incubating the apoprotein with synthetic chromophore (Fig. 1a) and molecular oxygen in the absence of Ca^{++} [8].

Heterogeneous Nature of Aequorin

Although commonly spoken of as if it were a single substance, aequorin is
in fact a family of closely related substances or isoproteins[9]. The various
isoaequorins are best resolved by isoelectric focusing, and can be shown by
this technique to have isoelectric points between 4.2 and 4.9. In pooled
extracts of <u>Aequorea</u> as many as 12 separate luminescent bands have been
resolved[10]. The question naturally follows as to whether this heterogeneity
reflects genetic differences among the individual specimens of <u>Aequorea</u> in the
pool, or whether all specimens contain the same mixture of isoaequorins. Isoelectric focusing gels run on extracts of single large specimens of <u>Aequorea</u>
have given a partial answer. As many as eight luminescent bands have been
distinguished in such gels, making it clear that the multiple bands of

470

pooled samples are not primarily a reflection of genetic heterogeneity. What is not yet clear is whether all specimens of Aequorea contain all of the iso-species of aequorin, and in the same proportions. It is difficult to load the gels from single specimens heavily enough to detect all of the luminescent bands present, and the fact that it has been possible to detect more luminescent bands in gels from pooled samples may reflect nothing more than that those gels were the more heavily loaded.

The chemical basis of the heterogeneity of aequorin is not known. The various isospecies have the same molecular weight by SDS gel electrophoresis[9], and NH_2-terminal analysis reveals no heterogeneity in amino acid sequence[11]. The emission spectra of the various isoaequorins are at least superficially the same. Although detailed kinetic studies have not been carried out on any of the separate isoaequorins, several observations suggest that differences may exist in the kinetics of their reactions with calcium. The most striking evidence of this sort is that not all samples of aequorin display the same pattern of decline of light intensity after they are rapidly mixed with a high concentration of Ca^{++}[12]. Fractions with an unusually slow rate of decline are sometimes collected early in the elution of aequorin from ion-exchange columns. Furthermore, the decline of luminescence of any given sample of aequorin is seldom strictly exponential, and the deviation from exponential behavior varies somewhat from sample to sample. This might very well reflect the presence of multiple isoaequorins in which the limiting rate constants are different. Clearly, it will be important to make a careful study of the properties of the individual isoaequorins. If they are found to differ in important respects, it may be necessary to resolve the photoprotein into its separate constituents before using it for certain types of work.

The Relation Between $[Ca^{++}]$ and Luminescence

There are two widespread misconceptions about the calcium concentration-effect relation for aequorin. The first is that over a wide range light intensity varies as the square of $[Ca^{++}]$. The second is that the square-law relation applies indefinitely with decreasing $[Ca^{++}]$. Until recently we accepted both of these notions, but in recent experiments[13,14] we have found that both are clearly incorrect. Figure 2 shows the calcium concentration-effect curve for aequorin determined under conditions of pH, salt concentration, and $[Mg^{++}]$ likely to be encountered in physiological experiments. Double logarithmic coordinates are required to display the results to best advantage. Both we[3,15] and others[16,17] have used semilogarithmic plots ($[Ca^{++}]$ on the log scale) at times in the past, but these show only the top two decades of the luminescent response. The full concentration-effect curve spans more than six orders of magnitude of light intensity, and some very interesting and instructive

471

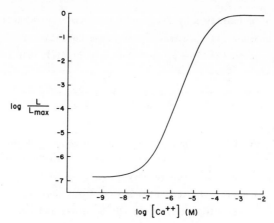

Fig. 2. Calcium concentration-effect relation for aequorin. Log-log plot derived from results obtained both with Ca-EGTA buffers and with dilutions of $CaCl_2$ (see ref. 13). Peak light intensity was measured when 10 µl of an aequorin solution was forcefully injected into 1 ml of the test solution. L/L_{max} is the ratio of peak light intensity in the test solution (L) to that in saturating $[Ca^{++}]$ (L_{max}). Solutions not buffered with EGTA were freed of contaminating Ca^{++} by passage through a column of chelating resin (Chelex-100) before the addition of $MgCl_2$ and $CaCl_2$. All solutions contained 150 mM KCl, 1 mM $MgCl_2$, 5 mM PIPES (piperazine-bis(2-ethanesulfonic acid)); pH 7.0, 21°C.

aspects of the relation are revealed only in the lower two-thirds of the curve. The full calcium concentration-effect curve for aequorin is sigmoid on a log-log plot (Fig. 2) and has three phases, which we will refer to as the calcium-independent phase, the logarithmic phase, and the saturation phase.

The fact that the calcim concentration-effect curve is horizontal at very low $[Ca^{++}]$ indicates that the calcium ion is not an absolute requirement for the luminescent reaction of aequorin. The calcium-independent luminescence has several properties that suggest it may result from the same basic reaction that is evoked by calcium, however. Neither type of luminescence is influenced by the removal of O_2. Factors (such as Mg^{++} and increased salt concentration) that inhibit the calcium-dependent luminescence also diminish the calcium-independent light. These observations are consistent with the idea that the role of calcium in the reaction is to greatly accelerate a process that takes place slowly in its absence.

The calcium-independent luminescence of aequorin was discovered as a direct outgrowth of attempts to generate calibration curves for the measurement of very low $[Ca^{++}]$. Previously it had been assumed, implicitly or explicitly, that the only limit to the detection of low $[Ca^{++}]$ with aequorin would be imposed by the difficulty of measuring very low light levels. In practice this frequently does serve as a limiting factor, but then it should be possible to extend the range of measurement by introducing more aequorin into the compartment of interest or by improving methods of light detection. The fact that the calcium concentration-effect curve levels off at very low $[Ca^{++}]$ presents an obstacle that cannot

472

be overcome by such means, however. It places an inescapable lower limit on the range of calcium concentrations that can be measured with aequorin.

In its logarithmic phase the calcium concentration-effect curve reaches a slope of about 2.5 (on double logarithmic coordinates). The steepness of the curve has both practical and theoretical implications. From the practical standpoint it means that differences in light intensity give an exaggerated impression of differences in [Ca^{++}], and that spatial patterns of light emission will be dominated by regions of locally high [Ca^{++}]. The theoretical implication is that multiple (at least 3) calcium binding sites must be involved in the acceleration of the luminescent reaction by Ca^{++} (see below).

In the saturation phase the rate of the luminescent reaction becomes independent of [Ca^{++}] again, presumably because all of the calcium-binding sites are occupied. Some step of the reaction subsequent to the binding of Ca^{++} must now limit the rate of the reaction. Factors that influence the sensitivity of aequorin to calcium do not, as a rule, alter this limiting rate.

The luminescence of aequorin is influenced by a number of environmental factors in addition to [Ca^{++}]. As one might anticipate, the rate of the luminescent reaction increases with temperature. This is true at all [Ca^{++}], but the Ca^{++}-independent component is influenced relatively more by temperature than is the rest of the curve (Fig. 3). Salt concentration has a marked effect on the sensitivity of aequorin to Ca^{++} (Fig. 4). As the concentration is increased the calcium-independent phase is depressed and the logarithmic phase is steepened and shifted to the right; the saturation phase is influenced only to the extent that it is abridged by the shift of the logarithmic phase to the right.

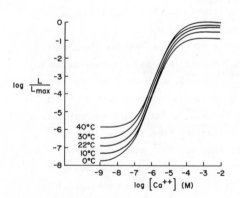 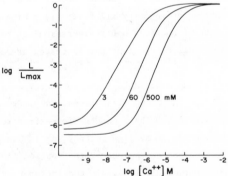

Fig. 3. Influence of temperature on the calcium concentration-effect curve for aequorin. Curves determined as in Fig. 2, but in 150 mM KCl, 5 mM PIPES, pH 7.0. All measurements are expressed as fractions of L_{max} at 40°C.

Fig. 4. Influence of salt concentration on the sensitivity of aequorin to calcium. Calcium concentration-effect curves determined as in Fig. 2 in the presence of 3, 60, and 500 mM K^+; each solution contained 2 mM PIPES, pH 7.0, 21°C.

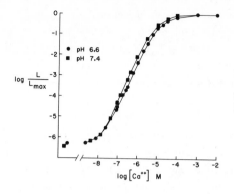

Fig. 5. Influence of pH on the calcium concentration-effect curve for aequorin. Curves determined as in Fig. 2 in 50 mM KCl, 5 mM PIPES, 22°C.

Moderate changes in pH have a relatively minor effect on the aequorin reaction (Fig. 5). Decreases in pH shift the logarithmic phase slightly to the right, but have virtually no effect on the calcium-independent or saturation phases.

Actions of Cations other than Ca^{++}

Many divalent and trivalent cations are capable of substituting isomorphously for calcium in a variety of chemical systems, including calcium-binding proteins. The ions that have been studied most extensively from this standpoint in biological systems are cations of Group IIa (especially Mg^{++}, Sr^{++}, and Ba^{++}) and some of the lanthanides. For the most part Mg^{++} and Ba^{++} have been found to antagonize the actions of Ca^{++}, while Sr^{++} often mimics Ca^{++} but is less effective[18]. The lanthanides may mimic or antagonize Ca^{++}, and are often effective at lower concentrations than Ca^{++} because their size, high charge density, and flexible coordination geometry all favor tight binding[18].

Most of the ions under discussion are capable of evoking aequorin luminescence. All of the lanthanides tested are more potent than Ca^{++} in this respect ($Tb^{+++} > Eu^{+++} \simeq Yb^{+++} \simeq Gd^{+++} \simeq Pr^{+++} > La^{+++} > Ca^{++}$), although the quantum yield and maximum achievable light intensity are in all cases somewhat less than with Ca^{++}. Of the Group IIa cations, Ca^{++}, Sr^{++}, and Ba^{++} evoke luminescence, but Mg^{++} does not. The concentration-effect relations for these ions and for Eu^{+++}, a lanthanide of average potency, are shown in Fig. 6. Sr^{++} and Ba^{++} are not only less potent than Ca^{++}, the maximal light intensity that they can evoke is also considerably less. In the latter respect they resemble pharmacological partial agonists, and like partial agonists they are capable of surmountably antagonizing the actions of Ca^{++}. Mg^{++} decreases the calcium-independent luminescence and serves as a pure antagonist to the actions of Ca^{++}. It does not evoke luminescence in any concentration, but shifts the calcium concentration-effect curve to the right without changing its slope or the maximum light intensity

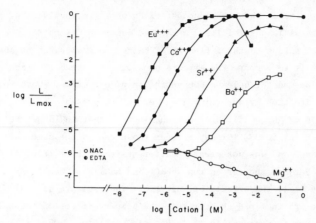

Fig. 6. Comparison of the actions of various cations on aequorin luminescence. Concentration-effect curves for Eu^{+++}, Ca^{++}, Sr^{++}, Ba^{++}, and Mg^{++} were all determined at 21°C by diluting ultrapure salts of the respective ions in a solution of 150 mM KCl, 5 mM PIPES, pH 6.5, that had been freed of Ca^{++} by passage through a column of chelating resin (Chelex-100). The point marked NAC indicates the level of luminescence evoked by the diluent alone; the point marked EDTA indicates the level after 1 mM EDTA had been added to the diluent to chelate any residual calcium. Ordinates are all normalized to the peak light intensity obtained with saturating $[Ca^{++}]$.

achievable. In the presence of 150 mM KCl, the addition of 10 mM Mg^{++} shifts the curve about 1 log unit to the right[14].

The concentration-effect curve for Eu^{+++} is parallel to that for Ca^{++} except at very high concentrations (> 1 mM), when the peak light intensity and quantum yield tend to drop off. In the latter circumstance the rate of decline of luminescence becomes very much more rapid than in saturating $[Ca^{++}]$. This, together with the reduced peak light intensity and quantum yield, suggests that a Eu^{+++}-induced inhibition or inactivation may develop simultaneously with the luminescent reaction at high $[Eu^{+++}]$.

Increasing salt concentrations have been found by various workers[3,16,17] to shift the calcium concentration-effect curve for aequorin to the right (see Fig. 4). Moisescu and Ashley[17] have reported that NaCl is more effective than KCl in this respect, and that a 1:1 mixture of NaCl and KCl is more inhibitory than the same total concentration of either salt alone. They concluded that the inhibitory effects of NaCl and KCl were due to competitive binding of Na^+ and K^+ at the calcium-binding sites of the photoprotein. We have done comparable studies and found that 150 mM NaCl does displace the calcium concentration-effect curve slightly more than 150 mM KCl, but that the curve determined in a mixture of 75 mM NaCl and 75 mM KCl falls between those run in 150 mM NaCl and in 150 mM KCl. The question follows as to whether the inhibitory effects of NaCl and KCl result from a specific competition between the monovalent cations and Ca^{++}, or from nonspecific modulation of Ca^{++}-binding by electrostatic shielding of negatively charged binding sites. The high concentrations of salts involved certainly make the latter a plausible possibility. A related question has to do with the possible role of anions in the influence of salts. One approach to answering these questions is to compare the effects of salts of a number of other monovalent cations and anions on the calcium concentration-effect curve. We have made a start in this direction by determining curves in the presence of KNO_3 and LiCl. The curve in 150 mM KNO_3 was indistinguishable from that in 150 mM KCl. The curve determined in 150 mM LiCl lay between those run in the same concentrations of KCl and of NaCl and was not clearly different from either (Fig. 7). Although the difference between the curves run in KCl and NaCl has been found consistently and is probably real, the effects of KCl, LiCl, and NaCl are so similar (cf. Fig. 6) that we are inclined to attribute the primary effect of the salts of monovalent cations to nonspecific electrostatic shielding. More salts will have to be studied before this conclusion can be drawn with confidence, however.

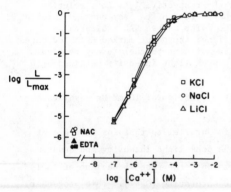

Fig. 7. Comparison of the effects of salts of three monovalent cations on the sensitivity of aequorin to calcium. Calcium concentration-effect curves were determined in 150 mM solutions of KCl, NaCl, and LiCl, each containing 5 mM PIPES, pH 7.0, at 21°C. All calcium concentrations were established by simple dilution. For this reason the lower points on the curves are slightly too high (note difference between points marked NAC and EDTA – abbreviations defined in legend to Fig. 6).

Role of Sulfhydryl Groups in the Luminescent Reaction

Shimomura et al.[1] and Shimomura and Johnson[7] have provided evidence that the integrity of sulfhydryl groups is essential for the luminescent response of aequorin to the addition of calcium ions. Our studies have confirmed this and indicate that the reaction of aequorin with any of a number of sulfhydryl-modifying agents not only leads to complete loss of calcium-evoked luminescence but also enhances calcium-independent light emission. Maleimides abolish the response to calcium, apparently irreversibly. Dithiobisnitrobenzoic acid, sodium tetrathionate, and methylmercuric carbonate also inactivate aequorin, but in the case of these reagents the reaction may easily be reversed by incubating the inactivated protein with a thiol (e.g., dithiothreitol). An example of this sort of interaction is shown in Fig. 8. In experiments of this type, it is usually not possible to fully restore luminescent activity, possibly because a significant fraction of the chromophore is oxidized in connection with the marked increase in calcium-independent light emission. At present we do not know whether the total amount of light

476

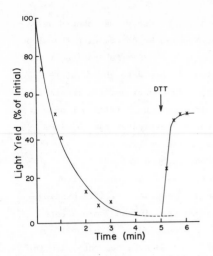

Fig. 8. Inactivation and reactivation of aequorin by sulfhydryl-modifying agents. The calcium-dependent light yield of aliquots of an aequorin solution was measured at various times after the addition of 10^{-4} M dithio-bisnitrobenzoic acid. After 5 min dithiothreitol (DTT) was added to a final concentration of 10^{-3} M.

emitted before the addition of the thiol corresponds to the deficit in the re-covery of the response to calcium, or whether some "dark" reactions account for part of the inactivation of aequorin. Nor do we have any explanation of why the sulfhydryl group is key to luminescence. Data derived from fluorescence spectro-scopic studies suggest that the –SH group lies in or near a hydrophobic pocket in the protein and probably close to the chromophore, which is also in a hydro-phobic pocket.

Models of Reaction Mechanisms

Results such as those shown in Figs. 2–5 have required two major adjustments in our thinking about the function of calcium in the luminescent reaction of aequorin. First, the role of calcium in the luminescent reaction probably should be re-garded not as that of an indispensable trigger, but as that of a co-factor capable of greatly accelerating a process which takes place slowly in its absence. Second, the steepness (slope 2.5) of the logarithmic phase of the calcium con-centration-effect curve indicates that each aequorin molecule must possess at least three interacting sites at which calcium ions can bind to contribute to the acceleration of the luminescent reaction. The interaction between the sites could be a truly cooperative one, in which the binding of Ca^{++} to one site in-creases the affinity for calcium of the remaining sites on the same aequorin molecule. In that case it is not necessary to assume that the effects of binding Ca^{++} to the separate sites are other than simply additive. Alternatively, the steepness of the calcium concentration-effect curve could result from an inter-action among the binding sites such that three sites on the same aequorin mol-ecule must simultaneously be in a particular state for luminescence to occur. Although the latter form of interaction is not in the strict sense of the word a cooperative one, the end result is much the same. In theory, one should be able to distinguish between the two sorts of interaction by determining the slope of the curve relating the amount of calcium bound to $[Ca^{++}]$. The results of such an experiment are not yet available, and even if they were their interpretation would be subject to a major uncertainty. That derives from the fact that measurements of calcium binding must by their nature be carried out on spent aequorin, and there is no assurance that the properties of the calcium-binding sites do not change in the course of the luminescent reaction. Nevertheless, the experiment is an important one that needs to be done.

The calcium-independent luminescence of aequorin is not accounted for by conventional schemes (e.g., 16, 17) in which the occupancy of specific binding sites by calcium is viewed as being an indispensable step in the luminescent reaction. To be tenable, such schemes must now be supplemented with the assumption that the luminescent reaction can also proceed by a separate path in which calcium is not involved. A model in which a single mechanism was capable of accounting for the whole calcium concentration-effect curve would be far more appealing. Two-state models have characteristics that suit them well to situations of this sort[19,20], and we have recently shown that such a model describes the calcium concentration-effect curve for aequorin remarkably well[13]. The model is based on the following assumptions:

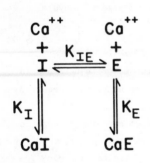

(1) Each aequorin molecule has multiple (at least three) binding sites, each of which can exist in two states (denoted E and I, for effective and ineffective).

(2) The E and I states of the binding sites are interconvertible, and are in equilibrium (equilibrium constant $K_{IE} = \frac{[I]}{[E]}$). The transitions of the individual sites on an aequorin molecule are assumed to occur independently (i.e., there is no true cooperativity).

(3) Ca^{++} has a much higher affinity for the E state than for the I state of the binding site. The binding constants for the two states are K_E and K_I.

(4) The light-emitting reaction (Fig. 1) is initiated whenever three binding sites on the same aequorin molecule are simultaneously in the E state (with or without calcium bound to them).

In the two-state model the binding of calcium is not viewed as producing any specific change in the aequorin molecule that would "trigger" the luminescent reaction as in the conventional occupancy model. Rather, by binding preferentially to sites in the E state, calcium tends to increase the proportion of sites in that state (E + ECa). This, in turn, increases the probability that three sites on a given molecule will be in the E state simultaneously. Needless to say, with appropriate values for the various constants, the model can be made to fit the full calcium concentration-effect curve within experimental error. Particularly attractive features of the two-state model are

(1) that the calcium-independent light is an intrinsic feature of the proposed mechanism,

(2) that the effects of partial agonists like Sr^{++} and Ba^{++} can be accounted for simply by assuming that these ions have a lower selectivity than Ca^{++} for the E state of the binding site, and

(3) that the effects of pure antagonists like Mg^{++} (including the influence on the calcium-independent light) are completely described by assuming that the antagonists bind preferentially to the I state of the binding site. Unfortunately, too little is known about the calcium-binding sites of aequorin to permit an assessment of the plausibility of this or any other model from the standpoint of protein structure.

Potential Applications of Aequorin

To date, most studies on aequorin have been carried out by investigators primarily interested either in the mechanism of bioluminescence or in using the photoprotein as a calcium indicator (for detailed review see ref. 3). We feel that aequorin also has great potential as a model system for the study of many questions of a more general nature. The outstanding attribute of aequorin in this respect is that light is one of its reaction products, and light can be measured with extraordinary ease and sensitivity. As a result, one can easily and precisely measure reaction rates over the more than 6 orders of magnitude spanned by the calcium concentration-effect curve. It is difficult to imagine another sort of biochemical system in which this would be possible. It seems likely, therefore, that aequorin will prove useful as a tool for the investigation of such subjects as

(1) the nature of the binding of calcium and other ions to specific sites on proteins,

(2) the nature of nonspecific interactions of ions with proteins,

(3) the functions of amino acid side chains in catalysis,

(4) the mechanisms by which protein structure can be perturbed, and

(5) theories of drug-receptor interaction and of protein-ligand interaction.

ACKNOWLEDGEMENTS

This work was supported by USPHS grant HL 12186. Parts of it were done during the tenure of a British-American fellowship of the American Heart Association and the British Heart Foundation (to D.G.A.) and a research fellowship from the Minnesota Heart Association (to F.G.P.). The authors gratefully acknowledge use of the facilities of the Friday Harbor Laboratories, University of Washington.

REFERENCES

1. Shimomura, O., Johnson, F. H., and Saiga, Y. (1962) J. Cell. Comp. Physiol. 59:223–239.

2. Shimomura, O. and Johnson, F. H. (1966) Bioluminescence in Progress, ed. by F. H. Johnson and Y. Haneda, pp. 495–521, Princeton University Press, Princeton, NJ.

3. Blinks, J. R., Prendergast, F. G., and Allen, D. G. (1976) Pharmacol. Rev. 28:1–93.

4. Cormier, M. J., Wampler, J. E., and Hori, K. (1973) Fortschr. Chem. Org. Naturst. 30:1–60.

5. Cormier, M. J., Hori, K., Karkhanis, Y.D., Anderson, J. M., Wampler, J. E., Morin, J. G., and Hastings, J. W. (1973) 81:291–298.

6. Cormier, M. J. and Hori, K. Personal communication.

7. Shimomura, O. and Johnson, F. H. (1969) Biochemistry 8:3991–3997.

8. Shimomura, O. and Johnson, F. H. (1975) Nature (London) 256:236–238.

9. Blinks, J. R. (1971) Proc. 25th Int. Cong. Physiol. Sci. 9:68.

10. Blinks, J. R. and Harrer, G. C. (1975) Fed. Proc. 34:474.

11. Prendergast, F. G. and Mann, K. G. In preparation.

12. Van Leeuwen, M. Unpublished observation.

13. Allen, D. G., Blinks, J. R., and Prendergast, F. G. (1977) Science 195:996–998.

14. Allen, D. G. and Blinks, J. R. In preparation.

15. Blinks, J. R. (1973) Eur. J. Cardiol. 1:135–142.

16. Moisescu, D. G., Ashley, C. C., and Campbell, A. K. (1975) Biochim. Biophys. Acta 396:133–140.

17. Moisescu, D. G. and Ashley, C. C. (1977) Biochim. Biophys. Acta 460:189–205.

18. Williams, R. J. P. (1970) Q. Rev. Chem. Soc. (Lond.) 24:331–365.

19. Janin, J. (1973) Prog. Biophys. Mol. Biol. 27:77–120.

20. Colquhoun, D. (1973) Drug Receptors: A Symposium, ed. by H. P. Rang, pp. 149–182, Macmillan, London.

Isolation, Properties and Function
of a Calcium-Triggered Luciferin Binding Protein

Milton J. Cormier and Harry Charbonneau
Bioluminescence Laboratory, Department of Biochemistry
University of Georgia, Athens, GA., USA

SUMMARY

An acidic Ca^{2+} binding protein has been isolated from crude extracts of *Renilla reniformis*, a coelenterate whose bioluminescence is initiated by a signal from a nerve net. This protein has been purified 6000 fold to homogeneity and, when isolated in the absence of Ca^{2+}, contains one mol of coelenterate-type luciferin non-covalently bound to the protein. Thus it is termed a Ca^{2+} triggered luciferin binding protein ($BP\text{-}LH_2$). The molecular weight of $BP\text{-}LH_2$ (18,500), the iso-electric point (4.3), the number of high affinity Ca^{2+} binding sites (2; $K_d = 1.4 \times 10^{-7}$M) and the amino acid composition data indicate possible similarities to other Ca^{2+} binding proteins such as muscle troponin C, brain modulator protein and aequorin. In the presence of Ca^{2+}, $BP\text{-}LH_2$ will transfer its bound luciferin to luciferase with subsequent light production. $BP\text{-}LH_2$ thus serves as the terminal link between nerve excitation and the bioluminescence flash through Ca^{2+} which apparently acts as a second messenger.

INTRODUCTION

Bioluminescence in the marine coelenterate, *Renilla reniformis*, is initiated by a signal from a nerve net[1,2]. The intracellular luminescence may be initiated upon tactile or electrical stimulation of the animal resulting in concentric waves of greenish luminescence. The rapid onset and decay of this *in vivo* luminescence suggest that it is under fine control[3]. The discovery and partial purification of the protein which links the nerve signal to bioluminescence was reported earlier[4] but it has now been purified to homogeneity and its characteristics are reported here. It is an acidic Ca^{2+} binding protein which, when isolated in the absence of Ca^{2+}, contains one mol of coelenterate-type luciferin non-covalently bound to the protein. Thus it is termed a Ca^{2+} triggered luciferin binding protein ($BP\text{-}LH_2$).

Luciferin in solution is readily autoxidized but $BP\text{-}LH_2$, in the absence of Ca^{2+}, provides an environment in which luciferin is stable. Under these conditions $BP\text{-}LH_2$ will not react with luciferase to produce light[4]. The addition of Ca^{2+} to $BP\text{-}LH_2$ results in the transfer of luciferin from $BP\text{-}LH_2$ to luciferase. This is followed by a luciferase catalyzed oxidation of luciferin to produce oxyluciferin, CO_2 and light[13].

The physicochemical characteristics of $BP\text{-}LH_2$ are described and compared with those of a number of Ca^{2+} binding proteins previously reported[5-8]. Although a number of the characteristics of BP-LH$_2$ are shown to be similar to those of the Ca^{2+} activated modulator proteins[8-12,24,25] it is also apparent that BP-LH$_2$ falls into a different class of proteins from the functional point of view.

METHODS

Assay of BP-LH$_2$ - BP-LH$_2$ is quantitated by measuring the total light yield from the luminescent reaction of BP-LH$_2$ with purified *Renilla reniformis* luciferase[13]. The total photon yield is directly proportional to the total BP-LH$_2$ present in the reaction mixture and the ratio of total photons to total protein (specific photon yield) is analogous to specific activity. The total photon yield is obtained by measuring the luminescence of the light reaction in a photometer calibrated in absolute photons with the luminol light reaction described by Lee et al.[14]. The reaction is performed by injecting 5-20 µl of BP-LH$_2$ solution into .53 ml of assay buffer containing 10 mM Tris-HCl, 10 mM CaCl$_2$ (pH 7.5) and approximately 5 ug/ml purified luciferase at room temperature (23-25°C).

Purification of BP-LH$_2$ - This protein has been purified from crude extracts of *Renilla reniformis* approximately 6000 fold to homogeneity by a combination of methods which include adsorption of contaminating proteins on aluminum hydroxide gel, anion exchange chromatography, gel filtration chromatography and an affinity chromatography step (p-benzyl oxyaniline Sepharose) previously described[13]. The details of this purification are being reported elsewhere. BP-LH$_2$ was found to be homogeneous by a number of criteria which include disc and sodium dodecyl sulfate (SDS)-polyacrylamide gel electrophoresis, isoelectric focusing and sedimentation velocity and equilibrium studies. BP-LH$_2$ activity was found to be associated with the only protein band found in the disc gels. Approximately 1.7 mg of pure protein having a specific photon yield of 2 x 10^{15} hν/mg, are obtained per kg of *Renilla* providing an overall yield of 55%.

Physicochemical studies - Ultracentrifugation studies, determination of Stokes radius, amino acid analyses and electrophoretic techniques were performed as previously described[13]. Carbohydrate content was examined using the phenol sulfuric acid procedure[15] and the periodic acid fuchsin sulfite staining method following polyacrylamide gel electrophoresis[16]. Phosphate content was determined using the method of Chen et al.[17] with the ashing procedure described by Ames and Dubin[18]. Protein concentration was determined by the biuret method described by Goa[19].

RESULTS

Physical properties - The physicochemical data obtained from BP-LH$_2$ are shown in Table I. The physical data shows that BP-LH$_2$ is a highly acidic globular protein having a molecular weight of approximately 18,500. The protein is composed of one single polypeptide chain, since the molecular weight obtained for the native protein by sedimentation equilibrium is in agreement with the molecular weight based on SDS gels. The isoelectric point of 4.3, obtained from analytical isoelectric focusing gels, is consistent with the high mobility of the protein on 10 or 15% polyacrylamide gels and its relatively tight binding to ion exchangers.

TABLE I: PHYSICOCHEMICAL PROPERTIES OF BP-LH$_2$

$S^{\circ}_{20,\,W} \times 10^{13}\ s^{-1}$	2.32 ± 0.02
Isoelectric point	$4.3\ \pm 0.1$
$\varepsilon^{0.1\%}_{276nm}$	1.3
Luciferin content	1 mol/mol protein
Molecular weight:	
Sedimentation equilibrium 16,900-18,500	
SDS gel electrophoresis 20,000	

Chemical composition - Table II gives the amino acid composition of BP-LH$_2$ as well as four other Ca^{2+} binding proteins for comparative purposes. As expected from its acidic nature, 27% of the amino acid residues of BP-LH$_2$ are either aspartate or glutamate. In addition, the protein is characterized by one histidine residue, 3 half-cystines and methionine residues, phenylalanine content twice that of tyrosine, and the absence of tryptophan residues. No trace of tryptophan appeared in triplicate analyses of samples hydrolyzed for 24 hrs. in the presence of thioglycolic acid. The phosphate content of ashed samples of BP-LH$_2$ was less than 0.2 mol phosphate/18,500 g protein. No carbohydrate was detected in BP-LH$_2$ samples. Using the specific photon yield and the quantum yield (5.5%) of the *in vitro* bioluminescence reaction[13], one can calculate the content of coelenterate-type luciferin. On this basis, BP-LH$_2$ contains one mol luciferin/18,500 g protein.

Spectral properties - The absorption spectrum of BP-LH$_2$ is shown in Fig. 1. The protein has two characteristic absorption maxima at 276 nm ($\varepsilon^{0.1\%}_{276\ nm}$ = 1.31) and 446 nm ($\varepsilon^{0.1\%}_{446\ nm}$ = 0.47). The visible band at 446 nm arises from the coelentrate-type luciferin which is bound to the protein.

The uncorrected fluorescence emission (λmax = 332 nm) on excitation at 290 nm is also shown in Fig. 1. This fluorescence has been tentatively assigned to tyrosine, since the protein contains no tryptophan. Furthermore, the emission from free tyrosine in solution is near the emission of the protein when the two spectra are obtained on the same instrument under identical conditions. In addition to the near UV fluorescence, there is a fluorescence emission at 520 nm (not shown) due to the luciferin content. The spectral distribution of this emission, and its maxima is very similar to that obtained from luciferin in methanol solution.

483

TABLE II: AMINO ACID COMPOSITION OF BP-LH$_2$ AND COMPARISON
TO OTHER Ca^{2+} BINDING PROTEINS

	BP-LH$_2$	Bovine Brain[8] Modulator Protein	Rabbit Skeletal[6] Muscle Troponin C	Hake[7] Parvalbumin	Aequorin[5,22]
Lysine	12	8	9	12	13
Histidine	1	1	1	1	4
Compound X		1			
Arginine	8	7	7	1	6
Aspartic Acid	24	24	22	13	21
Threonine	9	12	6	5	8
Serine	8	5	7	5	8
Glutamic Acid	20	29	31	10	20
Proline	4	2	1	0	6
Glycine	10	12	13	12	13
Alanine	19	12	13	19	11
Half-Cystine	3	0	1	1	3
Valine	10	8	7	4	8
Methionine	3	10	10	1	4
Isoleucine	8	8	9	7	9
Leucine	14	10	9	8	11
Tyrosine	4	2	2	0	6
Phenylalanine	8	8	10	10	7
Tryptophan	0	0	0	0	4

Calcium binding properties of BP-LH$_2$ - The Ca^{2+}-binding properties of BP-LH$_2$
have been studied using the intrinsic fluorescence which presumably arises from
tyrosine residues. Figure 1 shows that the fluorescence intensity at 320 nm (uncorrected) increases 4 fold on the addition of saturating amounts of Ca^{2+}. This
fluorescence change can be reversed by the addition of EDTA to Ca^{2+} treated BP-LH$_2$.
The increase in fluorescence intensity is accompanied by a shift of the emission
maxima 7 nm to the blue. The fluorescence emission at 520 nm (uncorrected)
increases 5 fold with the addition of Ca^{2+}. The 520 nm emission in the presence
of Ca^{2+} is not stable since the luciferin molecule of BP-LH$_2$ is now free to undergo oxidation. This oxidation results in the gradual loss of intensity at 520 nm
with a concomitant increase in fluorescence intensity at 420 nm due to luciferin
degradation. Because of the instability of the 520 nm fluorescence, we have used
the stable tyrosine fluorescence to determine the Ca^{2+} binding parameters of
BP-LH$_2$. A solution of BP-LH$_2$, which was prepared free from EDTA, was titrated
using small aliquots of Ca^{2+} standards. The titration resulted in a gradual
hyperbolic increase in fluorescence intensity with increasing Ca^{2+} concentration.

The fluorescence increase was saturated at $1-2 \times 10^{-5}$ M Ca^{2+}. A plot of fractional saturation of fluorescence versus the ratio of total Ca^{2+} concentration to protein concentration[20] indicated the existence of two Ca^{2+} binding sites. The best fit of this data was achieved by using a K_d of 1.4×10^{-7} M.

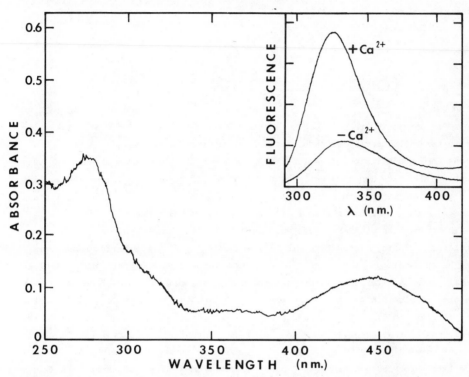

Fig. 1. Absorption spectra of BP-LH$_2$. The insert illustrates the fluorescence change that occurs upon Ca^{2+} addition (2×10^{-3} M) to BP-LH$_2$ on excitation at 290 nm.

We also examined the affect of Ca^{2+} on the rate of the luciferase catalyzed bioluminescent oxidation of BP-LH$_2$. A 3.8 fold increase in reaction rate was observed when Ca^{2+} concentrations were varied from .25 μM to 10 μM. The point of half saturation occurred at 2.0×10^{-6} M Ca^{2+}; a similar value was observed during the titration of tyrosine fluorescence. Thus we have concluded that the Ca^{2+} binding sites involved with the enhancement of tyrosine fluorescence are the same sites involved in triggering the bioluminescent reaction.

As shown in Table III, other Ca^{2+} binding proteins involved in the regulation of intracellular processes also have two high affinity Ca^{2+} binding sites with dissociation constants similar to those of BP-LH$_2$.

TABLE III: COMPARATIVE Ca^{2+} BINDING DATA

Protein	K_d (M)	Stoichiometry
Bovine Brain Modulator[8]	1×10^{-6}	2
	8.6×10^{-4}	2
Troponin C[21]	4.8×10^{-8}	2
	3.1×10^{-6}	2
BP-LH$_2$	1.4×10^{-7}	2

DISCUSSION

It is noteworthy that in order to obtain BP-LH$_2$ in its native state, i.e. with its attached luciferin, extreme caution was required to maintain Ca^{2+} free conditions throughout its purification. Thus if BP-LH$_2$ is chromatographed in the presence of Ca^{2+} its non-covalently bound luciferin is removed[4]. Therefore Ca^{2+} contamination at any stage in the purification would have resulted in the isolation of a Ca^{2+} binding protein whose function would have remained unknown.

The properties of BP-LH$_2$ have much in common with a number of calcium binding proteins including muscle troponin C[21], aequorin[5,22,23] parvalbumin[21] and brain and heart modulator proteins[8-12,24,25]. As indicated in the text, and in Tables I-III, these similarities in properties include amino acid composition, molecular weight, isoelectric point and the presence of two high affinity Ca^{2+} binding sites. The increase in tyrosine fluorescence observed upon binding of Ca^{2+} to BP-LH$_2$ is analogous to that observed when Ca^{2+} binds to troponin C and parvalbumin[26-28]. Recent chemical and physical evidence suggest that the bovine brain and heart modulator proteins and troponin C have evolved from a common ancestor[8,11]. Because of the chemical and physical similarities of BP-LH$_2$ to aequorin as well as to the modulator proteins and troponin C it is tempting to speculate that all of these proteins may have common ancestral origins. Structural conservation in these proteins may be related to the mechanism of Ca^{2+} binding. In fact, Kretsinger and Barry[29] have predicted a three-dimensional structure for muscle troponin C in which the ligands for Ca^{2+} involve the side chains of numerous residues.

Although there may be structural similarities between the above mentioned Ca^{2+} binding proteins it is also evident that troponin C and the modulator proteins perform different biochemical functions. In this same sense BP-LH$_2$ appears to represent a functionally different class of Ca^{2+} binding protein. For example, BP-LH$_2$ does not exhibit Ca^{2+} dependent activation of phosphodiesterase although there are relatively high levels of this modulator protein in crude extracts of *Renilla**. On the other hand, BP-LH$_2$ serves as the terminal link between nerve

---

*We thank Dr. F. L. Siegel for making these measurements for us.

486

excitation and bioluminescence in *Renilla*. The role of BP-LH$_2$ in this process is now well understood as outlined below.

Function of BP-LH$_2$ - The probable *in vivo* function of BP-LH$_2$ is best understood in terms of its interaction with the isolated components of the *Renilla* bioluminescence system. Three membrane associated proteins, which occur in equimolar concentrations in *Renilla*, have been isolated and the role of each in bioluminescence is now known. One of these proteins (luciferase) is an oxygenase which catalyzes the oxidation of coelenterate-type luciferin resulting in the formation of oxyluciferin, CO_2 and blue light[13]. A second protein, termed green fluorescent protein, is responsible for the green *in vivo* bioluminescence by acting as a specific energy transfer acceptor[30,31]. This process occurs by specific interaction of the green fluorescent protein with luciferase to form a complex which results in non-radiative energy transfer[31]. A third protein, the Ca^{2+}-triggered luciferin binding protein (BP-LH$_2$), serves as the terminal link between nerve excitation and the bioluminescence flash through Ca^{2+} which apparently acts as a second messenger. The relationship between these three proteins is illustrated in the scheme below:

BINDING PROTEIN — LH$_2$ + Ca^{2+} ⇌ Ca — BINDING PROTEIN — LH$_2$

Ca — BINDING PROTEIN — LH$_2$ + E ⇌ E — LH$_2$ + Ca — BINDING PROTEIN

E — LH$_2$ + O$_2$ + GFP ⟶ OXYLUCIFERIN* — E — GFP + CO$_2$
(MONOANION)

OXYLUCIFERIN* — E — GFP $\xrightarrow{\text{ENERGY TRANSFER}}$ OXYLUCIFERIN — E — GFP*
(MONOANION)

OXYLUCIFERIN — E — GFP* ⟶ OXYLUCIFERIN — E — GFP + LIGHT (509 nm)

LUCIFERIN (SYNTHETIC)   OXYLUCIFERIN MONOANION

LH$_2$ = LUCIFERIN   GFP = GREEN-FLUORESCENT PROTEIN   E = LUCIFERASE

In the absence of Ca^{2+}, BP-LH$_2$ will not produce light with luciferase because its bound luciferin is unavailable. The binding of Ca^{2+} to BP-LH$_2$ allows the transfer of luciferin to luciferase resulting in a bioluminescent oxidation. In fact, an *in vitro* mixture of the three proteins illustrated in the above scheme will produce a flash of green luminescence upon the addition of Ca^{2+}. The color of the light is identical to that observed *in vivo*.

In vivo, BP-LH$_2$, luciferase and green fluorescent protein are apparently sequestered within a membrane and thus protected from Ca^{2+} in the resting state[3,34]. In fact, these proteins can be isolated prepackaged within a membrane bounded vesicle. These vesicles, termed lumisomes, have been isolated and highly purified[3]. Lumisomes contain BP-LH$_2$, luciferase and green fluorescent protein and will produce a flash of green light in the presence of Ca^{2+} under appropriate conditions. Ca^{2+} will initiate a flash of light from a suspension of lumisomes either by hypotonic rupture of the lumisomal membrane[3] or upon establishing a Na^+ gradient across the lumisomal membrane[32-34] with Na^+ being isotonic on the inside of the lumisomes and low in concentration on the outside.

A plausible model which can account for the above observations is illustrated below:

CALCIUM: THE LINK BETWEEN NERVE EXCITATION AND BIOLUMINESCENCE

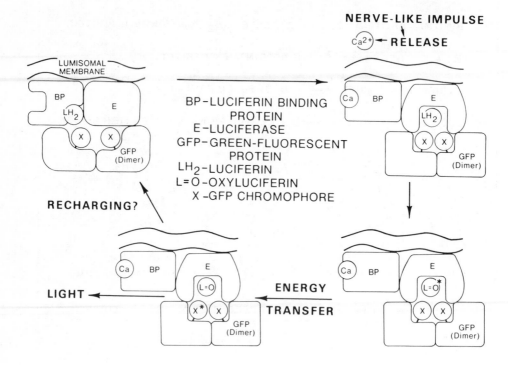

NERVE-LIKE IMPULSE

Ca^{2+} ← RELEASE

BP—LUCIFERIN BINDING PROTEIN
E—LUCIFERASE
GFP—GREEN-FLUORESCENT PROTEIN
LH$_2$—LUCIFERIN
L=O—OXYLUCIFERIN
X—GFP CHROMOPHORE

In this model BP-LH$_2$ is the terminal link between nerve excitation and bioluminescence with Ca^{2+} acting as a second messenger[35]. Membrane depolarization could result in a rapid and transient influx of Ca^{2+} across the lumisomal membrane turning the bioluminescence system on. A rapid resequestering of Ca^{2+} could turn the system off.

ACKNOWLEDGEMENTS

This work was supported by grants from the NSF(BMS 74-06914) and the ERDA(AT 38-1-635).

REFERENCES

1. Harvey, E.N. (1952) Bioluminescence, pp. 168-180, Academic Press, New York.

2. Nicol, J.A.C. (1975) J. Exp. Biol. 32, 619-635.

3. Anderson, J.M. and Cormier, M.J. (1973) J. Biol. Chem. 248, 2937-2943.

4. Anderson, J.M., Charbonneau, H. and Cormier, M.J. (1974) Biochemistry 13, 1195-1200.

5. Shimomura, O. and Johnson, F.H. (1969) Biochemistry 8, 3991-3997.

6. Collins, J.H., Potter, J.D., Horn, M.J., Wilshire, G. and Jockman, N. (1973) FEBS Lett. 36, 268-272.

7. Pechère, J.F., Capony, I.P. and Demaille, J. (1973) Syst. Zool. 22, 533-548.

8. Watterson, D.M., Harrelson, W.G., Keller, P.M., Sharief, F. and Vanaman, T.C. (1976) J. Biol. Chem. 251, 4501-4513.

9. Lin, Y.M., Liu, Y.P. and Cheung, W.Y. (1974) J. Biol. Chem. 249, 4943-4954.

10. Teo, T.S. and Wang, T.H. (1973) J. Biol. Chem. 248, 588-595.

11. Stevens, F.C., Walsh, M., Ho, H.C., Teo, T.S. and Wang, J.H. (1976) J. Biol. Chem. 251, 4495-4500.

12. Wang, J.H., Teo, T.S., Ho, H.C. and Stevens, F.C. (1975) Adv. Cyclic Nucleotide Res. 5, 179-194.

13. Matthews, J.C., Hori, K. and Cormier, M.J. (1977) Biochemistry 16, 85-91.

14. Lee, J., Wesley, A.S., Ferguson, J.F. and Seliger, H.H. (1966) In Bioluminescence in Progress, Johnson, F.H. and Haneda, Y., Ed., Princeton, N.J., Princeton University Press, p. 35.

15. Dubois, M., Gilles, A., Hamilton, J.K., Rebens, P.A. and Smith, F. (1956) Anal. Chem. 28, 350-356.

16. Zacharius, R.M., Zell, T.E., Morrison, J.H. and Woodlock, J.J. (1969) Anal. Biochem. 30, 148-152.

17. Chen, P., Toribara, T. and Warner, H. (1956) Anal. Chem. 28, 1756-1758.

18. Ames, B.N. and Dubin, D.T. (1960) J. Biol. Chem. 235, 769-775.

19. Goa, J. (1958) Scand. J. Clin. Lab. Invest. 5, 218.

20. Halfman, C.J. and Nishida, T. (1972) Biochemistry 11, 3493-3498.

21. Kretsinger, R.H. (1976) Ann. Rev. Biochem. 45, 239-266.

22. Blinks, J.R., Prendergast, F.G. and Allen, D.G. (1976) Pharmacol. Rev. 28, 1-93.

23. Ward, W.W. and Cormier, M.J. (1975) Proc. Natl. Acad. Sci. USA, 72, 2530-2534.

24. Brooks, J.C. and Siegel, F.L. (1973) J. Biol. Chem. 248, 4189-4193.

25. Brostrom, C.O., Hwang, Y.C., Brenkenridge, B.M. and Wolff, D.J. (1975) Proc. Natl. Acad. Sci. USA, 72, 64-68.

26. Winter, M.R.C., Head, J.F. and Perry, S.V. (1974) Calcium Binding Proteins Drabikowski, W., Strzelecka-Golaszewska, H., and Carafoli, E., eds., Elsevier North-Holland, Amsterdam, p. 109.

27. VanEerd, J.P. and Kawasaki, Y. (1972) Biochem. Biophys. Res. Comm. 47, 859-865.

28. Burstein, E.A., Permyakov, E.A., Emelyanenko, V.I., Bushueva, T.L. and Pechere, J.F. (1975) Biochim. Biophys. Acta 63, 1-16.

29. Kretsinger, R.H. and Barry, C.D. (1975) Biochem. Biophys. Acta 405, 40-52.

30. Wampler, J.E., Hori, K., Lee, J.W. and Cormier, M.J. (1971) Biochemistry 10, 2903-2909.

31. Ward, W.W. and Cormier, M.J. (1976) J. Phys. Chem. 80, 2289-2291.

32. Anderson, J.M. and Cormier, M.J. (1976) Biochem. Biophys. Res. Comm. 68, 1234-1241.

33. Henry, J.P. (1975) Biochem. Biophys. Res. Comm. 62, 253-259.

34. Cormier, M.J., Lee, J. and Wampler, J.E. (1975) Ann. Rev. Biochem. 44, 255-272.

35. Rasmussen, H. (1970) Science 170, 404-412.

Aberrant 35S-Methionine Incorporation and Calcium Uptake in Dystrophic Pectoralis, Cardiac and Brain Membrane Vesicles

Earl M. Ettienne and Robert H. Singer

Departments of Physiology and Anatomy

University of Massachusetts Medical Center

Worcester, Massachusetts   01605

INTRODUCTION AND SUMMARY

Avian muscular dystrophy is an autosomal recessive genetic disease character-ized by early hypertrophy and loss of function of the pectoralis major.[1] The dis-ease is progressive, ultimately resulting in atrophy and heavy lipid deposition.

Previous investigators have noted a decrease in the ability of the dystrophic sarcoplasmic reticulum to concentrate Ca^{2+}.[2] More recently, Sabbadini, et. al.[3], have shown an abnormal calcium uptake in avian dystrophic sarcoplasmic reticulum. They indicated, using freeze-fracture techniques, that a $90\overset{o}{a}$ particle of the vesi-cle membrane exhibited a decreased population and suggested that they might be the ATPase involved in calcium transport.

Our studies confirm the earlier observations of a decreased rate of Ca^{2+} up-take and Ca^{2+} binding capacity of dystrophic fragmented sarcoplasmic reticulum vesicles which are isolated from both embryonic and adult pectoralis. In addition, we have found a quantitative deficiency in a 65,000 and 12,000 dalton component of the dystrophic FSR at the time of myoblast fusion by measuring ^{35}S-Methionine in-corporation into the SR coupled to high resolution polyacrylamide gel electrophor-esis. Analysis of total tissue calcium by atomic absorption spectroscopy revealed major differences in the total calcium content between normal and dystrophic mus-cle.

Similar analyses of heart and brain from dystrophic embryonic animals showed that calcium uptake, storage and total content was opposite to that of the pector-alis muscle; however, with progression of the disease in the adult animals, cal-cium uptake and storage eventually reflected the pattern of the pectoralis FSR.

MATERIALS

The pectoralis, heart and brain were excised from 14-20 day-old chicks and adult dystrophic animals and normal controls.

The minced tissue was homogenized for 40-90 seconds in a Tekmar homogenizer using 4 volumes (w:v) of 0.3M sucrose and 10mM imidazole, pH 7.0. The homogenate was centrifuged in an S.S.-34 Sorval rotor at 17,000 g's for 20 minutes to pellet tissue fragments and large organelles. Mitochondrial contaminants of the superna-tant were further removed by staining through four layers of cheesecloth. After spinning at 34,000 g's (S.S. 34) for 20 minutes, the pellet was resuspended in

0.6M KCl and 10mM imidazole, pH 7.0 for 15 minutes to remove contaminating contractile proteins. A final spin at 198,000 g's for 20 minutes effectively pellets the
microsomal fraction representing the internal membranes. The pellet was resuspended in 0.3M sucrose to a final concentration of 1 to 5 mg per ml and used for either
further purification or for the various assays.

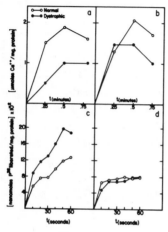

FIGURE 1 Ca^{2+} uptake sarcoplasmic
reticulum from (a) adult and (b)
embryonic animals. ATPase measurements are given in (c) and (d).

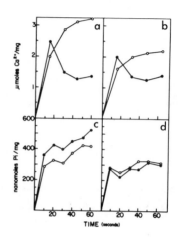

FIGURE 2 Ca^{2+} uptake in (a)
adult heart SR and (b) homogenate.
ATPases are given in (c) and (d).

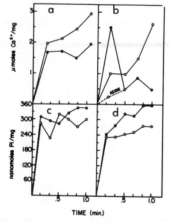

FIGURE 3 Ca^{2+} measurement
in (a) adult and (b) homogenates brain vesicles.
ATPases are given in (c) and (d).

METHODS AND RESULTS

Measurement of fast uptake of calcium: The rates of calcium uptake by the
various purified reticulum fractions were measured by adding vesicles to a reaction medium of 5mM MgCl$_2$, 5mM ATP, 100mM KCl, 10mM K-oxalate, 10mM imidazole at
pH 7.0, 0.2mM EGTA and 0.05μcuries/ml ^{45}CaCl$_2$. Optimum rates are achieved when
the CaCl$_2$ concentration of the medium was adjusted to 10μM/mg SR protein.

492

The initial fast uptake of calcium by the embryonic dystrophic pectoralis sarcoplasmic reticulum was greater than control values by as much as 30%. Adult dystrophic sarcoplasmic reticulum exhibited an initial rate which was 30% of the normal (Figure 1a, b).

Calcium uptake by vesicles isolated from adult dystrophic heart show initial values 25% greater than controls. However, there is a subsequent unloading of the dystrophic sarcoplasmic reticulum reaching steady state values which are less than 50% of controls (Figure 2a).

Membrane vesicles extracted from adult dystrophic brain show, steady state values for Ca^{2+} transport which are 30% below controls. Initial fast uptake of calcium by the dystrophic vesicles is 17% below controls (Figure 3a).

Assay for Ca^{2+} stimulated ATPase activity: Calcium stimulated ATPase activity was determined by monitoring at 10 second intervals ^{32}P liberated from ATP^{32} in a reaction medium of 100mM KCl, 5mM $MgCl_2$, 5mM ATP, 10mM K-oxalate, 10mM imidazole, pH 7.0, 0.2mM EGTA and 10µM $CaCl_2$/mg SR protein in 0.01µcuries/ml ATP^{32}.

The initial rate of ATP hydrolysis by the adult dystrophic sarcoplasmic reticulum was more than 50% greater than control values. This difference is reflected as a 70% drop in efficiency of the dystrophic system compared to control values (Figure 1c). The ratio of Ca^{2+} transported to ATP hydrolyzed is 1:2 in the dystrophic SR; normal controls exhibited a ratio of 1.8:1 ATP hydrolysis in the embryonic normal and dystrophic showed no differences. However, the dystrophic SR showed an initial efficiency 50% greater than normal controls (Figure 1d).

In the isolated adult heart vesicles, the initial calcium transport efficiency was 4.71 to 1 ATP hydrolyzed in normal controls. The dystrophic system was 10% less efficient at 4.28:1 (Figure 2c). Adult brain dystrophic vesicles exhibited a transport efficiency 25% below that of normal controls.

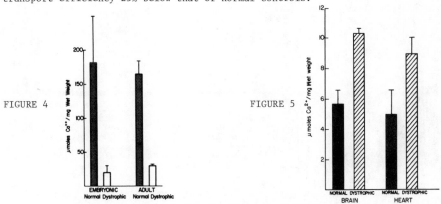

FIGURE 4 FIGURE 5

Measurement of total tissue calcium in pectoralis, brain and heart by atomic absorption spectroscopy.

Measurement of total tissue calcium: We first measured total bound calcium simply by excising the pectoralis and washing in phosphate buffered saline. The

493

tissue was tared, hydrolized and incinerated at 700ºC. Total calcium was then measured in small aliquots with an atomic absorption spectrometer. Extracellular calcium was replaced by pre-incubation of the excised tissue in Lanthanum Chloride.

When values for the intracellular tissue calcium from the normal embryos were compared to values for dystrophic embryonic tissue, they were found to exceed the dystrophic by a factor of 9:1 (Figure 4). Similar differences were observed between adult normal and dystrophic tissues, where the measured ratios were 5.6:1. These ratios were consistent whether the tissues were pretreated with lanthanum to remove extracellular calcium or left untreated. In the adult brain and heart tissues, the dystrophic tissues were found to exceed the normal controls in intracellular total calcium by as much as 2:1. These results are consistent with unpublished observations of an enhanced calcium accumulation by embryonic brain and heart vesicles (Figure 5).

^{35}S-Methionine incorporation in muscle cell cultures: Duplicate synchronous myoblast cultures of normal and dystrophic chick pectoralis have been incubated in a methionine-free, ^{35}S-Methionine enriched medium. Following six hours of incorporation of the label and encompassing the fusion of cells to form myotubes, the cultures were washed, harvested and the sarcoplasmic reticulum co-extracted with normal muscle carrier. 50µg samples of the purified sarcoplasmic reticulum were loaded onto gradient slab gels. Gels sliced into 1mm thick sections, HCl-hydrolyzed and counted (Figure 6). It is evident that a component running at 55,000 to 65,000 daltons has incorporated a greater amount of the label in the dystrophic cultures. Our tentative conclusion is that the higher rate of incorporation is reflective of a higher rate of turnover of the protein. This conclusion is partially substantiated by the observation that coomassie blue staining of protein at these molecular weight ranges after SDS-PAGE analysis is less in the dystrophic.

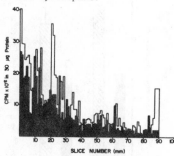

FIGURE 6
Histogram of counts following ^{35}S-Methionine incorporation into SR proteins from embryonic muscle cell cultures.

References

1. Asmundson, V. S. and Julian, L. M. (1956), J. Hered. 47, 248-252.

2. Peter, J. B., Fiehn, W., Nagatomo, T., Adniman, R., Stempel, K. and Bowman, R. (1974), Exp. Concepts in Musc. Dyst. II, A. H. Milhorat, Ed., American Elsevier Publishing Co., Inc., New York, (1974)

3. Sabbadini, R., Scalese, D. and Inesi, G. (1975), FEBS Letters, 54 (1):8.

Calcium-Mediated Regulation
of Intracellular Human Chorionic Gonadotropin in Hela$_{65}$ Cell Cultures

Robert J. Fallon, Nimai K. Ghosh, and Rody P. Cox
Division of Human Genetics, Departments of Medicine and Pharmacology
New York University Medical Center
550 First Avenue, New York, N.Y. 10016, U.S.A.

SUMMARY

Sodium Butyrate induces HeLa cell cultures to ectopically produce large quantities of the glycopeptide hormone, human chorionic gonadotropin (HCG). Calcium ions modify this response by increasing intracellular HCG levels while slightly reducing the hormone secreted into medium. Colcemid, a well-characterized inhibitor of microtubular assembly, also increases intracellular HCG and reduces the extracellular level. Calcium ions appear to modulate the secretory activity of HeLa cells and these effects may be mediated through an effect on the microtubular system.

INTRODUCTION

Calcium ion occupies a central role in cellular regulation and adaptation to changing environments, as shown by its ability to modulate secretory, proliferative, and contractile activity in diverse cell systems[1-4]. Its capacity to bind tubulin in vitro[5] and to induce depolymerization of microtubules[6,7] has prompted a search for evidence that cytoplasmic calcium activity is a physiologic regulator of microtubule assembly and disaggregation. Human granulocyte chemotaxis appears to be regulated by intracellular calcium levels[8]. Calcium ionophores affect mobility and distribution of cell surface receptors in transformed fibroblasts, and decrease Concanavalin A-induced agglutination of these cells[9]. Calcium has been implicated in many microtubule-dependent functions: axopodium extension in the mould Actinosphaerium cichorni[10], ciliary beating in mussel gills[11], and morphological change in HeLa cells exposed to sodium butyrate[12]. HeLa cells grown in medium with sodium butyrate assume a fibroblastic shape within hours and this morphological change appears to be a prerequisite for HCG production[13,14]. Calcium ionophores were reported to prevent butyrate-mediated changes in cell shape[12]. Therefore, an investigation of the role of calcium in regulating HCG synthesis and secretion was initiated. Present evidence indicates that calcium ions interfere with secretion of HCG in HeLa cells exposed to sodium butyrate. These results are consistent with the current belief that calcium ions may be important in vivo regulators of microtubule function.

RESULTS

Table 1 shows the changes in intracellular and extracellular HCG mediated by sodium butyrate and the effect of adding calcium chloride. In all experiments

TABLE 1

EFFECT OF CALCIUM ON SECRETION OF HUMAN CHORIONIC GONADOTROPIN BY HELA$_{65}$ CELLS

EXP. NO.	HCG LEVELS (mIU/10^6 CELLS)			% INTRACELLULAR		
	CON$^\theta$	SB	SB + CaCl$_2$	CON	SB	SB + CaCl$_2$
1	74.1(4.9)*	804.2(21.2)	70.(84.2)	6.2	2.6	54.6
2	91.1(11.)	269.7(99.0)	207.8(103.9)	10.8	26.8	33.4
3	4.8(.36)	74.2(12.9)	47.1(17.5)	7.0	14.8	27.1
4	11.6(.42)	87.0(4.2)	70.4(10.1)	3.5	4.7	12.8
5	133.3(3.8)	403.4(25.5)	426.6(72.4)	2.8	5.9	14.5
6	43.6(10.3)	175.6(64.6)	183.0(149.2)	19.3	26.6	44.9

* EXTRACELLULAR HCG LEVELS (INTRACELLULAR VALUES IN PARENTHESES)
θ (KEY: CON- CONTROL(NO ADDITIONS); SB- SODIUM BUTYRATE 4mM,48 hours; CaCl$_2$- CALCIUM CHLORIDE 10mM,48 hours.)

- HeLa$_{65}$ cells were grown in Waymouth medium containing 10% fetal calf serum and antibiotics as previously described[15]. HCG was quantitated by radioimmunoassay with antiserum against β-subunits of HCG(Institute of Bio-Endocrinology,Montreal) to insure immunochemical specificity. ^{125}I-HCG-β is employed as the tracer antigen and a second antibody,goat anti-rabbit γ-globulin, is used as the precipitant of radioactive antigen-antibody complexes.

the intracellular levels of HCG are increased by calcium ions and this increase is significant at the 95% confidence level when analyzed by paired one-tailed t-test (p<0.05)[16]. At the same time the extracellular levels of HCG are decreased in four of six experiments (table 1), and the total HCG levels are not significantly changed (.50>p>.40).

Colcemid, a known inhibitor of microtubule assembly, also increases intracellular HCG levels and slightly reduces the amount secreted into medium. These effects are similar to those observed with calcium.

DISCUSSION

Potent homeostatic mechanisms resist experimental modulation of cytoplasmic calcium concentration, estimated to be in the range of 10^{-6} to 10^{-8}M in normal, resting cells.[17]. This value can be raised experimentally by inhibition of the active mechanisms responsible for the efflux of calcium ions from the soluble compartment[18], or by inducing influx of Ca^{2+} ion by ionophores[19]. This latter procedure is accompanied by inhibition of protein synthesis[20]. The higher extra-

TABLE 2

EFFECT OF COLCEMID ON SECRETION OF HUMAN CHORIONIC GONADOTROPIN BY HELA$_{65}$ CELLS

HCG LEVELS (mIU/10^6 CELLS)			% INTRACELLULAR		
CON	SB	SB + COLCEMID(1.5mM)^X	CON	SB	SB + COLCEMID(1.5mM)^X
2.2(.058)*	35.1(2.62)	27.1(4.21)	2.6	6.3	12.4

*(SEE TABLE 1)   X(COLCEMID ADDED FOR THE FINAL 16h. OF A 48h. EXPERIMENT)

cellular calcium levels, exceeding those found in vivo, are used to saturate available intracellular reservoirs. The finding that the percentage of intracellular HCG is elevated in all experiments despite similar total hormone levels strongly suggests that the enhancement of intracellular HCG levels is due to decreased secretion rather than increased synthesis. It is conceivable that any agent preferentially interfering with the process of secretion may mimic the effects of calcium ions. The results obtained with colcemid are in conformity with this contention. The intervention of calcium ion in the secretory process may occur at many steps, such as interference with protein packaging in granules or fusion of vesicle and plasma membrane. However, since increased concentrations of this cation may serve an anti-microtubular function, calcium ion may enhance intracellular HCG in HeLa$_{65}$ cell cultures by inhibiting microtubule-dependent secretion of this glycopeptide hormone.

ACKNOWLEDGEMENT

This research was supported by NIH grant GM 15508. R.J.F. is a Medical Scientist Trainee supported by NIH grant GM 07308.

REFERENCES

1. Ebashi, S. and Endo,M. (1968) Prog. Biophys. Mol.Biol. 18 123.
2. Dulbecco,R. and Elkington,J. (1975) Proc.Natl.Acad.Sci. U.S. 72 1584
3. Hellman, B. (1976) FEBS Lett. 63 125.
4. Boynton,A.L. and Whitfield, J.F. (1976) In Vitro 12 479.
5. Hayashi,M. and Matsumura,F. (1975) FEBS Lett. 58 222.
6. Weisenberg,R.C. (1972) Science 177 1104.
7. Borisy,G.G.,Olmsted, J.B., Marcum, J.M. and Allen,C. (1974) Fedn.Proc. 33 167.
8. Gallin,J.I. and Rosenthal,A.S. (1974) J.Cell Biol. 62 594.
9. Poste,G. and Nicholson,G.L. (1976) Biochim. Biophys.Acta 426 148.
10. Scliwa,M. (1976) J.Cell Biol. 70 527.
11. Satir,P. (1975) Science 190 586.
12. Henneberry,R.C., Fishman,P.H. and Freese,E. (1975) Cell 5 1
13. Ghosh,N.K. and Cox,R.P. (1976) Nature 259 416.
14. Ghosh,N.K., Rukenstein, A.,and Cox, R.P. Biochem. J., in press.
15. Deutsch,S.I., Silvers, D.N., Cox,R.P.,Griffin, M.J. and Ghosh,N.K. (1976) J.Cell Sci. 21 391.
16. Goldstein,A. (1964) Biostatistics: An Introductory Text, Macmillan, New York.
17. Rasmussen,H. (1970) Science 170 404.
18. Lehninger,A.L., Carafoli,E. and Rossi,C.S. (1967) Adv. Enzymol. 29 259.
19. Reed,P.W. and Lardy,H.A. (1972) J. Biol.Chem. 247 6970.
20. Bottenstein,J.E. and deVellis,J. (1976) Biochem.Biophys.Res.Commun. 73 486.

Evidence for Separate Active Mitochondrial Efflux Mechanisms for Ca^{2+} and Mn^{2+}

Karlene K. Gunter, Randy N. Rosier, Donald A. Tucker and Thomas E. Gunter

Radiation Biology and Biophysics
University of Rochester; Rochester, NY 14642

INTRODUCTION

Ca^{2+} uptake into metabolically inhibited mitochondria may be induced by setting up an internally negative membrane potential without oxidation of substrate or hydrolysis of ATP[1]. There is evidence[2,3] that influx of Ca^{2+} or Mn^{2+} into mitochondria normally occurs down an electrochemical gradient maintained by electron transport via a "uniport" mechanism, but evidence also exists[4-8] that this uniport system is not the only mitochondrial Ca^{2+} or Mn^{2+} transport mechanism.

While Ca^{2+} and Mn^{2+} influx have been shown to be via the same mechanism[9], efflux of these ions takes place via a separate mechanism[10]. An active component of Ca^{2+} and of Mn^{2+} efflux has been postulated[4,10]. This hypothesis may help explain data in the literature relating to internal to external $[M^{2+}]$ ratios across the mitochondrial membrane[4-6], ruthenium red effects[4,7] effects of low levels of uncouplers on Ca/O ratios[11], and uptake of Ca^{2+} into submitochondrial particles[12]. Further evidence is presented here, supporting the existence of operationally distinct, active efflux mechanisms for both Ca^{2+} and Mn^{2+}.

MATERIALS AND METHODS

Efflux experiments were carried out by adding EGTA, the reagent to be tested, and the reagent plus EGTA to parallel samples of a mitochondrial suspension (prepared following Schnaitman and Greenawalt[13]) after uptake of 80 nmol/mg protein of Ca^{2+} or Mn^{2+}. Samples were stirred continually. 1 or 2 ml aliquots were withdrawn at preset times and the mitochondria separated from suspending medium with either Millipore filtration or centrifugation. Supernatants or filtrates were counted for ^{45}Ca or ^{54}Mn using techniques described previously[4].

Submitochondrial particles (SMP's) were made by sonication of mitoplasts prepared by techniques similar to those discussed by Schnaitman and Greenawalt[13]. A number of tests of sidedness of these SMP's[14] indicated the presence of a mixture of right side out and inside out vesicles. Consequently, techniques were developed for selectively energizing inside out or right side out vesicles. 12 mg/ml cytochrome c was added at the time of sonication. Cytochrome c external to the vesicles was removed by KCl wash. Inside out vesicles were energized by addition to the suspension of ascorbate and phenazinium methyl sulfate (PMS), which is membrane permeable and transfers reducing equivalents from ascorbate to cytochrome c. Experiments using the fluorescent carbocyanine dye, di $O-C_3-(5)$, indicated inside positive membrane potentials after this form of energization. Right side out vesicles are energized by addition of ascorbate and cytochrome c.

Since reducing equivalents from ascorbate can't cross the membrane in this case, only right side out vesicles are energized. Fluorescence data indicates inside negative membrane potentials after this form of energization.

RESULTS

The mechanism(s) of divalent cation efflux can be studied either by direct study of efflux or by a study of influx into inside out SMP's.

Efflux In studies of effects of metabolic inhibitors on the rate of efflux of Mn^{2+} or Ca^{2+} from mitochondria, it was found that Mn^{2+} efflux was as much as 50% slower in the presence of a metabolic inhibitor (2 mM KCN) plus the chelating agent EGTA as in the presence of EGTA alone, while Ca^{2+} efflux was faster where metabolic inhibitor was present.

While the rates of both Mn^{2+} and Ca^{2+} efflux in the presence of EGTA were affected by uncouplers, different characteristics were observed. Over the range of CCCP concentration studied (up to 50 μM), Mn^{2+} efflux rates in the presence of EGTA and CCCP were slower than those with CCCP alone. Mn^{2+} efflux rates in the presence of CCCP plus EGTA were markedly slower than those with EGTA alone over the range of CCCP concentration from approximately 0.15 to 15μM.

The effects of CCCP and EGTA on Ca^{2+} efflux rates appeared to depend on how long Ca^{2+} was held before spontaneous efflux occurred. Where rapid spontaneous efflux occurred, CCCP in the .01 to .05 μM concentration range slowed net Ca^{2+} efflux to a minimum rate whether or not EGTA was present. In the case where Ca^{2+} was held for at least 20 min. before spontaneous efflux, in a CCCP concentration range just less than that necessary for complete uncoupling (0.5 to 1.2 μM), the rate of efflux for the CCCP plus EGTA case was less then that with CCCP alone.

Uptake in Submitochondrial Particles Preliminary results of studies of uptake in selectively energized vesicles are shown in Fig. 1. Uptake of either Mn^{2+} or Ca^{2+} into these

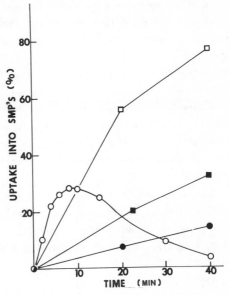

Fig. 1: Percent uptake of Ca^{2+} (circles) and Mn^{2+} (squares) into right side out (open symbols) and inside out (ISO) (filled symbols) energized vesicles versus time. SMP's (20% ISO) were suspended in 135 mM mannitol, 45 mM sucrose, 24 mM Hepes (pH 7.2) and .5 mM Na succinate. At time zero 750 nmol M^{2+}/mg SMP protein (i.e. very approximately 250 nmol M^{2+}/mg mito protein) and 1.3 μM antimycin A were added.

499

vesicles required the presence of substrate and either ADP or ATP. Mg^{2+} was required for Ca^{2+} uptake but not for Mn^{2+} uptake. FCCP, CCCP, ruthenium red, and KCN inhibited uptake into vesicles energized in both the right side out and inside out sense, while oligomycin had no effect on uptake.

DISCUSSION

The inhibitory effects of metabolic inhibitors and uncouplers on Mn^{2+} efflux can be understood in terms of the active efflux hypothesis. If the energy for an efflux pump is derived from substrate oxidation, then the action of the pump can be inhibited by a metabolic inhibitor blocking the source of energy or by an uncoupler dissipating the energy through a parallel pathway. The uncoupler could also act as a site specific inhibitor of Mn^{2+} efflux.

Lack of observed slowing of Ca^{2+} efflux with metabolic inhibitors does not prove that there is no efflux pump for Ca^{2+}. The amounts of inhibitor used certainly affect membrane potential. Perhaps the decrease in membrane potential compensates for the effect of the inhibitor on the pump and causes these effects to be unobservable. The effect of small amounts of uncoupler on Ca^{2+} efflux suggests a site specific inhibition in this case. These results suggest that the observation of Stucki and Ineichen[11] that Ca/O ratios increase in the presence of low levels of uncoupler is due to uncoupler inhibition of an efflux mechanism.

The preliminary results on Ca^{2+} and Mn^{2+} uptake in inside out energized SMP's made from liver mitochondria, confirm in a new mitochondrial system the results of Loyter[12] et al., who studied heart mitochondria. The data suggest that the source of energy for this uptake is substrate oxidation.

REFERENCES

1. Scarpa, A. and Azzone, G.F. (1970) *Eur. J. Biochem. 12*, 328.
2. Selwyn, M.J., Dawson, A.P. and Dunnett, S.J. (1970) *FEBS Lett. 10*, 1.
3. Rottenberg, H. and Scarpa, A. (1974) *Biochemistry 13*, 4811.
4. Puskin, J.S., Gunter, T.E., Gunter, K.K. and Russell, P.R. (1976) *Biochemistry 15*, 3834
5. Massari, S. and Pozzan, T. (1976) *Arch. Biochem. Biophys. 173*, 332.
6. Azzone, G.F., Bragadin, M., Pozzan, T. and Dell'Antone, P. (1976) *BBA 459*, 96.
7. Sordahl, L.A. (1974) *Arch. Biochem. Biophys. 167*, 104.
8. Crompton, M., Capano, M. and Carafoli, E. (1976) *Eur. J. Biochem. 69*, 453.
9. Vainio, H., Mela, L. and Chance, B. (1970) *Eur. J. Biochem. 12*, 387.
10. Gunter, T.E., Puskin, J.S., Gunter, K.K. and Russell, P.R. (1977) *Biophys. J. 17*, 253a., also submitted for publication.
11. Stucki, J.W. and Ineichen, E.A. (1974) *Eur. J. Biochem. 48*, 365.
12. Loyter, A., Christiansen, R.O., Steensland, H., Saltzgaber, J. and Racker, E. (1969) *J. Biol. Chem. 244*, 4428, *ibid* 4422.
13. Schnaitman, C. and Greenawalt, J.W. (1967) *J. Cell Biol. 38*, 158.
14. Rosier, R.N., Gunter, K.K. and Tucker, D.A., Manuscript in preparation.

Calcium-Binding Proteins
Extracted from Erythrocyte Membranes

S. Tsuyoshi Ohnishi
Department of Biological Chemistry, Department of
Anesthesiology, Hahnemann Medical College,
Philadelphia, Pennsylvania, 19102.   U. S. A.

We have previously reported that a troponin-c type calcium binding protein can be extracted from human erythrocyte membranes[1,2]. However, we have further found that at least two Ca^{++}-binding proteins exist in the membranes. During the studies of erythrocyte actin(eA) from human red blood cell[2,3], it was found that the superprecipitation of the hybrid of crude eA and rabbit skeletal myosin (rM) was Ca^{++}-sensitive(Fig.1(A)), while that of purified eA and rM was not(Fig. 1(B)). Since this suggested the existence of regulatory proteins in the membrane, we attempted to extract such proteins. Ghosts prepared from outdated blood bank blood were treated with acetone and extracted with 10 mM EDTA, 1 mM DTT and 5 mM Tris-HCl(pH 8.5). The extract was applied to a DEAE cellulose column, and then eluted by NaCl gradient as shown in Fig.2. The fraction pointed by an arrow A had the maximum Ca^{++}-binding capacity as measured by the murexide method[4].

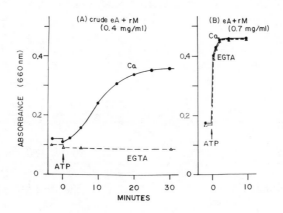

Fig.1. Superprecipitation of (A) crude eA + rM, and (B) purified eA + rM. Concentrations of added ATP, Ca and EGTA were all 0.5 mM. 50 mM KCl, 2 mM $MgCl_2$. pH 7.0 and 25°.

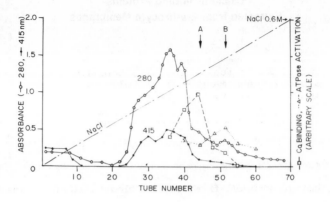

Fig.2. Gradient elution(0 - 0.6 M NaCl, pH 6.1) of ghost extract
through DEAE cellulose column.

This fraction was further purified by CM cellulose column to give two proteins,
one with a slight yellow color(Ca^{++}-binding protein), and the other without
color(non Ca^{++}-binding protein). The name of ossanin is proposed for the former
protein to signify that "bone" is found in the "ghost" (os(latin) = bone). As
shown in Fig.3, the yellow color comes from an absorption peak around 415 nm,
which is similar to that of hemichrome. Various attempts to remove this color
from ossanin were unsuccessful. The molar ratio of (heme(monomer)/ossanin) is
estimated to be 0.5 - 1.0. The functional role of heme in ossanin is unknown.

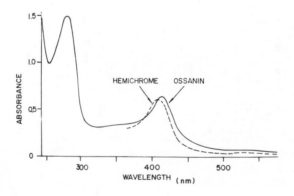

Fig.3. Absorption spectrum of ossanin and hemichrome. Hemichrome was
prepared by denaturing hemoglobin with SDS.

502

A possibility is that ossanin is the site of binding for hemoglobin (for example, Heinz body) on the membrane. The Ca^{++}-binding constant of ossanin is 4×10^6 (M^{-1}) in the absence of Mg^{++}, and 1×10^5 in the presence of 2.5 mM Mg^{++}. The colorless protein seems to be a tropomyosin-like protein (ossamyosin ?). Molecular weights of subunits of these proteins were similar to those of muscle troponin and tropomyosin except for 23,000 subunit of ossanin, which is 18,000 daltons in muscle troponin (Fig.4).

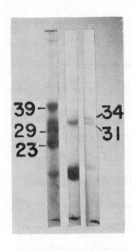

Fig.4. SDS gel electrophoresis of (left) ossanin, (middle) hemoglobin, and (right) tropomyosin-like protein. Samples were migrated from top to bottom through 12 % gels (pH 6.4). The markers near the bottom show the position of the tracking dye, BPB. Numbers show the molecular weight in 1,000 daltons.

The Mg-Ca-ATPase of ghosts (5) was activated by either fraction A or B. However, the activation by the fraction B is greater as shown in Fig.2. Muscle tropomyosin also activated the ATPase. According to the SDS gel electrophoresis, the fraction B contains a protein with the molecular weight of 18,000. This protein was further purified by DEAE column chromatography in urea, and was found to have similar properties as troponin-c, i.e., low phenylalanine content as shown by low 280 nm absorption, and the Ca^{++}-dependent increase of tyrosine residue fluorescence (excitation 276 nm, emission 306 nm). Involvement of these proteins in the Ca-pump (6) as well as in the maintenance of mechano-chemical properties of erythrocyte membranes is being studied.

(1) Ohnishi, T., Abstract of the 10th International Congress of Biochemistry (Hamburg, 1976) 277
(2) Ohnishi, T., British J. Haematology (1977) 35, 453
(3) Ohnishi, T., J. Biochem. (1962) 52, 307
(4) Ohnishi, T., Masoro, E. J., Bertrand, H. A., and Yu, B. P., Biophysical J. (1972) 12, 1251
(5) Luthra, M. G., Hildenbrandt, G. R., and Hanahan, D. J., Biochim. Biophys. Acta, (1976) 419, 164 ; Bond, G. H., and Clough, D.L., Biochim. Biophys. Acta, (1973) 323, 592
(6) Schatzman, H. J., and Vincenzi, F. F., J. Physiol.Lond. (1969) 201, 369; Lee, K.S. and Shin,B. C.,J. Gen. Physiol.,(1969) 54, 713

Cytochemical Localization
of Calcium in the Chick Allantoic Epithelium

Clara Varady Riddle[*]
Zoology Department
Oregon State University
Corvallis, Oregon U.S.A.

The chick allantois has classically been considered as the
storage organ for the waste products of the embryo. Recent elabora-
tion of the physiological properties of the allantoic epithelium has
established its role in the conservation of ions and water. The
active absorption of sodium[1,2], and perhaps calcium[3], have been demonstrated. Ultrastructural studies have shown that the allantoic
epithelium has a cellular heterogeneity[4] and has the structural
characteristics of ion and water transporting epithelia[5]. The
allantoic epithelium of the chorioallantoic membrane of 15-17 day
embryos consists of three cell types: (1) granular cells with
numerous granules; (2) mitochondria rich cells with an abundance of
mitochondria and a tuft of long microvilli; and (3) basal cells with
a paucity of organelles and no luminal exposure.

Calcium was localized differentially in the three cell types
using two separate cytochemical techniques. The pyroantimonate-
osmium tetroxide reaction (PAO) was utilized for the localization of
cations[6]. Calcium chelation controls indicate that calcium is the
cation precipitated by pyroantimonate. The intensity of the reaction varies according to the cell type. Basically, excluding
nuclear deposits, the granular cells have a large amount of
deposits; the mitochondria rich cells have from none to moderate
amounts; and the basal cells have none (Figure 1). In the granular
cells deposits occur along the cytoplasmic side of the plasma mem-

---

[*]Present address: Anatomy Department, Temple University Medical
School, Philadelphia, Pennsylvania, 19140 U.S.A.

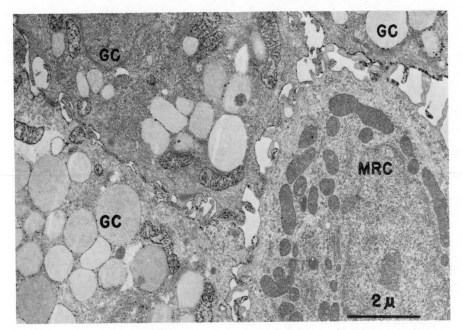

Figure 1. Slightly tangential section of granular cells (GC) and mitochondria rich cell (MRC) after PAO fixation. In the granular cells pyroantimonate deposits are on the cytoplasmic side of plasma membranes and in the intramembranous spaces of mitochondria.

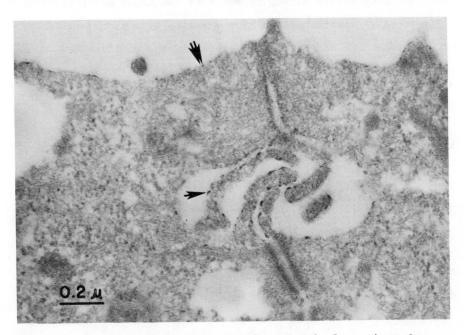

Figure 2. Lightly stained cross section of apical portion of granular cells after glutaraldehyde-calcium fixation. The faint electron opaque deposits on the lateral membranes (arrow) are larger than on the apical membranes (double arrows).

brane and generally, the apical membranes have smaller deposits than the basolateral membranes. Deposits are also present in the intramembranous spaces of mitochondria, on the membranes of granules, and in the nuclei. In some mitochondria rich cells there are a few deposits on the inner plasma membrane surfaces, in the intramembranous spaces of mitochondria, and in the nuclei. Basal cells are devoid of deposits except in the nuclei.

Calcium was also localized by the calcium-glutaraldehyde technique of fixation in 2.66% glutaraldehyde in 0.067 M s-collidine buffer, pH 7.3 with 10 or 90 mM $CaCl_2$[7]. Very faint electron opaque deposits are present along the cytoplasmic side of the entire plasma membrane of all the cell types. The apical membranes of the granular cells and the mitochondria rich cells have smaller deposits than their respective basolateral membranes (Figure 2). The deposits on the entire plasma membrane of the basal cells are also small. When glutaraldehyde-calcium fixation is followed by PAO post-fixation, the pattern of deposition is the same as when glutaraldehyde-calcium fixation is used alone.

The role of calcium in this ion transporting epithelium has yet to be defined; however, these results suggest that calcium is involved differentially in some functions of the various cell types.

REFERENCES

1 Garrison,J and Terepka,A (1972) J. Membrane Biol. 7,146-163
2 Moriarty,C and Hogben,C (1970) Biochim. Biophys. Acta 219,463-470
3 Moriarty,C (1973) Exptl. Cell Res. 79,79-86
4 Coleman,J and Terepka,A (1972) J. Membrane Biol. 7,111-127
5 Riddle,C (1976) Doctoral thesis. Oregon State University
6 Hardin,J, Spicer,S and Greene,W (1969) Lab. Invest. 21,214-224
7 Oschman,J and Wall,B (1972) J. Cell Biol. 55,58-73

(Ca²⁺-Mg²⁺)ATPase Activator of Red Blood Cells:
Decreased Response in Sickle Cell Membranes

Frank F. Vincenzi and Rama M. Gopinath
Department of Pharmacology, University of Washington
Seattle, Washington   98195

INTRODUCTION

A molecular abnormality has been known in sickle cell anemia for sometime but the mechanism of sickling is poorly understood. Sickle cells have been reported to show increased red blood cell (RBC) membrane Ca^{2+} permeability[1], increased RBC Ca^{2+} content[2], and decreased active transport of Ca^{2+} [1,3]. Active transport of Ca^{2+} across the RBC membrane has been associated with a $(Ca^{2+}-Mg^{2+})$ATPase and a cytoplasmic protein activator of the $(Ca^{2+}-Mg^{2+})$ATPase has been described[4]. The present work describes a decreased response to cytoplasmic activator of $(Ca^{2+}-Mg^{2+})$ATPase of membranes from RBC's of patients with sickle cell anemia.

MATERIALS AND METHODS

Membranes were prepared by hypotonic hemolysis[5] of RBC from outdated blood bank blood (normal) and from blood of patients with sickle cell anemia (HbSS). By modification of hemolysis conditions, it is possible to prepare membranes with little or no bound endogenous cytoplasmic activator (I20 membranes) or with sufficient bound activator to maximally activate the $(Ca^{2+}-Mg^{2+})$ATPase (I310 membranes)[6]. Except as noted, I20 membranes were used for the work presented here. ATPase activities were defined and assayed as described previously[5]. RBC cytoplasmic activator was partially purified according to the method of Luthra et al.[7]. Preparation and separation of irreversibly sickled cells was carried out according to Fischer et al.[8].

RESULTS AND DISCUSSION

As shown in Table 1, $(Ca^{2+}-Mg^{2+})$ATPase and $(Na^{+}-K^{+}-Mg^{2+})$ATPase activities of normal and HbSS membranes did not differ significantly. The Mg^{2+}-ATPase activity of HbSS membranes was significantly higher than normal. In the presence of RBC cytoplasmic activator (provided by a hemolysate of normal cells) the $(Ca^{2+}-Mg^{2+})$-ATPase activity of normal membranes increased about 3-fold. Activity of HbSS membranes increased less than 2-fold.

We tested the effect of membrane-free hemolysate of normal and HbSS RBC on membranes from normal and HbSS RBC. The results (Table 2) were the same whether normal or HbSS hemolysate was used as a source of cytoplasmic activator. Thus, cytoplasmic activator is present in HbSS cells and is not defective. These results also suggest that hemoglobin S does not interfere with the effect of cytoplasmic activator although they do not rule out an effect of hemoglobin S bound tightly to the HbSS membrane.

TABLE 1

ATPase ACTIVITIES OF NORMAL AND HbSS RBC MEMBRANES:
EFFECT OF HEMOLYSATE OF NORMAL CELLS

Membranes	Mg^{2+}-ATPase	$(Na^+-K^+-Mg^{2+})$ATPase	$(Ca^{2+}-Mg^{2+})$ATPase
Normal I20	0.24 ± 0.16	0.60 ± 0.20	1.20 ± 0.37
Normal I20 + hemolysate[†]	0.22 ± 0.15	0.65 ± 0.20	3.63 ± 1.03
HbSS I20	0.59 ± 0.30*	0.52 ± 0.35	1.00 ± 0.40
HbSS I20 + hemolysate[†]	0.50 ± 0.09*	0.59 ± 0.40	1.93 ± 0.49**

Values represent µmoles Pi mg protein^{-1}·hr^{-1}, mean of five experiments ± S.E. HbSS samples from three different patients.

* Significantly different from normal membranes, P < .05, ** P < .01.

[†] Membrane-free hemolysate of normal RBC's as the source of cytoplasmic activator (0.2 ml hemolysate/ml).

To rule out the possibility that binding sites on HbSS membranes were somehow inaccessible to exogenous cytoplasmic activator, we prepared I310 membranes. It was shown by Farrance[6] that I310 membranes have high $(Ca^{2+}-Mg^{2+})$ATPase activity because endogenous activator binds to the membrane during hemolysis. $(Ca^{2+}-Mg^{2+})$-ATPase activity of HbSS I310 membranes was less than that of normal I310 membranes (not shown). In other experiments (not shown), apparent affinities of normal and HbSS membranes for partially purified cytoplasmic activator were shown not to differ. Neither did the apparent affinity of the ATPases for Ca^{2+} differ.

A question arose as to whether the smaller response of HbSS membranes might be due to some irreversibly sickled cells (ISC's) in the population. Basal $(Ca^{2+}-Mg^{2+})$ATPase activity of ISC's was not different from that of non-ISC membranes of HbSS blood (data not shown). The influence of activator was also not significantly different comparing ISC and non-ISC HbSS membranes.

The above results are preliminary and have not been shown to be constant or specific for sickle cell anemia (as contrasted, for example, with other hemolytic anemias). Nevertheless, it appears that $(Ca^{2+}-Mg^{2+})$ATPase of HbSS membranes may have a limited capacity to respond to the cytoplasmic activator. An associated decrease in the capacity of the RBC Ca^{2+}-pump would be compatible with suggestions

TABLE 2

$(Ca^{2+}-Mg^{2+})$ATPase OF NORMAL AND HbSS RBC MEMBRANES: EFFECT OF NORMAL OR HbSS HEMOLYSATE

Membranes	$(Ca^{2+}-Mg^{2+})$ATPase
Normal I20	0.93
Normal I20 + Normal Hemolysate[†]	3.48
Normal I20 + HbSS Hemolysate[†]	3.29
HbSS I20	0.80
HbSS I20 + Normal Hemolysate[†]	1.49
HbSS I20 + HbSS Hemolysate[†]	1.18

Values represent μmoles Pi mg protein$^{-1} \cdot$hr^{-1} (N = 2). HbSS samples from two different patients.

[†] Membrane-free hemolysate of normal or HbSS RBC's as the source of cytoplasmic activator (0.2 ml hemolysate/ml).

that intracellular accumulation of Ca^{2+} is involved in the sickling process.

Supported by NIH Grant AM-16436.

REFERENCES

1. Palek, J., Church, A., and Fairbanks, G. (1976) in Membranes and Disease (L. Bolis, L.J.F. Hoffman, and A. Leaf, eds.). Raven Press, N.Y. pp. 41-60.
2. Palek, J. (1974) Proceedings of the First National Symposium on Sickle Cell Disease (J.C. Hercules, A.N. Schecter, W.A. Eaton, and R.F. Jackson, eds.). pp. 219-221.
3. Schrier, S.L. and Bensch, K.G. (1976) Membranes and Diseases (L. Bolis, L.J.F. Hoffman, and A. Leaf, eds.). Raven Press, N.Y. pp. 31-40.
4. Bond, G.H. and Clough, D.L. (1973) Biochim. Biophys. Acta. 323, 592-599.
5. Farrance, M.L. and Vincenzi, F.F. (1977) Experientia, in press.
6. Farrance, M.L. (1977) Ph.D. Thesis, University of Washington.
7. Luthra, M.G., Hildenbrandt, G.R., and Hanahan, D.J. (1976) Biochim. Biophys. Acta. 419, 164-179.
8. Fischer, S., Nagel, R.L., Bookchin, R.M., Roth, E.F., and Tellez-Nagel, I. (1975) Biochim. Biophys. Acta. 375, 422-433.

AUTHOR INDEX

510

SUBJECT INDEX